You're the Professor, What Next?

Ideas and Resources for Preparing College Teachers

The Committee on Preparation for College Teaching

A joint committee of the American Mathematical Society, the Mathematical Association of America, and the Society for Industrial and Applied Mathematics established by the Joint Policy Board for Mathematics.

Donald W. Bushaw, Washington State University
Bettye Anne Case, Florida State University, Chair
Robert H. McDowell, Washington University
Richard S. Millman, California State University, San Marcos
Robert R. Phelps, University of Washington
Michael C. Reed, Duke University (through 1989)
Richard D. Ringeisen, Clemson University and Old Dominion University
Stephen B. Rodi, Austin Community College
James G. Simmonds, University of Virginia
Theodore A. Vessey, St. Olaf College (through 1987)
Guido L. Weiss, Washington University

Research Associate: M. Annette Blackwelder, Florida State University

You're the Professor, What Next?
Ideas and Resources for Preparing College Teachers

Bettye Anne Case, Editor

A Project of
The Committee on Preparation for College Teaching
American Mathematical Society   Mathematical Association of America
Society for Industrial and Applied Mathematics

Editorial Advisors:
Donald W. Bushaw
Robert R. Phelps
Richard D. Ringeisen
Stephen B. Rodi

MAA Notes Number 35

Published and Distributed by
The Mathematical Association of America

The illustrative sketches were created by Henry Frandsen, University of Tennessee. (See pages i, 2, 32, 38–41, 46, 52, 60, 66, 71, 77, 82–84, 86, 148, and 166.)

All articles which appeared previously in other publications are reprinted here with permission. We thank the authors and the original publishers for this permission to reprint.

The seals or insignia of universities which appear on the first pages of Chapters 6 through 13 are used with permission of those institutions.

A grant from the US Department of Education through its Fund for the Improvement of Postsecondary Education (P116B00184) partially supported this publication and the project it reports. Despite any copyright notice on this page, permission is granted to any Federal agency or to investigators working on Federal grants to reproduce or borrow freely from all original material in this volume.

General staff support for this volume was provided by the Department of Mathematics of the Florida State University; typesetting in TeX was by Melissa E. Smith, Senior Art Publication Production Specialist.

©1994 by the Mathematical Association of America

ISBN 0-88385-091-5

Library of Congress Catalog Number 94-74101

Printed in the United States of America

Current Printing

10 9 8 7 6 5 4 3 2 1

MAA Notes and Reports Series

The MAA Notes and Reports Series, started in 1982, addresses a broad range of topics and themes of interest to all who are involved with undergraduate mathematics. The volumes in this series are readable, informative, and useful, and help the mathematical community keep up with developments of importance to mathematics.

MAA Notes

1. Problem Solving in the Mathematics Curriculum, *Committee on the Teaching of Undergraduate Mathematics,* a subcommittee of the Committee on the Undergraduate Program in Mathematics, *Alan H. Schoenfeld,* Editor
2. Recommendations on the Mathematical Preparation of Teachers, *Committee on the Undergraduate Program in Mathematics, Panel on Teacher Training.*
3. Undergraduate Mathematics Education in the People's Republic of China, *Lynn A. Steen,* Editor.
5. American Perspectives on the Fifth International Congress on Mathematical Education, *Warren Page,* Editor.
6. Toward a Lean and Lively Calculus, *Ronald G. Douglas,* Editor.
8. Calculus for a New Century, *Lynn A. Steen,* Editor.
9. Computers and Mathematics: The Use of Computers in Undergraduate Instruction, *Committee on Computers in Mathematics Education, D. A. Smith, G. J. Porter, L. C. Leinbach, and R. H. Wenger,* Editors.
10. Guidelines for the Continuing Mathematical Education of Teachers, *Committee on the Mathematical Education of Teachers.*
11. Keys to Improved Instruction by Teaching Assistants and Part-Time Instructors, *Committee on Teaching Assistants and Part-Time Instructors, Bettye Anne Case,* Editor.
13. Reshaping College Mathematics, *Committee on the Undergraduate Program in Mathematics, Lynn A. Steen,* Editor.
14. Mathematical Writing, by *Donald E. Knuth, Tracy Larrabee, and Paul M. Roberts.*
15. Discrete Mathematics in the First Two Years, *Anthony Ralston,* Editor.
16. Using Writing to Teach Mathematics, *Andrew Sterrett,* Editor.
17. Priming the Calculus Pump: Innovations and Resources, *Committee on Calculus Reform and the First Two Years,* a subcomittee of the Committee on the Undergraduate Program in Mathematics, *Thomas W. Tucker,* Editor.
18. Models for Undergraduate Research in Mathematics, *Lester Senechal,* Editor.
19. Visualization in Teaching and Learning Mathematics, *Committee on Computers in Mathematics Education, Steve Cunningham and Walter S. Zimmermann,* Editors.
20. The Laboratory Approach to Teaching Calculus, *L. Carl Leinbach et al.,* Editors.
21. Perspectives on Contemporary Statistics, *David C. Hoaglin and David S. Moore,* Editors.
22. Heeding the Call for Change: Suggestions for Curricular Action, *Lynn A. Steen,* Editor.
23. Statistical Abstract of Undergraduate Programs in the Mathematical Sciences and Computer Science in the United States: 1990–91 CBMS Survey, *Donald J. Albers, Don O. Loftsgaarden, Donald C. Rung, and Ann E. Watkins.*
24. Symbolic Computation in Undergraduate Mathematics Education, *Zaven A. Karian,* Editor.
25. The Concept of Function: Aspects of Epistemology and Pedagogy, *Guershon Harel and Ed Dubinsky,* Editors.
26. Statistics for the Twenty-First Century, *Florence and Sheldon Gordon,* Editors.
27. Resources for Calculus Collection, Volume 1: Learning by Discovery: A Lab Manual for Calculus, *Anita E. Solow,* Editor.

28. Resources for Calculus Collection, Volume 2: Calculus Problems for a New Century, *Robert Fraga,* Editor.

29. Resources for Calculus Collection, Volume 3: Applications of Calculus, *Philip Straffin,* Editor.

30. Resources for Calculus Collection, Volume 4: Problems for Student Investigation, *Michael B. Jackson and John R. Ramsay,* Editors.

31. Resources for Calculus Collection, Volume 5: Readings for Calculus, *Underwood Dudley,* Editor.

32. Essays in Humanistic Mathematics, *Alvin White,* Editor.

33. Research Issues in Undergraduate Mathematics Learning: Preliminary Analyses and Results, *James J. Kaput and Ed Dubinsky,* Editors.

34. In Eves' Circles, *Joby Milo Anthony,* Editor.

35. You're the Professor, What Next? Ideas and Resources for Preparing College Teachers, *The Committee on Preparation for College Teaching, Bettye Anne Case,* Editor.

MAA Reports

1. A Curriculum in Flux: Mathematics at Two-Year Colleges, *Subcommittee on Mathematics Curriculum at Two-Year Colleges,* a joint committee of the MAA and the American Mathematical Association of Two-Year Colleges, *Ronald M. Davis,* Editor.

2. A Source Book for College Mathematics Teaching, *Committee on the Teaching of Undergraduate Mathematics, Alan H. Schoenfeld,* Editor.

3. A Call for Change: Recommendations for the Mathematical Preparation of Teachers of Mathematics, *Committee on the Mathematical Education of Teachers, James R. C. Leitzel,* Editor.

4. Library Recommendations for Undergraduate Mathematics, *CUPM ad hoc Subcommittee, Lynn A. Steen,* Editor.

5. Two-Year College Mathematics Library Recommendations, *CUPM ad hoc Subcommittee, Lynn A. Steen,* Editor.

These volumes may be ordered from:
The Mathematical Association of America
1529 Eighteenth Street, NW
Washington, DC 20036
800-331-1MAA   FAX 202-265-2384

You're the Professor, What Next?

Table of Contents

Ideas and Resources for Preparing College Teachers

Part I — In the Beginning

These chapters explain how the AMS-MAA-SIAM Committee on Preparation for College Teaching came into being and describe its early work. Statistics and anecdotes are included which reflect the varying teaching climates and demands in doctoral departments. There is analysis of survey data that the committee gathered to inform its recommendations.

Chapter 1 – The Committee on Preparation for College Teaching 2

Introduction 2

Anecdotes 4

First Report 7

A Dialogue — What do future college teachers need? What is feasible in Ph.D. programs? 9
Bettye Anne Case ◇ Richard Millman ◇ Stephen Rodi ◇ Ivar Stakgold

Chapter 2 – The Survey Data *Bettye Anne Case and M. Annette Blackwelder* 14

Part II — Shared Responsibilities

This section emphasizes that preparing graduate students and orienting new faculty for their tripartite faculty role are responsibilities shared by doctoral and hiring departments. A general model and some pragmatic suggestions are provided for departments developing a plan. Mathematicians seeking the same goals outside a structured program are given guidance.

Chapter 3 – Getting Started 32

Whose Responsibility? *Bettye Anne Case* 32

From Grad School to Tenure: How to Support Young Mathematicians *Stephen Rodi* 34

Chapter 4 – The Professional Seminar 38

Chapter 5 – A Broad Mathematical Preparation 42

Mathematical Content and Teaching *Guido L. Weiss* and *Stephen B. Rodi* 43

Part III — Programs in Doctoral Departments

These eight university mathematical sciences departments cooperated in the project assisted by the Fund for the Improvement of Postsecondary Education (FIPSE). The program mentors individually developed activities suited to the experiences and needs of their advanced doctoral students. Special seminars and courses devoted to broad-based professional growth were established. The seminars emphasized many facets of college-level teaching.

Chapter 6 – The University of Cincinnati *Edward P. Merkes* 46

Chapter 7 – Clemson University *Joel V. Brawley* 52

Chapter 8 – Dartmouth College   *Kenneth P. Bogart ◇ Marcia J. Groszek* 60

Chapter 9 – The University of Delaware   *Ivar Stakgold ◇ Judy Kennedy ◇ M. Zuhair Nashed* 66

Chapter 10 – Harvard University   *Daniel L. Goroff* 71

Chapter 11 – Oregon State University   *Dennis Garity* 77

Chapter 12 – The University of Tennessee   *Henry Frandsen ◇ Sam Jordan* 82

Chapter 13 – Washington University   *Steven G. Krantz ◇ Robert McDowell* 86

Part IV — Invited Papers

These chapters include a variety of essays contributed for this book. Some were written by mathematicians about specific aspects of programs to broaden the training of graduate students or to orient new faculty; others consider the state of the profession more generally and give helpful information for new professors.

Chapter 14 – Preparing New Faculty 90
   Clifton Corzatt & Paul D. Humke ◇ Mary W. Gray ◇ Diane Herrmann ◇ Herbert E. Kasube
   Edith H. Luchins ◇ Richard D. Ringeisen ◇ Pat Shure ◇ Richard Tapia ◇ Janice B. Walker

Chapter 15 – Views and Experiences 106
   Manuel P. Berriozábal ◇ Donald W. Bushaw ◇ Donald R. Cole ◇ Stephen A. Doblin
   Ed Dubinsky ◇ Michael Freeman ◇ Leonard Gillman ◇ Bruno Guerrieri
   Stephen G. Krantz ◇ Brian Lekander ◇ Eleanor G. Palais ◇ Wanda M. Patterson
   T. G. Proctor, Susan J.S. Lasser, Mary Ellen O'Leary & Robert W. Snelsire
   A. Selden & J. Selden ◇ Barbara E. Walvoord ◇ Ann E. Watkins ◇ Roselyn E. Williams

Part V — Resources

This is a rich supply of information, references and resources which can be used for personal growth or to organize programs that prepare faculty for the full range of their academic responsibilities.

Chapter 16 – Professional Societies and Governmental Agencies 148

Chapter 17 – Materials and References 155

Appendices

Those interested in teaching in two-year colleges should consider carefully the recommendations in Appendix A. In addition to the essays written specifically for this book, short articles from a number of journals and newsletters have been selected and are copied in Appendix B with permission. The wide range of topics and opinions help an individual reader sample ideas from the mathematical community.

Appendix A – Recommendations of the American Mathematical Association of
   Two-Year Colleges   *Gregory D. Foley* A-1

Appendix B — Reprints of Selected Short Papers B-1

   Careers and Graduate Education B-I-1

   Social Issues B-II-1

   Undergraduate Education B-III-1

Index of Authors I-1

Preface

Most new PhDs from American doctoral mathematical sciences programs move on to teach in colleges and universities. While the excellence of their mathematical education is recognized around the world, systematic preparation for teaching and other professional functions distinct from research has often been lacking. This book is designed to improve this situation by providing a convenient collection of materials for use in efforts to improve preparation for teaching in mathematics. It can be used as a reading resource for the college teacher preparation components of a doctoral program, as a reference manual for mathematicians already embarked on teaching careers, or simply for browsing. Some graduate students and faculty members will want to read it straight through.

The Committee on Preparation for College Teaching hopes that many mathematicians are already convinced graduate departments should play an expanded role in preparing new PhDs for all aspects of their future careers and that activities like those described in this book will become an integral part of doctoral and postdoctoral programs in mathematics. The reports and resources we present can help senior mathematicians enhance their graduate programs or more effectively mentor junior mathematicians both before and after the PhD. Careful planning and preparation for the teaching and service aspects of a productive academic life are important.

The philosophy and work of the Committee on Preparation for College Teaching are described in the first chapter while the committee's recommendations are woven throughout the text. Later chapters analyze surveys and describe the efforts of the eight doctoral departments that participated in our FIPSE grant project. (Most of the illustrations were created by Henry Frandsen to portray the intents and activities of the participating program at the University of Tennessee. He generously permitted us to illustrate our thinking at many places in the book with his art and drew additional sketches at our request - beginning with the keynote sketch introducing Chapter 1.) The Committee appreciates the many thoughtful essays written for this book. They provide a rich mosaic of the best in thought about teaching college mathematics. The articles reprinted in the appendices are from an initially larger collection which we were generously given permission to reprint by authors and original publishers. We regret that space considerations did not permit us to reprint all of those articles. (The annotated Table of Contents and the Author Index are provided to facilitate use of this material.)

The committee's work was influenced in a number of ways: By the discussions resulting when we sponsored presentations at professional meetings; by colleagues who responded with suggestions after a limited distribution of some of these materials; by mathematicians early in their careers who have told us things they would have wished for; and by the climate of thought represented in publications and in private communications within the mathematical community during the last five years. We thank each of those who helped to shape our thinking and thus enabled us to write this book.

We leave many questions unanswered or partially answered, both practical and philosophical. Some are addressed in our *First Report* (see "Realities" page 8 and "Open Questions" page 9). Some are being considered through other efforts within the professional societies; for instance the Committee on Professional Recognition and Rewards — like this committee established by the Joint Policy Board for Mathematics — has asked questions like "What are appropriate manifestations of scholarly work by mathematics faculty?" and thereby opened discussion about appropriate research emphases for doctoral students. Some are long-standing: How can communication be improved between departments producing PhDs and employing departments? Some questions we hope are temporary, such as those resulting from the general downturn in academic hiring.

We appreciate the efforts of all the people who contributed to the production of this book. The office staff at Florida State University helped throughout the project, and crucial staff support was provided at times at Austin Community College, the University of Texas at Austin, Washington State University and the University of Washington. Brian Lekander, the Program Officer at FIPSE for our project, not only gave sound guidance and provided information to assist the project, but he attended a meeting, spoke, and contributed an essay for this book. We were assisted by Elaine Pedreira and the MAA staff personnel and by the copyreaders, June and Richard Kraus. Melissa E. Smith prepared the manuscript copy in TEX, cleverly overcoming a multitude of technical challenges.

As editor, I owe some special thanks: Colleagues at both the University of Texas in Austin, where I spent the first semester of preparation of this manuscript, and at Rice University provided advice and arranged communications support which facilitated the work. As this effort goes to press, I wish to express my personal gratitude to Don Bushaw, Bob Phelps and Steve Rodi for the commentary each contributed and their many thoughtful suggestions and critical appraisals at every stage of this production and, in recent months, on a daily basis. They, and Annette Blackwelder, on whom I depended heavily, were involved with both coordination of the site projects and in the production of this book. Their imprints are on virtually every page. My family and my professional associates in other endeavors have been patient when this project demanded priority.

Bettye Anne Case

Tallahassee, Florida

April 1994

You're the Professor, What Next?

Part I

In the Beginning

Chapter 1 – The Committee on Preparation for College Teaching

Chapter 2 – The Survey Data

These chapters explain how the AMS-MAA-SIAM Committee on Preparation for College Teaching came into being and describe its early work. Statistics and anecdotes are included which reflect the varying teaching climates and demands in doctoral departments. There is analysis of survey data that the committee gathered to inform its recommendations.

The Committee on Preparation for College Teaching

Introduction

Anecdotes

First Report

A Dialogue

Introduction

In asking "What else do I need?", the new PhD above highlights in a few words the charge of this committee. All aspects of the preparation of doctoral students for academic careers are of concern to the committee. The foremost goal has been to make recommendations, suggest and test structures, and prepare material designed to improve teaching competence and smooth the transition from doctoral student to junior faculty. Means for implementation of the desired mathematical and pedagogical goals necessarily will vary among doctoral programs. Typically this will begin with some form of teaching assistant (TA) orientation and training. But such programs are incomplete when compared with what is discussed in this book. The litmus test of the difference is simple: In this project the focus is on the graduate student's future responsibilities as a faculty member rather than the immediate problem of preparing the graduate student to do a better job in tomorrow's class as a TA. It is desirable that doctoral students have opportunities to confront professional issues early and forge answers to questions in conjunction with peers and seasoned, knowledgable, and effective faculty members of the departments in which they are studying. A secondary benefit for the doctoral student is the development of close relationships to more faculty than those in the student's specialty.

A lot of attention is given in this book to providing information about the diversity of academic institutions in which mathematics is taught and to encouraging adequately broad preparation for a variety of employment destinations. It is important that the graduate student be able to adjust to whatever type institution ultimately provides employment. Dramatically different proportions of research, service, and teaching are expected in different institutions. Graduate students need to know that. And hiring departments need to make their expectations in these matters clear. Newly appointed faculty should not be treated as subjects about to take a test without being given syllabus and text. If for no other reason, the expense of recruiting faculty and the difficulty of replacing them should encourage departments to do everything possible including providing ample, even generous, support to make sure that new junior faculty make a successful transition from graduate school.

In addition to providing material for use with graduate students, this book provides information and resources which will make it easier for mentors of new faculty to help them in the years preceding tenure. Many of the essays, prepared for a primary audience of advanced graduate students and their faculty mentors, will be equally helpful for postdoctoral instructors and other junior faculty.

In this introductory chapter, we give some information about the environment in which the Committee on Preparation for College Teaching began work. The committee members decided early to undertake the collection of information to give guidance to its recommendations. A survey of deans indicated reasonable satisfaction with preparation of junior faculty members in mathematics but pointed to the need for better communication among all departments preparing new mathematics faculty. A universal statement was that most new faculty had received little, if any, pedagogical training and almost always required a period of adjustment from graduate school to the frequently remedial nature of their new course loads.

The committee decided to seek funding for collecting additional information and for disseminating that information. Without such dissemination, these efforts would likely reach only a narrow audience and be useful only on a small scale. The committee was awarded a grant from the Fund for the Improvement of Postsecondary Education (FIPSE, Department of Education) that allowed it to accomplish both of these goals and also to assist eight pilot projects for advanced graduate students in doctoral departments, thus implementing the committee's ideas in an experimental setting.

These efforts together constitute the funded project "Preparing the College Mathematics Teachers of the Future" and are described in several chapters which follow. Chapter 2 reports statistical information which supports earlier data and anecdotal information contained in the committee's early actions and recommendations. Chapters 6 through 13 contain detailed reports from the eight experimental sites. Finally, more information useful to those who want to run similar programs in their graduate departments is reprinted in Appendix B; additional resources and references are described in Chapters 16 and 17.

The project had to take into account the great diversity of graduate mathematical science programs in the United States[1]. Some dimensions of this diversity are sketched in the following list, adapted here from a publication of the American Association for Higher Education which is listed in the **Selected Bibliography** of Chapter 17 (see Case and Huneke):

(1) More than 200 departments offer the PhD degree; some are in mathematics as traditionally defined, and others are in statistics, operations research, or applied mathematics. Some 600 departments offer the master's degree in mathematics.

(2) For many reasons, departments vary widely in the kinds of teaching roles assigned to graduate students. At some universities, usually private, graduate students do not account for much of the teaching. For example, Princeton University and the University of California, Berkeley, are ranked closely with regard to the quality of their mathematical research faculty but the responsibilities assigned their teaching assistants are very different (for an explanation of "rankings," see **Summary** reference [G], Chapter 2). Dartmouth College and Florida State University graduate students also experience very different teaching worlds, though the mathematics departments are similarly ranked. State institutions, such as Ohio State University, the University of Texas at Austin, the University of Michigan, the University of Wisconsin, and Purdue University, all highly ranked, must demand more productivity from their teaching assistants — and typically give them more early help in the mechanics of teaching — than do many private schools.

(3) Regional accreditation agency requirements affect assignments of duties to teaching assistants; for example, institutions in the Southeast are not permitted to assign full classroom responsibility until a graduate student has successfully completed eighteen graduate hours in mathematics.

(4) Some orientation programs for teaching assistants begin the summer before duties begin, with stronger programs continuing for several semesters. Most de-

---

[1] More information about the environment and programs in doctoral departments is available in publications listed in Chapter 17, e.g., *Educating Mathematical Scientists* (Board on the Mathematical Sciences), *Graduate Education in Transition* (Conference Board of the Mathematical Sciences), and *A Challenge of Numbers* (National Research Council).

partments require graduate students to participate in TA training activities; in some, however, many or all activities may be voluntary.

The committee members actively sought from colleagues information and opinions related to the charge. Some of this is framed in the second section of this chapter as *Anecdotes*. This section is included as a reminder that there are some serious problems to be overcome by mathematicians who are having problems with their teaching, and that they need help to do this.

The third section of this chapter, *First Report*, formed the intellectual basis of the committee's subsequently funded FIPSE proposal. The programs in eight departments partially assisted by the project (described in Chapters 6-13) put into practice and tested the recommendations of this report. The charge to the committee and its initial understanding of that charge are explained in the First Report. It may be well to mention that, because statistics and operations research appeared to have been excluded in the charge and formulation of the committee, no such separate departments were included among the pilot sites although some of the pilot site departments included these areas.

The committee ran panel discussions at several national meetings and sponsored two sessions of invited papers. Those sessions were a two-way street: the committee collected information that helped the FIPSE project and formulation of recommendations; the information disseminated by the presentations was available for immediate application. Most of the essays of Chapters 3, 4, 5, 14, and 15 had their genesis in these paper sessions. The last section of this chapter, *A Dialogue*, includes comments made by three speakers at a committee presentation early in the work of the funded project. This was a well-attended session, and participants took advantage of the opportunity of open microphones to join the discussion.

Anecdotes

Editor's Note: *The word pictures below deal with adjustments of new teachers to their students (and vice versa). The path is not always smooth. The real incidents on which these are based occurred at about the time the committee began this book. The authors have disguised the principals for this article, and, for this reason, we have not identified these authors. We do, however, appreciate their thoughtful contributions. It is our hope that awareness of these sorts of awkward situations may help others to move confidently to find solutions when faced with similar problems.*

Several of the incidents occurred when faculty or students were from or appeared to be from other cultures, and there are overtones of xenophobia. This is especially unfortunate when directed toward mathematicians, since American

mathematics has been much strengthened by past waves of immigrant scholars. At this time, over half of the US mathematics PhD degrees are awarded to foreign nationals, so it is economically imperative for protecting the investment in their educations that a good number of them be assimilated into our faculties. See also Chapters 1, 2, 4 and some other material in Case, Keys to Effective Instruction (MAA Notes 11 which is reviewed in Chapter 17 of this book). The intent of that work was to help graduate students and other part-time instructors in doctoral departments, but some of the information and suggestions there for international teaching assistants may find wider application by beginning faculty members and their mentors.

Passing the Torch

Some years ago, we had a graduate student who, though somewhat reticent, was sometimes given to outrageous remarks and unexpected social blunders. The faculty felt he showed the kind of poise and preparation in practice lectures that justified putting him into a course of his own. However, his first teaching experience was a disaster, and it fell to me, as advisor to graduate students, to try to pick up the pieces. Bill (not really his name) wanted to go into a teaching career, but I had to tell him that, under ordinary circumstances, he would not teach on his own again as a graduate student. In talking with him, I became convinced that he had failed at all aspects of teaching — from preparing his lectures to designing his exams, from interacting with students to choosing their homework.

I offered him the chance to serve as an apprentice in my discrete math course, sharing the course planning, helping choose the homework, helping make up the exams, and sharing the lecturing. My two conditions were that he had to agree in advance that I and I alone had the final say on when he was ready to progress from one stage of responsibility to the next, and that he had to observe me in everything I did in teaching the course. As part of deciding when he was ready to lecture, we agreed he would give practice lectures to me the night before he planned to give them in class, and I would critique them and decide whether he could give them the next day. We talked a lot about how silly remarks meant in fun could alienate a class member; about the value of having worked an example to the final number before trying it on the blackboard; about how to respond positively to the apparently most silly questions; and about what preparing a lecture actually means. I showed him my lecture notes; I showed him how I would develop the notes by trying out alternative approaches; how I would practice from the notes on sticky points; and then how I would use the notes only on those occasions when I wanted to state the numbers in an example exactly the way I planned them, or to jog my memory if something seemed to me to be out of place.

After about three weeks, he gave his first lecture, and I sat there and took notes - not on his subject, but on his style. We went back to my office for an hour right afterward and went over my notes in detail. He tried again a few lectures later, and this time the debriefing took less than a half hour. From then on he took over half the lectures, and, by the second-to-last week of the quarter, I told him he no longer needed me to help him prepare, but I recommended that he still go to the room where we worked in the evening, to practice.

In the last week of the quarter, I left town for two of his lectures, confident that the class was well served. The student course evaluations were positive, giving Bill a sense of accomplishment. However, my real understanding of just how far he had grown came months later when we attended the lecture of a famous mathematician, giving a talk aimed at undergraduates. The lecture was full of insight, and brought its subject essentially to the level of the students, but nonetheless I could tell it fell flat. As we walked out of the room together, Bill said to me "That could have been a great lecture if only he had prepared his examples in advance and practiced." He was right, and, though still a graduate student himself, he was ready to reach out to offer help to another.

Asians and Racist Rumors

Case 1: When I served as assistant department chair, it was my lot to admit students at registration time to classes essential to their progress but closed due to high preregistration numbers. I am generally reluctant to increase a full class section if any other solution can be found for the student. In a semester when we had three sections of Calculus III, the two-day sections were heavily enrolled, but the evening section had ample space. When a student appeared with a concern about her ability to return regularly from a school practicum assignment in time to make the day classes for which she was preregistered, I suggested the evening section, which had as instructor Professor A from my first anecdote. The student requested transfer to the other day section, and I refused on the grounds that this section was simply too full and there was another solution available for her problem. She burst into tears and admitted she was scared to death of Professor A. Her roommate had told her that no one could understand him and most people flunked his classes, and she felt entitled to a "real American" like Professor C, who was teaching the other day section. She was quite surprised to be told that despite having fair skin and blue eyes, Professor C was born in Eastern Europe and is not a "real American." In fact, Professor A has been in the USA several years longer than Professor C. Professor A is a naturalized citizen and Professor C is a resident alien. Both speak in accented, but understandable and correct, English. Professor C did a PhD in topology at a highly prestigious university and is certainly not "easy." The student left, still in tears and still convinced her roommate's assessment of Professor A was correct.

Case 2: In summer session two years ago, an un-

usually weak group of students happened to sign up for my section of calculus. Many of them were really students from other colleges, but their parents lived nearby and they were trying to pick up mathematics credits in the summer to transfer back to their primary institutions. The status of these students meant they had not pre-tested for placement. If they had been pre-tested, several of them would probably been placed in an algebra course. The first exam was about the same as those given with reasonable results in prior semesters, but it was disastrous for many of these students. The only "A" was earned by the one Chinese student in the class. Afterwards, when the students compared notes, several irate young men confronted me, wanting to know why I was so unfair. Didn't I know that those Chinese and Japanese people have different brains from us and isn't it unreasonable to expect so much from Americans? They cited the success of Japanese industries as proof of their argument. I was not sympathetic and several of them withdrew.

What Performance Can I Expect From Students?

Professor T began teaching in a tenure-track appointment with enthusiasm and confidence from being named "best TA" and from strong teaching practice as a graduate student. He was aware, even dismayed, that a number of his new students were not well prepared and seemed likely to fail courses if the course requirements were those with which he was familiar. Without asking for information or advice, he "adjusted" his expectations.

Some colleagues quickly became aware that students who passed Professor T's course often had difficulty in other and subsequent courses. However, his (informal) faculty mentor felt it would be intrusive to comment. By late in Professor T's third year engineering faculty were complaining loudly. The statistical differences between his sections and most others was well established. Finally, "official" action was taken to correct the matter and Professor T received the guidance which could have been given him much earlier.

Professor T's failure to see a problem and seek advice in this and other correctable matters — and the timidity of faculty members who wanted to help him — could have resulted in a less happy outcome. As it worked out, Professor T was awarded tenure in the last year in which consideration was possible.

If I Use Good English Will My Students Understand Me?

Case 1: Professor A is Korean, a former mathematics instructor at a major Korean university who later came to a midwestern state university to do a PhD in analysis. He was one of our first foreign-born professors and as such was subject to close scrutiny when he joined our department in the early seventies. We were satisfied with the public lecture he gave as part of the interview process, but we still wanted to be sure that students would understand his speech. As a member of his review committee, I visited a graduate class he was teaching in real variable analysis. He presented a lengthy proof of a theorem. When he had filled three panels of the chalkboard, he suddenly went back and changed several of the verbs in the proof. He then turned to the class and said, "Excuse me, I forgot I was working in the subjunctive." As he continued, several graduate mathematics students could be seen asking each other what in the world a subjunctive was. I left the room so I could laugh. We appointed Professor A.

Case 2: Professor B is Chinese. He first came to our college from Taiwan as a master's-level graduate student and worked for me as a tutor for basic skills students in a remedial program for entering freshmen. His English was not the best, but he was a good learner and a former high school teacher in Taiwan. He was sympathetic and good at seeing student difficulties, so was quite popular as a tutor. We hired him after he completed a PhD. I visited him as a member of his review committee in a section of a freshman mathematics course for liberal arts students. During the lecture he referred to the arithmetic mean, properly placing the accent on the third syllable of arithmetic. Two members of another ethnic group who were students in the class thought this was a funny Chinese pronunciation of arithmetic and literally laughed at him. He showed some embarassment at their response and hurried on. I knew both students from another program, so I cornered them after class. I told them about how *arithmetic* is pronounced when used as a noun (with the accent on the second syllable) and how it is pronounced when used as an adjective (with the accent on the third syllable, as Professor B correctly used it). They were quite shocked. I then asked them, "Why do you believe I have told you the truth about how mathematical English is pronounced, but you don't trust Professor B to pronounce it correctly?" They immediately replied that he didn't look like someone they should trust, and we had a serious discussion about the implications of that statement.

Do As I Say...

We arrange for a tenure-track faculty member to go to an early class each time a graduate student is in charge of that student's own course or section of a multisection course. I remember when a colleague went to the first lecture of a student about whom we had worried somewhat after some practice lectures. When I asked him how the student did, the reply was, "He had about ten students there and he gave a fine lecture. But he didn't ever really look at the students. He didn't know they had tuned out two minutes into the lecture, and he just went on and gave that lecture. Ten students, and he just gave that lecture! He even asked if there were any

questions, and when there weren't – because they didn't know what was going on – he went on and gave that lecture."

How does this happen? Perhaps in this way: I can remember being impressed as a graduate student with the professor who managed to condense each chapter of Kelley's *General Topology* into a single lecture. It was exciting — frightening but exciting — to see him deliver these one-hour masterpieces, and to realize when I went over my pages of notes that it was really all here, the essence of Kelley's entire book, condensed into the first few weeks of a functional analysis course. But just because I was thrilled as a graduate student doesn't mean that that style is appropriate for the undergraduates I teach. This is something we try to teach our graduate students early on, but it clearly does not always get through to them. In a graduate course our goal is often to convey the most material possible to a motivated audience through the only medium available, a lecture. Thus our students, impressed as they may be with our prowess, take this as an example of the best teaching they can remember, and it becomes a prototype for them. If there was ever a time for "Do as I say, not as I do," this is it. If we believe we can't afford the time as we teach graduate courses to show students how we would teach undergraduates, we must do the second best thing – we must tell them how to teach undergraduates, ask them to observe our own efforts with undergraduates that model varied teaching methods, and act as mentors for their early experiences.

Cultural Educational Expectations

Professor D is from India. He came to our college via a Canadian PhD program in which his major advisor was also an Indian. His language usage was fairly good, but he had constant arguments with his students. The basic problem seemed to be that Professor D had no idea what an American high school was like and didn't care to find out. He expected a student background which scarcely exists in this country and was bitterly disappointed in most of his students. He was counseled extensively by his colleagues, but to no avail. He was discharged at the end of his third year for his refusal to adapt and constitutes my only example of a foreign-born faculty member who did not work out. He accepted a contract with a second college, then broke it the last week before classes began to take an industrial computing position.

If the Shoe Fits...

This old refrain is used to title a report[2] about a fascinating series of skits, "Micro-Inequities," which have been presented at national joint mathematics meetings.

[2] The skits were originated by Sue Geller and Pat Kenschaft. The quotation is used in a column by Barry Cipra reprinted in *Winning Women Into Mathematics*, p. 3 (see Kenschaft, Chapter 17).

Through playacting, actual incidents which occurred in the preceding months in the lives of women mathematicians are portrayed "for laughs" in a manner which does not belie their seriousness. (An example too simple and happening too frequently to deserve a skit is: A student asks a secretary, "Where is Dr. A?" The secretary answers "Dr. A is down in 210" — an office twenty feet down the hall. The student, arriving at the door of 210, sees Dr. A at her computer terminal and looks bewildered. One of three comments will usually follow: "Where is Dr. A?" or "Uh, Mrs. A?" or (least frequently) "Are you Dr. A?". The third response is not, alas, voiced by the majority of students — even by the majority of women students, and even less often does the student confidently begin, "Dr. A, I was told that you can advise me about...".) The book *Winning Women Into Mathematics* is full of good information about this general situation.[3]

First Report
How Should Mathematicians Prepare for College Teaching?[4]

Background of This Study

In 1987, AMS, MAA, and SIAM representatives requested that the MAA establish a committee on Preparation for College Teaching; Guido Weiss was the initial chair. The committee's charge is:

> *To make recommendations concerning the appropriate preparation of college teachers of mathematics, taking into account the varieties of institutions of higher education, the diversity of the mathematical sciences, the backgrounds and career interests of college students, the impact of computers, and insights from research on student learning.*

Consideration of these topics by our discipline is timely. The "professor as teacher" and the centrality of teaching to the university and college mission are high profile topics today in publications like *The Chronicle of Higher Education* and in the popular press (e.g., *Profscam* by Sykes). Mathematicians of previous generations took great interest in teaching and considered it to be an important part of their profession. In biographical articles (such as those of American mathematicians in *The College Mathematics Journal*) one senses a natural assumption by young, but soon to be prominent, mathematicians that teaching several sections of undergraduate students was a principal reason they were hired at their colleges and universities. They taught with an ap-

propriate eagerness and enthusiasm both because that is what was expected of them and because they instinctively knew that this was the field where the seed corn of the next generation of mathematicians would be nurtured.

As regards teaching today's undergraduates, many faculty do not see an encouraging picture. There are only a few well prepared students who work hard; the declining numbers of American mathematics graduate students and undergraduate majors add further discouragement. The full participation of the faculty and graduate students in the exciting explosion of mathematical research in this country, a very good thing in itself, often diverts energy and attention from teaching. In particular, it may make it more difficult to encourage faculty interest in activities better preparing their graduate students for future faculty responsibilities. There are serious problems in this country's overall educational system which produce students often ill prepared to benefit from 'effective" college teaching. But putting real effort into enhancing the quality of college and university teaching is a constructive action for mathematicians, in the hope that other segments of the academic and educational establishment also will do their part.

The committee feels that a certain restoration of balance is needed. Before facing the complex academic situation today, it is appropriate and necessary that a component of a graduate student's preparation be aimed at effective teaching; assuring this is the doctoral department's responsibility. Our investigations and dialogue have implications for undergraduate as well as graduate programs and indicate need for involvement of a broad group of our colleagues.

Key Principles

The committee's recommendations center on three ideas; two apply to all graduate students in mathematics and the third concerns implementation.

I. Graduate education should not be limited to specialization in narrow areas related to thesis topics or current areas of research. The student needs to be prepared to meet a wide range of professional responsibilities. Narrowness is shortsighted. Breadth of knowledge forms the background for teaching a variety of courses, for advising students, and for recognizing different kinds of talent in mathematics. Further, the increasing interrelation of ideas at advanced levels of research makes it imprudent to be ignorant concerning major branches of mathematics. Besides being the mark of an educated person in the discipline, a broad background is necessary if one is to be open to new mathematical ideas. It is essential if one is to be discerning when evaluating peers, hiring faculty, organizing curricula, and managing resources in support of student learning.

II. There should be systematic attention to issues and excellence in the teaching of mathematics. This attention includes giving positive reinforcement to teaching assistants and junior faculty who are performing well,

---

[3] See also references of the ***Summary*** data report of Chapter 2: [B], [Ha],[He], [J], [S], and [W].

[4] Reprinted with permission from *Notices* of the Amer. Math. Soc. 36:10 (December 1989), pp. 1343–1345.

helping those who are teaching inadequately, emphasizing that teaching is important, and providing a basis of information for graduate students who may be faced with very different teaching circumstances as faculty.

III. Doctoral departments have the responsibility for assuring the principles above. Long-term success depends on individual efforts blended and fortified by collegial resolve. In particular, a doctoral department agreeing that broad mathematical exposure is desirable will devise strategies to be sure all graduates have seen a wide spectrum of advanced undergraduate and graduate topics.

As regards the teaching of mathematics, many graduate departments have an orientation and training program for teaching assistants which does a fine job of training in the survival skills needed for their duties in the atmosphere of the graduate institution. But more is needed to support the lifelong challenges of academic mathematics. The mathematics department seeking to assure a resource base for future teaching will want each graduate student, after some teaching experience, to be involved in an explicit and serious introduction into what is known about effective teaching. This will be developed by role models from its faculty; senior faculty will stimulate a respect for teaching which is emulated by graduate students and junior faculty.

Departments that acknowledge these needs and accept the responsibilities will find creative and effective means, perhaps widely varying, to succeed.

Realities

The Committee recognizes serious difficulties in implementing these principles. The very good model of graduate preparation for research excellence which focuses on deep understanding of a problem and then spirals out to a broader knowledge related to that problem appears to discourage early "breadth" in graduate programs. Limitations on resources may check good ideas and intentions in the absence of strong and innovative resolve. There are many problems in our educational system which impede progress but are beyond the Committee's charge or influence. Here are a few: (1) There is a limit to the length of time a graduate student is supported. (2) Graduate students and junior faculty sometimes see a negative incentive to spend time on teaching: Absence of student complaints may generate adequate teaching ratings while time spent on teaching may be considered "stolen" from research. (3) There is often perceived to be real difficulty in evaluating teaching effectiveness. (4) The current small numbers of US citizens entering graduate programs and consequent recruiting competitiveness limit departments from unilaterally adding requirements, however worthy. (5) Even when there is departmental commitment to the principles of mathematical breadth for graduate students and of providing a resource basis for the demands of future teaching, it is unlikely there is faculty unanimity.

Recommendations

The Committee has some ideas for implementing programs to achieve the proposed goals. But the Committee has no complete paradigm to offer.

Undergraduate Programs. Traditionally, mathematics faculty were expected to be able to teach almost all undergraduate mathematics courses. The general understanding and intellectual resources of faculty members were usually adequate for planning and effectively teaching courses they had not taken. This unifying tradition, not present in other sciences, is challenged as the variety of mathematics undergraduate offerings increases. It is worth saving. The Committee sees several ways undergraduate programs can be more fully utilized to assure stronger and broader faculty preparation. First, the individual with more mathematics courses and experiences at the undergraduate level has a head start toward mathematical breadth, given a fixed time for the graduate program. Also, advanced undergraduate courses in the graduate department can play a dual role. They strengthen the baccalaureate degree of mathematics majors, while presenting information at the right level of sophistication for filling lacunae in the knowledge of graduate students. Graduate programs may take advantage of such undergraduate courses to complement their "core" programs.

Resources for Teaching. The Committee suggests that exposure to a broad spectrum of mathematics be linked explicitly to the goals of excellence in teaching. The structure of this double-edged effort might be that of a semester- or year-long course which would include both seminar talks on mathematical topics and consideration of the issues and literature in the teaching of mathematics.

Talks would be presented by graduate students to one another in an organized setting and moderated by a senior faculty member who is both familiar with the mathematics and sensitive to issues of style and presentation. The mathematical topics of the seminar would be determined by the strengths — and gaps — in members' backgrounds. Some examples of the kind of knowledge that might be shared are: dynamical systems and chaos; the story of the classification of finite groups; the $P = NP$ complexity problem; the use of the computer as an experimental device for the pure mathematician; controversy about computer based proofs (the four color problem); special functions as a background for harmonic analysis; the vocabulary of modern applied statistics; the distinction between algebraic and analytic number theory; surprising new applications of pure mathematics in the applied sciences; the importance of probabilistic models in applied mathematics.

Graduate students would deal both with content and with an audience of peers depending on them for understanding. Videotaping, probably familiar to the graduate students from their "survival training" for TA duties, might again be used. Resulting discussions about the effectiveness of the presentation would provide motivation for introduction to some of the literature in higher education. The organization of the seminar experience in itself would model the hypothesis that content should not be divorced from presentation. An introduction to the reputable and stimulating literature on teaching provides a basis for the lifelong learning and improvement characteristics of a good teacher.

Student as Teacher

The Committee knows that it is not possible to decide which graduate students will someday become teachers. The availability of quality teaching for the next generation is of vital interest to industry as well as a national priority. This is a strong encouragement for industrial, government and academic support of activities related to future teaching; and there are still the usual valid arguments in favor, e.g., that good communication skills are needed in industry as well as teaching.

Most graduate students are happy to do some teaching as part of the requirements for their departmental stipends and they work hard at it. Their personal attitudes usually go well beyond work for pay; they want to teach well, and genuinely enjoy it. A balance of duties during graduate school which includes some teaching provides an experience base of value to all students.

Open Questions

Some issues that deserve additional investigation and consideration are: examples of useful literature on effective teaching; the appropriateness and desirability of graduate programs other than the research doctorate for some two- and four-year college teachers; appropriate undergraduate preparation; the role of probability and statistics for mathematicians; the impact on mathematicians of the computer revolution and resulting implications for changes in mathematical training.

Summary

Our goal is that mathematicians be broadly knowledgeable about their subject and about teaching, that they be interested in teaching, including its challenges, and that all professors respect and strive for sound teaching. We believe that departments must take seriously their role in preparing teachers for college classrooms; considerable effort and innovation may be necessary to do this within the available resources. College and university administrators are increasingly concerned about teaching effectiveness. There is a danger that they may respond to political or parental pressure by implementing simplistic measures that will further discourage young faculty. It is up to the mathematical community to present them with well thought out proposals to insure that our faculty are well prepared and effective, in the hope that solutions can be found to meet the needs of students and teachers which at the same time support the basic principles of academic freedom and recognize the value of scholarship.

Improvement, to be lasting and effective, comes best from within. The time seems to be right to recapture some of the earlier enthusiasm for undergraduate teaching. New outlets for creative pedagogy are opened by wide availability of computers and calculators for computations and graphics; the NSF is putting resources into undergraduate instruction; some well publicized steps are being taken toward improving cooperation between mathematics departments and mathematics education programs. Mathematicians need to grab this ring of change firmly as it comes around.

Teaching as a coequal part of higher education's professorial responsibilities is a grand and noble tradition and the committee believes that teaching and scholarly creativity form a symbiotic relationship.

A Dialogue
Paradigm Meets Reality
What do future college teachers need?
What is feasible in PhD programs?

Joint Mathematics Meetings,
San Francisco, January 16, 1991

Bettye Anne Case, Organizer and Moderator

Good morning!

You've been promised here this morning a dialogue, and you'll have your part in it... and we hope we'll have a lively dialogue. The Committee on Preparation for College Teaching has a three-year FIPSE-funded project, "Preparing the College Mathematics Teachers of the Future," and serves as Advisory Board to the project. A number of the members are here with us other than the two panelists on the committee, Steve Rodi and Rich Millman; if you'd stand a second... Don Bushaw, Rich Ringeisen, and Guido Weiss. I wanted you in the audience to identify the members of the committee because your input to any of us is important, and we each want to hear your ideas.

As a result of this project, we hope to learn more about — and specifically how — graduate programs can convey need for lifelong learning and breadth of mathematical understanding and the attitudes and specific information needed to support future productive careers in the academic world. This would include attention not only to pedagogy but to the attitudes and the climate as regards teaching. We are also interested in and are encouraged by our funding agency to look closely at questions of success in graduate programs of underrepresented groups... in other words, "pipeline" issues and the mesh between undergraduate and graduate programs. We will be conducting a survey later this year and hope to be able to provide helpful information which is not now available in the mathematical community.

Let's move to our four eager panelists and let them present some ideas to you — I'll summarize at the end, and then we'll open the floor mikes. Steve Rodi and Richard Millman are going to lean somewhat toward the consumers' view — the needs view — and Ivar Stakgold and Bus Jaco will look at the producers' side, or what graduate programs can realistically do.

Stephen Rodi

In the division of labor for this panel, I was assigned the job of talking about the need for a project like the one our committee is organizing. In many ways, this is the easiest job of any panelist. I do not have to make a case. Others are already doing it for me.

Daily newspapers, the *Chronicle of Higher Education*, *UME Trends*, NSF reports, state legislatures, and federal government analyses and documents all are full of the same message: colleges and universities need to pay more attention to undergraduate teaching. Just two examples are the recently published Boyer report *Scholarship Reconsidered: Priorities of the Professoriate* from the Carnegie Foundation for the Advancement of Teaching, and recommendations in one section ("Research and Teaching") from Lynne Cheney's report *Tyrannical Machines: A Report on Educational Practices Gone Wrong and Our Best Hopes for Setting Them Right* issued by the National Endowment for the Humanities.[5] If you want something more readily at hand, try the article in the Education Section of the December 10, 1990, Newsweek. Its subtitle: "Universities are rediscovering the virtues of undergraduate teaching."

The simple fact is that a rather dramatic imbalance exists between what graduate faculties (and hence graduate students) see as important in the professional preparation of PhDs in mathematics and what most of those same new PhDs actually will be doing for most of their professional lives. As a profession, we need to be responsible citizens and respond to this national chorus of voices crying for attention to this imbalance. To accomplish such a goal, we need to sensitize graduate faculties to the needs of undergraduate education and to their responsibility in training the next generation of teachers who will meet these needs. We need to create (or re-create) a system which recognizes and rewards those within mathematics who apply their skills and talents to these educational problems. We need to broaden the notation of scholarship in our community to afford respect to those who devote much of their work to mathematical exposition and pedagogy.

As Boyer persuasively writes, there are four components of scholarship, all of which are important and need to be recognized: the discovery of new knowledge, the integration of knowledge, the application of knowledge, and teaching. And, as he forebodingly comments, we are at this very moment "training another generation of faculty who will affect the quality of campus life for years to come."

Some of you know that for ten years I was in training as a future Jesuit priest before I found my final resting place here in mathematics. The way we now train graduate students, as they get ready to step forth into their first professional positions and to face undergraduates all over the nation, is comparable to some of my seminary classes of long ago in which we discussed the fine points of the medieval theology of causality as propaedeutic to missionary work among the working classes in Central America. Somehow, the means did not fit the end!

In adjusting means to ends, I have some recommendations about how graduate programs need to treat graduate students and also some recommendations for the programs themselves and their home institutions. First, here are recommendations that apply to students in PhD programs in mathematics.

Every graduate student, at the time of acceptance into a graduate mathematics program, should receive from the receiving department an apologia for our profession which emphasizes that the student is entering a profession in which teaching of undergraduates is an important and respected component. This apologia should make the point that in all cases anyone who wants to pursue our profession should be willing, able, and at least occasionally even eager to spend part of his/her career teaching undergraduates and should be ready to accept some preparation for undergraduate teaching as part of the graduate program. The purpose of such an apologia would be to set the tone and parameters of our profession right at the beginning as people enter the profession's "novitiate" in graduate school. It makes clear that all of us in the profession must be protectors of the mathematical seed corn, of the next generation of human beings who will practice mathematics in our society and who will become our colleagues as we age.

Actual teaching should be required of all graduate students while they are in graduate school. Successful teaching should be a condition of receiving the degree. In addition, toward the end of the graduate degree program, all students should be required to participate in a "proseminar" of the kind our committee has proposed. Each graduate program should design such a seminar (or comparable experience) to make best use of its own resources. But, in all cases, successful completion of the seminar with its concentration on issues related to teaching, should be a requirement of getting the PhD degree, a requirement as essential as passing qualifier examinations and defending the dissertation.

The committee has outlined elsewhere various models this seminar might follow. These are models only and we want many institutions to develop others. But in one way or another all of them should include readings on the history and nature of pedagogy; study and discussion of teaching techniques; practice teaching with group critique; discussion of the educational system which pre-

---

[5] See Chapter 17, **Selected Bibliography**, for both references.

pares students to come to college (so that future assistant professors can get a realistic perspective on the backgrounds, learning styles, and educational needs of undergraduates); and an introduction to the many components of our profession which can support their work as teachers (MAA, AMATYC, COMAP, UME Trends, and multiple others).

Graduate programs should recognize outstanding teaching by graduate students with awards and in other ways. This tells the student early in his/her professional career that such activity is respected and is worth devoting time and effort to. And, at least in some cases, the graduate department might want to establish internship programs with nearby community colleges or local high schools as part of the training process for their students. Such internships are a common part of the training of professionals in other areas and should be given more serious considerations in mathematics. And one must exorcise from our consciousness the fallacy that dissertation directors who praise the teaching skills of their students in letters of reference are really sending coded message about the student's lack of research ability. Let outstanding references about teaching become a point of pride among us, not a kiss of death.

Now let me mention some attitudinal and practical changes that need to take place in PhD-granting departments and in their home institutions.

PhD departments need to take genuine pride in building a reputation for graduating students who are outstanding teachers. They need to see this feather in their cap as no less significant than graduating outstanding research students. They need to commit resources (people, space, time, money) to the parts of their program which develop their students as teachers. And that commitment should be comparable to the commitment which promotes traditional research. Graduate departments need to take the lead in our community in redefining the notion of scholarship. They need to respect the organization and transmission of knowledge as much as its discovery. They need to be leaders in saying that the ability to organize and distribute knowledge is a gift in its own right as important to our profession, to the work of the university, and to society as any other. PhD-granting departments and their institutions need to re-discover that the German model of the university is not the only one, was not the first model to be introduced in the US, and may not be the one best suited to meeting the needs of American society.

If we are to give fresh importance to undergraduate teaching, we also need to modify how new junior faculty are treated when they step into their first faculty positions. They need to be encouraged to make their mark by becoming known as teachers. Universities and departments must eliminate any hint that such a commitment will negatively affect their chances of tenure or promotion or salary increases or access to the perks of travel or financial support. They need to be given sab-

baticals to develop as teachers as readily as they are to write books. And, if a junior faculty member devotes a number of years to outstanding work as a teacher, that person needs to be provided release time later for research activity.

In short, all mathematics departments, but especially those among the PhD granting departments who set the style for our profession and are the touchstones against which we all measure our success, need to create an atmosphere, training programs, and reward structures in which teaching undergraduates has the highest of priorities and is accorded the highest respect. We hope that the work of this committee will help move us all down this important road.

Thank you.

Richard S. Millman

We have been asked to be controversial and stir up debate, so I will see if I can do that. I was looking at my teaching evaluations for the last quarter and decided that this was all silly stuff. After all, what do these students know about what I am trying to teach? What do they know about the subject matter? What do they know about methods of delivery? The answers are obvious. Thus, I will not pay attention whatsoever to teaching evaluations because after all, damn it, I have standards. My courses are demanding, and grades were not high. That's the reason I get such low ratings.

The problem with the stream of consciousness above is that there is literature on student evaluation of teaching. None of it claims that student evaluation of teaching is the most wonderful thing in the world, but it does show, for example, that students do not give you a high evaluation because they expect a high grade. It does not say that the more you run and jump around, the better evaluation you are going to get. It is a different kind of phenomenon.

It is not the particular question about how you measure teaching that I want to address. Rather, I want to emphasize that there is, in fact, some written work on the subject of how effective student evaluation of teaching is. We should acquaint our graduate students with that literature, and we should teach our junior faculty about it. We ourselves as senior faculty should also know about it. There were certainly no seminars when I was in graduate school in 1966-71 that talked about such things. We were expected to pick up good teaching by osmosis. We now recognize that quality teaching is far too important to leave to osmotic pressure.

The times have changed (as Steve Rodi put it so well). Faculty are beginning to think about the issues that students will face both in their intellectual future and in their careers. That is one thing that we absolutely have to talk about with our students. We have to work with our graduate students and ask them to remember what it is that people are going to do when they finish their mathematics degree. Not everyone is going to become a research mathematician. There are articles that are written that ask industrial mathematicians what do they do. The answer is, "I write." That is something that most practicing mathematicians don't realize. When we are working with our students who are not going to go on to graduate school, we have to make sure there is a track for those people who are going to go on to the "real world" also. We have to be acquainted with the pedagogical literature. We have to ask where our students are going.

We also must broaden the notion of scholarship. Let me mention one or two things along these lines, as much ink has been devoted recently to these ideas. In particular, Ernest Boyer of the Carnegie Foundation has been a leader in this movement. Scholarship has to be defined very broadly. I remember a colleague of mine who was writing a book and had not gotten tenure before the book was done. The tenure committee asked how much that should count. The response of the tenured mathematics faculty was, "anybody can write a book". For those of you who have not done it, try it some time. It is very difficult and time consuming. Furthermore, a good book is a significant contribution to the literature and is important. (On the other hand, I would certainly advise untenured people not to take that risk right now.) The value of written works which are not research articles is something we need to discuss with the entire senior faculty of all institutions. The appropriate notion of scholarship will depend on the university, but it may mean something other than research papers.

There are many publics to whom we teach, and all of them are important. I think we have been focusing too much on the obvious mathematical needs. We are attuned to them. We feel that we can throw our worst teachers at those folks in business. After all, they don't even need trigonometry! The elementary education, secondary education kids — well, let's see. There must be faculty who have never written an article. We can go ahead and call them math education specialists and have them teach that group. After making those kinds of teaching assignments, should we then wonder why the K-12 education is the way it is or why the Business School doesn't understand us? We must carefully tailor our teaching schedule to the interests of our faculty and then reward quality teaching to any audience. All audiences are important audiences. On a personal note, when I taught lots of calculus, it was always business calculus by my choice. I really enjoyed working with that group of students. When there are differences in learning and differences in audiences, there have to be differences in teaching methods.

There has to be cultural sensitivity. The "proseminar" that we have talked about in our committee is an important idea because it can also present examples of the work of Uri Triesman at UC Berkeley, Manny Berriozabal of the University of Texas at San Antonio, and Carolyn Mahoney of Cal State University–San Marcos (and others) about the differences in learning that

go on. An example of cultural sensitivity follows. Suppose that you have 45 undergraduate proctors helping in some of your beginning math courses, and they are all white. Furthermore, you have about 10% African-Americans in your classes. There has to be a cultural sensitivity on the part of the math department that you should make a special effort to get undergraduate proctors who are African-American. It is very, very hard to get minority faculty. Getting minority undergraduates who have finished enough mathematics so they can proctor elementary mathematics is nowhere near as difficult. Not only will we provide some role models, but maybe we will encourage some minority students to go into teaching! Teaching has become, unfortunately, a very static sport. We teach the way we were taught, and it doesn't take a long time to realize that this procedure results in a stationary point in the process. We need to make sure that new ideas such as symbolic manipulation, cooperative learning, group learning, writing to learn mathematics, *et cetera*, get introduced. They don't have to be followed as long as we know there are other teaching strategies that are available out there.

Another problem lies with academic administrators. We need to build into the reward system the policy that if people are going to devote their time (as Uri Triesman, for example, did) to helping with some of the mathematical/social problems, or to addressing different teaching needs or audiences, then they will be rewarded. The dean, academic vice president, and president should brag about how well such faculty have done. They should emphasize how important it is to get grants from the education directorate of NSF as well as the research directorate. They should say how important it is to get recognition in the community for the teaching activities of the department.

I have taken a much bleaker view of what is happening than I really believe. As Steve says, the times, they are a-changing. I believe they are changing for the good. On the other hand, I was ordered by our fearless leader to talk about the needs, and so that is what I have done. I will finish up by saying that I really am optimistic that we mathematicians are working hard on our teaching and paying much more attention to curricular issues.

Ivar Stakgold

My remarks today are in three parts: (1) A description of the mismatch between graduate programs and available jobs; (2) A discussion of the qualities desirable in a college teacher; (3) An outline of a change in our graduate program at the University of Delaware.

(1) Our present graduate education is not particularly geared to the existing job market. The mismatch is worse if we take into account lost opportunities — potential jobs that might be available to mathematicians but which our students cannot fill because of inadequate education and attitudes. To paraphrase

Jim Glimm: The things that mathematicians are uniquely qualified to do are not greatly in demand in industry. The available mathematical jobs in industry can be done by physicists, engineers, computer scientists, and operations researchers as well as by mathematicians. If we want our students to compete for these jobs, we have to provide them with a different type of graduate education than at present.

I believe most of our PhD programs are basically similar although the quality of students may vary from one university to another. Incidentally, there is a Board of Mathematical Sciences study under way of ten selected successful programs — they may well arrive at different conclusions than mine.[6] The goal of the typical PhD program is to prepare a student for a junior faculty or postdoc position at one of the around 150 research universities. The centerpiece of the program is the research dissertation intended as an apprenticeship in research. What sort of success have we had? The research establishment in the US is second to none, but the majority of our graduates do not publish beyond their PhD. I conclude that the reviews are mixed. Most of the PhDs do not spend the major part of their career in an environment where research is a primary requirement.

Accepting for the moment that our present PhD program is good preparation for a research career, I doubt whether it is suitable as preparation for college teaching. The often narrow focus of a research dissertation does not provide the kind of breadth and scholarship useful to a college teacher.

(2) What then are the qualities needed in a college teacher? Let me mention three:

◇ An understanding of the main strands of mathematics and some of their connections. A college teacher must be equipped to teach a variety of courses.

◇ An appreciation of the role of mathematics in civilization. There is something remarkable in that mathematics has attracted many great minds throughout history and has become part of our humanistic tradition. Equally important is an understanding of how mathematics contributes to science, engineering, computing, and many other fields. Mathematics is the intimate companion to science and technology. Computing is no longer confined to applied mathematics but is beginning to play an important role in many branches of pure mathematics. (Note, for instance, the new *Journal of Experimental Mathematics* which will explore experimental evidence for reasonable and promising conjectures in mathematics.)

◇ Enthusiasm and intellectual curiosity that will stimulate students in service and major courses. My idea of a college teacher is someone who reads and enjoys the *Monthly*, the *Magazine*, the *Intelligencer*, the *Notices*, the *UMAP Journal*, and *SIAM Review* (or at least a few of them!).[7] May I point out a particularly fascinating article on the oscillations of suspension bridges in the Dec 90 issue of *SIAM Review*. In it the authors, Joe McKenna and Al Lazer, present a very plausible explanation for the collapse of the Tacoma Narrows Bridge.

(3) I would now like to say a few words about a recent change in one aspect of our graduate program at the University of Delaware.

All non-thesis master's students (this is the vast majority and includes all those going on to the PhD) will be required to take two "thematic seminars", each carrying two hours of credit. One seminar will be offered each spring and themes will vary from year to year. Both pure and applied mathematicians are required to take the same seminars. The topic for spring '91 is optimization.[8]

Some features of these seminars:

(a) To show the interconnections among branches of mathematics, including some that are not normally included in our curriculum.

(b) The seminars are theme and problem oriented. We want to show how various fields of mathematics enter in the analysis and solution of a class of problems or how a general area in mathematics contributes to other fields.

(c) Computing (both numerical and symbolic) is to be an integral part of every seminar.

(d) Small-group learning with sharing of ideas among students. Promotion of interaction in class as well as group projects.

(e) Students will give presentations. Teaching skills will be stressed. The additional breadth provided by the seminars should be useful to prospective college teachers.

Besides optimization, future themes might include convexity, fixed points, dynamical systems, probabilistic methods, number theory, Fourier methods.

I am pleased to acknowledge the intellectual support given to this project by Bettye Anne Case and the financial support of FIPSE.

---

[6] See the **Selected Bibliography**, Chapter 17, for reports by the Board on Mathematical Sciences and the Conference Board of the Mathematical Sciences, the latter chaired at that time by Stackgold.

---

[7] See **Journals and Newsletters**, Chapter 16.

[8] See Chapter 9 for a description of activities subsequently developed at the University of Delaware.

Chapter 2

The Survey Data

Despite much surveying (as any departmental administrative assistant will quickly assure you), meager description is available of either the graduate student cohort or doctoral programs. The information which was collected and analyzed in this project can be used both as a predictive device for planning future strategies and a warning of corrections needed in graduate education. This was obtained through a two-stage survey designed jointly with the AMS-MAA Data Committee. The demographic information was in the same form as the Data Committee's "Survey of New Doctorates." Since our first survey was mailed as a piggy-back to that familiar survey, the response rate was high. (See Tables 1 and 2 of the *Summary* below.) In the first survey, departments reported all graduate students who had entered during the preceding twelve months, which included academic 1990-1991. The second survey was sent in the summer of 1992, the earliest time which would reflect persistence by graduate students through two academic years. In addition to demographic data, the survey included items directly related to the concerns of the Committee on Preparation for College Teaching such as departmental entrance and doctoral examination expectations and the teaching climate in the department and university. Because of the differences between departments, analysis is based on groupings of departments of similar type; the demographic data is categorized by race, sex, and citizenship characteristics. So that it would be widely and quickly available, the analysis was published in two papers in the *AMS Notices*. Those papers are reprinted below. The more recent, which includes the follow-up information, is denoted *Summary*; the analysis of the earlier reporting is denoted *First Survey*.   BAC

Doctoral students and persistence in programs. The doctoral student cohort, and, separately, those entering in 1990-1991, are described by race, sex, and citizenship characteristics. (See *First Survey*.) The entering group is followed for two years, and the resulting information about persistence is reported along with some information about whether those leaving programs in less than two years had earned a degree. (See *Summary*.) Precise answers about persistence to the PhD degree would require a long-term tracking project of at least ten years. Despite the abbreviated tracking time, some reliable conclusions are possible. In particular, it is clear — both in the entire doctoral student cohort and in the entering group which was tracked — that the numbers of students from the ethnic groups now underrepresented among mathematicians are so small that the junior faculty a few years hence will not include a significantly higher proportion from these groups than now.

Departmental expectations. There is a tabulation about the undergraduate courses expected or required as entering preparation for beginning graduate students. This information, helpful to graduate departments, is essential for undergraduate advisors and potential graduate students. The topics, format, and timing of preliminary or qualifying examinations for doctoral candidacy are described to give further information about departmental expectations. Items indicating mathematical breadth have implications about both departmental expectations and the preparation for future college teaching. Direct indication about teaching is provided by questions about graduate student teaching activities, teaching awards, and the importance of teaching qualifications in hiring decisions.

Summary:

Doctoral Retention, Departmental Expectations, and Teaching Preparation*

Bettye Anne Case and M. Annette Blackwelder

This report is based on two surveys conducted by the Committee on Preparation for College Teaching, a joint Committee of the AMS, the Mathematical Association of America (MAA), and the Society for Industrial and Applied Mathematics (SIAM). The surveys were framed jointly with the AMS-MAA Data Committee and funded partially through the Fund for the Improvement of Postsecondary Education (FIPSE) of the US Department of Education. The first survey was reported in the Notices, May/June 1992, pages 412–418.

Introduction

The mathematical sciences community, through departmental efforts and professional society and government agency initiatives, is currently considering many facets of college mathematics teaching. Among these are the current climate and attitudes regarding teaching in doctoral departments, the composition of the group of graduate students, and the retention of members of underrepresented groups in doctoral programs. Studies over many years have shown that the pool of new doctorates in mathematics, which later forms the pool of junior faculty members, does not reflect population patterns [BMS, CBMS, CT, D, NRC]. To improve the effectiveness of teaching for all students, a first step is to examine the need for and availability of role models in the professoriate from various societal groups.

At its inception in 1987, the Joint Committee on

Preparation for College Teaching felt that a paradigm of mathematical breadth and teaching-related activities was lacking in doctoral departments. Little information was available about the citizenship, sex, and race characteristics of doctoral students, and there was none about retention. Thus a proposal was made to FIPSE for funding to assist the development of some example programs and the implementation of a two-stage survey of doctoral departments. FIPSE encouraged sequential surveying for evaluation purposes. The first survey was the subject of a previous report in the *Notices* [CB]. Now the Committee is pleased to report here to the mathematical community the only sequential surveys tracking a particular beginning doctoral student cohort to determine retention rates over the critical first two years.

The wider project assisted by FIPSE centered on the development of organized activities late in doctoral work to ease the transition from pre-faculty to junior faculty. The activities are designed to encourage graduate students to engage in career-long learning and self-evaluation in three areas: mathematical knowledge, teaching, and service. FIPSE has supported the initiation of such programs at the University of Cincinnati, Clemson University, Dartmouth College, the University of Delaware, Harvard University, Oregon State University, the University of Tennessee, and Washington University. The project has shifted into publication activities and presentations and will document what these departments have done.

In its early survey of the literature, the Committee found that collected material for pre- and junior faculty about professional development in the three areas cited above was not available. The need for such information was reinforced by the experience of the faculty mentors of the site programs. A book, *You're the Professor, What Next?*, in the "MAA Notes and Reports" series is intended for those soon to enter or having recently entered the professoriate, as well as the senior faculty who serve as their mentors [C, 1994].

The Surveys, Department Groups, and Response Rates

The demographic information reported in the Committee's first survey [see CB] included race, sex, and citizenship characteristics for all doctoral students and, separately, for the cohort who entered their current department in 1990–1991. The second survey determined retention and degrees awarded at the end of their second academic year in the department for the previously reported 1990–1991 entrants. The first survey also included items about what departments expected of students and the departmental teaching climate. These items were expanded and clarified in the second survey, and this report consolidates the information about teaching from the two surveys.

Persistence through the first two years of graduate school is reported with some information about whether those leaving earned a degree. At the end of aca-

* Written as part of the College Teaching Project. Reprinted by permission of the American Mathematical Society from the *Notices of the AMS* 40:7 (September 1993) pp. 803-811.

demic year 1990–1991, the departments in Groups I, II, III (doctoral mathematics departments), and Group Va (doctoral applied mathematics departments) were asked to provide sex, race, and citizenship data both for all graduate students and, separately, for those entering their programs in the calendar year beginning June 1, 1990. Group IV, comprising separate departments of statistics, and Group Vb, comprising departments of operations research and management science, fell outside the Committee's charge [CT], which focused on mathematics teaching. (See [G] for an explanation of departmental groupings.)

At the end of academic year 1991–1992 the departments were asked to indicate how many of the 1990–1991 entrants remained in their programs in good academic standing. Several departments that did not respond to the first survey took this second opportunity to participate in the retention study and provided data about their 1990–1991 entrants for both time periods. Table 1 shows the number of departments providing demographic information at each phase of the surveying.

Table 1. Responses about retention

	Initial Data	Retention Data
Group I	38 of 39	32
Group II	42 of 43	37
Group III	67 of 88	55
Group Va	12 of 17	11
All Groups	159 of 187	135

Departmental expectations are reflected in the preparation that is expected of incoming graduate students and in the nature of preliminary or qualifying examinations for doctoral candidacy. The survey asked about such expectations, in addition to indicators of mathematical breadth, graduate student teaching duties and preparation for those duties, and additional teaching-related activities. Questions about teaching awards and the importance of teaching qualifications in hiring decisions provide some indication of departmental attitudes toward teaching. The overall responses for teaching-related information appear in Table 2; usable responses vary from one item to another.

Table 2. Responses about teaching in doctoral programs

Group I	39 of 39
Group II	43 of 43
Group III	73 of 88
Group Va	12 of 17
All Groups	167 of 187

A number of points should be made about the information on teaching-related activities. The data constitutes virtually a census on many items from Groups I and II due to the high response rates. Some of the departments in Group III are very small or award few PhD degrees [G]. These surveys may have seemed less meaningful to them, decreasing their response rate. Five departments from Group Va did not provide data. Anecdotal information indicates that at least one of those programs is no longer active, but some are known to be strong programs. In some of these seventeen departments, the graduate students do not teach. The intent of this survey may have been misinterpreted in such departments, causing a lack of response. Finally, characteristics are generally reported as percentages in the tables which follow, so they may be applied to the extent that the user has confidence that they project accurately over all the departments of a group.

Precise answers about persistence to the PhD degree would require a long-term tracking project of at least ten years. Despite the abbreviated tracking time, some reliable conclusions are possible. In particular, it is clear—both in the entire doctoral student cohort and in the subgroup tracked through two years—that the numbers of students from the groups underrepresented among mathematicians are too small to achieve balance in the professoriate. The raw baseline numbers about some minorities are so small that giving percentages is misleading. In addition, these findings predict that rates of persistence through graduate school will be lower for underrepresented groups than for white males. The data is not surprising, since it validates prevailing anecdotal information. These figures are sufficient to conclude that the junior faculty a few years hence will not include a significantly higher proportion of underrepresented ethnic groups than now. The findings are a serious concern for faculty who seek to assure an effective professoriate in mathematics.

Two-Year Retention of Doctoral Department Entrant from Academic Year 1990-1991

In 1992 each US doctoral department reported the following information about the group of graduate students entering that department in the calendar year beginning June 1, 1990: degree held at entry, percentage of US citizens, and numbers and percentages of members of underrepresented minorities. The percentages and numbers were also given for all graduate students enrolled in the departments in academic year 1990–1991 [see CB]. The second survey shows (see Table 3) how many of the 1990–1991 entrants remained in the same department or had received a degree from that department at the end of academic year 1991–1992. The degree awarded was not specified on the survey. Since the time period was two years or less, it is likely that most entrants with only bachelor's degrees (Table 4) were awarded master's degrees.

Table 3. Retention in doctoral programs by department groups

	Total Number Reported	Remaining	Awarded Some Degree	Left Without Degree
Group I	922	82%	5%	13%
Group II	756	81%	3%	16%
Group III	923	78%	3%	19%
Group Va	122	74%	6%	20%
All Groups	2723	80%	4%	16%

Degree at Entry into Doctoral Program. The degree held at entry into the doctoral program was recorded, and, unsurprisingly, more entrants holding only bachelor's degrees had left within two years with new degrees (presumably master's degrees) than had those already holding a master's degree. (See Table 4.) In all department groups the percentage remaining in the doctoral program is highest for entrants holding the master's degree.

Table 4. Retention of bachelor's only holders vs. master's holders

	Bachelor's Entrants Reported	Remaining	Awarded Some Degree	Left Without Degree
Group I	671	81%	6%	13%
Group II	529	80%	4%	16%
Group III	674	77%	4%	19%
Group Va	85	71%	8%	21%
All Groups	1959	79%	5%	16%

	Master's Entrants Reported	Remaining	Awarded Some Degree	Left Without Degree
Group I	182	85%	3%	12%
Group II	227	85%	1%	14%
Group III	249	80%	1%	19%
Group Va	37	81%	none	19%
All Groups	695	83%	1%	16%

If the subgroup of US citizens with bachelor's degrees only is compared to that with master's, most trends are similar to the cohort without regard to citizenship, including a higher percentage remaining from master's entrants. The percentages of those dropping out are consistently a bit higher for US citizens, averaging 17.5% rather than 16% as above. The most significant difference is noted in Group Va where 28% of the US citizen students who entered with master's degrees left graduate school without additional degrees.

Sex of Entrants. The "champagne glass" shape of

the graph by age group representing numbers of women studying mathematics is well known [BR, D, L, NRC]. Nearly 50% of the US citizen undergraduate mathematics majors are women. However, the entering graduate student cohort drops to 34% women, and only 22% of new doctorates were awarded to women in 1992 (up slightly from 20% in 1991) for Groups I, II, III, and Va combined. (This is based on the raw data of [D]; see also [G, J].) One also observes even lower percentages of tenured women in prestigious departments [B, HA, HE, S, W]. These are comparisons at a fixed time, hence of different age groups.

Although the percentage of doctorates awarded to US citizen women has not changed much for a number of years, the reported data gives hope that the percentage of women may rise in three to five years: 33% of the entering 1991–1992 US citizen cohort remaining after two years were women, down less than 1% from percentages of women in the entering group.

Whether one looks at the entire group in Table 5 or US citizens only in Table 6, the percentage leaving within two years without a degree is higher for women than for men when all departments are combined. (This casts doubt on the folklore which holds that the greater attrition of women students in doctoral programs is due to the fact that many of them choose to leave graduate school after receiving master's degrees.)

Table 5. Retention of women vs. men

	Women Reported	Remaining	Awarded Some Degree	Left Without Degree
Group I	226	79%	7%	14%
Group II	249	78%	3%	19%
Group III	337	78%	3%	19%
Group Va	37	59%	3%	38%
All Groups	849	78%	4%	18%

	Men Reported	Remaining	Awarded Some Degree	Left Without Degree
Group I	696	83%	4%	13%
Group II	507	83%	3%	14%
Group III	586	77%	4%	19%
Group Va	85	80%	7%	13%
All Groups	1874	81%	4%	15%

Percentages by sex were reported earlier for all graduate students and for the 1990–1991 entering cohort [CB]. Using the database of the second survey, the percentages of women in that US citizen entering cohort for Groups I, II, III, and Va were 26%, 35%, 39%, and 30%, respectively. In Group I the two-year retention of women is better than for men, although it may be significant that the entering percentages of US citizen women are lowest in Group I.

Table 6. Retention of US citizen women vs. men

Women	Reported	Remaining	Awarded Some Degree	Left Without Degree
Group I	134	81%	5%	14%
Group II	164	74%	2%	24%
Group III	223	78%	4%	18%
Group Va	22	59%	none	41%
All Groups	543	77%	3%	20%

Men	Reported	Remaining	Awarded Some Degree	Left Without Degree
Group I	381	78%	5%	17%
Group II	301	82%	3%	15%
Group III	342	79%	4%	17%
Group Va	51	78%	8%	14%
All Groups	1075	80%	4%	16%

Table 7. Retention of US citizen vs. others

US Citizens Reported		Remaining	Awarded Some Degree	Left Without Degree
Group I	515	79%	5%	16%
Group II	465	79%	3%	18%
Group III	565	78%	4%	18%
Group Va	73	73%	5%	22%
All Groups	1618	79%	4%	17%

Other Citizens Reported		Remaining	Awarded Some Degree	Left Without Degree
Group I	407	86%	4%	10%
Group II	290	85%	4%	11%
Group III	358	76%	3%	21%
Group Va	48	75%	6%	19%
All Groups	1103	82%	4%	14%

The data for Group Va shows a much larger difference between women and men than that for other department groups. Women (US citizen and others) drop out of Group Va programs without a degree at a rate three times that of men. On the other hand, the rate at which men drop out of Group Va departments is lower than the average for all groups. It should be kept in mind that the average number of students in Group Va departments is significantly lower than in Groups I and II, and that the lower response rate for Group Va departments may mean percentages do not project as accurately (see also the section, "The Surveys, Department Groups, and Response Rates," above).

Retention and Citizenship. The national media have often trumpeted the low percentage of US citizens among new doctorates in mathematics, engineering, and the physical sciences. For mathematics these percentages were 45% in 1991 and 44% in 1992. (This is based on the raw data collected for [D].) There is reason to believe that these percentages may soon rise to reflect higher percentages of US citizens in more recent entering classes. The report of the first survey indicated that 56% of all doctoral department students are US citizens [CB], with even higher percentages of entering students. The percentage of US citizens among the 1990–1991 entering cohort after two years is 59%. The retention information of Table 7 shows that rates for *US citizens* and *others* are not as different as anecdotes sometimes imply; the greater numbers of US citizens in the present doctoral cohort, combined with this persistence rate, may mean higher percentages of US citizens among new doctorates in a few years. (A reasonable anecdotal explanation for the higher dropout rate for noncitizens in Group III departments is that some international students come first to those departments and then, with improved English and perhaps US master's degrees, are accepted at Group I or II departments.)

Underrepresented Minority Doctoral Students Among US Citizens

The total numbers of graduate students from some ethnic minority groups are so small that analysis is difficult. For a more complete discussion of the reported ethnic minority students among all graduate students of 1990–1991 see [CB]. For the first survey a total of 4677 graduate students were counted in 143 doctoral departments. Of this total only 344 were from the following underrepresented minority groups: Black (African American), 134; American Indian, Eskimo, Aleut (Native American), 11; Mexican American, Puerto Rican, other Hispanic (Hispanic), 199.

Since anecdotal evidence indicated that members of these three ethnic minority groups dropped out more often than other graduate students (white, Asian, and unclassified race), the retention data was compared (see Table 8.) When the rates for the department groups are examined separately, however, minority drop out rates appear higher than that for other students in Groups II and III and lower in I and Va. (Again, it is suggested that the information above about rates of response and the usable responses on these items be kept in mind.) Of course it is not possible to project persistence to the doctorate on the basis of this two-year retention information, but the raw numbers of new doctorates in 1991 (eleven) and 1992 (eight) are interesting to note [D] because the reporting departments indicate about seventy individuals of the 1990–1991 entering cohort remained in programs after two years. (There is no information available about entering numbers when the current individuals awarded new doctorates, totaling nineteen, began their studies.)

Table 8. US citizens leaving doctoral programs without a degree

		Group I	Group II	Group II	Group Va	All Groups
Total Combined Minority*	Number Reported	23	30	40	9	102
	Remaining	83%	67%	65%	78%	71%
	Awarded some degree	4%	none	3%	11%	3%
	Left without degree	13%	33%	32%	11%	26%
Other**	Left without degree	16%	17%	17%	24%	17%
Usable Responses		32 of 39	37 of 43	55 of 88	11 of 17	135 of 187

*African American, Native American, or Hispanic.

**Asian and Other (includes white; does not include Unknown Race).

Since there are so few students in the entering 1990–1991 graduate cohort from each of the three underrepresented minority groups, the reports retaining the breakdowns by degree at entry and sex are given in Tables 9, 10, and 11. The usable responses on this item for Groups I, II, III, and Va are, respectively, 31, 37, 55, and 11.

Table 9. African American entrants to doctoral departments during academic 1990-1991:
Number by June 1992
(remaining; awarded some degree; left, no degree)

	Men Bachelor's Only	Men Master's	Women Bachelor's Only	Women Master's
Group I	(5,0,0)	(1,0,1)	(3,1,0)	none
Group II	(5,0,3)	(2,0,0)	(4,0,3)	none
Group III	(8,0,6)	(8,0,0)	(4,0,1)	(1,0,2)
Group Va	(2,0,1)	none	none	none
All Groups	(20,0,10)	(11,0,1)	(11,1,4)	(1,0,2)

Table 10. Native American entrants to doctoral departments during academic 1990-1991:
Number by June 1992
(remaining; awarded some degree; left, no degree)

	Men Bachelor's Only	Men Master's	Women Bachelor's Only	Women Master's
Group I	none	none	none	none
Group II	(0,0,1)	(1,0,0)	(1,0,0)	none
Group III	none	none	(1,0,0)	(1,0,0)
Group Va	none	none	none	none
All Groups	(0,0,1)	(1,0,0)	(2,0,0)	(1,0,0)

Table 11. Hispanic entrants to doctoral departments during academic 1990-1991:
Number by June 1992
(remaining; awarded some degree; left, no degree)

	Men Bachelor's Only	Men Master's	Women Bachelor's Only	Women Master's
Group I	(5,0,1)	(2,0,0)	(2,0,1)	none
Group II	(4,0,2)	(2,0,1)	(1,0,0)	none
Group III	(1,1,2)	(1,0,1)	(1,0,1)	none
Group Va	(3,1,0)	(1,0,0)	(1,0,0)	none
All Groups	(13,2,5)	(6,0,2)	(5,0,2)	none

Academic Expectations and the Teaching Climate

On each survey, departments provided information about entrance and candidacy requirements. From these, departments with two or more separately structured doctoral programs may be identified. Other items report mathematical breadth in doctoral programs, specific preparation for teaching, and some indicators of the departmental climate regarding teaching. Where the first survey revealed a need, items were clarified; some identical items remained. On a few items it is interesting to compare the results from the surveys, although neither the time interval nor the differences noted are large enough to reliably indicate trends.

Departmental Options and Requirements. A number of the Group I, II, and III departments have two or more distinct doctoral programs or options. The reported programs are rather generally classified here as (traditional) *mathematics, applied mathematics,* or *collegiate mathematics education.* The existence of distinct programs is implied when a department reports two or more options by either differing entrance expectations or differing doctoral preliminary or qualifying examinations. Usable responses for the purpose of reporting special program options by department groups were: I (31 of 39), II (38 of 43), and III (69 of 88).

Some Group I, II, III, and Va departments list doctoral program options in *statistics.* Since the Departments of Statistics (Group IV) are not included because of the charge to the Committee, options denoting statistics alone are not compiled. (See [CT] and [G].) In doctoral departments of mathematics or applied mathematics which include statisticians, statistics sometimes appears as a program option jointly with another topic (e.g., applied mathematics and statistics) or as one topic which may be tested on the preliminary examination for either traditional or applied mathematics options.

As expected most departments in Groups I, II, and III report what may be termed a traditional (or pure) mathematics program; the five exceptions are among the smaller doctoral programs in Group III. (See the *Notices* articles referenced in [G].) Among the programs classified as traditional, some include applied mathematics or statistics as one of several testing areas which may be selected for preliminary or qualifying examinations. The next most frequently observed separate programs usually have titles including the terms "applied" or "numerical" mathematics, and they show similarities to programs of Group Va departments. A total of twenty-five departments report that such programs have different entrance expectations, different preliminary examinations, or both. (See Table 12.) The information about these options is shown as an added group called Applied Options (AO) in some subsequent tables.

Table 12. Numbers of Group I, II and III departments with applied mathematics options (AO) similar to Group Va programs

	I	II	III
different entrance expectations only are reported	2	5	4
different preliminary examinations only are reported	2	0	1
different entrance expectations and examinations are reported	0	6	5

Recent publications about doctoral programs indicate that there are programs (or perhaps options within departments) which emphasize collegiate mathematics teaching [BMS; also see CBMS]. In the survey returns, three departments indicate such programs: one department each in Groups II and III which also have a traditional program; and one from the smaller programs of Group III which does not list a traditional program. These programs are titled "Undergraduate Mathematics Pedagogy" or "Mathematics Education". In light of the [BMS] report, more such programs would be expected to have been reported. There are probably a few other such programs among reporting departments which were not specified (one is known anecdotally), and there may be departments with such programs among those which did not respond to the survey. The total number of such programs is very small.

Entrance Expectations. The opportunity was provided to report uniform entry recommendations or re-

quirements for the department or to give separate sets of recommendations for various program options. Undergraduate advisors will want to use this information carefully so that students keep many possibilities open [CU]. The two collegiate mathematics education programs in departments which also have traditional mathematics programs indicate that the entrance expectations are similar or identical to the traditional program.

The list of undergraduate courses[1] was clarified after the first survey. The second survey asked, for specific courses listed below, the following two questions, each with a yes/no option:

Are there courses beyond multivariable calculus that you strongly recommend for students before entering your doctoral program(s)?

Are students who have not taken these courses allowed or expected to take ("make-up") these courses after admission to your program(s)?

An analysis follows by course, and some conventions and abbreviations will be used. Since the variation appears significant between department groups and types of options, overall percentages (averages) are not given.

Recommended Undergraduate Preparation

SR: Strongly Recommended. Note: A possible misinterpretation may involve timing; the item specifies "SR before entering" but some departments marking "not" SR may require make-up for accepted students who do not have it.

MU: Make-up enrollment. Note: Some possible misinterpretation on this item may have occurred; some departments marked "yes" to mean that these are the courses for which make-up is *required* (while make-up might be allowed on others).

AO: The Groups I, II, and III departments reporting applied program options similar to those of Group Va.

(x,y,z,u,v): A characteristic (such as percentage of SR responses) is given for, respectively, the departments of Groups I, II, III, Va, AO.

Real Analysis I (Adv. Cal. I) (38,43,66,12,22): SR for all of Groups I, II, III, and AO; SR for 92% of Group Va. MU (61%, 79%, 77%, 62%, 100%).

Real Analysis II (Adv. Cal. II) (37,43,63,12,19): SR: (95%, 91%, 92%, 75%, 74%). MU: (74%, 91%, 89%, 67%, 100%).

Modern Algebra I (38,42,62,8,17): SR for almost all of Groups I, II, and III with MU, respectively, 63%, 71%, 80%. SR for half of the applied programs for Groups Va and AO; MU for all.

◇ Perhaps a trend is developing toward the necessity in undergraduate preparation for both Real Analysis and Modern Algebra: several more departments strongly recommend them, and fewer permit make-up than in the earlier reporting.

---

[1] [It was implicit in the survey wording that a "course" is one semester.]

Linear Algebra I (29,39,59,11,22): SR in almost all programs with lower make-up rates in Groups I, II, III (32%, 54%, 53%) than in the applied programs for Groups Va, AO (75%, 73%).

Linear Algebra II (21,32,41,6,15): Many more departments marked neither yes or no on this topic than those above. SR: (62%, 53%, 49%, 50%, 67%); MU: (44%, 87%, 68%, 100%, 100%).

Ordinary Differential Equations (27,34,55,10,21): SR: (81%, 67%, 62%, 90%, 95%). MU: (50%, 74%, 66%, 33%, 80%).

Complex Variables (27,36,53,9,17): (74%, 47%, 49%, 78%, 71%). Groups I and Va schools are less likely to allow make-up. MU: (69%, 93%, 95%, 60%, 100%).

◊ For both Ordinary Differential Equations and Complex Variables the general pattern of SR and MU is similar, with most Group I, Va, and AO schools SR and fewer Group I and Va schools allowing MU.

Partial Differential Equations (21,30,48,7,20): SR: (29%, 37%, 15%, 86%, 90%); MU: (60%, 91%, 100%, 80%, 100%).

Topology or Geometry (28,35,53,4,14): SR: (71%, 46%, 49%, none, 29%);MU: (80%, 93%, 100%, N/A,100%).

Numerical Analysis (25,29,50,10,21): SR: (4%, 13%, 24%, 90%, 90%). MU is permitted in most cases.

Probability and Statistics (30,33,57,9,18): SR: (13%, 9%, 28%, 77%, 72%); MU: (75%, 100%, 100%, 80%, 89%).

◊ For both probability and statistics and numerical analysis there is a clearly heightened importance in applied programs.

Other: An open question was asked to determine other Strongly Recommended courses. Only six departments responded at all on this line. One, which could not be entered as a usable response for any of the above courses, said "at least six mathematics courses beyond Calculus III". Two listed computer science, one listed combinatorics, and two included discrete mathematics topics with statistics or programming.

Preliminary or Qualifying Examinations. These items were carefully reworked after the first survey. The simply stated questions of the first survey were not sufficient to reflect the varied and complex formats that exist. Examinations given in two stages, for example, could not be adequately described on the first survey. The new items were:

o **Format:** (oral and written, oral only, written only)

o **Timing:** (Usually completed no more than __ years after admission as a graduate student.)

o **Content:** (All students write the same exam which covers (list areas); Students have some choice of areas (describe).)

On the second survey departments were asked to provide information separately for program options (e.g., traditional and applied mathematics). The variety of structure and content formats reflected in this more complete information does not lend itself easily to consolidation with information from the first survey. Usable data was reported for Groups I, II, III, Va, and AO from, respectively, 30, 36, 53, 9, and 14 departments. (See Table 12 and the discussion preceding it about the AO programs within Group I, II, and III departments.)

Examination Format: Twenty formats were reported of the possible combinations of one- or two-stage exams, either or both of which may be written or oral or a combination of the two, and with a choice of topics or fixed content. Five formats were most reported, totaling 90 of the 142 usable responses. With numbers of reportings these are as follows:

Two-stage: written with choice, then oral with choice (28)
Two-stage: written with fixed content, then oral with choice (22)
One-stage: written with choice (15)
Two stage: written with fixed content, then oral and
         written with choice (13)
One stage: written with fixed content (12)

How many departments reported one-stage exams and how many two-stage? Group I, II, III departments split, respectively, 12–18, 8–28, 19–34. Only two Group Va departments reported one-stage exams, and one of those was composed of both oral and written parts. AO programs split 6–8. Generally the second stage of examination is likely to be closely related to the research area.

Examination Content: The majority of departments permit some student choice of areas. For one-stage exams, 18 of 47 departments report fixed content. For two-stage exams, 91 of 95 allow a choice of area, though about 40 of these may have a first-stage with no choice. Almost all Group Va and AO exams involve topic choice. It should be noted that many traditional programs specify a choice of areas which includes applied mathematics or statistics in addition to real and complex analysis, algebra, topology, and geometry. In applied departments or program options, mastery of traditional areas at an upper-division undergraduate level, particularly for analysis or algebra, is often implied and is sometimes stated. Topics reported are analysis, algebra, numerical analysis, methods of applied mathematics, probability, and, less frequently, differential equations, fluid dynamics, operations research, mathematical ecology, and computer science.

Timing of Examinations: For each group of departments, one-stage exams are most frequently given two or three years after admission, although the range of reports was one to five years. The most frequent time for the first of two-stage exams is after two years, though after one or three years is not unusual. The second of two-stage exams is most often given after the third or fourth year, with a range of second through fifth years.

Breadth of Mathematical Preparation. The earlier recommendations of the Committee on Preparation for College Teaching [CT] say that each graduate student, regardless of employment aims, should become familiar with a wide range of mathematical topics. Furthermore, the Committee believes that the responsibility for assuring this breadth lies with doctoral departments [CT, p.1344]. A later report, which resulted from a conference on graduate education, states, "A PhD program to prepare for college teaching would stress broadly based scholarship rather than narrowly focused research..." [CBMS]. Another report places responsibility for these goals jointly on graduate and undergraduate departments [CU, p.4,18, 20].

After the first survey, specificity was added to the item on breadth. Respondents were asked to indicate aspects of their doctoral programs that support a goal of mathematical breadth. About 80% of Group I, II, and III departments, and 50% of Group Va, indicated that they supported this goal through breadth of required courses in the department.

A collection of courses called a *minor* is required in about 30% of Group I and II departments, 15% of Group III departments, and, a much higher, 60% of Group Va departments. A required minor outside the department is most common in Group Va. In Groups I, II, and III a minor is most often a concentration in an area other than the research area within the department, or else there is choice of minor area within or outside of mathematics.

An open item asking for other breadth requirements elicited few responses beyond mention of preliminary examinations, language examinations, or courses which that particular department considers nonstandard but which are routinely reported in a number of similar departments. Another item of the survey was designed to show various activities which provide special preparation for college teaching. One of the options under this item was, "Talks in topical mathematical seminars where breadth is the intent." This activity was reported in Groups I, II, III, and Va by, respectively, 52%, 37%, 47%, and 20%. There were a few other useful suggestions:

o After the qualifying exam is passed, a survey paper on an approved subject is required within a three-week time period.

o Students in mathematical research groups interacting with other departments are encouraged to participate in joint activities with those departments.

o Students are required to attend colloquia.

o An internship is required in government or an industrial research laboratory; there is a required computer project.

o A minor oral in the form of a one-hour talk on a topic outside the area of the dissertation is required.

Factors Related to Teaching

The Committee believes that many activities and attitudes which make up the general environment and scientific life of a doctoral department influence students when they become junior faculty. Actual teaching duties are a direct influence, so the survey asked about orientation, training, duties, and supervision of Teaching Assistants (TAs). Even in departments where graduate students do not have teaching assignments, there are many opportunities for the faculty to help them prepare for future academic careers. (See also [BMS, BR, [C,1994], CH, CBMS, CT].) Since a wide variety of different activities were reported on the open questions of the first survey, a special attempt was made in modifying the second survey items to prompt respondents to indicate those activities while retaining the open item. Two items were included as indirect indicators of the teaching climate: awards to faculty and/or TAs; and the importance of teaching qualifications in the department's postdoctoral and other junior faculty appointments.

Teaching Assistants. The training, monitoring, and duties of graduate students who are involved in teaching-related activities were reported on two items of each survey. Some of the items are similar, and, for those, combined data can be presented (e.g., first survey data was retrieved for a few departments which did not respond to the second). On other items the clarifications in wording lead to a more accurate picture, especially for the Group Va departments. The combined data is almost a census of active doctoral programs, with 167 usable responses of 187 potential respondents.

First, departments had the opportunity to indicate whether graduate students involved in these activities are designated as TAs or by a similar title. Only six departments of 167 reported that they do not have this designation. Of those six, four went on to indicate specific teaching-related duties of their graduate students. (Three are in Group Va and are likely of the model where financial aid is termed a fellowship, but there are some duties.) Of the two departments reporting no TAs and no duties, one is a graduate department associated with a consortium of undergraduate colleges. For such departments anecdotal information indicates that graduate students often teach in the associated undergraduate colleges; the employment is similar to that of a part-time or adjunct instructor.

There is a significant increase in some teaching preparation and monitoring activities since the mid-1980s; this is documented by a previous survey of departments in Groups I, II, and III [C,1984]. It may be significant to note that on the first of the two present surveys (1991) the percentages reporting faculty observations of TAs (81%, 83%, and 72% for Groups I, II, and III, respectively) were somewhat higher than the combined 1991–1992 data. (Anecdotally, it is reported that current economic problems in departments have caused cutbacks on this and other desirable activities.) For Group Va departments there is no information from the 1980s; on the present surveys, 67% report orientations, 56% report faculty observations of TAs, and all report necessary English language training.

Table 13. Percentage of departments reporting teaching assistant orientation programs

	Group I	Group II	Group III
This report	92%	95%	86%
1987 [C, 1989, p. 37]	79%	86%	76%
1985 [C, 1989, p. 37]	66%	88%	64%

Table 14. Percentage of departments reporting English language training when necessary

	Group I	Group II	Group III
This report	100%	95%	76%
1987 [C, 1989, p. 36]	72%	70%	71%
1985 [C, 1989, p. 36]	60%	71%	68%

Table 15. Percentage of departments reporting faculty observations of teaching assistants

	Group I	Group II	Group III
This report	74%	83%	67%
1987 [C, 1989, p. 38]	62%	67%	67%

Table 16. TA duties and activities
(see also Tables 13, 14 and 15)

	Group I	Group II	Group III	Group Va	All Groups
Undergraduate teaching:					
Teach own class	82%	90%	86%	33%	82%
Problem sections for faculty lectures	87%	83%	58%	67%	72%
Tutorial sessions	63%	55%	51%	67%	56%
Computer or tutoring laboratories	32%	45%	53%	42%	45%
Substituting	21%	33%	31%	33%	29%
Evaluation methods:					
Videotaping	55%	48%	29%	22%	40%
Student evaluations	89%	90%	88%	78%	88%
Peer observations	26%	35%	17%	none	23%
Grading	92%	95%	96%	92%	94%
Examination design	58%	58%	60%	25%	56%

Doctoral students are assigned to teach undergraduate courses with varying amounts of responsibility and supervision. Four such levels were identified on the survey, and departments indicated a fifth activity, that of occasional substitute teaching for professors; these are reported in Table 16. Sometimes there are departmental examinations in courses at or below the calculus level; an item, "participating in examination design," shown in Table 16 was intended to elicit evidence of graduate student participation in making such examinations or in a formal discussion of test design.

Preparation for Future Professorial Responsibilities. One item was provided to elicit information about additional activities which prepare advanced graduate students for the role of college professor. The first op-

tion on this item was "nothing further", and only 15% checked this. The option "activities through other departments to enhance teaching abilities" (e.g., courses in mathematics education) was reported in only 7% of the departments. Mathematical talks given by doctoral students are a common activity, reported by 80% of the departments. Some talks are specifically designed to provide mathematical breadth for the audience and are reported above. Other talks are given in disciplinary research seminars. A professional seminar of the type recommended by the Committee on Preparation for College Teaching [CT] and assisted through the Committee's project [C,1994] explicitly provides an organized set of such activities. In addition to the eight project sites, fourteen departments report such seminars; these are somewhat more frequent in Group I departments.

Teaching Awards. There are university-wide and departmental awards reported for faculty and for TAs, with some departments reporting both kinds of awards. The existence of such awards can be an incentive to graduate students who teach. They may also be indicators to graduate students of the general interest in and importance of teaching in the institution and department. In the case of university-wide awards there is generally no assurance that any recipients will be from mathematics. The reporting is shown in Table 17.

Table 17. Teaching Awards

	Group I	Group II	Group III	Group Va	All Groups
For TAs	71%	85%	47%	67%	64%
For faculty	71%	80%	82%	58%	77%

The Importance of Teaching when Faculty are Hired. An identical item on each survey was intended to provide some indication about the weight departments place on teaching activities. The item asked, "In the view of your department chair, how important was evidence of teaching preparation and experience during graduate school in the decisions which led to appointment of entry-level assistant professors during the last three years?" Because this item is an opinion of an individual, it is reasonable to expect in some cases a significant change in reporting when a new department chair is elected or appointed. For the 119 departments responding to this item on both surveys, about half did change their responses. More of the changes were toward greater importance attached to teaching when hiring new faculty. The percentages of those who changed in their answers indicating {more,less} for Groups I, II, III, and Va were ({35%,19%},{40%,20%},{27%,22%},{29%,14%}).

A scale from 1 (not important) through 5 (very important) was provided. The time interval between the surveys was too small to expect that a trend could be perceived. For the two surveys the total number of departments responding on this item is 156. When the data is combined (using the second survey when both were usable) the responses by Groups I, II, III, and Va, and over all departments were, respectively, 2.9, 3.5, 3.8,

2.9, 3.5. See Figure 1.

FIGURE 1. Importance of teaching in hiring, by department groups

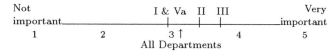

All Departments

References

[B] Lynne Billard, "The Past, Present, and Future of Academic Women in the Mathematical Sciences," *Notices of the AMS* **38**:7 (September 1991) pp. 707–714.

[BMS] Board on Mathematical Sciences, National Research Council, *Educating Mathematical Scientists: Doctoral Study and the Postdoctoral Experience in the United States*, National Academy Press, Washington, DC, 1992.

[BR] William G. Bowen and Neil L. Rudenstein, *In Pursuit of the PhD*, Princeton University Press, 1992.

[C,1989] B.A. Case, Editor. *Keys to Improved Instruction by Teaching Assistants and Part-Time Instructors*, MAA Notes **11**, 1989, Mathematical Association of America.

[C,1994] B.A. Case, Editor. *You're the Professor, What Next?*, to appear in the MAA Notes and Reports series.

[CB] B.A. Case and M.A. Blackwelder, "The Graduate Student Cohort, Doctoral Department Expectations, and Teaching Preparation," *Notices of the AMS* **39**:5 (May/June 1992) pp. 412–418.

[CBMS] Conference Board on the Mathematical Sciences, *Graduate Education in Transition*, Washington DC, 1992. Also reprinted in: *Notices of the AMS* **39**:5 (May/June 1992) pp. 398–402 and *Newsletter*, Assoc. for Women in Mathematics, (July/August) 1992, pp. 22–28.

[CH] B.A. Case and J.P. Huneke, " Programs of Note in Mathematics," *Preparing Graduate Students to Teach*, American Assoc. of Higher Educ., 1993, pp. 94–104.

[CT] "How Should Mathematicians Prepare for College Teaching?" *Notices of the AMS* **36**:10 (December 1989) pp. 1344–1346.

[CU] CUPM, *The Undergraduate Major in the Mathematical Sciences* (Bettye Anne Case, Chair, Subcommittee on the Undergraduate Major), MAA, 1991. Reprinted in: *Heeding the Call for Change*, MAA Notes **22**, 1992, pp. 225–247.

[D] All of the following Annual AMS-MAA Survey reports are in the *Notices of the AMS*. Donald E. McClure: 1992 First Report, **39**:9 (November 1992) pp. 1027–1033; 1991 Second Report **39**:6 (July/August 1992) pp. 573–581; 1991 First Report, **38**:9 (November 1991) pp. 1086–1122; 1990 Second Report, **38**:5 (May/June 1991) pp. 441–419; 1990 First Report, **37**:9 (November 1990) pp. 1217–1250. Edward A. Connors: 1989 Second Report, **37**:6 (July/August 1990) pp. 658–665; 1989 First Report, **36**:9 (November 1989) pp. 1155–1188.

[G] **Groups I and II** include the leading departments of mathematics in the US according to the 1982 Assessment of Research-Doctorate Programs conducted by the Conference Board of Associated Research Councils in which departments were rated according to the quality of their graduate faculty. **Group I** is composed of thirty-nine US departments with scores in the 3.0–5.0 range. **Group II** is composed of forty-three US departments with scores in the 2.0–2.9 range. **Group III** contains the remaining US departments of mathematics reporting a doctoral program. **Group IV** contains US departments (or programs) of statistics, biostatistics, and biometrics reporting a doctoral program. **Group V** contains US departments (or programs) in applied mathematics/applied science, operations research, and management science which report a doctoral program. **Group Va** is applied mathemat-

ics/applied science; **Group Vb** is operations research and management science. [These findings were published in *An Assessment of Research-Doctorate Programs in the United States: Mathematical and Physical Sciences*, edited by Lyle V. Jones, Gardner Lindzey, and Porter E. Coggeshall, National Academy Press, Washington, DC, 1982. The information on mathematics, statistics, and computer science was presented in digest form in the April 1983 issue of the *Notices of the AMS*, pages 257–267, and an analysis of the above classifications was given in the June 1983 *Notices of the AMS*, pages 392–393. Also see the April 1988 *Notices of the AMS*, pages 532–533.]

[GE] Julius Getman, *In the Company of Scholars: The Struggle for the Soul of Higher Education*, University of Texas Press, Austin, 1992.

[Ha] Jenny Harrison, "The Escher Staircase," *Notices of the AMS* **38**:7 (September 1991) pp. 730–734.

[He] Claudia Hension, "Merging and Emerging Lives: Women in Mathematics," *Notices of the AMS* **38**:7 (September 1991) pp. 724–729.

[J] Allyn Jackson, "Top Producers of Women Mathematics Doctorates," *Notices of the AMS* **38**:7 (September 1991) pp. 715–719.

[L] D.J. Lewis, "Mathematics and Women: The Undergraduate School and Pipeline," *Notices of the AMS* **38**:7 (September 1991) pp. 721–723.

[N] Richard Neidinger, "Survey on Preparation for Graduate School," *Focus* of the MAA **8**:4 (Sept 1988) p.4.

[NRC] National Research Council Committee on the Mathematical Sciences in the Year 2000, A Challenge of Numbers; People in the Mathematical Sciences, National Academy Press, Washington, DC, 1990. (See also, *Notices of the AMS* **37**:5 (May/June 1990) pp. 547–554.)

[S] Alice T. Schafer, "Mathematics and Women: Perspectives and Progress," *Notices of the AMS* **38**:7 (September 1991) pp. 735–737.

[W] Articles and responses about women mathematicians in prestigious departments: (1) Paul Selvin, "Profile of a Field – Mathematics – Heroism is Still the Norm," Science **255** (13 March 1992) pp. 1382–1383. (2) Letters and commentary related to the Selden article: Science "Women in Mathematics," Science 257 (Letters, 17 July 1992) pp. 309–310; President's Report (Carol Wood), *Newsletter*, Assoc. for Women in Mathematics **22**:5 (Sept/Oct 1992) p. 1 and Letter from Joan Birman, *Newsletter*, Assoc. for Women in Mathematics **22**:6 (Nov/Dec 1992) pp. 3–4. (3) "Women in Mathematics Update" Science 257 (July 17, 1992) p. 323.

Acknowledgments. The authors especially thank for their assistance, in addition to all the representatives of reporting departments: Dave Lutzer and other members of the AMS-MAA Data Committee; Allyn Jackson of the AMS staff; the other members of the AMS-MAA-SIAM Committee on Preparation for College Teaching (which is chaired by Case): Donald Bushaw, Robert McDowell, Richard Millman, Robert Phelps, Richard Ringeisen, Stephen Rodi, James Simmonds, and Guido Weiss. Technical support was provided by the Department of Mathematics, Florida State University, and special thanks go to Melissa E Smith, the department's senior art/publication production specialist. The research was partially supported by Department of Education FIPSE Comprehensive Grant P116B00184.

First Survey:
The Graduate Student Cohort,
Doctoral Department Expectations,
Teaching Preparation

Bettye Anne Case and M. Annette Blackwelder[†]

The following reports on a survey conducted by the Committee on Preparation for College Teaching, a joint committee of the AMS, the Mathematical Association of America (MAA), and the Society for Industrial and Applied Mathematics (SIAM), chaired by Bettye Anne Case. Both authors are members of the Department of Mathematics, Florida State University.

Introduction

As requested in 1987 by the presidents of the AMS, MAA, and SIAM, the joint Committee on Preparation for College Teaching was established with the following charge:

> To make recommendations concerning the appropriate preparation of college teachers of mathematics, taking into account the varieties of institutions of higher education, the diversity of the mathematical sciences, the backgrounds and career interests of college students, the impact of computers, and insights from research on student learning.

A description of the committee's early work and recommendations is found in its 1989 First Report [CT]. Finding no existing models that embodied their recommendations, the Committee secured funding from the Fund for the Improvement of Postsecondary Education, US Department of Education, to conduct a survey and to promote adoption of its recommendations. The project currently involves development of programs in eight doctoral departments.* as well as the survey. This article presents findings from the first stage of the survey.

The wider scientific community is concerned that future faculty members be representative of the various US citizen population groups. It is well known that women, and Blacks and Hispanics of both sexes, are underrepresented in the mathematical community relative to their proportion in the population. Annual data continues to show the same pattern of underrepresentation among new doctorates [D]. Further, anecdotal information suggests that persistence through graduate school is lower for underrepresented groups than for white males, and for US citizens than for noncitizens. Because graduate students will supply most of the new entrants to the professoriate, efforts toward change must address underrepresentation within the graduate student population.

Description of the Survey and Responses

The Committee's survey was designed to provide information about the race, sex, and citizenship characteristics of the graduate student cohort and about the preparation for future teaching and attitudes about teaching in graduate departments. The survey, formulated with the assistance of the AMS-MAA Data Committee, has two stages. First, a questionnaire was sent to doctoral departments in May 1991 with the Data Committee's Annual Survey of New Doctorates; the results of that survey are presented here. The second stage is a follow-up survey, to be conducted during Spring 1992, to collect information about retention in doctoral programs of the entering 1990-1991 graduate student cohort.

Doctoral departments are divided into six groups [G]. Groups I, II, III, and Va fall within the purview of the Committee's charge and hence were included in the survey. The excellent response rates, shown in Table 1, lend confidence to the data.

TABLE 1. Departmental response rates

Group I	36 of 39
Group II	41 of 43
Group III	65 of 86
Group Va	11 of 16
Total	153 of 184

The data about New Doctorates presented in this report is derived from the Annual AMS-MAA Survey [D] and is presented for comparison purposes.

Sex, Race, and Citizenship Data: The survey sought to classify students into six racial categories: Asian, Pacific Islander; Black; American Indian, Eskimo, Aleut; Mexican American, Puerto Rican or other Hispanic; None of the Above; Unknown. There were four classifications for citizenship: US; Canada; Other; Unknown. Each department was asked to provide this information for women students and for men students who first entered the department's program between June 1, 1990, and May 31, 1991, and whose highest degree was a bachelor's. The departments also provided race and citizenship information for students entering in that time interval with a master's degree, and for all full-time graduate students enrolled. Findings about graduate students will provide a baseline for retention data to be collected in the second stage of the survey.

Academic Expectations; The Teaching Climate: The survey included questions about the expected preparation of incoming graduate students and about preliminary and qualifying examinations. Information was also collected about teaching duties assigned to graduate students in the department, preparation for those duties, and any additional activities in preparation for a future professorial role. Questions about teaching awards and the importance of teaching qualifications in hiring deci-

---

[†] Reprinted by permission of the American Mathematical Society from the *Notices of the AMS* 39:5 (May/June 1992) pp. 412-418

* University of Cincinnati, Clemson University, Dartmouth College, University of Delaware, Harvard University, Oregon State University, University of Tennessee, and Washington University.

sions were included as indicators of the importance attached to teaching and the teaching climate in departments.

US Citizen and Underrepresented Population Groups Among Graduate Students and New Doctorates

Education Level of Doctoral Department Entrants: Most entrants—74% of the 2,805 reported by surveyed departments—into doctoral programs hold only the equivalent of a US bachelor's degree. Of the US citizens entering doctoral programs, 83% (of 1,693 reported) hold only a bachelor's degree. For Group I departments, this percentage is higher still, 88%. The remainder of reported entrants hold the equivalent of a US master's degree.

TABLE 2. Percentages of doctoral department entrants holding only a bachelor's degree

	US Citizens	Non-US Citizens	All Entrants
Group I	88%	67%	79%
Group II	81%	53%	70%
Group III	81%	55%	73%
Group Va	74%	58%	68%
All Dept Groups	83%	59%	74%

US Citizens: The percentage of US citizens among new doctorates has been stable for several years around 45% [D], but nearly 70% of entrants to doctoral departments with only a bachelor's degree are US citizens. Table 3 shows that 62% of the Doctoral Department Entrants are US citizens—a fairly high percentage compared to the 56% among All Graduate Students and the 45% among New Doctorates. Although these numbers may appear to promise a larger proportion of US citizens among future crops of doctorates, the situation is not so simple. For example, anecdotal reports from some departments indicate that the proportion of US citizens who persist to the doctorate is significantly lower than that of noncitizens. Retention will be investigated in the second stage of the survey.

(It should be noted that the numbers for All Graduate Students include, for the reporting departments, their 1991 doctoral recipients. The cohorts of lines 1 and 2 do not overlap and are combined for line 3; also, cohorts of lines 3 and 5 are included in that of line 4.)

TABLE 3. Percentage of US citizens among:

1. 1990-1991, Doctoral Department Entrants
   with Bachelor's Degree: Groups I, II, III, Va   69%
2. 1990-1991, Doctoral Department Entrants with a
   Master's degree: Groups I, II, III, Va.   41%
3. 1990-1991, Doctoral Department Entrants with either a
   Bachelors' degree or a Master's degree:
   Groups I, II, III, Va.   62%
4. 1990-1991, All Graduate Students: Groups I, II, III, Va 56%
5. 1991 New Doctorates: Groups I, II, III, Va   45%

Women Among US Citizens: With respect to their

proportion in the workforce, women are underrepresented among mathematical scientists of all types. Statistical and anecdotal information about this underrepresentation is well known and is collected in several papers and their references in the *Notices*, September 1991 [B, Ha, He, J, L, S], as well as in the previously cited statistical reports [D].

This underrepresentation is strongly exhibited among faculty and students in the mathematics doctoral departments comprising the population of this survey—Groups I, II, III, and Va. The Annual AMS-MAA Surveys report for Groups I-V that 24% of the US citizen new doctorates are women. However, within Group IV, the analogous figure is 38%, and within Group Vb it is 44%. Removing these department groups from the data results in a drop to 20%, as seen in line 4 of Table 4 (see next page). (Note that there is no overlap between lines 1 and 2, and that lines 1, 2, and 4 are included in line 3.)

TABLE 4. Percentage of women among US citizens

1. 1990-1991, Doctoral Department Entrants
   with Bachelor's Degree: Groups I, II, III, Va   35%
2. 1990-1991, Doctoral Department Entrants
   with Master's Degree: Groups I, II, III, Va   31%
3. 1990-1991, All Graduate Students: Groups I, II, III, Va 31%
4. 1991 New Doctorates: Groups I, II, III, Va   20%

As with the citizenship data, the numbers in Table 4 might be taken to predict eventual increases in the percentages of women among US citizen doctorates and faculty. But, again, anecdotal information—combined with the stability of the percentage of women US citizens among new doctorates for a number of years—invites caution. For whatever reason, the proportion of women who persist to the doctorate is significantly lower than that of men. The second stage of the survey will address the question: "What percentage of the entering 1990-1991 students persisted to a third year of graduate study or have received a master's degree?" For those who obtained a master's degree and then left a doctoral department, the more subtle questions about "intent" (to complete only a master's or to go on for the doctorate) versus leaving with the master's degree as a "consolation prize" will not be answered.

Underrepresented Minority Groups: Some ethnic minority groups comprise a smaller proportion among mathematicians and mathematics graduate students than in the general population; this pattern remains even when one looks at women only or men only. Table 5 combines the data from departmental Groups I, II, III, and Va and gives both reported numbers and percentages for three such US citizen minority groups. Table 6 reports Groups I and II data. Even allowing for nonreporting departments, the numbers are very small: out of a total of 4,677 doctoral students in 143 departments, only 344 are members of these minority groups.

Because the numbers are so small, one must be wary of generalizations, but a few comments can be made. The significantly higher proportion of Hispanics among "all" graduate students than among "entering" graduate students stands out. (The weight of the Hispanic data in the combined data for these minorities causes the same effect in the Totals.) This pattern—which is also found when the data is examined by departmental Group and is especially marked in Group III—does not have favorable implications about the number of Hispanics receiving doctorates in the future. Another unexpected note: for the applied departments of Group Va, the percentage of "entering" Black graduate students is significantly higher than for "all" Black graduate students; perhaps good recruiting will lead eventually to more doctorates awarded to Blacks.

TABLE 5. Numbers and percentages from underrepresented minorities among US citizen graduate students

	Entering Grad. Students with BAs Degree Only	Entering Grad. Students with MSs Degree	All Entering and Continuing Grad. Students	New PhDs
Black				
Number	49	14	134	8
Column %	3.5%	4.7%	2.9%	2.3%
Amer. Indian, Eskimo, Aleut				
Number	3	2	11	2
Column %	0.2%	0.7%	0.2%	0.6%
Mexican Amer., Puerto Rican or Other Hispanic				
Number	24	6	199	4
Column %	1.7%	2.0%	4.3%	1.2%
Total of Above				
Number	76	22	344	14
Column %	5.5%	7.4%	7.4%	4.1%
Usable Responses	146/184	146/184	143/184	173/184
Reported US Citizens of Known Race	1384	296	4677	343

It is interesting to determine whether the proportions of men and women for the minority groups follow those that in each of the graduate student cohorts. For Blacks, the breakdown of men and women is very close to that for the entire US citizen cohorts, although in departmen-

tal Groups I and II a larger percentage of Black graduate students entering with a bachelor's degree are women— 50%. Also, Eskimos-Indians have a higher percentage of women in each of the cohorts; six men and five women were reported among All Graduate Students, and three women and two men among Entering Graduate Students. For Hispanics, the proportion of women entrants holding only bachelor's degrees follows that predicted by the population, but otherwise there are proportionately fewer women among Hispanics than generally. There were two Hispanics reported among new doctorates — one man and one woman.

TABLE 6. Underrepresented minorities among US citizen graduate students in Groups I and II

	Entering Grad. Students BAs Only		Entering/Continuing Grad. Students	
	I	II	I	II
Black				
Number	8	17	31	42
Column %	2.0%	4.2%	1.9%	3.2%
Amer. Indian, Eskimo, Aleut				
Number	0	2	1	6
Column %	0.0%	0.5%	0.1%	0.5%
Mexican Amer., Puerto Rican or Other Hispanic				
Number	7	7	56	30
Column %	1.8%	1.7%	3.5%	2.3%
Total of Above				
Number	15	26	88	78
Column %	3.8%	6.4%	5.4%	5.9%
Usable Responses	33 of 39	41 of 43	34 of 39	39 of 43
Reported US Citizens of Known Race	398	404	1616	1319

Entrance and Continuation Expectations and Characteristics Related to Teaching

One portion of the survey was devoted to questions about the preparation expected of entering graduate students, continuation requirements for students, preparation for teaching, and the teaching climate in departments.

Entrance Requirements. Entrance requirements or recommendations for upper-division undergraduate courses are far from universal. The survey asked departments whether or not certain courses were "strongly recommended" as background for entering graduate students, and whether these courses could be made up in graduate school. In some cases, the "strongly recommended" courses were actually required for entrance. Several departments indicated that graduate students are accepted only as master's candidates until deficiencies in background are made up. Results about the categories of courses surveyed are indicated below.

Real Analysis. Real Analysis I (Advanced Calculus I) is "strongly recommended" in 96% of doctoral mathematics departments overall, with 80% reporting they will admit students and let them make-up the course after enrolling. (The corresponding percentages for Real Analysis II are 91% and 86%.) It is interesting that *all* of the Group Va (applied) departments responded that Real Analysis I is "strongly recommended," and only 70% will allow make-up after entrance to graduate school.

Modern Algebra. Predictably, few Group Va departments "strongly recommend" Modern Algebra I, but 93% of Group I, II, and III departments do. This percentage drops to 73% when it comes to Modern Algebra II; make-up is allowed in 90% of Group I, II, and III departments combined, but by only 81% of Group I departments.

Probability and Statistics. For Groups I, II, and III only 21% list Probability and Statistics as "strongly recommended," and in Group I only one department does so. However, in Group Va there is stronger positive response: 64% of the departments indicate "strongly recommended." Probability and Statistics have been included repeatedly for many years in professional society curriculum recommendations [CU] as essential in undergraduate major programs. The Committee on Preparation for College Teaching feels that some study of Probability and Statistics at the upper-division undergraduate level is necessary for the broad foundation in mathematical topics important to undergraduate teachers [CT, p. 1346].

Additional Courses. Answers reflect differences between "pure" and "applied" programs. Under the category *Applied Mathematics*, 75% of Group Va departments indicate "strongly recommended courses"; only 23% of the other departments mark this category. The *Applied Mathematics* courses most often listed are Numerical Analysis and Partial Differential Equations. These courses and other courses in the general area of applied mathematics were sometimes listed under *Other*.

The courses listed in the *Other* category most frequently as "strongly recommended" are Topology and Complex Variables (Complex Variables was also sometimes listed under *Applied Mathematics*). Linear Algebra and Ordinary Differential Equations were reported by a number of departments, and this confusion led to a change in the second stage of the survey, in which the ambiguous term "upper-division courses" will be re-placed by "courses coming after single and multivariable calculus." There will also be provision for a separate listing when two programs within a department have different entrance expectations.

The Preliminary Examination. As with entrance expectations, the areas examined may differ by choice of program within a department; on the examinations, there may or may not be a choice of topics within a program. Again, responses found in this first stage of the survey will help to clarify the questions for the second stage. Respondents were asked to say whether the format was oral or written, indicate when the examination needs to be completed, and describe the content of the examination.

The usable responses on this item were high. Here are some interesting highlights.

⋄ Only two departments did not report a preliminary examination. (One said there was none, while one did not answer the item.)

⋄ A "written only" examination is more common in Groups I and III, but in Groups II and Va both a written *and* an oral examination are the more common testing requirement.

⋄ The examination is typically taken in the second or third graduate year.

⋄ Generally, on the written portion of the examination, at least two areas are included, with three or four areas most typical.

⋄ In about 60% of the departments, some choice of testing area is available to the students.

Content of preliminary examinations vary, because they appropriately reflect the research interests of the faculty. Therefore, it is not possible to list a set of nonoverlapping areas, topics, or concentrations from which departmental testing areas are taken. Samples of responses indicate in some cases student choice of topic:

⋄ all of complex analysis, real analysis, topology, analysis on manifolds, linear algebra, and abstract algebra;

⋄ three from algebra, applied mathematics, complex analysis, real analysis, or topology;

⋄ from a department with two doctoral programs: Pure Mathematics: algebra, real and complex analysis, topology; Applied Mathematics: analysis, numerical analysis, PDE and numerical PDE, fluid dynamics, and another topic chosen from research area;

⋄ two from analysis, algebra, linear algebra, combinatorics, or probability.

Mathematical Breadth. As indicated earlier in the discussion about undergraduate mathematical preparation, the Committee on Preparation for College Teaching considers it desirable that graduate students become familiar with a wide range of mathematical topics at the advanced undergraduate level:

Graduate education should not be limited to specialization... Breadth of knowledge forms the background for teaching a variety of courses, for advising students, and for recognizing different kinds of talent in•mathematics. Further, the increasing interrelation of ideas at advanced levels of research makes it imprudent to be ignorant concerning major branches of mathematics. [CT, p. 1344]

Accomplishing this goal is a joint responsibility of undergraduate and graduate departments [CU, Case 1991, pp. 4, 18, 20] and may not be specifically reflected in doctoral requirements. To provide a basis for later Committee recommendations, the survey attempted to ascertain the extent to which doctoral programs appear to encourage mathematical breadth. Answers indicate that:

◇ a number of departments consider that their preliminary or qualifying examinations, over several areas, accomplish this goal;

◇ few departments require a minor outside of the department—even in applied mathematics only 42% do so; and

◇ 80% believe "breadth" is reflected through their areas of required courses.

Preparation and Duties of Teaching Assistants.
The Committee does not consider "TA training" sufficient preparation for college teaching, and such training is, in fact, the primary concern of another committee. Nevertheless, the Committee does see TA training as an important part of the total preparation. In 42% of Group Va (applied) departments, graduate students are not involved in teaching at all, or else have responsibility only for assisting professorial faculty and grading in upper-division courses. Various TA training mechanisms were listed, and the most interesting findings concern orientation programs (Table 7), English language training (Table 8), and faculty observations of TAs (Table 9). Data from a previously published survey [C] about departments in Groups I, II, and III indicates an increase in all three activities since the mid-1980s.

TABLE 7. Percentage of departments reporting teaching-assistant orientation programs

	Group I	Group II	Group III
1991	94%	95%	85%
1987 [C, p. 37]	79%	86%	76%
1985 [C, p. 37]	66%	88%	64%

TABLE 8. Percentage of departments reporting English-language training

	Group I	Group II	Group III
1991	94%	90%	75%
1987 [C, p. 36]	72%	70%	71%
1985 [C, p. 36]	60%	71%	68%

TABLE 9. Percentage of departments reporting faculty observations of teaching assistants

	Group I	Group II	Group III
1991	81%	83%	72%
1987 [C, p. 38]	62%	67%	67%

Student evaluations are common, but other components of TA training and supervision are less frequently used. Although peer observations by other graduate students have been used and considered effective for a number of years [C, p. 36], the reported incidence is small. Overall, 28% of programs reported videotaping of classes, with incidence in Groups I and II at 40%. Videotaping of mock classroom sessions is thought to be much higher. (For example, non-specific "Uses videotape" queries in 1985 and 1987 were answered positively in Groups I, II, and III by over half the reporting departments.)

Teaching Activities of Graduate Students.
Individuals typically teach their own classes, with varying levels of supervision, in departments other than the Group Va (applied) departments. Duties of graduate students almost universally include grading and student contact in recitation or tutorial sections, if not in classes with full responsibility. Over half of the departments indicated that graduate students occasionally act as substitute teachers for regular faculty. Although graduate students often teach courses in which department-wide examinations are used, 73% (excluding Group Va) nevertheless report that their graduate students gain experience in the designing of examinations.

Additional Items Regarding Teaching.
The Committee recommends that all third- and fourth-year graduate students be involved in "systematic attention to issues and excellence in the teaching of mathematics" [CT, p. 1345]. The grant to the Committee through the Fund for the Improvement of Postsecondary Education provides for pilot programs to design "professional seminars" and other such mechanisms. These activities would target advanced graduate students and are in addition to TA training, which focuses primarily on an immediate concern for competent teaching in the department.

The survey asked departments to list additional aspects of their program that prepare advanced graduate students for undergraduate teaching, and there were few responses: 92% of departments listed nothing. Of the few responses, involvement in seminars in their mathematical specialty was mentioned most frequently. Several others mentioned the opportunity to teach sections of classes beyond single-variable calculus with full classroom responsibility. One department responded that students are encouraged to take a course in the Science Teaching Department.

TABLE 10. Departments making teaching awards

	Group I	Group II	Group III	Group Va
TA awards	50%	66%	31%	none
Faculty awards	17%	36%	17%	13%

The last two items of the survey were included to provide some indication of the departmental climate regarding teaching or attitude toward the importance of teaching (Table 10). One item concerned departmental teaching awards; since there are university- or college-wide awards programs that may not have been reported in these responses, the second stage of the survey will ask about such awards.

The last item on the survey was intended to provide some indication about the importance the departments attached to teaching activities. The departments were asked: "In the view of your department chair, how important was evidence of teaching preparation and experience during graduate school in the decisions which led to appointment of entry-level assistant professors during the last three years?"

A scale from 1 (Not important) through 5 (Very important) was provided. The weighted mean answer was 3.5 with the responses by Groups I, II, III, and Va, respectively, 3.1, 3.4, 3.8, and 2.9.

FIGURE 1. Importance of teaching in hiring, by department groups

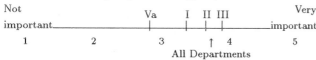

References

See **Summary** references above. [C] of this paper is [C, 1989] of the **Summary** references.

Data Analysis and Presentation: In addition to the AMS-MAA Data Committee assistance, the survey forms were prepared, distributed, and collection coordinated by Monica Foulkes and the AMS staff. Allyn Jackson of the AMS staff provided helpful editorial assistance. Technical support and other assistance were provided by the Department of Mathematics, Florida State University, and special thanks go to Melissa E Smith, the department's Senior Art/Publication Production Specialist. The research was partially supported by P116B00184, Comprehensive Fund, FIPSE.

You're the Professor, What Next?

Part II

Shared Responsibilities

Chapter 3 – Getting Started

Chapter 4 – The Professional Seminar

Chapter 5 – A Broad Mathematical Preparation

This section emphasizes that preparing graduate students and orienting new faculty for their tripartite faculty role are responsibilities shared by doctoral and hiring departments. A general model and some pragmatic suggestions are provided for departments developing a plan. Mathematicians seeking the same goals outside a structured program are given guidance.

Chapter 3

Getting Started

Bettye Anne Case

Stephen B. Rodi

Whose Responsibility?

Bettye Anne Case

Mathematicians packing before departure for their first regular faculty job may need to select from among their own capabilities in quite different proportions, depending on the destination. They will be able to fill the right luggage if, well in advance of the departure date, each acquires a balance of capabilities and understandings related to the tripartite faculty role.

How can this be assured? Who has the obligation to see that a person has the right components to pack for a specific new job? In the earliest recommendations of the Committee on Preparation for College Teaching, we placed this responsibility on doctoral departments. (See the **First Report** in Chapter 1.) After three years' working with the experiments of this project, I would say nothing to diminish the necessity and urgency for doctoral department efforts. In addition, it is now clear that there is enough responsibility to spread around: doctoral students themselves must begin habits of self-evaluation and professional development; and hiring departments, in both postdoctoral and tenure-track cases, should provide caring mentors and possibly formal activities. The early acceptance of this personal responsibility will serve mathematicians well through the changes of a career in academia. Hiring departments will be better served when their departmental and college priorities and needs are understood and hence may be met. Junior faculty members who have had the opportunity in their doctoral or postdoctoral programs for a structured introduction to a broad range of information about pedagogy and faculty obligations are able to adjust more easily to serve their new students.

But joint responsibility is sometimes perceived as belonging to no one. If you have come this far reading or browsing in this book, and if you find yourself saying "Well, yes, someone should..." then please consider this anecdote: When new to departmental administrative work, I asked the university police for advice about an incident. After the threat of a more serious problem had abated, the officer in charge thanked me for calling; I replied that I was not sure when to call. He said "If you ask yourself 'Should I call?' then do it!" If now you find yourself saying "Should I do something about this?", the answer is "yes." I hope you will find in this book many things to help you properly acquit your perceived responsibility.

What can and should a doctoral department do to help its graduate students or postdoctoral fellows? To use an analogy with the upbringing of our natural children, a department should assume responsibility to assure at least two types of nurturing.

◇ The parent role: Individual mentoring in research (where the thesis director is often referred to as the mathematical parent) and perhaps also in teaching;

◇ The school role: An organized activity led by senior mathematicians which anticipates future responsibilities with a broad view. (See in particular Steve Rodi's essay in this chapter, and information about the Professional Seminar in Chapters 4-14.)

The organized activity to meet this responsibility may share some techniques with the orientation of teaching assistants but differs in that its focus is on the lifelong careers of the graduate students.

There is bad news and good news about the support, financial and personal, needed for success. The bad news is that money probably cannot make a program successful unless it is considered valuable by a reasonable number of the department's faculty and doctoral students. The good news is that a small core of faculty who believe it will be valuable can make a difference by their enthusiasm. The financial costs are not great, but faculty time must be allocated; it will not endure long if it is an "off load" activity for either the professor or the doctoral students.

What might developing mathematicians do to help themselves? Some helpful activities, in or outside a formal program, are perhaps obvious but are important enough for quick mention: Read (the essays, reprints and references of this book are a strong start); go to professional meetings; learn about areas of mathematics other than your research interest; talk with other graduate students and faculty about teaching methods and controversies; ask to observe classes of professors or graduate students who are trying new methods or are considered especially effective; try new ideas in your own classroom if you are teaching; develop your interaction and communication skills with peers and students; develop contacts who can help you learn more; and think a lot about all of this.

At this writing the Young Mathematicians Network (loosely patterned after the more established Young Scientists Network) offers valuable shared information through an e-mail newsletter and encourages collegiality and a broad view of the profession through informal communications.

If you are a doctoral student or postdoc who is given the opportunity to participate in an organized activity of the type of the professional seminars described in this book, then you are fortunate. It is very practical (and perhaps is open to cynical abuse) to note that this activity is for the future job seeker what a prime-time photo op is to a politician. It lends seriousness and legitimacy to your claims of interest in teaching, and the teaching portfolio which can then be assembled gives substance to those claims. If you do not have the benefit of an organized program, you may arrange for yourself the benefits of some of these activities. Especially in that case, you may want a teaching mentor in addition to your research director; with some thought you will be able to identify a faculty member who is known to be a good teacher and who you think will want to help you.

What must wait until after hiring? What are the responsibilities of hiring departments? "Preparation" for college teaching is really a lifelong activity; it must continue past the award of the doctorate at least to the point of tenure consideration. Hiring departments, especially those who hire new PhDs as assistant professors or postdoctoral fellows, sometimes treat them as though they are being tested. Unfortunately, unwritten rules sometimes exist, and should be taken into account. The novice faculty member must, therefore, do everything within reason to discover what is expected, both explicated and not. Well-meaning senior faculty sometimes feel strongly that junior persons' freedom is abused if they are advised in certain choices they might make. The shortsightedness of this "sink or swim" policy is apparent when you consider the negative effects on students and the expense of recruiting, replacement, and other administrative implications.

The hiring department may be very different from the departments (undergraduate and graduate) of the new PhD's experience. The newly appointed person may not know just what is expected, and it may be that, even if former mentors from the doctoral department are consulted about a problem, they cannot help. The only answer seems to be that hiring departments should assure that each new faculty member has a mentor regarding each type of new responsibility. Some interesting ethical questions arise as to the role of this mentor (or mentors) at decision junctions such as retention and tenure. How much obligation is there to be an advocate, or to remain neutral? How much of the mentoring relationship is confidential? Though there was a lively discussion among ten of us at a recent meeting of mathematical sciences department chairs, definitive answers were not reached. All agreed that mentoring was essential and that these and other practical and ethical considerations must be kept in mind.

What can individual senior faculty members do to help? The most effective mentors to work with advanced doctoral students or new PhDs who are thinking about their future academic roles are professors who in their careers have shown accomplishment and interest in effective teaching, productive scholarship, and unselfish service activities. If the numbers of developing mathematicians in the department justifies, senior faculty can encourage setting up suitable programs, possibly including both advanced graduate students and newly appointed faculty. In either situation, senior faculty members have a responsibility to inform themselves about the resources available for help. This book and the sources it lists are a good foundation; it is important to stay abreast of current situations that affect developing mathematicians. (The information through the Young Mathematicians Network is also useful for senior faculty acting as mentors.)

The doctoral or hiring departments can and should build a bridge for mathematicians in transition toward the professoriate. Individuals must give close attention to their preparations if there is to be a safe crossing.

From Grad School to Tenure:
How to Support Young Mathematicians

Stephen Rodi

Some Introductory Comments. I am one of a small group of people who for more than two years sat through almost all the discussions (electronic and face-to-face) which made up the committee's work and also read carefully all the site reports which you will find in Part III. I see my job here as making a sort of meta-analysis, trying to pick out some common threads that cut across all this material but perhaps are not explicitly the purview

of any one report.

How should you use the information in this volume? I strongly recommend beginning by reading the eight doctoral department reports in Part III. I found these reports extraordinary, first in that they exist at all and then for the range of information they provide for imitation. These essays describe the work of real pioneers. Not many doctoral programs in mathematics have paid attention to training graduate students to be teachers. Many have had short training sessions for teaching assistants, frequently concentrating on the adjustments required by foreign national assistants. But, prior to the effort of these eight programs, few departments set out systematically to make matters of undergraduate instruction, breadth of mathematical learning, and future professional responsibilities (other than those of research) an explicit and even required part of the training of every graduate student. I suspect such a requirement will become more and more common as the value of the work of these eight programs becomes apparent.

And, like the travelers who marked out the Oregon trail westward a century ago, these pioneers have marked out a practical and useful route. The individual project directors thought carefully about what they wanted to achieve and were observant and flexible in making midcourse corrections as needed. They were also perceptive in making meta-analyses of their own programs. In addition to containing a great deal of specific information that will be immensely useful to anyone who wants to go forth and do likewise, the essays are reflective on why the program directors followed certain courses of action and are honest in evaluating what worked and what did not work.

One reader of these site reports commented that they should be shortened because "there was too much repetition in them." I and others disagreed, and so they remained essentially as submitted. The similarities, in my opinion, are the most important indicators. Quite independently, sometimes in very similar language, the reports say, "This is what is really needed," or, "This really worked," or, "I would do this differently next time." That is exactly the kind of information potential new program directors need. The commonalties you the reader find in the site reports give the best clue to what might be a *sine qua non* in your program.

Points of Focus of A Proseminar for Graduate Students. Here are some of the points of focus I decided were important as I read the site reports and listened to the discussions and debates that took place within the Committee on Preparation for College Teaching.

"Preparation is critical." Successful programs of the kind described do not happen by chance. They require careful thought in advance about practical problems (e.g., when are graduate students free? how can this be done without over-burdening their schedules?) and careful structuring of a plan for the group to follow. These seminars are the sort of activity, related to ped-

agogy and general issues of professionalism, for which a faculty member might be tempted to "wing it" week after week. After all, there are many other demands on one's time.

It is important to recognize that very few graduate mathematics faculty have the instant expertise to "wing it" successfully in this setting for very long. And those who have the talent would do a better job with advanced planning. Students quickly pick up the lack of purpose and focus in a seat-of-the-pants approach. The psychological result will be a lessening, not a heightening, of the importance graduate students give to the professional issues around which the seminar is organized. Some of the experimental site directors were senior in their departments, accomplished mathematicians, and savvy in the ways of our profession. But their reports make clear that detailed advance planning and regular attention to the seminars made all the difference as regards success.

"Departmental support is a necessity." These programs take time, money, space, and other resources. Except for time and the money associated with the project director's salary, the resource requirements are surprisingly small. The project reports suggest a variety of strategies for getting the necessary support. The most important single factor may be provision of time for the director to do the job well by reassignment to this project in place of other teaching duties. One should emphasize that such a seminar is another form of teaching, equally as important for graduate students as specialized topics in mathematics, and hence is a course for which deans and others need to provide funds. If departments, colleges within universities, and universities are serious about responding to issues of instruction and professionalism within mathematics programs, now so often discussed in the media, they will have to provide money to support their rhetoric — but far less money than one might fear. All in all, a great deal of good is done relatively cheaply.

When high moral arguments fail to convince a graduate faculty member, department chair, or dean to "chip in some funds," we might persuade such persons with a reminder that graduate student experiences such as those cited in the eight experimental programs in Part III frequently make the difference for those seeking a first position. Graduate programs which provide preprofessional support are likely to develop a noticeable edge in placing their students over those which do not.

"Do it in the mathematical sciences and involve senior faculty." The enhancement programs described in Part III should not be "farmed out." The mathematics department should take primary responsibility for conducting and planning such programs but should consider calling on appropriate colleagues elsewhere in the university (second-language experts, professional development specialists, or professors of education) for assistance at appropriate times. If participants both inside and outside are carefully selected, the results

may include very good programs and very productive relationships. The purpose is to change the culture within mathematics itself, to have mathematicians themselves recognize the full range of their professional responsibilities, and to deal with them as mathematicians. Hence, within the graduate mathematics program itself, issues of instruction, breadth, and every aspect of professionalism should be part of the training of new mathematician. These are an essential component of what students should experience as neophyte mathematicians in their graduate school novitiate.

Such goals probably cannot be achieved without involving senior graduate mathematics faculty in developing and teaching the kinds of seminars proposed by the committee. These mathematicians are, by their very status, models for the next generation. They add credibility by their involvement. They have the wide experience needed to enrich the program. Finally, because they have already reached tenure and perhaps full professorship, they are at a stage in their professional life where they can devote attention to such matters without jeopardizing their chances for advancement by diverting professional attention from their own research (a reality one must face). Of course, these program directors should be eager to draw on talent elsewhere in the university, including colleagues in mathematics education and elsewhere in the College of Education, for various components of their seminars.

A report from the Joint Committee on Professional Recognition and Rewards will be issued through the American Mathematical Society about the same time the book you are now reading comes off the press. I am the only person who served on that committee and the Committee on Preparation for College Teaching. It has been very interesting to observe over the past two years how these two committees support each other in defining more broadly the role of the mathematician as a professional. While maintaining the primary and special status of research in the life of a mathematician, both committees have recognized in their own ways a broad range of other matters that should be a natural part of a mathematician's professional life. The report of the Joint Committee on Professional Recognition and Rewards reinforces the importance of direct involvement by mathematicians in these areas and should be in the material studied by participants in the proseminars recommended by the Committee on Preparation for College Teaching.

"Deliberately establish an atmosphere of openness and equality." By this I mean an atmosphere in which faculty and graduate students begin to interact as professional colleagues who can learn from each other. Mathematics has tended to be a solipsistic profession in which individual thinkers wrestled alone, *manus a manu*, with difficult abstract problems. The issues on which proseminars concentrate frequently are more social in implication and require a higher level of group solution. By having faculty and graduate students in-

teract during the seminars, one hopes to develop an easier acceptance of group problem-solving throughout the student's professional career. Not only does this kind of interchange lead to better solutions, it also is becoming essential in the consultation that occurs everywhere between mathematics and other disciplines. It also characterizes the teamwork which the corporate world prizes more highly every year. Many mathematicians will work in this corporate world and academic mathematicians will send most of their non-PhD majors to it.

"Be patient and flexible." The experimental reports in Part III do not claim to have "gotten it right" on the first try. They report modifications from Year I to Year II and even adjustments within a semester's plan. They did not expect instant perfection and would admit they did not get it. They recognized that they were breaking new ground, and they were perceptive enough to recognize and address the flaws in the design of their programs. But most, I think, would also say that they got a lot closer to a very good program in one or two iterations than they might have expected in advance.

The site mentors also recognized the importance of letting the program grow naturally. For example, it might be rare to make such a seminar an absolute requirement for all PhD degrees in the first year of its implementation. But it should not be unexpected that, over some period of years, as faculty and graduate students alike observe the advantages of such a seminar, it would become a formal part of the program for the majority of graduate students.

"Pay attention to the nitty-gritty." As we hear so often these days in debates on balanced budgets, national health care, and NAFTA, the devil is in the details. It is no different here. Seminars like those described in Part III run smoothly when someone worries carefully about reference books in a departmental library, a supply of provocative articles to stimulate discussion, taping machines to record and review teaching segments, and scheduling of experts from within the department or university to enrich the presentations. Let us remind you the reader that the Committee on Preparation for College Teaching hopes to have already done some of this for you in this book. Part V and Appendix B in particular provide you with a substantial bibliography (part of which has been annotated) and an at-hand supply of articles for which you have to search no further than these covers. The bibliography, of course, points you to many other sources, some of which arrive in your mail box regularly, like journals of the professional societies and publications about undergraduate mathematics instruction. Think of Part IV as a seminar session in which a dozen or more individuals reflect on the issues that will arise in your own seminars. Finally, the reports in Part III are full of hints about how to organize the small details that may determine the success or failure of your own efforts.

Some Comments About New Assistant Profes-

sors. The Committee on Preparation for College Teaching set out to describe experiences graduate students should have as part of a broad-based professional training. After the eight experimental programs in Part III were set up and anecdotal evidence began to filter back from these and other sources, it became clear that the charge to the committee was too narrow in one aspect. It should have included a consideration of the professional development needs of new assistant professors.

At present, follow-up work with assistant professors is as largely unattended to as the broader needs of graduate students themselves. Without it, the efforts begun in graduate school may be undone by lack of cultivation. This is especially important for those new assistant professors who did not have the opportunity for a proseminar or similar experience while in graduate school.

Many young, new assistant professors have only vague ideas of what is expected of them other than publishing enough research to satisfy a tenure committee. Occasionally, they are even unaware of the full implications of this publishing requirement or of how to go about meeting its demands. They frequently are unsophisticated about the social and collegial aspects of departmental life and "fitting in" to this new world. This inexperience generally is overlooked in graduate students, but it can quickly become a hindrance for someone on tenure-track.

What is needed is an organized mentoring program for new assistant professors for at least the first two years of employment and perhaps all the way to tenure. Where there are sufficient new hires to justify it, such mentor programs for new faculty could in part be group-oriented. Departments could have a series of seminars organized for newly hired assistant professors in which information about university procedures, tenure expectations, teaching duties, local resources, and professional expectations are laid out. In particular, an opportunity could be provided for recently hired assistant professors to meet regularly among themselves — or with some wise departmental elders present for consultation — to discuss issues important to their intellectual and social adaptation. Where the number of new hires is small, basic information of the kind described should be available at least in a manual or other written form.

In all cases, however, perhaps more constructive than these group efforts would be establishing a true one-on-one mentor relation between a new assistant professor and a senior faculty member. What is envisaged here is an interaction that goes beyond advice on research questions and publication, although that would be part of the intent. In fact, a new assistant professor might have two mentors, if a single person could not meet all the individual's needs. One mentor might work seriously in the same research area as the assistant professor, and the other would be a personally interested guide who could take an active interest in the new faculty member's adjustment to other aspects of academic life.

Some departments, as part of the tenure or promotion

process, appoint a senior faculty member to be an advocate for the candidate before the review committee. This role could be expanded so that the advocate would take a personal interest in the new colleague from the moment of hiring and actually help guide the assistant professor to the point of tenure.

The kind of support described here obviously can be very beneficial to a new faculty member. But it also makes good sense from a departmental perspective. It increases the chances that effort and expense invested in hiring will be rewarded in the long run with a permanent staff member. It also enhances the quality of a new employee's contribution to the department's overall mission. In particular, as more women and minorities enter academia as a result of long and careful cultivation by many mathematicians, departments have an especially important obligation to see that these new faculty members are supported all the way to tenure review.

The Committee on Preparation for College Teaching decided that the material gathered for this book is potentially as useful a resource in establishing mentoring programs for new assistant professors as in running proseminars for graduate students. In fact, the committee chose some of the materials in the book precisely with that purpose in mind. Indeed, in the absence of a mentor, the committee hopes new assistant professors can use the book to inform themselves about a broad range of professional issues beyond their research interests. Hence, the committee describes and advertises this volume as appropriate for both graduate students and new faculty.

Chapter 4

The Professional Seminar

Faculty Duties

Many new and not-so-new faculty members have viewed the varied demands of the job with some dismay. As mathematicians contemplate the array of competing obligations and the range of demands over the years of a career in college teaching, they have good reason to look bewildered. Doctoral students and postdoctoral fellows can be helped by experienced faculty members to develop an integrated and balanced sense of these responsibilities. The thesis director may try to give doctoral students some ideas about what their future employment may involve, but the development of a strong research program forms the basis of that relationship. We recommend that in a planned and organized way departments provide for new mathematicians a forum to discuss and learn about the diverse duties of the professoriate; this activity should be considered part of the assignment of at least one respected, usually senior, faculty member.

What is a Proseminar?

We will describe a seminar for mathematicians which will help pre- or junior faculty store information and experiences to draw on in the future. Other professions (e.g., medicine) have similarly intended seminars often called professional seminars. The site programs of this project (see *Part III*) used a shortened form of that descriptive title; proseminar. We sketch below some characteristics of proseminars which are successful in easing the upcoming transition for their participants. They are then able to profit from valuable additional experience, study, and professional activities directly related to teaching.

A proseminar might have both advanced graduate students and postdocs as participants. Most of the suggestions and activities below are suitable for such a mix, or lend themselves to easy adaptation. We think it is important to emphasize that, though some activities may be similar, the proseminar is distinguished from teaching-assistant orientation or training by its goal; it is not directed at the teaching productivity needed in the doctoral department. It is, of course, an almost inevitable secondary benefit that graduate students currently teaching will teach with more interest and effectiveness. Indeed, this may be the factor that sells some faculties and administrators on supporting the activity.

A Generic Model

The exact form, style, scope and content of a proseminar will depend on its institutional setting, participant needs and interests, and its faculty leadership. As efforts to execute the resulting plans proceed, enhancements especially appropriate to the goals or backgrounds of the local participants will evolve. Examples of some combinations of these possibilities and of local adaptations

appear in the site reports in this book. Seven of those site programs included most of the features we describe below. (The "thematic seminar" at the University of Delaware was somewhat different. Its stated main goal was to provide a model for participants of the way mathematical ideas from various sub-areas may be related and contribute to a mathematical theme. Despite this difference in primary goal, many ideas about teaching methods and professionalism were modeled or discussed in the context of learning about the mathematical theme.)

We shall now sketch, if only as a base for discussion, elements of a proseminar. We shall mention goals, activities, supporting technical considerations, and benefits. No one departmental program is likely to encompass all of these elements. While acknowledging the vitality of well-considered local variations, we believe that a proseminar that has most of the following characteristics will yield the greatest benefits.

Goals: What motivates the choice of proseminar activities?

◇ The primary purpose of the proseminar is to give the participant a strong push in development as a professional teacher of undergraduate mathematics.

◇ The proseminar is aimed at preparation for a career in academia and at building a foundation for continuing to learn how to teach mathematics, not merely for current duties, e.g., as a teaching assistant or postdoctoral instructor. It focuses on the broad demands and the resources available to support an academic career.

◇ It is important to evaluate and meet in the proseminar the needs of graduate students who are being supported during doctoral work by research appointments but who intend future academic careers.

◇ The nature of the activities of the proseminar encourages, even necessitates, a level of cooperation and interaction among the participants which is not typical of graduate mathematics courses. It may well be the first time some women, minority, and international students have the sort of confidence in being part of the community of mathematicians which comes more easily to many members of the majority culture.

◇ All the participants will become more aware of the needs of students from groups underrepresented in mathematics.

◇ The proseminar provides an opportunity for most graduate students to have as mentors other highly respected faculty members in addition to the professor supervising their research.

◇ The proseminar acquaints the participants with some of the best thinking and current information about undergraduate teaching in general and especially in mathematics. It provides an introduction to some of the most important relevant books, reports, periodicals, professional organizations and governmental agencies.

◇ The proseminar enhances the participants' appreciation of the productive interplay of research, teaching, and service in college faculty careers so that they have some idea of what may be expected of them wherever in the academic world they go.

◊ Above all, the proseminar instills a sense of the importance of being serious about teaching at all levels, and of doing it well.

Activities: What can be done to assist students to reach the proseminar's goals for them?

◊ The proseminar draws on illustrative mathematical material within the typical undergraduate curriculum. It is an opportunity to broaden and strengthen the participants' mathematical knowledge.

◊ Participating students visit advanced undergraduate classes taught by professors who are good teachers and report on what they learn from doing so; an internship can sometimes be arranged. Reciprocally, classes being taught by proseminar participants can be critiqued by professors.

◊ The activities of the proseminar are designed to exemplify good teaching. (We practice what we preach.)

◊ Participants are introduced to the useful literature on teaching in general, and on teaching mathematics in particular. The introduction includes browsing, exposure to especially significant sources, and group discussion. Visitors from other parts of the university may make valuable contributions to the proseminar's involvement with this area.

◊ Activities are designed to provide information and enhance sensitivity so participants will be able to help and retain students from underrepresented groups. There is an opportunity to serve several societal goals by incorporating a cogent plan for helping future faculty member works effectively with these students.

◊ Discussions about thorny questions of professional ethics may avert potential problems in the future or make it easier to handle them if they later arise.

◊ The participants will be much interested in examples (perhaps about, or involving visits by, recent graduates of the program) to illustrate the varying expectations in beginning academic positions. Information about the wide range of faculty responsibilities and varying proportions of the research, teaching and service assignments come to life in this way. Care must be taken that activities do not overly focus on "how to job-look," but the natural concern and interest about finding a job can be used to advantage.

◊ Most of the participants who will be applying for teaching jobs will be happy to have guidance in developing a teaching portfolio as part of the proseminar activities. They may also take keen interest in videotapes and critiques. More generally, participants should become acquainted with some of the better means of evaluating students, instructors, and programs, and some of the issues involved in using them. (See also the essay by Don Bushaw, Chapter 15.)

MODELING GOOD TEACHING

◊ Participants will be interested in knowing about professional societies and governmental agencies. Many publications of these organizations are used in the proseminar. It may be possible to arrange for a group of participants to attend a national or sectional mathematics meeting.

Mechanics: What technical considerations are related to the smooth and effective conduct of the proseminar?

◊ The participants are not beginning graduate students.

◊ The seminar is a graded course (perhaps Pass-Fail) on the student's transcript. Twenty-five or more 75-to-90-minute time periods, flexibly used, are needed. Time may also be spent individually or in groups in computer labs, observing faculty and/or peer classes, in libraries, attending professional meetings, and so on. Such a mix may simulate, on a small scale, the life of a mathematics professor.

◊ The seminar is counted as teaching on the faculty load in the same manner as other courses and appropriate resources similarly provided. (Traveling to professional meetings or discussing similar programs with colleagues is especially helpful.)

◊ The proseminar emphasizes the interplay of research, teaching, and service activities in the role of all college and university faculty members, albeit in varying proportions; as a consequence, the mentor(s) at each site have, among them, strong reputations in each of these roles.

◇ A departmental collection of resource books on teaching and publications of mathematical organizations is important for faculty and graduate students. Local bookstores may be persuaded to keep some of the more important items of these kinds in stock.

◇ Budget: The seminar costs after some initial setting-up are not significantly different from those provided generally in support of a course, but are perhaps a different mix. Copying and regular additions to the pedagogy collection will not cost very much but probably more than analogous expenses for other courses. University or departmental resources are probably available if videotaping is planned. It is helpful to provide the seminar director with a small entertainment budget to encourage the development of collegiality among the participants. (One department with strong resources furnishes a pizza lunch once a week!)

Motivations for the department: If doctoral departments are asked to provide this experience for their advanced students and/or postdoctoral fellows, are there benefits for the department in addition to those for the individual participants?

◇ Graduate student recruiting may be enhanced. The student who is already interested in undergraduate college teaching will be attracted by a description of interesting activities; others will be attracted by the general "we care" attitude which pervades departments when such programs are in place.

◇ Participants will do more effective teaching for their remaining time in the doctoral department, and some may be given more responsibility during their last couple of years. This may have an indirect effect of lightening some faculty loads to compensate for the time the program costs and will further enrich the graduate students' preparation for their careers.

◇ Participants serve eagerly as peer mentors for new graduate students or postdocs; in some situations they can be more effective than faculty in working with these developing mathematicians who are only a few years behind them on the same path.

◇ Proseminars are likely led by one professor or a small number of professors who are assigned the proseminar as a course. As course directors, these professors are likely to involve other colleagues as guest presenters or as mentors of individual graduate students. This association of people who may not have previously worked closely together can have as a valuable side-effect enhanced faculty collegiality among those not in the same mathematical area.

◇ Some deans or central administrators will likely be pleased with this activity. This is an opening to help them be better informed; support based on understanding and sympathy (which is very possible in this case) is likely to be the most stable. Of course, if they like the activities for reasons in addition to the department's primary goals, so much the better.

◇ It feels good to "do the right thing."

Local Interpretation

We have described above many features which may contribute to developing a strong proseminar, and the following chapters detail eight successful programs. However, there is no ideal project to clone or replicate. "Best" is a program you are able to direct which is most helpful for your participants.

Chapter 5

A Broad Mathematical Preparation

The Joint Committee on Preparation for College Teaching has given much thought to the question of what it should say about the mathematical content of doctoral programs. We have considered changes in the mathematical landscape and shifting connections among parts of mathematics as well as with other disciplines. Earlier reports concerning preparation for college teaching*, written more than twenty years ago, were more specific about content than would, we believe, be justifiable today. The committee unanimously endorses overall breadth of mathematical training, which the future collegiate faculty member can obtain in experiences that cover both the undergraduate and graduate years. We hope that our advocative position on breadth will be helpful when graduate or undergraduate major curricula are discussed and also for those who seek to be well-rounded mathematicians.

The chapter concludes with an essay written by two members of the committee. The version here was influenced by comments from other members and by reports from the example programs described in Part IV of this book. The philosophy of this essay is reflected in the committee's *First Report* (reprinted in Chapter 1).

Mathematical breadth. The committee believes that it would be impossible to specify a comprehensive list of essential mathematical topics that would not soon be made obsolete by rapid changes in the vitality and usefulness of the separate topics. However, breadth will better prepare students to face the challenge of future change in the mathematical environment as well as in teaching expectations and administrative responsibilities.

---

* A "Beginning Graduate Program in Mathematics for Prospective Teachers of Undergraduates," Committee on the Undergraduate Program in Mathematics, 1969. Reprinted in *A Compendium of CUPM Recommendations* Vol. 1, two volumes of such reprints published by the MAA in the early 1970's after cessation of NSF funding to CUPM.

Technology. The charge to the committee includes "...taking into account... the impact of computers". It is our impression that most graduate students in mathematics now acquire, or have acquired, reasonable competence along these lines. We believe that all graduate students should have easy access to computers or computer terminals and should acquire familiarity with electronic mail, mathematical typesetting and the uses, including teaching applications, of some standard mathematical and statistical packages. They should be familiar with the most commonly used operating systems and should have some experience with the types of personal computers used in undergraduate education. In addition, they should be familiar with graphing calculators and their use as a teaching tool.

Editor's Note: Attribution is difficult for the materials in this chapter since all members of the committee contributed ideas. The essay which follows is the thought of its two authors who are on the committee. Parts of an earlier version if this essay were endorsed by the committee through incorporation in its **First Report**.[2] (See Chapter 1.)

Mathematical Content and Teaching

Guido L. Weiss and Stephen B. Rodi

We feel that graduate students should become acquainted with a wide spectrum of mathematical topics by the time they obtain the PhD degree. Consequently, graduate education should not be restricted to a narrow area related to a thesis topic or a current area of research. Narrowness of this kind is short sighted. A student should be prepared to meet a large variety of professional responsibilities related to future mathematical research and other aspects of professional activities.

There are many reasons why we espouse this philosophy. Let us present a few of them. Experience has shown that there exists a considerable interrelation of ideas at advanced levels of research; thus, a broad knowledge of mathematics is a necessity if one desires to advance in research by taking advantage of this interrelation. Moreover, it is expected that each member of a mathematics faculty is capable of teaching essentially all undergraduate courses. This unifying tradition is not present in other disciplines and provides an important versatility that mathematics departments need for carrying out their heavy teaching duties. (In practically every university the mathematics department leads in the number of registered students that it teaches.) This capacity is being challenged by the increasing variety of mathematics undergraduate courses that have been introduced in recent years because of the ever increasing number of students who enter mathematics courses from other disciplines: economics, actuarial science, physical sciences, special branches of engineering, medical and biological sciences, statistics, education, and behavioral sciences. Mathematics professors owe it to these students to have some knowledge about the uses of mathematics in these various fields. The "mathematization" of various disciplines has been growing rapidly in recent years and there are good reasons for believing that the tempo is increasing.

We see, therefore, that a graduate program should produce young mathematicians with a backgrounds broad enough to enable them to embark on successful careers.

What is sometimes called the "core" curriculum that is now in effect in most American graduate programs in mathematics goes a long way to provide students with the breadth we believe is needed. A typical program involves "qualification"[1] in three or four areas. This usually is achieved by taking one-year courses in these subjects and then by passing corresponding qualifying examinations. An example is a course in algebra, one in analysis, another in differential geometry, and one in probability and statistics. Other subjects are often substituted for the ones just mentioned. Moreover, the qualifying process may include some form of examination or a presentation by the student in a more specialized area related to the subject of the thesis; in many departments more than one such presentation is required. Another common requirement is a demonstrated proficiency in a language other than English. In addition to all this, almost all departments offer advanced courses in areas that are of particular interest to individual members of the faculty.

We believe that it is a fair conclusion that a student who has obtained the PhD degree in a department that demands such requirements is, indeed, equipped with a broad knowledge of mathematics and has demonstrated ability to perform original research. For this reason many agree that the typical American PhD is a "more solid" degree than the doctoral degree administered by mathematics departments in most other countries. In fact, several departments outside the US have recently set up programs that are very much influenced by the typical American PhD program in mathematics we have just described.[2]

Despite these very positive aspects of American graduate education, we feel that more should be done in graduate programs in mathematics in order to improve the preparation of the young mathematician for their careers. Most PhD mathematicians lead their professional lives in institutions where teaching undergraduates is a most important activity. Moreover, most recipients of a PhD degree hope that they can find a position in such an institution. Thus, a successful graduate program should prepare the student to become an effective teacher at the undergraduate level and to become acquainted with the various features of an undergraduate program in mathematics. Let us address this question.

We propose, therefore, that, in addition to what we have described above as "typical" features of a graduate program, further efforts be made to prepare the student to become an effective teacher at the undergraduate level. This can be done in various ways. For example, each student may be required to participate in a semester or year-long seminar in which the student or faculty make presentations which focus on showing the relationship of a particular subject to other parts of mathematics. These presentations might also concentrate on the problems students might encounter in learning the material. Another approach could involve "model" lectures by a faculty member, together with discussions on the merits (or faults) in the presentation. The purposes of these efforts are to allow the student to concentrate both on how to make a good representation and on how to learn a new topic in mathematics. Thus, this is another way one can add to the breadth of the future undergraduate teacher.

Finally, we would like to point out that, with imagination and good will, much could be done in each department of mathematics to enrich the graduate program along the lines we have described. Each professional mathematician should be able to present a series of lectures, addressed to beginning graduate students, that describe and explain salient features of the areas of the mathematician's research. For example, one of us has given lectures on the basic ideas that involve Fourier Series and how they can be used to explain a considerable amount of topics in Analysis. These lectures were addressed not only to beginning graduate students but to undergraduate majors in mathematics at several different institutions. Topics such as partial differential equations, the theory of distributions, basic facts about bounded linear operators, ergodic theory, and probability theory are only a few of the topics that can be associated with Fourier series with a minimal background required from the listeners. We cannot give details of these lectures in this presentation, but we welcome inquiries on this topic.

Let us close with a list of topics that can provide considerable understanding of the breadth and scope of mathematics to an audience that has the preparation of the normal graduate student entering in a graduate program in mathematics:

(1) Finite groups and the fact that they have been classified;

(2) Galois theory and associated geometric constructions;

(3) Aspects of Algebraic and Differential Geometry and their differences;

(4) The theory of special functions and how they arise in several different contexts;

(5) Probability theory and its applications and ties with other branches of mathematics;

(6) The basic differences in the analysis of one, two, three and higher dimensional spaces.

(7) NP Completeness and its implications for discrete math and optimization; in particular, the study of the "$P = NP$" problem and its ramifications.

These are just a few examples of topics that can provide the stimulation and breadth we discussed. We mention them since we are aware of presentations involving them that have been very successful. We encourage all interested parties to develop materials in these and other subjects that could be used to enrich a graduate program along the lines we have discussed.

Notes

[1] See also the data reported in Chapter 2, **Summary** and **First Survey**. (Specific descriptive information about these procedures which amplifies our examples is given there.)

[2] Through a FIPSE project, the committee sponsored special programs at eight universities that were designed to carry out activities of the type we have just described. See Part IV Chapters 6-13 for a description of each of these programs (University of Cincinnati, Clemson University, Dartmouth College, University of Delaware, Harvard University, Oregon State University, University of Tennessee, and Washington University in St. Louis). These descriptions provide information about inclusion of a breadth of mathematical topics in the context of doctoral department activities, and evaluations of the effectiveness of the activities.

You're the Professor, What Next?

Part III

Programs in Doctoral Departments

Chapter 6 — The University of Cincinnati

Chapter 7 — Clemson University

Chapter 8 — Dartmouth College

Chapter 9 — The University of Delaware

Chapter 10 — Harvard University

Chapter 11 — Oregon State University

Chapter 12 — The University of Tennessee

Chapter 13 — Washington University

These eight university mathematical sciences departments cooperated in the project assisted by the Fund for the Improvement of Postsecondary Education (FIPSE). The program mentors individually developed activities suited to the experiences and needs of their advanced doctoral students. Special seminars and courses devoted to broad-based professional growth were established. The seminars emphasized many facets of college-level teaching.

Chapter 6

The University of Cincinnati

Edward P. Merkes

An Overview. Our Department has awarded a PhD degree since early in this century. Presently, the doctoral program has about forty full-time, mostly international, students with an equal distribution of males and females. In preparation for teaching their own courses, the advanced degree candidates are encouraged to enroll in the regularly scheduled year-long course, "The Proseminar in the Teaching of College Mathematics." The purpose of the course is to emphasize the teaching and service aspects of a professional career in mathematics. The course includes discussion of and experience with a diversity of alternative teaching techniques, some learning theory with application to attracting minorities to mathematics, training in the use of technology in the classroom, and an exposure to a wide variety of mathematical topics and applications. Our goal is to produce a broadly educated PhD who is aware of the character and mission of many kinds of institutions and the cultural differences of the students that they serve.

Other Mentor: Lowell Leake

"Mathematics contains much that will neither hurt one if one does not know it nor help one if one does know it."

Although often paraphrased today, this is a quotation from almost three centuries ago by J. B. Mencken, *De charlataneria eruditorum*, II, 1715. Perhaps after 277 years we, as teachers of mathematics, at least have begun to overcome this myth by making the subject more relevant for students in specialized areas through a selection of applications that are more specifically discipline-oriented. Our halls, nonetheless, still echo with the proclamations that calculus is calculus whether it be for engineers, for business students, for physical science majors, for biologists, for economists, or even for mathematicians.

To some degree our graduate students, especially the international students, seemed to view mathematics as a discipline unburdened by any applications outside those of the physical sciences. One important aspect in the training of future college teachers of mathematics, therefore, is to familiarize each with some non-traditional applications and with some sources for such applications. There is an ever-growing collection of literature precisely on such unusual uses of mathematics that range from the most elementary to the sophisticated. If we are to renew interest, appreciation, and enthusiasm for mathematics by today's students, meaningful illustrations of its efficacy in a wide spectrum of disciplines are a step worth taking!

Our department's curriculum does not include precalculus courses. Consequently, there are few classes in any quarter taught solely by graduate students. However, each graduate student has full responsibility some time for a calculus or elementary statistics course. Initially, our graduate students receive a week-long orientation to their duties. Unless experienced as teachers in this country, they are used as tutors, assistants to instructors, or graders. They receive various materials about teaching and discuss the content. Each quarter Professor Lowell Leake works to improve the teaching performance of the TAs who teach their own courses by first videotaping a session. The TAs review the tape and comments in writing on their own presentation. Later, the TA and Professor Leake meet privately to discuss ways to improve the instruction. When it is necessary, the process iterates until both are satisfied. (A more detailed description by Leake will appear in a forthcoming issue of *Primus*.) Henceforward, the faculty member to whom the assistant is assigned monitors that assistant's progress as a teacher.

Our first attempt to train advanced graduate students in teaching more advanced undergraduate mathematics began in the summer of 1991. We initiated an informal course that met twice a week in two-hour sessions over a five-week period. There were ten students in the group representing such countries as China, Colombia, India, Korea, Poland and, with only one representative, the USA. The initial sessions discussed undergraduate education in each student's native country including comments about teaching methods as well as curriculum. In most of these countries, the curriculum for a mathematics major generally follows the more classical pattern common in our country some years ago. If one seeks the intersection of the teaching styles for the courses taken by this heterogeneous group of nationalities, it was the pure uninterrupted lecture, often for more than an hour, sometimes augmented weekly by a small group discussion or problem session. When asked if they maintained attention for such long lecture sessions, the reply was, "Yes, because we had to if we hoped to stay in college." (They also admitted they were aware of the content of each lecture and studied it before the presentation, a practice seldom if ever found in this country.) It was not a surprise that these students did not defend this style and were quite receptive to discussions of alternate methods of teaching advanced undergraduate courses.

This experience proved to be the model introduction for our seminar each successive autumn. The students began to know each other by name as well as background. At the close of the introductory sessions we distributed a delightful essay, written by a faculty member, H. David Lipsich, on four of his "best" teachers, each of whom used an effective yet different pedagogical technique. (You may request a copy.) The graduate students ranked in writing the four teachers and discussed their ranking in a subsequent session. This simple exercise seemed to awaken their interest in teaching!

After these and other cursory discourses on general instructional methods, I gave an intentionally flawed 30-minute lecture on "What is Linear Programming." It included some feeble attempts to engage orally this group of advanced students who were generally unfamiliar with the topic. The audience was informed in advance that numerous principles for good teaching were to be violated during the lecture. (The exercise is designed to help overcome a reluctance of international students to be openly critical of a professor.) Afterward, each person in the group evaluated the lecture. They identified all the intentional flaws, along with certain unintentional ones. As an assignment, each participant listed in writing some of the weaker and some of the stronger points of the presentation. To encourage and guide the criticism, we supplied each student with a form often used by the Department of Mathematical Sciences at the University of Cincinnati to evaluate the teaching of our TAs (Appendix 1).

This type of critical evaluation of presentations constituted the foundation for the remaining sessions of the summer course. Each student selected for a lecture/discourse presentation a topic that could be a component in, or at least a supplement to, an advanced undergraduate mathematics course. Although initially the international students were reserved, the audience became more constructively critical with each session and the lectures grew increasingly more interactive and polished. The course ended with a full session devoted to a college professor's responsibilities beyond those of teaching and specialized research.

From that summer experience participants learned to place greater emphasis on teaching in general and on particular teaching styles, as evidenced by a noticeable improvement in the mechanics and clarity of the lectures they presented. Since advanced graduate students frequently lecture in departmental courses, this summer course demonstrated that additional attention to communication as well as content can lead to improved teaching.

A formal course sequence entitled *Proseminar in the Teaching of College Mathematics* was approved by the Department of Mathematical Sciences and the University of Cincinnati by the end of the summer of 1991. It was offered for credit for the first time during the 1991-92 academic year. The author of this article was the "instructor," although a better title is "facilitator." Since almost all of our advanced students in the PhD program are foreign nationals, we permitted a few American-born students seeking a master's degree to participate in the proseminar. The latter students, many of whom wish to teach in community colleges, read and presented summaries of sections found in a general book on effective teaching. For this we selected *The Craft of Teaching: A Guide to Mastering the Professor's Art* by Kenneth E. Eble. In addition, all the students read and reviewed the proposed faculty roles cited in the report of the Committee on the Mathematical Sciences in the Year 2000, *Moving Beyond Myths, Revitalizing Undergraduate Mathematics*. Many of the suggestions in this report are in unfamiliar phrases, so there was need to amplify and illustrate such statements (or jargon, or vocabulary) as

◇ Learn about learning,

◇ Explore effective alternatives to lecture and listen,

◇ Involve students actively in the learning process.

First Quarter. After initial introduction to the literature, the remainder of the quarter to a large degree was devoted to the task of discovering ways for undergraduate students to play a more active role in their own education.

Five styles of teaching were illustrated and discussed by the facilitator and the participants. Each student selected an application of a topic from an advanced undergraduate course. After consultation with a faculty member, the student made a presentation to the group using at least one of the alternative styles of presentation. The participants role-played the students in the particular course for the selected topic, and afterward they scrutinized orally, or in writing, the presentation for ways it might be improved.

The styles considered were: lecture/recitation; lecture with class interaction; lecture with a writing component; collaborative learning; and the unadulterated Moore Method followed by some modifications.

1. We began the discussion of teaching methods with the well-prepared *uninterrupted lecture*. No illustration of this prevalent style was necessary. The participants were quick to point out that, until a date in the very distant future when all mathematics classes at colleges and universities are relatively small, there is little chance that this method of instruction will vanish from the earth. Combined with small discussion groups and careful selection of visual aids, the proseminar members concluded that this method is not necessarily ineffective at least for the more motivated students. Indeed, it is the most widespread manner in which all the international students, and many of the students educated in this country, were taught mathematics in the past. Nevertheless, the group universally agreed the uninterrupted lecture method should be confined to large classes and then only if a small-group recitation component is an integral part of the course.

2. Next we discussed *strategies for promoting class participation* during a lecture. The group watched the tape of George Polya, *Let Us Teach Guessing*, to illustrate one such method. Each presenter of a topic thereafter prepared a lecture with relatively frequent places to engage the audience by asking a question or a sequence of questions. Initially, those who selected this method had presentations that were labored and stiff. In time, however, the interaction between the participants and the lecturers became very effective and natural as it often does in a normal smaller class setting.

3. The University of Cincinnati for a number of years has conducted workshops on *Writing Across the Curriculum* (WAC) and last year hired a Director for WAC, Dr. Barbara Walvoord. After asking the proseminar participants to review articles in *Using Writing to Teach Mathematics, MAA Notes #16*, we invited Dr. Walvoord to speak on the applications of WAC principles at a departmental colloquium. There was resistance during and after the presentation to the suggestions on how to use WAC techniques in the classroom. When asked how each studied mathematics, however, the participants generally responded that they learned by writing key ideas and by creating illustrations on paper. In a way, this pattern is used by WAC, using a notebook for reading assignments. In this notebook students record in their own language any points that require greater clarification as well as essential ideas. To introduce WAC more fully, we followed the colloquium with a number of graduate student lectures that incorporated an *in-class* writing component as opposed to a simpler "term paper" assignment. (Have you ever asked a calculus class to define the derivative after spending one or more class periods on the concept? Or have you ever asked an advanced calculus class to define the limit of a sequence using words and no symbols?)

4. The principles employed by Uri Treisman for *collaborative learning* were introduced by viewing his MAA videocassette. The participants attended, and reported on, some courses given by the department where collaborative learning techniques are practiced for minority students as well as for more heterogeneous groups of college freshmen.

5. A *modified Moore Method* was next introduced with reference to an article by D. Cohen (*Amer. Math. Monthly*, 89 [1982] 473-490). Essentially, the instructor presents the definitions and theorems on handouts. Following the theorems are suggestions or hints on how each might be proved. The students take turns presenting lectures that provide illustrations and detailed proofs while the instructor and the remaining students question and, at times, challenge the speaker. One would suspect that this method would prove popular with foreign nationals who often have language difficulties, although this did not appear to be the case.

6. Finally, we considered the *pure Moore Method*, first by watching the R. L. Moore videocassette of the MAA and then as outlined by F. B. Jones (*Amer. Math. Monthly*, 84 [1977] 273-278). A faculty member, who as a student was trained by this method, spoke about the pros and cons of the technique. To illustrate the ideas, even though all the participants had violated the method's rule by having read books on the subject, a brief set of definitions and perhaps nontraditional theorems in linear algebra was handed out in the penultimate class of the quarter. Students were asked to prove the theorems without consulting either the literature or their colleagues.

Some of the participants presented proofs during the final session. (This handout is included as Appendix 2.)

Second Quarter. The students of each specialty were encouraged to collaborate in preparing lectures that could be understood by an advanced undergraduate mathematics major on important aspects or applications in their particular field. The lecturers not only promoted class interaction but also required writing as a means for learning. This format along with themes selected to fill observed gaps in the background of the particular group of students, fosters breadth of knowledge over and above that obtained by the relatively standard qualifying examinations normally required of a PhD candidate. (The collection of students in the seminar that quarter, for example, did not know of Caratheodory's theorem on convex sets in R_n. This, along with related convexity theorems, became topics for special talks.) Excellent sources for some of the applied presentations are the SIAM publications *Mathematical Modelling: Classroom Notes in Applied Mathematics*, and *Problems in Applied Mathematics*, both edited by Murray Klamkin.

Third Quarter. After the projects of the previous quarter were completed, we began each two-hour session with a brief guest lecture usually by a faculty member. The subjects included: "How I improved my English;" "Using the overhead projector effectively;" "The best teachers of mathematics I have known;" "Graphing calculators for statistics and calculus courses;" "Visualization in 3D calculus;" "Spreadsheet mathematics;" "The math major's undergraduate curriculum; what it is and what should it be?;" "Using MAPLE in teaching and research;" "Why so few women in mathematics?;" "What to include in a vita;" "Building a teaching portfolio;" "The job interview;" "How NOT to give a Colloquium;" "Writing a review;" and "Using e-mail." A reading list usually accompanied each lecture. However, many sessions began with the students "brain storming" the topic. For example, the graduate students in the session on a job interview created a list of "do's and don'ts" after which the articles on the topic were suggested. (These articles are in the *AMS Notices*, 38 [1991], 891-894 and 39 [1992], 560-563.) Finally, we had students report on mathematical articles and articles about mathematics in such publications as *UME Trends, FOCUS, The American Mathematics Monthly, Mathematics Magazine, AMS Notices, SIAM News*, and *SIAM Review*.

As alluded to earlier, in the second full year of the offering of the proseminar we incorporated the first summer experience into that of the initial quarter. We built tutorials and scheduled laboratory sessions for the graduate students on the use of MAPLE. It consumed very little time for these bright students to become efficient in its use. We directed them to articles on computers in the classroom. A similar approach was followed for graphing calculators. We discovered how little history proseminar-participants know even in their specialities. Consequently, we required some historical talks about mathematics.

We plan to devote additional time this quarter to learning more about learning with some guest speakers from the Department of Psychology.

The proseminar also served to point out gaps in our initial TA training. In an experiment with a faculty member from the College of Education, for example, we discovered a wide difference among the participants when they graded quiz performances of other students. A picture and the background of the fictitious student were provided and, after viewing this information, each TA graded the person's quiz. The performance on the quiz was identical for each fictitious student, but the pictures and the background differed from student to student. To our dismay the grades ranged from 55 to 90%.

What should we do differently in the future? It would be better to slow the pace. The participants need more time for the development of pedagogical skills necessary to engage undergraduates as active participants in the lecture and to employ writing as a teaching tool for learning. There should be a greater emphasis on nontraditional applications of mathematics and meaningful use of modern technology. We need to include some open discussion about cultural differences and the nature of prejudice.

One approach that is not recommended is to devote too much of the time in the proseminar to a single theme. When this is done, the course becomes just another hurdle for the graduate students and enthusiasm wanes. Furthermore, a variety of topics, interspersed with discussions of teaching styles and professionalism, enables each period to be more self-contained. It fosters participation by students not necessarily registered for the proseminar when they find the topic of a particular session is of interest.

Did the participants benefit from the proseminar? During the final session they were surveyed to determine their reactions to the proseminar. We introduced a somewhat unusual evaluation form, mainly to determine whether the attitudes toward the teaching functions of these future college teachers had changed. (Improved pedagogical skills and awareness of alternative teaching styles were assumed to be a natural consequence of the course.) On our evaluation form (Appendix 3) last year, the universal positive reply to Question 2 about becoming more aware of the importance of teaching perhaps suggests that they did. The next item, however, indicates that there was indeed a benefit. After a quarter of the proseminar, the group's average had a greater percentage of the effort devoted to teaching over research, which is contrary to their reaction at the start of the course. Finally, the interaction demanded by the courses brought the students together. By way of illustration, one American student stated it was the first time in her two years as a graduate student that she got to know and talk about mathematical and nonmathematical issues with the Asian students. Overall, there is a much improved sense of cooperation among the students and the faculty mentors as well. Also, a number of the recitation sessions conducted by TAs have now become examples of collaborative learning.

In summary, international students often need assistance adjusting to the cultural diversity of this country. Once this is accomplished, they are at most an accent away from their American colleagues and just as receptive to alternative styles of teaching. For any group of graduate students in the mathematical sciences we recommend, from our experience, that a Proseminar in the Teaching of College Mathematics incorporate each of the following ingredients:

1. Have the students frequently read and report on articles about teaching and general articles about the state of the profession in various mathematical publications, e.g., *FOCUS, SIAM News, NOTICES, UME Trends.*

2. Have some students, usually people with undergraduate degrees from institutions in this country, read and report on at least one non-discipline general book on teaching at a college or university.

3. Have the students practice alternatives to the "lecture and listen" style of teaching. Have the students play the role of undergraduates when not the lecturers and play the role of peer evaluators after the presentations. Rotate the role of lecturer.

4. Permit the students to select a topic for one of their own presentations (with your approval). Require it to be at a level beyond calculus, and one that might be supplementary to a course in an undergraduate mathematics curriculum. Encourage applications of mathematics. Require advanced students to give introductory lectures on aspects of their specialities that can be understood by first-year graduate students in mathematics or in mathematical statistics.

5. Make the sessions "fun" rather than just another hurdle for the students.

6. Have guest lectures from faculty in other disciplines as well as members of the mathematical sciences community.

7. Intersperse the mathematical presentations with discussions on being an effective member of a faculty and the importance of other forms of scholarship in addition to research.

8. Develop a form to evaluate your success to be completed at the final session of the course.

Appendix 1.
Teaching Evaluation Form

Instructor, Date of Visit, Course, Number of Students Attending, Starting Time, Ending Time

1. How well was the instructor prepared for the class session?
2. How clear was the presentation?
3. How well were the topics motivated?
4. Was there an appropriate division between theory and example for the course?
5. How much enthusiasm was there in the presentation?
6. Was the pace of the presentation appropriate for the course?
7. Was it easy to hear and understand the instructor?
8. Was there evidence of eye-contact by the instructor with the class?
9. Comment on the instructor's blackboard techniques.
10. How many questions did the students ask during the class period, and how well did the instructor respond to these questions?
11. Did the instructor ever solicit the class to help?
12. How many questions did the instructor address to the class? Did the instructor provide adequate time for a response?
13. What generally was the reaction of the class to such questions?
14. Describe any additional effective interaction between the instructor and the students in the class.
15. Did the students appear attentive? For how long?
16. To what degree was the class atmosphere friendly?
17. Did the instructor ever use good analogies, humor, or storytelling to make an important point?
18. Were there any annoying mannerisms in speech or otherwise of the instructor?
19. Were assignments given and clearly described?
20. Please list ways you believe the instructor might improve the technique used for teaching this course.

Appendix 2.
Equations with Inequalities for Constraints

Consider the linear system $Ax = b$, where A is an $m \times n$ matrix of rank m and $b \in R_m$. We say x^* is a feasible solution of this system if $Ax^* = b$ and $x^* \geq 0$. (The inequality for the vector means all its components are non-negative.) A basic index set B is a subset of $\{1, 2, 3, \ldots, n\}$ consisting of m distinct integers such that the indexed column vectors of the matrix A corresponding to these integers form an invertible submatrix AB. A solution x^{**} is said to be basic if there is a basic index set B such that all the components of x^{**} that are non-zero have indices in B.

Theorem 1. *The system $Ax = b$, $x \geq 0$, has a solution if and only if it has a basic feasible solution.*

A nonempty set C in R^n is called convex if for any pair of points x and y in C the points $\{z|z = tx + (1t)y$ for some $t, 0 < t < 1\}$ are also in C. A point w of a convex set C is said to be an extreme point if $w = tx + (1t)y$ for x and y in C and some $t, 0 < t < 1$, implies $x = y$.

Theorem 2. *The set $C = \{x \in R_n | Ax = b, x \geq 0\}$, if nonempty, is convex, and has at least one extreme point. Furthermore, x^* in C is an extreme point if and only if x^* is a basic feasible solution of $Ax = b, x \geq 0$.*

Corollary. *There are at most $\binom{n}{m}$ extreme points of the set C above.*

An index $j \in \{1, 2, 3, \ldots, n\}$ is called a null index if the jth component is zero for each solution of $Ax = 0, x \geq 0$.

Theorem 3. *If $C = \{x \in R_n | Ax = b, x \geq 0\}$ is nonempty, then j is a null index if and only if there is a non-zero vector $\lambda \in R_m$ such that $\lambda^T A \geq 0$, $\lambda^T b = 0$, and the jth component of $\lambda^T A$ is strictly positive.*

Appendix 3.
Survey

1. Of the various teaching techniques illustrated this quarter, which of these would you attempt if teaching a mathematics course beyond the calculus? (If more than one choice, rank the first choice as 1, etc.): (a) uninterrupted lecture; (b) lecture/question; (c) lecture/writing; (d) collaborative learning; (e) modified Moore method; (f) pure Moore method; (g) none — instead, I would do $\mathcal{X}$. Reply:(a); (b); (c); (d); (e); (f); (g); $\mathcal{X}$.

2. Has the Proseminar helped you become more aware of the importance of the teaching component of being a mathematician (even in a non-academic type environment)?

3. In the chart below what percentage of your effort should be spent in each category if you are a faculty member of a(n): (i) four year college? [Research; Teaching; College and Community Service]; (ii) university that offers a master's degree as its highest degree in mathematics.[Research; Teaching; College and Community Service]; (iii) university that offers a PhD degree in mathematics.[Research; Teaching; College and Community Service].

4. If you are not a citizen of the USA, would you change the percentages should you return to the country where you received your initial college level education? How? (i) Research; Teaching; College and Community Service; (ii) Research; Teaching; College and Community Service; (iii) Research; Teaching; College and Community Service.

5. Which of the techniques of teaching should receive more emphasis in the proseminar?

7. What should we do to make the Proseminar more meaningful?

SIGNED (Optional)

Chapter 7

Clemson University

Joel V. Brawley

An Overview. Clemson University is the state-supported land grant university of South Carolina. The campus borders Lake Hartwell and is located in the northwest corner of the state in the foothills of the Blue Ridge Mountains. The Department of Mathematical Sciences has some 50 faculty members and around 100 graduate students.

The Clemson project on preparing PhD students for careers in college teaching was developed with the following goals in mind: (i) to help the students become more effective teachers; (ii) to help the students become more aware of the components and expectations of the profession; (iii) to broaden the students' mathematical perspectives; and (iv) to achieve the above three goals without imposing too much additional expense (in terms of time and money) on the students. These goals were met through the careful design of a professional seminar carrying three hours of graduate credit, the purchase of a small amount of video equipment, and the establishment of a special departmental library for the project.

Some Background Information. Before describing the Clemson project, let me say a little about myself and how I came to be involved in the project. I have been a university mathematics teacher for over thirty years. My formal training to be a professor came in 1960 when, as a first-year teaching assistant, I was assigned to teach a course in college algebra. I was given a book and syllabus for the course, and I was told the building, the room number, and the meeting times. Many of my contemporaries from other universities received about the same training.

Certainly, we have come a long way in TA training since those days. Most colleges and universities with teaching assistants put considerable effort into making sure they are prepared before putting them into the classroom. But not all of us who eventually go into college teaching serve as TAs. Further, being a TA is very different from being a professor. Until recently, with the instigation of programs like those described in this volume, very little formal effort was put into preparing mathematics PhDs for careers in college teaching.

There are several plausible explanations for this neglect. Krantz [8] offers the first three of the following four attitudes which have been observed: (i) good teaching is unimportant, (ii) the components of good teaching are obvious, (iii) the budding professor has spent a lifetime in front of professors, observing all kinds of teaching, so certainly this person knows the traits that define an effective teacher, (iv) good teachers are born and not made so not very much can be done. Krantz effectively argues that (i) is false and that (ii) and (iii) are true. The rub, he says, is that "sometimes the obvious has to be pointed out to us."

This article discusses a program at Clemson University whose goal is to prepare better mathematics PhD students for careers in college and university teaching. After some background information, the paper describes the various components of the Clemson project and then gives a class-by-class outline of a course taught in the fall of 1992. It is hoped that the paper will provide everyone considering a similar course at their university some ideas on ways to proceed.

Before undertaking the Clemson project, I leaned a little toward (iv) and sided at times with the great algebraist Emil Artin, who was known to be an exceptional teacher. In a 1960 address [1] in which he discussed what he thought should be included in a graduate algebra course, Artin said, "In closing allow me a few words on the pedagogical side of the course. We all know that the best planned course can be ruined by a poor presentation. It is my experience that nothing can be done about it. The first seminar talk of a student has always revealed to me his future teaching abilities. I have often tried to improve this by talking to the student and explaining to him his mistakes. I have had little or no success whatsoever. On the other hand, we know that with very good students notoriously poor teachers have had frequently remarkable success. I am therefore of the opinion that we may safely leave aside any consideration of this problem."

With Artin's remarks in mind and with the realization that I had never had any formal training in "how to teach," it was with some degree of anxiety and reservation, and after some "arm twisting" by my department head, that I agreed to be the site mentor for the project at Clemson University. Perhaps I ultimately agreed to be the site mentor because I believe in the philosophy: if one wants to be better at something, then through dedication and hard work one can be better.

Still, I had very little idea of how much improvement I could or should expect from the students relative to their classroom abilities. Even so, I felt we should focus on improvement. So it was during my first meeting with my initial group of students that I stated, partially with tongue in cheek but also with some seriousness, the following (mathematically simplified) goal for improvement: If the teaching ability of an individual student at the beginning of the project is represented by a number x in the closed interval $0 \leq x \leq 1$, (with 1 denoting perfection) and if the teaching ability of this same student at the end of the project is denoted by $f(x)$, then the goal is that $f(x)$ be at least the square root of x. Incidentally, I also pointed out that, if there are only two kinds of teachers, meaning the domain of the function is the set $\{0, 1\}$, then taking square roots does absolutely nothing, in which case Artin's perception and our project goal are not in conflict. I did say, however, that I felt the continuum $0 \leq x \leq 1$ to be the more appropriate domain.

As with all attempts at quantifying the subjective, it is difficult to judge whether or not we met our goal. However, from the beginning I was sure that $f(x)$ would be no smaller than x and that the difference $f(x) - x$ (i.e., the improvement) would depend not only on x but also on many other complex factors including desire, effort, practice, feedback, experience, innate talent, and so on. Now, after teaching the seminar twice, I am convinced, as were the students, that $f(x) > x$ for all x in these two classes; additionally, I believe that I personally am a better teacher because of my involvement.

Goals for the Clemson Project:

To help our students become more effective classroom teachers; to help our students be more aware of the many components and expectations of being a college professor; to broaden our students' mathematical perspectives; to achieve the above three goals in a manner that is not very expensive to the students (in either money or time).

These goals were met through the careful design of a professional seminar carrying three hours of graduate credit, the purchase of a small amount of video equipment, and the establishment of a special departmental library for the project.

The Clemson Professional Seminar. The primary component of the Clemson project is a one-semester professional seminar course carrying three hours of graduate credit. The single requirement for enrolling in the course is the completion of the doctoral preliminary examinations (which usually come after two or three years of graduate study), and registration is voluntary. The course has been offered the past two fall terms and will henceforth be offered every other year. Generally, each of the students in the class has spent several years as a TA, so most of them have considerable teaching experience. The class meets twice weekly for sessions of one hour and fifteen minutes each and has one or two evening socials during the semester.

The seminar, which meets about 28 times during the semester, can be roughly divided into three nearly equal components: (i) readings and discussions of issues of teaching and pedagogy and observation of experienced teachers; (ii) mathematical presentations and their critiques; and (iii) panel discussions on academic matters. There are also a final session, a course critique, and a wrap-up.

The readings, discussions, and observations of teaching come mostly at the beginning of the course to set the stage for the student presentations. In these early sessions the students gain a better understanding of various techniques, methods, and elements of effective college teaching, and they get some understanding of the different expectations found in college and university environments. Further, they establish a camaraderie with other class members to prepare them better to accept and offer the peer evaluation which comes later.

The primary textbook for the pedagogy sessions is [8]. This book is quite nice for evoking discussion and getting the students involved in analyzing and describing their own experiences. We also use several other sources for these sessions, including readings from [2], [3], [7], [9], [11], [12], and [13]. The first term, we used Eble [7] as the primary textbook, but I found the newly available Krantz book to be much better for our purposes because it is directed explicitly at mathematics students. (Eble, an English professor, directs his book toward all disciplines.)

As a part of the pedagogy portion of the course, the participants are required to visit the classes of some of our more experienced professors and report back their observations. Their instructions for these visits are to "focus on the teaching techniques and not the mathematical content." Of course, I obtain the permission of the professors prior to the visits, and each student visits a different professor. These visits provide for some interesting discussion. After we have spent considerable time discussing the elements of effective teaching (including such issues as testing, grading, discipline, advising, textbook selecting, et cetera) we start lecturing to each other. Each student gives two presentations — an hour lecture and a 20-minute presentation. I should mention that, in the spirit of "what's good for the goose is good for the gander," I myself give two presentations just as the students do.

The hour lectures are pitched at the senior or beginning graduate level and must be on an important mathematical topic not generally found in the student's proper course work. There are many such mathematical topics, and the students select their own topics, subject to my approval. I do pass out lists of possibilities (see Appendix 1), and I encourage the students to browse through our departmental library for possible topics. What is intended in these hour lectures is for some real teaching and some real learning to be going on and that the lectures broaden the students' mathematical perspectives.

The 20-minute presentation is expected to be something different, and we are somewhat flexible here; there may be: (i) instruction in calculator or computer use; (ii) a demonstration with a hands-on geometrical model or other mathematical object; (iii) a report on a conference; (iv) a brief introduction to a certain method of teaching or theory of learning; or (v) a short research presentation as would be done at a professional meeting. Thus, the lecture can be either mathematical or pedagogical and it can also involve team teaching.

Both of a student's presentations are videotaped, and a written evaluation of each is made by every class member. (See evaluation forms in Appendix 2.) Generally, the first 10-15 minutes of the period following a student lecture is used for a self-critique after the lecturer has had the opportunity to review the videotape and the peer evaluations. Then we go on with the next lecture. I go through the same lecture and evaluation process as the students.

The last third of the course is composed of panel dis-

cussions of various teaching-related issues. The panelists usually come from the faculty within the university (though not all from mathematics), but we also use outside expertise when available. Generally, each of these panel discussions is composed of a 45-minute panel presentation followed by a 30-minute round-table discussion. The exact topics we consider in these sessions depend, of course, on the availability of faculty and other outside panelists, but there is reasonable consistency from semester to semester. The last time I taught the course the panel discussions were on the following topics: (1) getting students involved in learning; (2) tenure, promotion, and the job market; (3) three faculty philosophies of university teaching; (4) industry, government, and the teaching of mathematics; (5) writing across the curriculum; (6) technology in the classroom; (7) miscellaneous professional responsibilities; and (8) women and minorities in mathematics. I also showed and we discussed the film *Let's Teach Guessing* by George Polya, a famous research mathematician who also did fundamental work on teaching.

Support for the Project. The project drew support from three sources. In addition to the FIPSE grant project by which this book is produced, I wrote and received a "Teaching Awards" grant from competitive funds within the university, and I received support from the department, not only by way of encouragement from my department head, but also by way of some travel funds, some equipment money, and the inclusion of the seminar as part of my regular teaching load. These sources allowed for the purchase of a camcorder with tripod and the establishment of the special departmental library. Also they provided some summer support for me to develop the course.

The faculty in my department has been very supportive. Everyone I have asked to help (and there have been many) has willingly done so; several have indicated their willingness to teach the course in the future.

Grading. Grading is on the usual scale (A, B, et cetera) and is based exclusively on participation and attendance; i.e., there are no written examinations, and no attempt is made to grade the final teaching performances. In fact, each student is told that the requirements for an A are to: (i) come to class; (ii) actively participate in the readings and discussions; (iii) give the required presentations; (iv) conscientiously fill out the appropriate evaluation forms; and (v) strive for self-improvement. Each of the students who has taken the course has satisfied these requirements and received an A. Although I usually give grades to distinguish between students, I believe that this class, composed entirely of post preliminary-exam students with proven mathematical abilities, is different and that the pressure to excel comes from other sources, namely, from oneself and from peers.

Expense. The course imposes very little financial expense on the students. While they do have to pay for the three hours' credit they receive for the seminar, they generally register for three fewer hours of research. There are no expenses for textbooks, as all books used in the course are borrowed, usually from the project's library. Additionally, the course is designed to interfere minimally with the student's other study; the only substantial time requirements come in attending class and in preparing for the presentations. I would estimate that, above this, two to three hours of work per week are required.

The fact that the course counts as part of my regular teaching load means it is not tremendously expensive in relation to my time; and, presumably, the added expense to the department and university helps us better fulfill our mission by producing better teachers as well as enhancing our own undergraduate instructional program because those taking the seminar usually teach another year or two at Clemson before entering the profession.

The Departmental Library. With the help of the supporting grants, we have created a departmental library of books and materials relating to the course. This library contains both books on pedagogy and books containing material suitable for lectures. For example, most of the MAA publications related to the teaching of mathematics have been bought for the library, as have been many of the MAA books on mathematics itself. We have obtained multiple copies of the books we use as textbooks (e.g., [7], [8], [11]) so that each student is loaned these books. New books are added to the library from time to time, often on recommendation of the students.

Student Evaluation of the Professional Seminar. The students in the project have been very enthusiastic about it. On the anonymous final evaluations, thirteen of the fifteen students gave the course an excellent rating, and the other two judged it as good. Some of the positive comments from these evaluations were: a valuable experience with candid discussions; I believe every member of the class improved as a lecturer; seeing ourselves on video was very instructive; this is an excellent course and should be required of all PhD students planning to enter the academic world; the panel sessions with the other professors were very helpful; I enjoyed hearing the faculty present their teaching philosophies; [the course] gave us a chance to discuss teaching techniques and methods; the perspectives on tenure, interviewing, and career advice were invaluable and gave me ideas that I doubt I would have picked up elsewhere; I will miss this course when it is finished.

The major negative comments came the first time I taught the course and pertained to the appropriateness of the Eble book [7] and the fact that the students felt they would profit from more panel sessions and pedagogy sessions. I thought these criticisms were valid, so in the second offering we used Krantz [8] and added several sessions.

Some Personal Reflections. In spite of my initial concerns and fears, I enjoyed developing and teaching this course. Although I spent considerable time thinking about the course and working on it before it actually began, I found that my preparation time for the early discussion sessions was greater than my preparation time for a normal first-time mathematics course. Perhaps this was because I am not as experienced with this kind of teaching as I am with the teaching of mathematics. Also, there is some effort in organizing the panel discussions. On the other hand, because I had no papers to grade and because much of the course involved student lectures, my total time commitment to the course during the actual semester itself was roughly equivalent to that for a standard course.

Both times I taught the course, the class developed a closeness not often found in normal mathematics classes. I treated the students as my less-experienced teaching peers, and, because the final course grade was of no concern (neither to me nor to them), I found it was easy to view them in this way. I asked for their help, and they gave it. Every student in the class took the assignments seriously, and during the two offerings there were very few absences, only one or two of which were unexcused.

It was clear from the peer evaluations that the students genuinely tried to help each other improve as teachers. Another example of cooperation was evidenced when a US student told a Chinese student, who regularly works on improving his accent, that he'd be glad to go over the Chinese student's videotape with him to point out the places where his pronunciation needed some work.

Other important items now apparent to me are these: students in their later years of graduate study with several years of teaching experience are in a unique position. Not only are they still in the "student mode" and are willing and anxious to learn, accept criticism, and improve, but they are also well qualified to critique the teaching of their peers. Moreover, their involvement in the whole process is as important as the involvement of the faculty.

I believe the students have left our professional seminar both better prepared for college teaching and with broader mathematical perspectives. Moreover, while the whole improvement process may be a difficult-to-measure function of many complex variables, I feel the results were well worth our efforts.

A Class-by-Class Summary.

Class 1. **Introduction**. At the first class, I passed out materials for the course including a syllabus, a description of the course, a list of possible lecture topics (see Appendix 1), and the books by Krantz [8], Schoenfeld [11], Courant and Robbins [6], and Eble [7]. (The first two books were given to the students and the others were loaned to them from the departmental library.) I described the history of the project and the goals and purposes of the course, and I also discussed several ways the current course would differ from the one given the first time we had offered it. I encouraged the class members to begin thinking of a topic on which they might give a one-hour lecture and also a topic for a 15-20 minute presentation.

Classes 2–4. **How to Teach Mathematics: Krantz [8].** (2) Starting with me, the class members each described their own background and indicated those people who had been their greatest influences, both mathematically and as model teachers. This took much of the period and was enjoyable to all. We had time to discuss only the preface to Krantz's book. Near the end of the class, I listed those Clemson mathematics professors who were willing to have their classes visited, and I asked the students to be thinking about which of these (or others) they would like to visit. The purposes of these visits, I explained, would be to examine the techniques of some experienced teachers.

(3) We agreed on which faculty classes the students would visit (subject to the faculty okays), and we then discussed Chapter 1. The students were very responsive and related many of their experiences (as they did throughout the course).

(4) We finished Chapter 1 and most of Chapter 2. Again there was considerable discussion on teaching techniques and it was clear the students had read their assignments.

Class 5. **Readings discussed.** We finished our discussion of Krantz [8] with Chapter 3, and talked a bit about using the book by Schoenfeld [11] as a source book for college mathematics teaching. I had now obtained all faculty okays for classroom visits, so I encouraged the class members to make their visits and to keep records of their observations for reports back to the class.

Class 6. **Getting students involved in learning.** This class was a presentation and discussion led by a professor in our Chemistry Department, Dr. Melanie Cooper, who has done considerable work with beginning TAs at Clemson. Her emphasis was on how to get the students involved in learning (primary emphasis on freshman and sophomore students). Major among her techniques were: (i) the use of small groups (e.g., group projects and group exercises); and (ii) the use of writing (e.g. "write 100-200 words describing something that you did not understand"). She also gave everyone a copy of a booklet [4] she had prepared for use by beginning TAs in our College of Sciences.

Class 7. **Tenure, promotion, and the job market.** Our department head, Richard Ringeisen, described in detail the tenure and promotion process and the spectrum of expectations found at various colleges and universities. He also talked about faculty evaluations at Clemson and at other places and encouraged the students to maintain, early in their careers, dossiers to include nearly every activity in which they are involved. The students were exposed to many new ideas that they had not previously considered, and they had a number of questions for our guest speaker.

Class 8. **Three different faculty philosophies of teaching.** Three departmental faculty members, each known to be an effective teacher, briefly outlined their careers and described some of their teaching techniques and philosophies and how these had evolved. (This term the faculty involved were Warren Adams, Doug Shier, and Herman Senter.) The fact that teaching is a personal thing was quite evident here because the philosophies espoused were var-

ied. Incidentally, the teaching done by the panelists ranged from the teaching of freshmen to the teaching of advanced graduate students.

Class 9. **Reports on classroom visits 1.** At the beginning of the period I passed out the lecture evaluation forms that we would use when we started "teaching" each other [see Appendix 2] and *The Quest for Excellence in University Teaching* [12] and *Micro-Teaching* [13]. I asked the class to read these for next time and to come prepared to spend a short time discussing changes we might want to make in the evaluation form. Then four of the students described the visits they had made to faculty classrooms, outlining the nice features of these visits and also describing, in some cases, how they might have done things differently.

Class 10. **Reports on classroom visits 2.** I began by listing the names of all members of the class (ordered randomly except that I went first) and asked for tentative lecture topics for the upcoming first lectures. The topics were:

(1) The Fundamental Theorem of Algebra — A Real Variables Proof,

(2) Lagrange Multipliers,

(3) An Introduction to Factorial Experiments,

(4) Diophantine Equations,

(5) An Introduction to Chaos Theory,

(6) What is NP-Complete?,

(7) The Number of Positions for Rubik's Cube,

(8) An Introduction to Game Theory.

We then heard descriptions of the remaining faculty classroom visitations and tried to make a general list of nice features we had observed as contrasted with a list of things we might have changed. Finally, we discussed the evaluation form which was generally agreeable to all, and we discussed several of the items in the Micro-Teaching handout [9].

Class 11. **Industry, government, and teaching mathematics.** Dr. Clark T. Benson, who has 25 years of university teaching experience and is now a mathematician at the National Security Agency, gave a presentation of his philosophy of teaching and discussed the importance of communication in government and industry as well as in teaching. I remember particularly one of the points he made in drawing an analogy between teaching and doing magic: "If a magician intends to make an orange disappear, he had better make sure that the audience sees the orange and then understands that it is gone." As teachers, he recommended our making sure the students understood clearly what was being discussed; i.e., "Show them the orange." There was much interesting discussion, and many other ideas were put forward.

Classes 12–19. **Lectures.** Lectures on topics chosen in Class 10 (see list above) were presented and critiqued.

Class 20. **Critique and film.** Let's Teach Guessing by George Polya [6].

Class 21. **Writing across the curriculum.** (Panelists: Cindy Harris and Robert Jamison from mathematics, Daniel McAuliff from the College of Engineering, and Arthur

Young, Professor of Technical Communication from the English department.) Each panelist was given 10-15 minutes to discuss experiences in communicating across the curriculum. Young talked about the national movement and gave an overall picture of what is happening at Clemson. McAuliff spoke about the program in engineering, its roots and evolution. Jamison told of some of the things he was doing in the upper-division undergraduate mathematics courses, and Harris described things she was doing in the lower-division undergraduate courses. Each of the panelists passed out illustrations used in their classes and gave us other materials to consider.

Classes 22–24. **Mini-presentations.** Each of the three speakers was given 15 minutes followed by a five-minute question-and-answer period, much like a session at a national meeting; thus, a speaker could run a minute or two over the allotted time. The goal was to communicate the material as though the audience were composed of beginning graduate students. A simple one-page evaluation form (see Appendix 2) was used for these mini-presentations.

The topics were:

1. Polynomial Functions on Matrices over a Finite Field.
2. Integral Global Optimization of Robust Discontinuous Functions.
3. An Algorithm for Function Optimization.
4. The Current Status of "Fermat's Last Theorem."
5. Computers in the Discrete Mathematics Classroom — A report on a DIMAC Workshop.
6. Orthogonal Canonical Forms of a Single-Input Linear Control System.
7. Generalized Lagrangian Duality
8. A Report on a Doctoral Colloquium at a Recent ORSA Meeting.

Class 25. **Technology in the classroom.** Panelists: William Hare, Clark Jeffries, Don LaTorre, Gil Proctor (all from the Mathematical Sciences Department). The panelist gave a personal perspective of how they had used various new technologies in their own classrooms and/or how they saw technology developing in the future.

Class 26. **Miscellaneous professional responsibilities.** (Panelists: Peter Nelson, Richard Ringeisen and Doug Shier, from the department.) The panelists were chosen in part because of their active involvement in many aspects of the mathematics profession; e.g., each one either was currently or had served on the editorial board of a journal (or journals). Among the topics discussed were (i) writing and refereeing articles for journals, (ii) writing and refereeing proposals, (iii) reviewing published articles and books, and (iv) various possibilities for committee work and other activities within the professional societies.

Class 27. **Women and minorities in mathematics.** Panelists were Renu Laskar, Frances Sullivan, Jennifer Key, and Calvin Williams (faculty members in our department). Each participant spoke of personal experiences in gaining a PhD in mathematics and in moving up in the profession and

each had suggestions for the class members. Laskar described, among other things, how difficult it was for her (as a woman) to get a college education in India and how students at times have tried to take advantage of her (without success) because of her accent. She encouraged hard work and dedication. Sullivan described a number of her educational experiences and some of the pressures she felt as an African American mathematics student growing up in the South. She recommended hard work and expressed her belief that we must demand much from students. But she also said that we should have compassion for them. Key told of the difficulties of being a research mathematician with family responsibilities (e.g., rearing children). She also said she believed she had not been discriminated against as a woman and that, ultimately, success depends not on gender but on the quality of the work one does. Williams, also a black growing up in the South some years later than Sullivan, related how his father, who died soon after Calvin had entered college, had pushed him to work hard and had encouraged him to concentrate on academics and not athletics (he had potential for a baseball career). Williams also stressed hard work and indicated he too felt one should demand much of students — with compassion. We class members agreed that this was an inspirational session.

Class 28. **Final discussion and critique of the course.** In the final class, I solicited class opinions about the course and asked for written evaluations of it. All of the participants expressed a belief that they had become better teachers because of the course, and they all commented on enjoying the Krantz book [8] and the discussion it evoked. Furthermore, they all indicated that they had gained many new perspectives on college-teaching.

Acknowledgements. Many people made contributions to the Clemson project by offering suggestions, pointing me to appropriate resources, participating on panels, allowing students to observe their teaching, and/or discussing with me various ideas concerning the project. At the risk of leaving some names out, I would like to express my appreciation to the following individuals: Warren Adams, Clayton Aucoin, Clark Benson, Annette Blackwelder, James Brannan, Thomas Brawley, Vincent Brawley, Bettye Anne Case, Melanie Cooper, Donald Bushaw, Lin Dearing, Paul Duvall, Vince Ervin, Bob Fennell, Chris Fisher, Mary Flahive, William Hare, Cindy Harris, Paul Holmes, Robert Jamison, James P. Jarvis, Clark Jeffries, David H. Johnson, John Kenelly, Mike Kostreva, Jennifer Key, Steven Krantz, Renu Laskar, Don LaTorre, Jeuel LaTorre, Daniel McAuliff, Robert McDowell, William Moss, Gary Mullen, Peter Nelson, James Peterson, Gil Proctor, Richard Ringeisen, Herman Senter, Roberta Sabin, Doug Shier, Frances Sullivan, Calvin Williams, Daniel Warner and Arthur Young. I would also like to thank the PhD students Marla Bell, Eric Bibelnieks, Tracy Bibelnieks, Karen Copeland, Mark Cawood, Judy Green, Robin Lougee-Heimer, Matt Myers, Franklin Shobe, Matthew Ten-Huisen, Craig Turner, Karin Vorwerk, Charles Wallis, Quan Zheng, and Deda Zheng who took the course and contributed greatly to its success. Their enthusiasm and their desire to know more about the academic world and to improve as teachers was evident.

References

[1] Artin, Emil, *Contents and Methods of an Algebra Course*, Tata Institute, 1960.

[2] Boyer, Ernest L., *Scholarship Reconsidered — Priorities of the Professoriate*, The Carnegie Foundation for the Advancement of Teaching, Princeton, 1990.

[3] Case, Bettye Anne (Ed.), A preprint of some materials to be published as *You're the Professor, What Next?* by the MAA in 1994.)

[4] Cooper, Melanie, *A Guidebook for Clemson University Teaching Assistants*, Clemson University, Department of Chemistry, 1991.

[5] Committee on Preparing for College Teaching, AMS-MAA-SIAM, "How should we prepare mathematicians for college teaching?" *Notices*, Amer. Math. Soc. 36: 1344-1346 (1989).

[6] Courant, R. and H. Robbins, *What is Mathematics?* Oxford Univ. Press, 1941.

[7] Eble, Kenneth, *The Craft of Teaching*, 2nd ed., Jossey-Bass, 1990.

[8] Krantz, Steven, *How to Teach Mathematics: A Personal Perspective*, American Mathematical Society, 1993.

[9] *Moving Beyond Myths — Revitalizing Undergraduate Mathematics*. National Research Council, National Academy Press, 1991.

[10] Polya, George, "Let's Teach Guessing", A videotape distributed by the Mathematical Association of America.

[11] Schoenfeld, A., *A Source Book for College Mathematics Teaching*, Mathematical Association of America, 1990.

[12] Sherman, T., L.P. Armistead, F. Fowler, M.A. Barksdale, and G. Reif. "The quest for excellence in university teaching," *Journal of Higher Education*, 48: 6-83 (1987).

[13] Vernon, Marilyn C., Micro-Teaching. (Transparencies used in a workshop given by the author who is with the Division of Continuing Education and Training, Federal Judicial Center, 1520 H. Street, N.W., Washington, D.C. 20005.)

Appendix 1 — Lectures and Lecture Topics

Each student is expected to give two lectures to the class. One of the lectures will be an hour lecture and the other 15-20 minutes. The hour lecture should be pitched at the advanced undergraduate or beginning graduate level and should involve, for the most part, material that the students have not explicitly studied in their proper course work. Each student is responsible for choosing a lecture topic and having it approved by the instructor. Below are some possible topics. The mini-lecture should be something a little different; e.g., it could be something such as (i) instruction in calculator or computer use, (ii) a demonstration with a hands-on geometrical model or other mathematical object,

(iii) a report on a conference, (iv) a brief introduction to a certain method of teaching or theory of learning, or (v) a short research presentation as would be done at a professional meeting. If requested, a little more time could be given for the mini-presentation which can be either on mathematics or on pedagogy and can involve team teaching.

Further Topics (beyond those chosen in Class 10)

1. Fractals
2. Entropy and Information
3. Random Number Generation
4. Cryptography
   - (a) Classical Cryptography
   - (b) Public-Key Cryptography
5. The Construction of the Number System
   - (a) From the Natural Numbers to the Complex Numbers.
   - (b) Completing the number line: From Q to R
     - (i) Cauchy Sequences and Valuations
     - (ii) Dedekind Cuts
   - (c) Algebraic and Trancendental Numbers.
     - (i) Construction of various transcendental numbers
     - (ii) e is transcendental and/or π is transcendental.
   - (d) The Fundamental Theorem of Algebra proved using complex variables
6. Infinite Cardinal Numbers.
7. Topics in Number Theory
   - (a) Continued Fractions
   - (b) Prime number generation
   - (c) Factoring methods (e.g., the Pollard ρ method)
8. How to Lie with Statistics
9. Simulation
10. How to Solve It.
11. Well-Ordering and Mathematical Induction.
12. Constructions with Compass and Straightedge.
13. Non-Euclidean Geometry.
14. Special Topics in Geometry or Topology.

Appendix 2 — Lecture Evaluation Forms

Form 1 (for the hour lecture): (More space is provided than is shown. There is a rating system of [1 — 2 — 3 — 4 — 5], "5" being best and "1" being worst. A comment area is also provided for each item.)

EVALUATION OF:

Lecture Topic:

1. Clarity: (Speaking voice and delivery, use of blackboards and/or other visual aids, lecture coherence, appropriateness of examples and level of presentation, et cetera)

2. Enthusiasm: (Was the instructor: "into" the lecture? Did the instructor show an enjoyment of the material and enjoyment in delivering it? Did the instructor project personal feeling for the material?)

3. Preparation/Organization: (Was the lecture well-organized, well-conceived, well-structured? Did you understand the objectives and were they met?)

4. Stimulation: (Were you "with" the instructor? Were you stimulated to think about the material during class? after class?)

5. Knowledge: (How do you rate the instructor's knowledge of the lecture material? How do you rate the instructor's interest in and concern for the material?)

6. Other Suggestions for Possible Improvement.

7. Did you learn anything?

Form 2 (for twenty-minute presentations):
FOLLOW-UP EVALUATION OF:

1. Topic:

2. Constructive criticisms (positive and negative). (You might wish to comment on such things as clarity, enthusiasm, preparation/organization, stimulation, breadth of and zest for knowledge, whether the lecture was on target with the audience, or other things you think would be helpful.)

3. If possible, compare this presentation with the first presentation given by the speaker by putting a mark on the scale shown or else, circle "Not comparable."

   Here,

   $2 =$ Clearly better than the first.

   $1 =$ Probably better than the first.

   $0 =$ About the same as the first.

   $-1 =$ Probably worse than the first.

   $-2 =$ Clearly worse than the first.

Chapter 8

Dartmouth College

Kenneth P. Bogart

Marcia J. Groszek

An Overview. Dartmouth College is a selective liberal arts institution with a fairly small graduate student body (about 5% of the undergraduate population.) The mathematics graduate program generally has a total of just over 20 students, approximately the same as the number of mathematics faculty. The typical mathematics graduate student is from the United States, has just received a BA, may have come from an undergraduate liberal arts institution, and has chosen to apply to Dartmouth partly because of an interest in teaching. About a third of mathematics graduate students are women, and only a very small number have been minority students. The retention rates for women and men are about the same. Dartmouth's PhD program in mathematics was conceived as a program to develop college and university teachers of mathematics—not teachers to the exclusion of research, but rather research mathematicians who were prepared to teach. The ingredients of the program aimed at teaching preparation were broad requirements over mathematical areas, a universal teaching requirement, and a progression of teaching responsibilities.

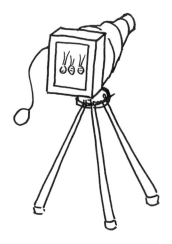

Brief History. Students learn to teach through experience in tutoring during the first two years of graduate study, a teaching seminar during the summer after the second year, and experience in subsequent years teaching courses under the supervision of faculty. We are currently developing an intensive teaching course, which will incorporate and greatly expand upon the current teaching seminar.

Mathematical breadth is provided by a certification system that requires knowledge in algebra, analysis, topology, and a fourth area of mathematics. Our low student-faculty ratio makes it possible to discuss professional development on an individual basis with thesis advisors, research seminar participants, and faculty course supervisors. Issues of professional development directly relevant to teaching may be discussed in the teaching seminar.

Formerly, course-work requirements included a "distributive requirement" covering at least 5 areas of mathematics with qualifying exams in algebra, analysis, topology, and a fourth area of the student's choice. The teaching requirement began with two years of leading discussion sessions, followed by a one-year break, and culminated in (1) a group effort in a multisection course, where the student was trained (indoctrinated) in teaching techniques in weekly meetings, followed by (2) a course in which the student was more independent.

Over the years, requirements evolved. We now have no formal course distributive requirements, though virtually all students satisfy the former ones. Qualifying exams are now in the form of certification (not necessarily by exam) by two faculty members of students' knowledge of the departmental syllabus in algebra, analysis, topology, and a fourth area of mathematics. The teaching requirement has also been modified: Discussion sections have been replaced by evening tutorials with graduate student tutors assigned a specific course for which they have done the homework. (This has the benefit of making "TAs" very popular with undergraduates.) Graduate students in mathematics now normally teach one course in the third and each succeeding year.

Tutoring rather than teaching results in serious concerns. First, our program has a serious commitment to developing excellent teachers, and we were concerned that doing away with discussion sections would leave our students less prepared than they would otherwise have been for teaching. Complaints from undergraduates reinforced these concerns. Further, our students faced a tight job market and we felt that a program designed to make them successful teachers would make their job searches easier. We responded to these with three initiatives:

1. We introduced a teaching seminar, conceived initially as a response to our local problems but quickly seen as a logical part of our commitment to the development of college and university teachers. The seminar is required of all students admitted to PhD candidacy and is given in the summer between their second and third years.

2. We introduced a teaching evaluation committee that would arrange classroom visits early each term to evaluate classes taught by graduate student teachers and new faculty, evaluating, mentoring, and returning as necessary.

3. We began giving instructors (including graduate students) student course-evaluation forms to use in their classes at the end of each term. In a more recent innovation, everyone is encouraged (and graduate student instructors are expected) to use a standard form that can be summarized in an electronic database and, before looking at them, to turn evaluations over to the department for entry into the departmental database. After the numerical responses are entered into the database and after grades are turned in to the registrar, the evaluation forms are returned to the individual instructors for their own use. Graduate-student instructors are encouraged to discuss their evaluations with faculty members.

Recently we have been working on the development of an intensive teaching course. This course will replace the current teaching seminar and will incorporate many of its features.

The Departmental Teaching Seminar. The teaching seminar has changed little in format over the years. The emphasis is on the preparation and delivery of lectures. Graduate students give two lectures each. Originally both were on on multivariable calculus, although now students are allowed to choose the subject of the second lecture while the first is from either multivariable calculus or calculus with algebra. The audience consists of their fellow students and three or more faculty

members who make up the teaching seminar committee for that summer. Each lecture is videotaped; it is also critiqued (without video replay) by the audience immediately afterwards. Students enroll in the seminar as a course, Math 147, graded on a credit/no credit basis. This course is required of all mathematics graduate students, and costs them nothing (except for the time commitment). Each summer, there are between three and six students in the seminar. Three or more faculty members serve on the teaching seminar committee and attend all seminars as one of their (typically two or three) committee/job assignments for the year.

The involvement of three or more faculty members has a major impact on the feedback; faculty do not hesitate to disagree in front of the students. This enhances the general "give and take" tremendously. This discussion can range over students' learning styles, issues of gender or race, blackboard style, interaction with students—the whole gamut of didactic technique. The videotaping is an essential ingredient. Students view their videotapes on their own; we typically recommend that each student invite another member of the seminar to view it at the same time. Students are also strongly encouraged to review their videotapes with a member of the speech department who does this for us on a volunteer basis.

At the end of the seminar the committee meets and advises the department chair on what kinds of teaching responsibilities each member of the seminar should get. When a student's second lecture is well crafted, encourages audience participation and is pitched at the right level, the committee recommends a standard beginning assignment — a section of calculus with algebra or a section of multivariable calculus. The majority of students get such a recommendation. If the second lecture and discussions demonstrate exceptional readiness, then the committee recommends an assignment of teaching independently in a relatively advanced course, usually the course the student wants or a related course. For the remaining students, the committee tries to recommend a course of action that will improve their teaching without assigning full teaching duties. Each such student is handled individually; it would be preferable to have a less subjective approach, but each case seems to call for special treatment. Typically such students do get a teaching assignment in the following year, but they may be asked to repeat the teaching seminar first.

Just the fact that not everyone in the seminar goes on to regular lecturing duties indicates that something more is needed. We've tried a number of ways of developing the additional expertise; the one that has the most promise of any we've tried is the teaching apprenticeship. This is based on the fact, well confirmed by studies made by specialists in education, that practice-teaching in the presence of a mentor who provides feedback is the single most effective part of the training of high-school teachers.

We have experimented with apprenticeships having a

varying degree of mentor-student interaction. The most striking experiment involved a student that the committee recommended for teaching with significant reservations. In his first course he received overall student ratings of poor or below-average, with about half of each. The advisor to graduate students explained to this student that without remedial action he would not get another opportunity to teach and, as a consequence, would not be able to get positive recommendations from the faculty about his teaching.

The two agreed that in the author's next appropriate course the student would serve as an apprentice; they drew up a careful agreement for intensive coaching and cooperative planning. In the following year when the student taught on his own, most of the teaching evaluations were evenly split between average and above average, with one below-average and several excellent ratings.

While we believe the teaching seminar is successful, this experience has convinced many of us that, by following the seminar with such apprenticeships, we can expect not only to remediate in situations such as those discussed above but also to develop superior teachers from good ones. However, the apprenticeship is still a very subjective approach, and it is not clear that other faculty member-student pairings would have the same effect. In order to find a more reproducible way to develop teaching skills, the department is now experimenting with a course that goes into more depth than the teaching seminar in order to reduce (to zero, we hope) the number of graduate students who need special treatment afterwards.

While the department has efforts under way to improve our method of developing effective teachers, we believe that the teaching seminar, followed by teaching experience, has resulted in PhDs who are significantly better teachers than the typical new PhD. While certain details of the seminar are specific to Dartmouth (for example, we operate year-round, so it is a convenience rather than an inconvenience to schedule the seminar in the summer when graduate student commitments to formal course work are lightest) we believe it is likely to prove successful elsewhere. Features of the seminar which we believe contribute most to its success are:

1. Involvement of three or more faculty members. This not only permits us to offer students different points of view, it encourages open discussion, influences the teaching of involved faculty, and demonstrates to the graduate students the high level of importance the faculty accords to teaching.

2. Videotaping, with contributions from both individual viewing and professional consultations.

3. Immediate discussion in depth. This discussion provides feedback when memories are fresh; research in learning demonstrates that immediacy of feedback is an important contribution to the learning process.

4. Far-ranging discussion. This can lead to topics far removed from lecture technique, such as gender, un-

dergraduate goals, and perhaps even the validity of lectures as the central organizational feature of a course.

5. Follow-up in the form of arranging a student's teaching in accordance with the committee recommendations and recognizing exceptional performance with exceptionally desirable teaching assignments.

6. Tying the first lecture to a course the graduate student is likely to teach. This gives the student a sense of immediate applicability.

7. Allowing the graduate student to "lobby" for a desired teaching assignment (and taking this lobbying seriously). This gives the students a sense of at least partial control over their futures and is a reward for doing well.

8. Keeping the lectures short and allowing the committee to set its own ground rules each year. This has made it possible to ensure faculty willingness to participate over the "long haul."

One feature of the Dartmouth environment that would not be easy to duplicate elsewhere is the combination of graduate student teaching load and graduate-student culture. Graduate students teach one ten-week course per year, and seem to feel an enormous sense of responsibility for their teaching. As a result, it is not unusual for a graduate student to spend two or three times as much time preparing for class as a faculty member would. In this way they develop very good habits, habits that may take a long time when first tried but don't take nearly so long when they are second nature.

All in all, the teaching seminar has been valuable for Dartmouth and its students, graduate and undergraduate; it has been easy to implement and maintain; and it is an accepted part of department life. We recommend it to other departments as part of a program to prepare graduate students for their roles as college and university teachers of mathematics.

The New Teaching Course. The new teaching course was inspired by several different factors. The first factor was the ongoing desire to continue to improve our graduate students' preparation for teaching. The second factor was the graduate students themselves. While most of them find the teaching seminar quite helpful, some were nonetheless unsatisfied with its limitations. They wanted more of an emphasis on aspects of teaching other than lecturing, and perhaps more time devoted to preparing to teach. The third factor was the growing nationwide sentiment that, in order to increase the mathematical understanding of diverse groups of students, we in the mathematical community need to change the way we teach.

Other programs that have gone beyond the departmental seminar for preparing graduate students to teach have generally taken one of two options. One is a teaching apprenticeship, the other a general learning-theory or teaching-methods course offered by an education department, psychology department, or teaching center. We saw several advantages to teaching a course within our own department. Unlike a general course, it would provide for learning and practice specifically in mathematics teaching and would encourage graduate students to consider teaching as an important professional activity by having them work together with active research mathematicians on questions of teaching. Unlike an apprenticeship program, it would provide a mechanism for changing the status quo in mathematics teaching by giving students a base of knowledge about learning and the opportunity to learn about and practice a variety of teaching methods.

It was not hard to realize that, like most mathematicians, we were not very well equipped either to change the way we ourselves teach or to help students learn to teach differently from the way we do. Most of us had neither been taught anything about learning theory nor been exposed to a variety of non-traditional teaching methods. To address this problem we did two things. We enlisted the participation of a colleague from the education department, and we prepared to spend a great deal of time reading all the relevant material we could find about learning theory, educational research, and classroom experimentation.

The course development project that followed was exciting, engrossing, and time-consuming. (The work was partially supported by a grant from the New England Consortium for Undergraduate Science Education.) Following top-down principles of course design, we began by trying to articulate our goals for the course. Our central goal was to produce effective teachers of mathematics. We immediately realized that in order to explicate that goal, we first had to understand our goals for mathematics students. A description of this process can be found in Appendix C. A great deal of reading, discussion, and debate prepared us for the next step, producing a list of subgoals that would contribute to effectiveness for mathematics teachers. Simultaneously, we considered the question of how we might evaluate our own effectiveness in meeting these subgoals. (See Appendix D.)

In summary,

◇ Mathematics students should experience learning and growth in four areas: viewpoint and motivation; knowledge and skill in subject area; ability to engage with mathematics; ability to learn and communicate mathematics.

◇ Teachers of mathematics should be able to: create and conduct an effective course; engage with students; develop as teachers; understand how students learn.

The content of the course we designed comprises four major areas: learning theory, with an emphasis on understanding how undergraduate students learn mathematics; teaching methods, with a concentration on the undergraduate mathematics classroom; learning environment, including questions of classroom dynamics, race, and gender; and professional development as a teacher of mathematics. The course design incorporates teaching methods that will provide a model for graduate students who are learning to teach, and a modular structure that will allow topics to be chosen in line with the needs of the graduate students and of the program.

A few specifics might give an idea of the nature of the course. In the area of teaching methods, we chose to concentrate initially on the use of lectures, writing assignments, and cooperative learning. Graduate students have the opportunity to experience all these methods as students and to practice using them as teachers. Depending on graduate student interest, later in the course any one of these methods can be explored in more depth or other methods can be considered. In the area of professional development, one of our areas of concentration is peer coaching. By learning how to observe each other's classes and helping each other think about how to become better teachers, graduate students will be prepared both to support each other's teaching through the remainder of graduate school and to share expertise with their colleagues throughout their professional lives.

The experience of our colleague from the education department, and the principles of learning theory we were incorporating into our course content, led us to the same conclusion: To help our graduate students learn to teach, we had to give them the opportunity to practice as they were learning. Since the teaching course will be offered during the summer at Dartmouth, we chose to arrange a practicum in the form of two two-week mathematics enrichment programs for local high-school students. These programs will be designed and taught by graduate students in the teaching course. Integral to this experience are supervision and feedback from faculty and structured peer-observation and coaching by graduate students working in pairs.

As this is being written, during the summer of 1993, we are offering the first pilot version of this course at Dartmouth. Like the teaching seminar it replaces, the course is required of all graduate students and is generally taken during the summer after the second year of graduate study. This scheduling fits in well with Dartmouth's system. As the program is in year-round operation, students are in residence during the summer; at the end of the second year, graduate students have generally just finished qualifying exams, and are in a good position to devote a significant amount of effort to teaching preparation. Four to six students is a typical number for the teaching seminar, and this year there are four students in the course. The teaching course is very intensive, meeting five days a week for at least an hour and a half, and students enrolled in it receive credit for two courses. Five weeks of the nine week summer term are devoted to classroom activities and four weeks to the practice-teaching experiences.

These specific details could easily be changed by another program intending to offer this or a similar course. We have designed the course with an eye toward flexibility. For example, rather than an intensive summer course with an attached practicum, the course could be offered as a weekly seminar throughout the graduate students' first year of teaching in the department. The modular structure allows some choice of topics to be covered. We plan to make a detailed course outline and reading lists available to other interested graduate programs.

During the next academic year, with some additional support from the New England Consortium for Undergraduate Science Education, we will evaluate the results of our pilot course offering. Appendix D outlines a diverse range of possibilities for evaluation. In particular, we have already collected some baseline data. We have years' worth of student evaluations of graduate-student instructors. (Our departmental evaluation form is included as Appendix A; a new supplementary form, designed to help evaluate the teaching course, is Appendix B.) We also have responses to a questionnaire on teaching which we gave to graduate students who had not taken this teaching course, and which we will give to graduates of the course.

Our goal is to prepare graduate students to be not only excellent teachers but leaders in the current movement to transform undergraduate mathematics education. It is too soon to evaluate our success. So far we can say only that all of us involved in the teaching course, students and faculty alike, are positive and enthusiastic. Not only do we find our ways of thinking about teaching transformed, but those of us who can compare our teaching before and after our involvement in this project find our classrooms transformed as well.

Appendix A

Course Evaluation: Department of Mathematics and Computer Science

Instructor, Course, Term, Dartmouth class

Instructions: Please rate each item. Unless instructed otherwise, please use the following scale: (1) for poor; (3) for average; and (5) for excellent. If you don't feel the question applies, then mark (0).

The Instructor

1. Gives well-organized lectures.
2. Explains clearly.
3. Uses the blackboard (or computer displays, etc.) well.
4. Makes it possible to ask questions in class.
5. Is available outside of class.
6. Made me think.
7. Overall rating.

Comments about the instructor:

The Course

8. was boring$\Longrightarrow$was interesting
9. pace too slow$\Longrightarrow$pace too fast
10. I learned nothing$\Longrightarrow$I learned a lot
11. less work than a typical course$\Longrightarrow$more work than a typical course
12. Overall rating of course.

The Text

13. too easy$\Longrightarrow$too difficult
14. confusing$\Longrightarrow$clear
15. irrelevant to course$\Longrightarrow$relevant to course
16. dull$\Longrightarrow$interesting

Comments about text:

The Homework

17. too easy$\Longrightarrow$too difficult
18. too little$\Longrightarrow$too much
19. no help$\Longrightarrow$helped me learn material

Comments about homework:

The Exams

20. too easy$\Longrightarrow$too difficult
21. too short$\Longrightarrow$too long
22. irrelevant$\Longrightarrow$fair test of course material

Comments about exams:

Written Comments

What were the best features of the course/instructor?

What were the worst features of the course/instructor?

What advice would you give to a fellow student considering taking this course?

Appendix B
Supplementary Course Evaluation Form

Instructions: Please rate each item according to the following scale: (1) strongly disagree; (5) strongly agree.

1. This course made me feel that mathematics is:
   (a) fun
   (b) interesting
   (c) useful
   (d) important
   (e) beautiful
2. This course improved my ability to answer questions about mathematics by:
   (a) learning from books
   (b) learning from teachers
   (c) learning from other students
   (d) thinking about it myself
   (e) thinking about it with other students
3. In this course I felt that:
   (a) it was okay to make a mistake in front of class
   (b) other students' participation in class helped me understand
   (c) my participation in class helped other students understand
   (d) other students helped me outside of class
   (e) I helped other students outside of class
4. After taking this course:
   (a) I have a better understanding of how all math fits together
   (b) I can visualize math problems better
   (c) I am better at proving things
   (d) I feel I mastered the basic ideas of the course
   (e) when I get stuck on a problem, I am more likely to view it as a challenge and not give up right away

Appendix C

Goals for Students in Mathematics Classes. We began our discussion of this course from the point of view of students in mathematics classes. What do we want students to get out of a math class? How can we teach in order to facilitate this? How can we measure our success in meeting these goals?

Detailed answers to the last two questions would depend on our research in mathematics education. We considered the first question from our perspective as mathematicians and reflective teachers of mathematics, and produced the following four broad goals for mathematics students.

First, students should learn a positive and perceptive attitude toward mathematics. They should have a coherent overview of the subject, and understand it in larger context. They should understand what makes mathematics, and particular areas of mathematics, interesting. They should appreciate math for its usefulness, challenge, and beauty. They should have a personal engagement with the subject and be confident of their own ability to think about and to do mathematics.

In any particular math course, students should learn a specific subject. They should have an overview, or a "mental map," of the subject, and they should understand why and how that area of mathematics works. They should be familiar with the body of material covered in the course and in particular with key points or cognitive thresholds of the subject. They should know and be able to work with and talk about the facts, theorems, computational skills and techniques, and applications included in the course.

Students should also learn methods and processes for doing and understanding mathematics. They should develop their geometric vision and intuition, their ability to make definitions and build mathematical theory, and their comprehension of and facility with the formal structures of mathematics. They should learn how to ask questions and formulate problems, how to approach a problem, how to get ideas. They should be able to engage in open-ended exploration and problem-solving. They should gain, from knowing how to "do math," an improved understanding of what mathematics is all about.

Students should learn how to function as part of a mathematical community. They should develop strategies for learning mathematics, both independently and cooperatively, using a variety of resources. They should become familiar with the language and standards of mathematical discourse. They should develop the skills of reading, writing, speaking and listening to mathematics, in order to both understand and communicate effectively.

In summary, mathematics students should experience learning and growth in four areas: viewpoint and motivation; knowledge and skill in subject area; ability to engage with mathematics; ability to learn and communicate mathematics.

Appendix D

Goals for Teachers of Mathematics. After discussing our goals for students in mathematics classes, and then doing a good deal of reading about teaching and learning mathematics, we began discussion of our teaching course. We began by asking ourselves what our goals were for the students in our teaching course, i.e., for prospective teachers of mathematics. In setting these goals we also considered the question of evaluation, broadly, in the sense of "What would constitute evidence that this goal was being reached?," and narrowly, in the sense of "What instruments could we use to measure success in this area?"

Our primary goal is that graduate students in this course become effective teachers of mathematics. For us this means that their own students should succeed in developing the skills and attributes we outlined as goals for mathematics students.

Among the ways we can measure this are: Compare the scores of students in their sections with those in other sections on common exams. Administer post-tests to students (again, of course, theirs and others) six months after the course ends, to test retention of course material. Examine records of course registration and choice of major to see whether students go on to take more mathematics courses or to major in mathematics and/or science. Examine student evaluation forms to see attitudes students express towards mathematics.

The four goals that follow are all subsidiary to this primary goal. Each is important in that it contributes to effective teaching.

Teachers should be able to create a course, in an intelligent and informed manner, and to conduct it effectively. Planning a course involves setting goals, choosing means of evaluation, planning a syllabus and course organization, choosing teaching techniques, and choosing motivational strategies. Teachers should know techniques and strategies in these areas and be able to choose those appropriate to the material, students, and larger context. They should also be able to evaluate a course in progress and make changes in response to their evaluation.

We will be able to measure our success by observing graduate students plan courses, making appropriate and considered choices from a range of options, and conduct them effectively. Among the instruments of observation we can use are graduate-student journals, written course plans, peer coaching, direct observation, videotapes, teaching evaluation committee reports, and student evaluations.

Teachers should value positive engagement with their students and know how to achieve it. We can assess our graduate students in this area through their journals, through student evaluations, and through direct observation by means of discourse tracking.

Teachers should know how to learn more about teaching, and want to do so. They should approach teaching challenges with confidence and a sense of adventure. They should have strategies for evaluating and improving their teaching skills. We can assess our graduate students in this area by examining their journals, observing the peer coaching process, and observing their participation in teaching seminar meetings.

Teachers of mathematics should know how students learn mathematics. We can test our graduate students' knowledge in this area directly. We can also observe their use of this knowledge in selecting teaching strategies and in responding to undergraduate learning problems. Instruments of observation can include their journals, our observations, and peer observations.

In summary, teachers of mathematics should be able to: create and conduct an effective course; engage with students; develop as teachers; understand how students learn.

Chapter 9

The University of Delaware

Ivar Stakgold

Judy Kennedy

M. Zuhair Nashed

An Overview. The University of Delaware grew out of a small academy founded in 1743. In 1867, it became part of the land-grant system, and was named a sea-grant college in 1976. The first PhD in mathematics was granted in 1968. The Department of Mathematical Sciences has 40 faculty members and 65 graduate students, of whom 15 are part-time.

A new element was introduced in the graduate programs in mathematics and in applied mathematics beginning with spring 1991: two semesters of participation in a "thematic seminar" whose theme changes from year to year. These seminars are less structured and more problem-oriented than the typical graduate course in mathematics. They are intended to show either how various fields of mathematics enter the solution of a problem or how an area of mathematics contributes to other fields. The seminars, by encouraging team work on designated projects, model an effective undergraduate teaching method with which the graduate students previously have little experience. Frequent oral presentations by students are an essential aspect of the thematic seminar.

Other Mentors: Thomas S. Angell, Cliff Sloyer, and Howard M. Taylor

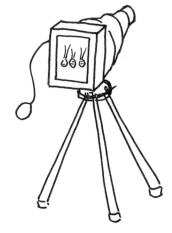

Background on University and Program. The University of Delaware with its 18,000 students is a state–assisted institution, which in our case means that 25% of the budget and about a third of our students come from the State of Delaware (whose population is 650,000). Most graduate programs began to be developed in the 1950's and the PhD in mathematics was first awarded in the late sixties. The program quickly became successful, turning out three or four PhDs per year and about twice as many master's degrees. Many of our students were (and still are) recruited from the Mid-Atlantic region, but by the early eighties we also saw an increase in the number of foreign students, although their fraction always remained below fifty percent and is now considerably smaller.

From its inception, the PhD program was modeled after those of the established institutions in the field, and our students found employment, in about equal numbers, in a) industry and government, b) universities where research is an important part of the hiring and promotion criteria, and c) colleges where teaching undergraduates is the principal responsibility. Typical academic placements – either initially or subsequently – included institutions such as VPI, Tennessee, Clemson, Vanderbilt, Georgia Tech, Georgetown, Iowa State, Minnesota, Toledo, Oakland, Montana State, Temple, Missouri, Smith, Lafayette, Vassar, Otterbein, Lycoming, Millersville, and various campuses of the SUNY college system.

Despite our reasonable track record, we have become aware of some gaps in our program. With a few exceptions, entering students are no longer so single-minded in their pursuit of a mathematical research career. A pop psychologist might attribute this to changing societal attitudes, to the influence of television, to the limited employment opportunities in research, or merely to a return to sanity. In any event, students and the general community are increasingly interested in preparation for teaching and for industry. The traditional orientation we provide our TAs is limited to getting the student ready to handle recitations in freshman calculus; for foreign students this includes language training and an introduction to the sociology of the American classroom. But as prospective college teachers our students would have to teach a wider range of courses and should therefore understand some of the connections between mathematical fields. In industry, collaborative work on projects is essential. Certainly the mathematical culture in industry is quite different from that in academia. With the encouragement of the project supported by FIPSE we developed the thematic seminars described below.

Participation and Administration: The Department of Mathematical Sciences revised its Master's degree programs in Mathematics and Applied Mathematics in 1990 to require two semesters of participation in a "thematic seminar" aimed at mathematical broadening and interconnections and improvement of teaching skills. Since for most students this is also the entry to our PhD programs, nearly all students are affected.

The first experience with the seminar portion of the program at the University of Delaware took place during 1991. This was scheduled before award of the FIPSE grant and does not take full advantage of pedagogical information which subsequently became available.

All non-thesis master's students (this includes all who go on to the PhD) are required to take two "thematic seminars" each carrying two hours of credit. One seminar is offered each spring semester with themes varying from year to year among some six to eight possibilities. Both the Mathematics and Applied Mathematics students take the same seminars. Each seminar carries two hours of credit (Pass/Fail); meetings are once a week for two hours.

The Thematic Seminar: The seminars are at the same time theme- and problem-oriented. We want to show how various fields of mathematics enter in the analysis and solution of a problem and, conversely, how a general area in mathematics contributes to other fields.

It is important that students think about interconnections among branches of mathematics, including some that are not normally included in our curriculum.

A thematic seminar has a number of features that distinguish it from other courses:

a) it is problem-oriented rather than a standard devel-opment of a branch of mathematics;

b) it shows either the wide range of applications of a mathematical area or how many branches of mathematics are used to solve problems in some thematic area;

c) it provides some historical perspective;

d) it encourages collaborative work and interaction among students and provides a model for small-group learning with sharing of ideas among students; it may provide an early introduction to research;

e) it fosters pedagogical skills by requiring students to present their work orally in class and encourages interaction in class;

f) computing (both numerical and symbolic) is an integral part of every seminar;

g) in the student presentations, teaching skills are stressed; the additional breadth provided by the seminars is needed for prospective college teachers;

h) other faculty members are involved in addition to the organizing mentor.

Sample "Themes" (Mathematical Topics) for Seminars: We give examples of themes that may be suitable for our seminar in future terms. These themes partly reflect the interests of our faculty and the orientation of the graduate program at the University of Delaware. In particular, most of the topics are in the areas of analysis and applied mathematics. Clearly different topics may be more suitable for other departments.

◇ Optimization

◇ Fixed Points: Brouwer's FP, Schauder FP, cones with order relation

◇ Dynamical Systems: Iteration of maps, chaos, fractals, Julia sets

◇ Probabilistic Methods: Geometric probability, probabilistic methods in combinatorics, statistical mechanics, Monte Carlo

◇ Approximation

◇ Number Theory: Algebraic, analytic and probabilistic methods

◇ Fourier Methods: Harmonic analysis, Fourier transforms, probabilistic interpretations, sampling, Fast Fourier transform, Hardy functions, connection with number theory and groups

Summary of Seminars: In Spring 1991, Angell, Nashed and Stakgold offered the initial thematic seminar on the topic of optimization. Other faculty members participated on occasion. For more details see Nashed's write-up below. In Spring 1992, Howard Taylor offered the Thematic Seminar on Aspects of Probability. The topics were random walks and Markov processes. One of the goals of the seminar was to tie these probabilistic ideas to the more familiar concepts of harmonic function, potential, capacitance, and others from classical analysis. In Spring 1993 (as this is being written) Judy

Kennedy is organizing a Thematic Seminar combining Fixed Points and Dynamical Systems. (See Kennedy's description below.)

Evaluation:

1. Students were enthusiastic about the teamwork aspect of the seminars. Quote: "The best aspect of the class was the teamwork. My group truly enjoyed working together ... It was particularly satisfying to start working with only a vague idea of the problem, combine individual ideas, and eventually arrive at a better understanding (and perhaps even a solution) of the problem."

2. Oral presentations were also regarded as a good experience by the students.

3. "The idea of making the seminar structure resemble the average college class as little as possible was a great idea, but, unfortunately, the intended structure sometimes went by the wayside."

4. Objections were raised to the logistics of getting the groups together and of the timing of project assignments.

5. Overall, the evaluations for the first two seminars were favorable, but there are clearly some adjustments that need to be made in better planning and execution.

Goals and Features of the Thematic Seminar.

A thematic seminar has a number of goals and features that distinguish it from traditional courses or seminars:

a. It aims at *mathematical broadening* and exploration of *interconnections* between the chosen theme and other areas of mathematics.

b. It is *theme-oriented*; it shows how apparently different problems (methods, frameworks, theories, etc.) influence each other and are often interwoven by common threads. Depending on the particular theme and emphasis, the seminar also shows either the wide range of applications of a mathematical area or how several branches of mathematics are used to solve problems in some thematic area.

c. It takes a *problem-oriented point of view*; yet it is not a "problem seminar." We are not preparing students for the Putnam or Olympiad. The seminar is problem-oriented within the framework of an intuitive approach (reasoning and formulation rather than technical methods and skills).

d. We abandon the usual lecture style, daily homework (of the "academic type"), and examinations. The seminar emphasizes the *exploratory approach* to the theme, rather than "reading all about it" from a selected list of textbooks. The seminar, however, does involve reading some expository articles related to its theme. Homework is replaced by "group projects."

e. The seminar *gets students involved*. It encourages collaborative work and interaction among students; they work in teams. The seminar provides a model for *small-group learning and sharing* of ideas (and difficulties) among students. It may also provide an early introduction to research; if this happens, it is a *by-product* rather than a principal goal of the thematic seminars.

f. The seminar aims at *improvement of teaching and presentation skills*. It fosters pedagogical skills by requiring students to present their work orally in class. Each project is presented by one team, but each team is required to work on all the projects; written reports on some of the projects are also required. The simulated class environment is that of a small team presenting its scientific findings to a critical "government agency," consisting in this case of the faculty members and the other teams of students who have also worked on the same project.

g. *Computing* (both numerical and symbolic) is an integral part of every seminar in the sense that *some* of the projects require computing. Again, this is a team effort which helps the less experienced students to learn from their more skilled peers on the same team.

h. The seminar provides some *historical perspective* on the chosen theme.

i. *The seminar gets faculty to interact and cooperate in the same project.* Typically, two to four faculty members are involved in addition to the organizing mentor. Thus the seminar may also provide a learning experience for the faculty!

j. *Intuition:* The seminar helps students to develop a feel for a "right way" and a "wrong way" of approaching a problem. This is partly accomplished by comparing the approaches of the different teams to the same project, and by the success and failure experiences.

k. The projects and team presentations help to *uncover hidden difficulties or deficiencies* that students may have. (As an example, one of the projects on optimization led to a boundary value problem for the differential equation $y'' - y = 0$. One student realized by inspection that $y(x) = Ce^x$ satisfies the equation, but was unable to match this solution to the boundary conditions. It turned out that he never had a course in differential equations in his undergraduate education!)

These are some of the main goals and features of the thematic seminar at the University of Delaware. To sum up, the two required semesters of thematic seminars (two distinct themes) are aimed at mathematical broadening and interconnections and improvement of teaching skills.

The seminars are both *theme-* and *problem*-oriented. We want to show how various fields of mathematics enter in the analysis and solution, and also to show how a general area in mathematics contributes to other fields. It is important that students think about *interactions* among branches of mathematics, including some that are not normally included in our curriculum. In student presentations, teaching skills are stressed; the additional breadth provided by the seminars is needed for prospective college teachers.

Finally, cooperative faculty effort is involved in each thematic seminar.

The Thematic Seminar, 1993, Judy Kennedy.

The subject for the 1993 thematic seminar was fixed points in dynamical systems. The aim of the seminar was to discuss (1) conditions under which fixed points arise; (2) how the existence of fixed points can be applied to problems in various branches of mathematics; and (3) generalizations of fixed points (such as periodic points and invariant closed sets). Topics covered include Brouwer's Fixed Point Theorem, Schauder's Fixed Point Theorem, Banach's Contraction Principle, Caristi's Theorem, Kakutani's Theorem, various applications of those fixed point theorems (such as Picard's Theorem), applications to integral and differential equations, fractal geometry applications, how periodic points and invariant Cantor sets arise in dynamics and the consequences of their presence (for example, Smale's horseshoe map, chaos, symbolic dynamics, and Sarkovskii's Theorem).

No systematic or complete coverage of the subject was attempted. Instead, the goals and tradition of the seminar were continued but with added stress on the active involvement of the students doing mathematics. In addition, the seminar explored the connections between various branches of mathematics, and students were required to make oral presentations of their conclusions and proofs to their classmates and instructor.

During the first half of the semester the students were asked to work **individually** on a list of theorems, problems and questions. The level of difficulty of the problems on the list varied from rather easy to quite difficult, but enough background information was provided to make the work reasonable to attempt. The students were also told that no one was expected to provide proofs (solutions) to all theorems (problems or questions) on the list, but rather all students were to find sufficient work that they could successfully complete. They were not to consult outside sources (books, papers, or other people). If they knew a solution from work done in other classes, they were sometimes allowed to present those solutions, but were not to represent that work as their own. Also, no words or results were to be used unless the presenter was willing to give definitions and careful explanations to the rest of the class.

During the second half of the semester the rules changed somewhat. The class moved to group work. Three projects and three related (and accessible) *American Mathematical Monthly* papers were selected. Project I concerned Schauder's Fixed Point Theorem and applications, along with the paper "A fixed point theorem for mappings which do not increase distances," W.A. Kirk, *Amer. Math. Monthly* **72** (1965) 1004-1006. Project II

concerned Kakutani's Fixed Point Theorem and applications, along with the paper "Analytic proof of the 'Hairy Ball Theorem' and the Brouwer Fixed Point Theorem," J. Milnor, *Amer. Math. Monthly* **85** (1978) 521-524. Project III concerned periodic points and Cantor sets in dynamical systems, along with the paper "Period three implies chaos," T.-Y. Li and J.A. Yorke, *Amer. Math. Monthly* **82** (1975) 955-992. Each student chose one project and spent the remainder of the semester working on that project, as well as reading the associated paper. The students choosing a given project and paper worked together as a team on that project. Six students chose Project I, four students chose Project II, and one student chose Project III. The varying sizes of the teams was not a problem since the larger a given team was, the more work was expected of that team. Teams then made reports, written and oral, on their projects, papers, and work completed.

The Thematic Seminar, 1991, M. Zuhair Nashed.*

In my Baltimore talk I reported briefly on the philosophy, scope, objectives and administration of a *thematic seminar* at the University of Delaware that is currently required for all first-year graduate students in the mathematical sciences. The seminar differs in essential ways from traditional methods of teaching courses; also it is neither a research seminar nor a problem-solving seminar, although it interfaces with some aspects of such seminars, as explained below.

The "Thematic Seminar" was introduced recently in the graduate program at the University of Delaware. Master's students in mathematics or applied mathematics were required either to write a master's thesis or to take additional courses. In both cases students were required to pass a written fundamental graduate examination. This latter requirement was recently replaced by the requirement that all non–thesis master's students (this is the vast majority and includes all those going to the PhD) take *two* "thematic seminars," each carrying two hours of credit, on a pass/fail basis. Meetings are once a week for two hours. One seminar will be offered each spring semester with themes varying from year to year. A list of some possible themes was proposed, but modifications are always possible. Both pure and applied mathematicians are required to take the same seminars. Topics in successive years will be chosen to encourage breadth, so the themes will be from different areas of mathematics.

Each seminar will have a faculty "organizing mentor" in addition to two or more faculty members who are also involved in the design and conduct of the seminar. The first offering of the seminar was in Spring 1991. "Optimization" was the mathematical theme. Ivar Stakgold

---

* From the Proceedings of an AMS Special Session at the 1992 Joint Mathematics Meetings in Baltimore organized for the AMS-MAA-SIAM Joint Committee on Preparation for College Teaching by Bettye Anne Case.

served as the faculty mentor, assisted by Tom Angell, Zuhair Nashed, and Cliff Sloyer. Some of the participating faculty members attended most, if not all, of the sessions and were actively involved in the discussions and critiques of the students' presentations. Each of these faculty members also was in charge of presentation of some topic within the theme of "optimization" and development of projects related to that topic.

Chapter 10

Harvard University

Daniel L. Goroff

An Overview. All of the dozen or so new graduate students who enroll each year in the Harvard Mathematics Department are PhD candidates with serious research interests. Sooner or later, however, nearly all will also find themselves teaching students who may have talents and goals quite different from their own. The department, therefore, takes quite seriously its responsibility to prepare graduate students for college teaching as part of their professional training. For this reason, the Mathematics Department has collaborated with the Derek Bok Center for Teaching and Learning for nearly ten years in running a comprehensive "Apprenticeship Program" that involves class visits, practice lectures, extensive feedback, guest lectures, and videotaping. Successful completion of such an apprenticeship or other substantial teaching experience is a prerequisite for enrolling in Math 301, a seminar course for graduate credit called "Theory and Practice of Teaching in the Mathematical Sciences." Whereas the training exercises that novice instructors carry out as apprentices help them find out about "how to teach," Math 301 is a master class that focuses more on "how students learn."

Other Mentors: Robin Gottleib, Joe Harris, and Barry Mazur

Math 301 challenges participants to develop a pedagogical philosophy of their own that is informed by and applicable to a wide variety of educational settings. In evaluating this seminar, participants are especially enthusiastic about our visits to local high-school classes, as well as our discussions of specific college classroom incidents and conundrums using the case method.

As mathematicians, we are trained to hide our scaffolding. The mathematical structures we build are not supposed to reveal traces of how they were constructed. Rather, we write as if setting delta equal to some strange function of epsilon was and should be immediately obvious. The final product is all that matters. Hunches, blind allies, and intuition were simply never needed, it

seems. As it is when we discuss our research results, so it is when we talk about our professional lives. We act as if how to be a mathematician is something people should just know. But perhaps young mathematicians especially could benefit from discovering that, on the contrary, explicit discussion about educational theory and practice can be respectable, interesting, and rewarding. This is what Math 301 sets out to prove.

Math 301: Structure and participants. Piloted in the fall semester of 1992, Math 301: Theory and Practice of Teaching in the Mathematical Sciences, met once a week at Harvard as a for-credit graduate seminar. Faculty members Daniel Goroff, Robin Gottlieb, and Barry Mazur were responsible for planning and leading this course. There were a total of nine regular participants. In addition to the graduate students from the Mathematics Department, one man from the Division of Applied Sciences and one woman from the Graduate School of Education also signed up to take Math 301 for credit. Several other graduate students who were not registered for the course attended selected sessions, too. Generally speaking, the group simultaneously had both diversity and the ability to work quite well together. About half were native speakers of English. All were highly motivated to improve their own teaching.

Some substantial teaching experience was a prerequisite for Math 301. In particular, those from the Mathematics Department were expected to have already completed our Apprenticeship Program for training graduate students as teachers. The elements of this apprenticeship have been devised, run, and refined over the course of almost ten years in cooperation with the Derek Bok Center for Teaching and Learning.

Specifically, each apprentice is paired with a coach who is already a successful calculus teacher. After observing and discussing several of the coach's classes, the apprentice prepares and delivers a practice lecture to a handful of volunteers from the coach's class. Based on feedback from the coach and these students, the apprentice revises this lecture. It then becomes the first of three lectures that the apprentice gives to the coach's entire class. The second is videotaped and reviewed by the apprentice in consultation with representatives of both the department's Committee on Instructional Quality and of the Derek Bok Center for Teaching and Learning. What the apprentice learns from this videotape is then implemented in the apprentice's third and final lecture. Thus, the Apprentice Program helps give prospective teachers the confidence and skills that will help them when facing a class of their own. All our graduate students must successfully complete the program before we assign them to teach calculus.

The fact that the participants in Math 301 had previously had at least as much teaching experience as that provided by the Apprenticeship Program was one of the factors contributing to the success of the seminar. In fact, all but one of the participants had progressed well beyond the Apprenticeship: most had taught their own classes several times, one was teaching concurrently, and the others had been leading recitation sections. In contrast with rank novices who tend to be either polemical about or oblivious to certain pedagogical issues, the graduate students in Math 301 based their work in the course on real educational problems, examples, and observations. This helped make the seminar particularly engaging not only to them, but to the faculty involved as well.

Cooperative Planning. The maturity and focus of those who signed up for Math 301 also meant that they had a clear idea of what they wanted to get out of the course. The faculty initially presented the requirements for Math 301 by setting out combinations of activities and assignments that students could choose to complete in order to pass the course. It soon became apparent, however, that there was little or no interest at all in several of the items on this menu. People felt they either knew enough about these already or could easily find out more when needed. The seminar therefore spent less time than anticipated on topics such as computers, career development, and professional activities other than teaching. We also cut down on invitations to guest speakers unless there was an explicit call for outside expertise, as when we took up adolescent psychology. Thus, the students and faculty in Math 301 took part in discussions and in shaping the course on an equal basis. As a result of these adjustments, participants did approach the topics and projects chosen with considerable energy, enthusiasm, and seriousness.

Assignments and e-mail feedback. Everyone's active involvement was also facilitated by the way we assigned background reading from the mathematics education literature each week. All participants were expected to produce short essays reacting to the reading assignment and relating it to their own teaching. Instead of just turning in this work at our meetings, everything was posted in advance to an electronic mail distribution list for the class. This allowed people time to reflect upon and respond to one another's ideas, leading to some lively exchanges both over the network as well as in class.

The e-mail messages quoted below convey some of the informal but earnest tone of this medium even if these have been modified slightly to protect anonymity or to improve grammar and readability. The point is that these graduate students definitely gained the permission and the vocabulary to carry on meaningful conversations about teaching, a topic often considered otherwise taboo. Moreover, they found out that such discussions can be engaging, rewarding, and "highly nontrivial."

Our first assignment, for example, was to study Bill Thurston's article on "Mathematical Education" from the September 1990 *Notices of the American Mathematical Society.* Because it is such a thoughtful and comprehensive survey by a highly respected mathematician, this became an organizing backbone for the seminar to which we referred repeatedly during the semester. From the beginning, everyone found something to seize upon. In initial e-mail postings, one student used the distinction Thurston draws between the depth and breadth of mathematics to criticize the national curriculum he had followed in his native country. Another challenged Thurston's statements about socially excluded groups. Picking up on what Thurston calls "the unanswered question in the air," a student ruminated about motivation, perspective, and, in his own words, "why we teach mathematics at the calculus level in such a different way from advanced courses." Still another offered refutation of Thurston's disparaging remarks about mathematical contests.

Key questions. To examine these and other issues more carefully required that we push beyond the question of "how to teach" addressed by the Apprenticeship Program and focus more on "how students learn." This became a central theme of the course. Anyone who might have thought there were simple answers to this question was quickly convinced otherwise. Ideally, Math 301 challenges participants to develop a pedagogical philosophy of their own that is informed by and applicable to a wide variety of educational settings. Here are specific examples of how we studied seven different questions about the nature of learning.

1. *How do students learn math in American high schools?* Visits to local public and private schools made a great impression on seminar participants and provided helpful insights about the expectations, the behavior, and the background of typical students. In preparation for these field trips, we studied the NCTM *Standards* and discussed their implications for college teaching. Most were intrigued and positive about such reforms. One reaction posted to e-mail stated that, "I do not like the tone of these *'standards'* that much. The practical aspects of mathematics is somewhat overemphasized so that they seem to talk about 'mathematical training' rather than 'mathematical education.' For example, I would like to find the word 'enjoy' somewhere in the section where they discuss the goals of mathematics education. Well, helping students to be 'self-directed learners' is mentioned later. But what is the use if they cannot enjoy it? After all, isn't demolishing the misconception that mathematics is difficult and dull the most important job to be done?"

In the event, they did find some of the high school classes tedious and mechanical. One team of observers described how "The teacher emphasized, both implicitly and explicitly, the importance of rules in calculus. While it is of course desirable to have students know the rules well, and to be able to take derivatives almost without thinking, we think that such a strong emphasis on rules may be counterproductive." People were concerned if not alarmed to realize that "When I teach calculus, I will be teaching it to students who have been through a course much like this one."

In contrast, everyone was enthralled by one nontraditional teacher whose interactive methods were often referred to throughout the rest of Math 301. People described this class visit as "exceptional" and "inspiring" in their e-mailed analyses. "Visiting the class was very educational for me," reported the student who had previously written about the rigid system in his country of origin, "The class was practically led by the students. The teacher almost never told the students what was right or wrong in advance; he let the students reach the conclusion themselves. He made good use of the board, putting on it practically all of the students' suggestions. His method was particularly effective for the purpose

of that class, where he was solving test and homework problems. I would think that if he were to teach new material, new theorems, and new definitions, he would have to modify his method somewhat. However, I believe with that atmosphere he would still be able to evoke maximum class participation.

"Perhaps the special ability of the students and the teacher or the size of the class made this so lively," he continues, "However, I think I could learn many things from this class that would be applicable to the way I lecture. First is to have an atmosphere that makes students comfortable participating, makes them not afraid of giving incorrect answers. Second is to make the students feel that the subject is exciting, something within their ability to learn, to master. When a student asked some questions whose explanations may have been beyond the scope of the class, the teacher did not discourage him by telling him that was something beyond his ability now, but rather he told him to think more about it."

Another student wrote about how he was fascinated to find that "Mathematics could be taught as something that is worked on together, as is the case in [this teacher's] class, rather than something that is 'handed down from a higher authority,' as some grade school teachers may have one believe." A foreign graduate student's e-mail described how "In my visits to the high schools, I paid special attention to the oral communication side. It is difficult to communicate orally in mathematics. . . But written communication has a tendency to flow one-way. So for active student involvement, oral communication must be emphasized. In this aspect, [the first teacher] is not so satisfactory. Just like our calculus lectures, many students pay more attention to what is written on the board. Naturally, there are not many questions from the students, not to say any suggestions. On the other hand, [the second teacher's] class is led by students' suggestions. There were so many questions that he had to ignore some. Just coming into the classroom, I noticed the small size of the board, suggesting more oral communication than writing. [The first teacher] on the other hand used his board effectively, with carefully chosen examples. But you cannot say the students were enjoying it. [The second teacher's] students were excited and enjoying the class."

2. *How do students learn calculus at colleges?* We compared and contrasted calculus exams from several different kinds of colleges and universities, in some cases dating back many years. Noticing the emphasis many recent tests place on calculators, applications, and ideas as opposed to traditional calculations, theorems, and proofs, we were able to identify current calculus reform trends and to debate their merits based on concrete examples.

Having studied the final examination on multivariable calculus from another university, a graduate student who was teaching the same subject wrote that "Compared with Harvard, this exam tests more calculational techniques...In a way it tests understanding of the subject, but I think it is possible that a student with reasonable understanding could miss many points and a student with poor understanding can get a good score by memorizing basic formulas combined with practice on old exams. Basically, all the problems involve just one-step logic. So it looks like a combination of exercises." About examination questions in general he concludes that "It seems there are three levels: first something like exercises, numerical formula type problems, which are, of course, an important part of the subject. Then there are geometric, pictorial problems. And finally a problem asking the students something more active. The third one is really ideal...E.g., how about 'What is Stokes' Theorem? Give a rough argument why you believe it is true. How would you use it? Make a problem of your own, where it can be used, and do it!' This may be too radical and, of course, difficult to grade. In a way, the real problem with examinations is of a practical nature. If we make clear what our goal in the course is, then it is easy to see what kind of problems will be 'ideal.' But usually, ideal problems are impossible to grade. So one must make a compromise."

3. *How does the way we test students affect the students' perception of what it means to have learned something?* With the various finals in front of us again, we examined the messages we send our students in our testing. As a way of trying to clarify our goals when teaching a given course, we speculated at length about what kind of test we would administer to students a year or two after taking calculus to see whether they really "got it." Curiously, all that seems clear is that the questions would be quite different from the ones we routinely ask.

These considerations led us to an interesting discussion of what constitutes a "conceptual question." One graduate student gave an example from an article he had found in the October 1992 *Notices of the American Mathematical Society* by David Gale entitled "Calculus Reform: Past, Present, Future (?)." We agreed that it is useful to recognize that not all qualitative questions are conceptual. Having recently studied the "Meno" dialogue of Plato in Math 301, one student drew an analogy to this point, with "the difference between Socrates and the Sophists: dialectic is more important than rhetoric; i.e., the most important thing is to engage the student." What is conceptual, like what is surprising, in mathematics depends very much on context. We discussed the importance placed by philosophers like Israel Scheffler on the educational role of surprise. In conclusion, one of the faculty presented a surprising and quite delightful proof of the Pythagorean Theorem.

4. *How do students learn from our expositions?* Faculty and students scheduled themselves to give special expository presentations. Most were delivered by the graduate students outside the regular Math 301 times: to other graduate students in the Department's "Trivial Notions Seminar;" for undergradu-

ates at the Math Club's weekly "Math Table;" or for groups of high school students. We made systematic arrangements for speakers to receive constructive feedback about their talks from fellow members of Math 301 and, in some cases, by surveying all the others present as well.

Some expository lectures were delivered to Math 301 itself, but were also designed to illustrate how topics can be adapted to suit various audiences. Thus, the assignment was not merely to familiarize our class with little-known but interesting and useful facts about Lagrange multipliers, for example. Rather, we specifically asked students to prepare three different presentations about Lagrange multipliers, say one for calculus students, one for economists, and one for advanced mathematics majors. Our class would pretend to be each of the different audiences in turn, then compare and contrast the treatments. Through this exercise, we gained insights about how to address groups appropriately depending on their level of experience and sophistication.

One source of subject matter on which graduate students from the Mathematics Department can draw is their minor thesis. This is a required expository paper to be completed in less than three weeks on a topic specified by the Department to have little or no overlap with the student's dissertation research. Along with the long-established "Basic Notions Seminar" given by faculty, it is another way in which the Department provides graduate students with a broad perspective on mathematics.

Perhaps because some students already feel rather satisfied with respect to mathematical breadth, the expository talks we assigned in Math 301 were the only aspect of the course about which anyone was unenthusiastic in the final evaluations. Along with many other positive comments, the one participant who complained wrote about how "The presentations made on parts of elementary mathematics took up far too much time for very little gain." Although few others would agree, the reason stated is that "most PhD students in math already have some experience giving talks to a variety of audiences."

5. *How do students learn from one another?* To help us reconsider the teacher's role in dealing with diverse sets of students, we watched and discussed a videotape produced by the Derek Bok Center for Teaching and Learning. Called "Thinking Together," it presents examples of interactive and collaborative learning strategies that Uri Treisman and others have proved remarkably successful for college science courses in general and for traditionally underrepresented groups in particular. People were especially interested in the film's portrayal of how Eric Mazur, a physics professor, is using conceptual questions and peer learning methods to improve measurably the performance of a large lecture class. Readings for this session included Professor Richard Light's undergraduate studies stressing the educational ef-

fectiveness of techniques such as frequent feedback and small group learning.

6. *What about the theory of student learning?* After reading about "Perry stages" and other accounts of cognitive development, we held a spirited discussion on adolescent psychology with Abigail Lipson, a representative of Harvard's Bureau of Study Counsel who specializes in working with math and science students. The class critiqued some of the advice she routinely gives undergraduates who are floundering in such courses. We also learned about what other techniques and resources we can draw upon to help such students. While William Perry's book *Forms of Intellectual and Ethical Development in the College Years* is concerned with how attitudes towards authority, belief, and truth in general develop during the college years, we found it fascinating to work out for ourselves how his scheme might shed light on the learning of mathematics in particular. This kind of psychological insight, for example, can make it easier to recognize, understand, and deal with the angry, fearful, and mechanical approaches to mathematics that afflict so many undergraduates.

7. *How do students learn in representative "case studies"?* We examined, discussed, and drew lessons from cases of realistic teaching incidents and conundrums. While our work on the "Meno" dialogue of Plato and on the local high schools also functioned as case studies in this way, we especially pondered a case in a series specifically written for such purposes by Abby Hansen. Entitled "Trouble in Stat 1b," it tells of a new and dedicated associate professor whose statistics class rebels when he tries to make them think.

Based on actual events, this three-part case includes many quotes from interviews with the professor, with members of his class, and with colleagues. Names and places are fictionalized in the write-up, which reads like a short story with many novelistic details included. The eight-page narrative of the first part of the case builds up to the professor's recalling in his own words how "As I finished writing Marie's equation, a woman's voice called out: 'Wait a minute — is that right?' It doesn't particularly matter who the speaker was — an average-aged, average-looking, bright woman student. Let's call her Joan. She was half out of her seat, waving one hand in the air, palm out, as if trying to flag down a train. I remember the tone of voice more than anything else. Joan's was furious. At the time, I was thinking that Joan's technical instincts were on target. But this was a discussion. The last thing I wanted was to kill it by either damning or blessing Marie's equation at this point. I stood there, wondering how to react to two things: Joan's 'wait a minute' — a direct order to stop the class — and her equally direct question: 'is that right?' "

Having read this in advance, everyone came to class with ideas about what had gone wrong leading up to this point, what to do now, and why. Participants were able to support their arguments and prescriptions by applying the pedagogical theory we studied to the concrete situation at hand. Later we passed out the short second and third parts of the case describing some of the havoc that actually ensued. Our discussion was especially engaging and productive because it was skillfully led by Kay Merseth, a professor at Harvard's Graduate School of Education who has much experience both with mathematics and with the case method. We even set aside time near the end of our meeting just to go over with her some of the discussion-leading techniques she uses.

The case method is increasingly popular in many disciplines, not only for teaching material to students, but also for training teachers about pedagogy. The American Association for Higher Education, for example, is sponsoring an initiative to collect such pedagogical cases along with supporting materials for the discussion leaders who might use them. At the moment, however, there are few cases that speak directly to teachers of mathematics. Producing good ones is a surprisingly involved task. The faculty of Math 301 have been researching and writing up the case of a young and well-meaning assistant professor whose miscommunication with his first calculus class resulted in hostility and, at one point, chaos. There is also an unexpected twist to this story which may make it of particular interest to international faculty.

This case study discussion and the high school visits were the two activities participants praised most in their evaluations of Math 301. "Overall, the seminar was a worthwhile experience," wrote one. "I learned a fair amount about college students, the people who I will soon be teaching. The discussions about various teaching techniques and their effectiveness were also useful; I may try some of these techniques. The most valuable parts of the seminar were those in which we studied how students actually learned and how teachers actually taught."

"The meetings were much fun, not boring" reads another participant's evaluation, "The approach through more fundamental issues rather than teaching skills was good. Teaching skills are important, but it is usually difficult to learn them by discussion meetings." A classmate concurs, "Overall I found Math 301 extremely interesting and useful...Thanks for a great seminar."

The final evaluation from another participant mentioned three lessons that were especially important: "The idea of a 'conceptual problem' is new to me. I was always aware that there are 'word problems' that could be answered, but I did not realize that some of these so-called 'real life problems' are after all, just the same as the 'straightforward' ones. I feel that a good problem need not have a fancy statement; if the problem has depth, it is much better than a trivial problem made complicated just by its wording." His second observation is that, "I have also reinforced my ideas about getting a class involved and have been trying this with the course I am now involved with as a teaching fellow."

Finally, he writes about how "I used to think of the job of a teacher is to teach a certain amount of material and to get the class interested in the material. Perhaps the job of a teacher is 'merely' to let the students learn. What I mean by this is the teacher may create an atmosphere which allows learning but the actual learning comes from the efforts of the students. An example is the interactive class we visited. The more that the students put into the class, the more they get out of it. The class is based on the students asking questions of themselves, and the teacher leads by discussing with the class, not lecturing to it."

Most participants similarly report that they are not only adjusting their current teaching practices because of Math 301, but that they also feel more knowledgeable and confident about future challenges they will face in the classroom. In view of this success, the Harvard Mathematics Department plans to continue offering such a course for graduate credit. Indeed, other departments have become interested in creating similar seminars for their graduate students.

While the next version of this seminar will be reorganized to benefit from our experience with this pilot project, there are also factors contributing to the success of Math 301 that should be maintained. These include: the constant exchange via e-mail of reactions to readings, lectures, or discussions; the full participation of faculty on an equal footing with the graduate students; and the ability to draw judiciously on resources such as the Derek Bok Center, the Bureau of Study Counsel, and the Committee on Preparation for College Teaching. Above all, Math 301 succeeds in demonstrating that future faculty can prepare to teach college well by addressing intellectually challenging and genuinely interesting questions about how students learn.

Acknowledgement: The author wishes to thank Robin Gottleib for reviewing a draft of this report and making several helpful suggestions. Dick Gross and Joe Harris were among faculty members contributing helpful ideas and encouragement.

Chapter 11

Oregon State University

Dennis Garity

An Overview. Oregon State University is the land grant university of the state and is Oregon's oldest state-supported college or university. The Department of Mathematics consists of about 35 professors and 45 graduate teaching assistants. Beginning mathematics graduate teaching assistants participate in an orientation session before they start teaching. This session includes an introduction to teaching duties and to the graduate program in mathematics. Each GTA gives a sample lecture, prepares a sample test, and gets feedback from faculty and more experienced graduate students during this orientation session.

A professional seminar for PhD candidates was introduced during the 1991-1992 academic year and was offered again during the 1992-1993 academic year. This professional seminar was designed to prepare PhD candidates for their future teaching responsibilities at colleges and universities, to increase the mathematical breadth of the candidates, and to foster their sense of professional and social responsibility. A key feature was the involvement of many faculty members from the mathematics department in describing their experiences with a variety of upper-level undergraduate courses. Another main focus was a detailed description of the responsibilities that the PhD candidates will have after they get their first jobs.

Other Mentor: Mary Flahive

A Professional Seminar for PhD Candidates at Oregon State University

Introduction. Oregon State University is one of three research universities in Oregon. Its student population is about 15,000, including approximately 3,000 graduate students. The mathematics department awards MS, MA, and PhD degrees. There are about 60 graduate students in mathematics, with about 50 receiving support as either graduate teaching assistants or graduate research assistants. About 30% of the mathematics graduate student population is female, and 72% of the students are US nationals. The graduate program in mathematics has been increasing in size in recent years.

The following table gives further information on the number of graduate degrees granted. (The numbers for 1993 are projections.)

YEAR	MS/MA	PhD
1993	7	2
1992	10	4
1991	11	4
1990	15	3
1989	8	1
1988	5	2
1987	6	2
1986	7	4

Graduate teaching assistants usually teach recitation sections for calculus or college algebra during their first year. More experienced graduate students often teach their own sections of calculus. An integral part of the department's orientation for new graduate students is a teaching workshop, in which faculty and more experienced graduate students participate. The workshop takes place in September during the week before classes begin.

On the first day, participants receive handouts on teaching and participate in discussions on teaching undergraduate mathematics. The use of graphing calculators in courses is explained, and the participants receive information on the background of students in our lower-division courses. At the end of the day, they receive a topic to present in a sample lecture, and they receive an assignment to prepare a practice test.

On the second day, the participants give a sample lecture and receive feedback from faculty and advanced graduate students. A discussion is held on composing tests, grading, dealing with problem students, academic honesty, and the university's sexual harassment policy. During the third day, the participants meet in small groups with faculty and advanced graduate students and discuss the sample tests they have composed. A discussion is held on teaching evaluations and on departmental teaching and the academic year.

During the year graduate teaching assistants hand out student evaluations, and a faculty member visits one of their classes and makes suggestions for improved teaching.

The Proseminar. The orientation described above prepares our students well for teaching lower-level mathematics courses, but does little to prepare them for the type of teaching responsibilities that they will have after receiving a PhD. This led to the formation of a professional seminar for PhD candidates designed to prepare them better for life as professors.

The seminar was first held during the 1991-1992 academic year and again during 1992-1993. It had weekly meetings. The principal audience consisted of mathematics graduate students who had completed the written examinations for the PhD degree. In addition to preparing the participants for their future teaching responsibilities, the seminar had two other purposes: to increase the mathematical breadth of the participants and to foster their sense of professional and social responsibility.

Alan Schoenfeld's *A Sourcebook for College Mathematics Teaching* (MAA, 1990), was the required text for the seminar. We supplemented it with department copies of many of the volumes of the MAA's *Notes* series and a bound collection of articles from various popular and academic sources, among them *UME Trends* and *The Chronicle for Higher Education*. Students compiled portfolios of articles they found informative. Schoenfeld's book was an invaluable resource. In addition to a very detailed bibliography, some of the chapters on which we focused contain information on instructional goals, MAA curriculum requirements, advising, and evalua-

tion.

The following are other key MAA *Notes* we used:

Number 16: Using Writing to Teach Mathematics, edited by A. Sterrett (MAA, 1990).

Number 9: Computers and Mathematics, edited by D. Smith, G. Porter, L. Leinbach, and R. Wenger (MAA, 1988).

Number 18: Models for Undergraduate Research in Mathematics, edited by L. Senechal (MAA, 1990).

Number 13: Reshaping College Mathematics, edited by L. Steen (MAA, 1989) .

One typical activity of the seminar had a faculty member from the department discuss some subject area which is taught at the upper undergraduate level. (Some examples are: advanced calculus, numerical analysis, abstract algebra, linear algebra, probability, and topology.) The visiting faculty member chose the topic. The discussion included some of the following: course prerequisites, content, level, teaching techniques, alternate teaching styles and other innovations (and whether they succeeded or failed), and comments on existing textbooks and software. During the first year, there were also presentations by the student-participants. These presentations often followed up one of the earlier faculty talks or presented a summary of information from some of our resource books. Presentations were given on writing in mathematics courses, visualization, technology in linear algebra, and a number of other topics. During the second year, partially on the basis of feedback from the students from the first year, the participants were asked to compile a bibliography or some other information that would be useful after they graduated. The mentors also presented additional material on other professional responsibilities that the students could expect to assume after they received their degrees. These presentations included: membership in mathematical and other professional societies; meetings of professional societies; the tenure process at different universities and colleges; committees on which faculty members serve; efforts to increase the numbers of women in mathematics; and some strategies in applying for jobs.

Plans for the future. Another component of the seminar was an evaluation of how the students' attitudes towards their teaching responsibilities changed over the course of the seminar. A proposal is being prepared to the mathematics department to offer this seminar every other year.

Winter Term 1991-1992

Mentors: Mary Flahive and Dennis Garity

The professional seminar for PhD candidates was offered for the first time during winter-term of the academic year 1991-1992. There were fourteen graduate students in the department who had passed their qualifying examinations at that time. Of this group, nine students registered for the seminar and one additional student sat in without registering. The seminar was scheduled for one to three credits and met once a week for one and a half to two hours each week during the term. It was usually divided into two parts. The first part was a presentation by a faculty member on some upper-level undergraduate course with a follow-up discussion. After a short break, the second part consisted of presentation by one of the graduate students either following up on a previous topic or presenting information from some of the resource material.

What follows is a week-by-week summary of the topics covered in the seminar.

Week one: The first part of the seminar was organizational. A list of resources was handed out, and the format of the seminar was explained. The participants were given the history of the development of the seminar and were told to purchase *A Source Book For College Mathematics Teaching*, edited by Alan Schoenfeld. They were also asked to keep notebooks of resources discussed in the seminar and to prepare topics for presentation. A list of possible topics was given.

The mentors together with the students developed a list of courses on which faculty members could speak. The list included courses titled Linear Algebra, Discrete Mathematics, Introduction to Proof, Differential Equations, Advanced Calculus, Numerical Analysis, Abstract Algebra, Number Theory, Probability, Topology, Complex Analysis, Statistics, Modeling, PDE, Introductory Analysis, History of Mathematics, and Geometry. For each of these courses it was suggested that the faculty member could discuss content, level, teaching techniques, uses of visualization, innovations and whether they succeeded or failed, possible textbooks and software, computer use, writing use, and current reform efforts.

Professor Garity then gave a presentation on the teaching of undergraduate topology. This was followed by a discussion of how to introduce some of the key concepts in topology, and how to choose texts for courses. The participants then described their undergraduate schools and discussed how they became interested in mathematics. The importance of personal encouragement of undergraduates by faculty was stressed.

Week two: A faculty member gave a presentation on teaching undergraduate probability courses. The seminar included the elements:

◇ Smaller colleges often have difficulty finding people to teach probability and statistics courses.

◇ Many graduate students and faculty say they do not know how to do the problems in the junior-level probability course.

◇ The junior-level course is not a popular course to teach, and students tend to find the course difficult because it is not algorithmic.

It was recommended that graduate students acquire a background in probability while they are still in graduate school. There was also a discussion of textbook choice, the use of technology, access to computers, and appropriate homework assignments.

The second part of the seminar was a presentation by one of the participants on the situation for women in mathematics. It included the elements:

◇ Women are more likely to be judged by their appearance than men. The atmosphere in the classroom is often different for women-instructors than for men-instructors.

◇ Statistics regarding women PhD recipients from the AMS *Notices* was presented.

◇ There was a discussion of "Winning Women into Mathematics."

◇ Possible reasons for the lack of women in mathematics include cultural expectations, the impression that mathematics is not feminine, family customs, the educational system, customs and practices within the mathematical community, and the availability of childcare.

Week three: A faculty member gave a presentation on teaching undergraduate numerical analysis. The presentation included the following points.

◇ Because of the rapid pace of developments in the field, the texts and literature have not completely caught up. Careful editorial work by instructors is needed with many texts.

◇ A balance of theory and experience with computers is needed so that it becomes possible to judge how often bad behavior occurs.

◇ It is desirable to have the students learn how to use the good available software and to be able to do some programming.

◇ Grading student computer programs can be extremely time-consuming and difficult.

The second part of the seminar was a presentation by one of the participants on computer use in linear algebra courses. The participant had developed a list of references and examples of how different people had used computers in such courses. A list of software packages that could be used in such courses was distributed. Some of the key points raised in the presentation were that computers:

◇ can provide tutorial assistance in such courses and can be a means for creating a self-paced learning environment;

◇ provide a tool for implementing illustrations and practical applications;

◇ can be a way of encouraging experimentation and discovery;

◇ can be used in the context of describing and illustrating mathematical concepts.

The third part of the seminar was a detailed description of the undergraduate mathematics curriculum at Oregon State University. This was followed by a discussion of the undergraduate major at different types of institutions.

Week four: A participant in the seminar presented a detailed summary of the book *Scholarship Reconsidered* by Boyer.

Some of the main points that were discussed were the following.

◊ The definition of scholarship has evolved over time, from religious to service to research. There is a need for each college to clarify its mission.

◊ There are four aspects of scholarship: discovery, integration, application, and teaching.

◊ Faculty members have different strengths. Creative contracts should be developed between institutions and faculty members, taking into account the mission of the institutions and the strengths of the faculty members.

◊ There are problems with overly specialized study. Scholarship needs to be redefined, taking the community into account.

This presentation led into a general discussion of the tenure and review process for faculty members.

The second part of the seminar was a presentation by Professor Flahive on mathematical demographics and on professional mathematical organizations. Demographics by D. C.Rung from the Fall 1990 *AMS Notices* were presented. These included the total number of undergraduate mathematics course enrollments (1,926,000), the number of full time mathematics faculty (19,411), and the number of BA and BS degrees in mathematics (19,191). The American Mathematical Society and SIAM and their publications were discussed. Other professional organizations, including MAA, AAUP, AAAS, AWM, and NAM, were mentioned.

The third part of the seminar was a presentation by a faculty member on the history of mathematics and science. A number of examples were given that illustrated the point that mathematicians tend to be bad historians. Topics discussed included: the works of Oresme (circa 1350 C.E.) and of John Philoponous (circa 551 C.E.); a brief history of the conflict between earth-centered and sun-centered models of astronomy; the development of the calendar; and the history of rigor in mathematics.

Week five: There were two presentations by participants in the seminar this time.

The first presentation gave a summary of postdoctoral opportunities. Much of the material presented came from October issues of the *AMS Notices*. The types of positions included named instructorships, postdoctoral positions at research institutes, foundation-supported postdoctoral positions, and positions in specific countries. There was also a discussion of sources of travel funds.

The second presentation was on the article "Reluctance to Visualize" in the *MAA Notes* volume on "Vi-

sualization in Teaching and Learning Mathematics." There was a detailed discussion of visualization versus symbolic manipulation. The presentation included a discussion of how mathematics classes and texts are organized with respect to visualization.

Week six: A faculty member gave a presentation on actuarial science. Options for mathematics students in actuarial science were discussed, and the importance of developing outside contacts was emphasized. The roots of probability in actuarial science were detailed. The Associate and Fellow levels in actuarial science were discussed, and the examinations leading to these levels were described.

A participant in the seminar then discussed writing in mathematics. Sources were an article in *UME Trends* from August 1991, *MAA Notes* number 16, and an interview with a faculty member at Oregon State University. Examples of major writing assignments were also given. These ranged from writing a chapter in a book to assembling a portfolio. Examples of shorter writing assignments were also given. One example was having students write down two sentences about a particular topic. Another was writing a letter to next year's students explaining the course. Another example was requiring weekly abstracts of the lectures and reading.

Another participant then presented some resources and opportunities for faculty members and for undergraduates. Faculty workshops and conferences listed in the *AMS Notices* were presented, and summer short courses for faculty were discussed. Good sources of information given were the *MAA FOCUS*, the *AMS Notices* and NSF announcements. Research experiences for undergraduates were also discussed.

Week seven: A faculty member gave a presentation on teaching advanced calculus. The goals for each term of our three-term sequence were discussed in detail. The advantages and disadvantages of a number of texts were discussed, and the participants gave their experiences with advanced calculus when they were undergraduates.

Professor Garity then gave a summary of other duties that faculty members have in addition to teaching and research. The variation in duties at different types of institutions was described. Some examples that were presented included committee work, student supervision, course development, and university service.

A participant in the seminar then gave a detailed introduction to the types of topics covered in complex analysis courses and provided an extensive bibliography and a number of course outlines.

Week eight: A participant in the seminar gave a presentation on visualization in conjunction with the use of graphing calculators. A handout was given with some sample problems that could be solved with various amounts of visualization. Participants then discussed their approaches to solving the problems. Some sample graphics programs on the calculators were demonstrated, and a discussion about how these calculators could be used in courses followed.

Another participant gave a presentation on opportunities for encouraging research by undergraduate students. A discussion followed about how to generate excitement about mathematics in undergraduates. Since this was the last meeting of the seminar, it ended with an evaluation of the seminar and with suggestions about changes that could be made.

Spring Term 1992-1993

Mentor: Dennis Garity

The professional seminar for PhD candidates was offered for the second time during spring term of the academic year 1992-1993. Most of the graduate students who had passed their written qualifying examinations had taken the seminar the previous year, so only three students signed up this time. There were two additional regular participants who did not officially register and two or three other participants who attended sporadically.

The seminar was again listed for from one to three credits and met once a week during the term. Because of student and mentor reaction to the previous year's seminar, a modified format was tried. The students had indicated that they got the most useful information from faculty presentations on specific courses and from discussions of their future responsibilities. A decision was made to have a single focus for each meeting and to have the student-participants prepare detailed bibliographies on some material rather than presenting material during the seminar.

A week-by-week summary of the seminar meetings follows.

Week one: This meeting was primarily organizational, and the format of the seminar and the expectations of the student participants were explained.

Week two: Professor Garity presented a discussion of teaching advanced calculus. The material presented and the discussion that followed were similar to that described above in week seven from the first year.

Week three: Professor Garity gave presentations on a number of different topics. First, there was a discussion of MAA resources including a short description of the *MAA Notes* series. A presentation was then given on various methods of teaching, including the lecture method, modified lecture method, Moore method, and self-paced method. The advantages and disadvantages of each were discussed. Participants related their experiences with various teaching methods. Finally, a short presentation was given on using writing in mathematics courses.

Week four: Professor Garity continued the presentation on writing from the week before. The participants described their experiences with writing both in mathematics courses and in other courses. The advantages and disadvantages of introducing writing in various ways were discussed.

Week five: Professor Garity and another faculty member led a discussion on the process of tenure at various institutions. The relative importance of teaching, research, and service, as well as departmental expectations for new faculty members were discussed. The participants showed a great deal of interest in these topics.

Week six: A faculty member gave a presentation on numerical analysis. The presentation was similar to the one described for the first year. Some additional experience had been gained by using the software program Matlab in a computer laboratory on campus. A detailed description of using this software and the accompanying laboratory was presented.

Week seven: A faculty member gave a presentation on teaching linear algebra using a computer laboratory and Matlab. Specific examples of computer assignments were presented and discussed.

Week eight: A faculty member and a finishing graduate student who had found a job gave a presentation on job-seeking strategies. This presentation generated detailed discussion, partly because of the current difficult employment situation facing new recipients of PhDs in mathematics. The importance of making personal contacts and of having all material ready early in the year was emphasized.

Week nine: The discussion on job seeking strategies from the previous week continued for a short time. Then an evaluation of the seminar and a discussion on how to change the seminar in the future occurred.

Evaluation. At the beginning and end of the seminar, each participant filled out a doctoral student survey on preparation for college teaching. The survey asked information about the participant's background and about attitudes towards teaching. In addition, the students were encouraged to submit written comments about the seminar, and these comments were considered in the modification of the seminar that took place during the second year.

During the second year, the participants again filled out the doctoral student survey and submitted their written comments.

Chapter 12

The University of Tennessee

Henry Frandsen

Sam Jordan

An Overview. The University of Tennessee, Knoxville, is the land-grant university of the state and its major state-supported PhD granting institution. The Department of Mathematics consists of about 45 professors, 50 graduate teaching assistants (GTAs), and 15 adjunct teachers. New mathematics graduate students normally participate in a departmental orientation session which includes an introduction to teaching duties and various aspects of the graduate program in mathematics. They take a one-credit-hour course in topics related to mathematics teaching during their first semester on the campus.

A three-credit-hour professional seminar for senior PhD students was introduced in the 1992-93 academic year. This seminar was designed to provide information and experiences for PhD students that would help them in the transition from the role of GTA to the role of tenure-track professor. Readings and discussions focussed on the non-research roles of college professors, the expectations of mathematics departments regarding new professors, the recent efforts by professional organizations to change the way that mathematics is taught in grades kindergarten through college, and the process of finding a teaching position.

The Department of Mathematics at the University of Tennessee has a standard PhD program in mathematics which dates back to 1948. Graduate students are supported with assistantships which require them to teach freshman and sophomore courses. Currently, due to accreditation regulations, graduate students are not allowed to have full responsibility for teaching courses for credit until they have at least 18 semester hours of credit in graduate-level mathematics courses. This restriction means that the responsibilities of first-year graduate students usually are limited.

All new teaching assistants receive some training in teaching before and during their first semester of graduate work. An initial orientation for the GTAs takes place over a period of three days just prior to the beginning of fall-semester classes. It provides a quick introduction to the department and its curriculum, the art of testing and grading, the process by which students drop and add courses and change sections, the process for reporting and prosecuting cheating cases, methods for the prevention of cheating, university policy on sexual harassment, an overview of the content of the courses which the GTAs will be teaching, the importance of dealing fairly but firmly with students, and some fundamentals of lecturing. The GTAs prepare and deliver sample lectures which are video-taped and critiqued. They also write exams which are critiqued.

The new GTAs are then required to take a full semester course in teaching techniques. This one-credit-hour course meets weekly and is team-taught by four mathematics professors. It includes more discussion of student-teacher conflict resolution, the prevention of cheating, course and lesson planning, test writing and grading, the undergraduate mathematics curriculum, and the university and departmental teacher evaluation processes and how they can be used to improve teaching performance. There are also discussions of the current efforts of the professional organizations MAA, AMS, SIAM, NCTM, and others to improve the teaching and learning of mathematics at the undergraduate level and some discussions designed to familiarize the GTAs with the use of calculators as teaching tools. This course progresses at a more leisurely pace than the orientation and treats topics in greater depth. It benefits greatly from the assistants having had some classroom instructional experience.

During their first year with full teaching responsibilities the new GTAs are required to have student evaluations of their classes reviewed by a professor from the Teaching Effectiveness Committee of the Department of Mathematics. After that the GTAs are largely on their own unless there are some serious complaints about their teaching.

In the fall of 1992 a new professional seminar for students who are about to complete their PhD degrees was initiated with the support of the joint AMS/MAA/SIAM Committee on Preparation for College Teaching. This seminar was designed to assist them in dealing with the non-research aspects of their careers as college professors. It attempted to acquaint the participants with the various research, teaching, and service responsibilities of a mathematics professor and the processes that will be used by colleges and universities to evaluate their work and make decisions about promotion and tenure — in short, how to succeed at being a mathematics professor. A second goal of this program was to provide a link between the University of Tennessee Department of Mathematics and Knoxville College, which is a nearby undergraduate liberal arts college with a mostly African-American student body. This objective was addressed by a series of presentations for undergraduate mathematics majors at both institutions by the PhD students.

The participants were introduced to issues in mathematics education from elementary school through graduate level, recent efforts of the professional organizations to improve mathematics education in the United States, and the roles and responsibilities of college mathematics faculty in these efforts. Faculty members currently involved in teaching experiments at the freshman level were invited to discuss their experiences.

Assigned readings included an article by the research mathematician W. P. Thurston (*AMS Notices* 37:7[Sept. 1990] 844-850) concerning the role of mathematics professors in improving school and college mathematics education, parts of the NCTM publication *Professional Standards for Teaching Mathematics*, the MAA monograph *Reshaping College Mathematics*, and the National Research Council's publications *Everybody Counts* and *Moving Beyond Myths*. Participants were also asked to read the book, *The Craft of Teaching*, by K. E. Eble (which provided some very controversial topics for discussion).

An important feature of the course was the creation

of a series of interactive presentations to undergraduate audiences led by mathematics students at the junior and senior levels at the University of Tennessee and at Knoxville College. The PhD students were asked to select topics based on *UMAP Journal* articles and to prepare presentations using methods such as those recommended by the NCTM *Standards*. (Presentations of materials found in UMAP modules were used to expand and to hone the teaching skills and to broaden the mathematical backgrounds of the participants.) The topics chosen had to be outside the area of the presenter's research and appropriate for the intended audience. Each topic selected was discussed with the entire class and approved by the faculty.

MODELING GOOD TEACHING

After selection of topics, a preliminary plan for each presentation was presented and discussed in class. The presentations (rehearsals) were given to the class, video-taped, and evaluated by the class. Using the feedback provided, the participants prepared the final versions which were given to the undergraduate audiences during the spring semester.

Another major activity of the course focussed on classroom teaching. Participants were asked to keep personal journals with observations about the classes they were teaching and to visit one or two classes taught by a colleague to look for techniques that could be "borrowed" to improve one's own teaching. These visits were documented with written reports that were supposed to be reflective evaluations of the visitor's own teaching (not critiques of the teaching being observed).

The professional seminar met once each week for two hours but carried three semester hours' of credit (graded A-F) to reflect the additional effort required for the lecture series. It was directed by Professors Frandsen and Jordan and was taken by five senior PhD candidates, each of whom had taught for the department for four or more years. The participants were three international students (all European males) and two domestic students (a white female and an Hispanic male).

The course was greeted with some resistance by both the participants and some faculty. It was considered a potential distraction from the more important work of writing a dissertation. The students generally had very conservative viewpoints about changes that are being called for by the national organizations. There were some heated discussions about teaching and the roles of the mathematics teacher and student, most of which never reached any closure or consensus. As the semester progressed, the participants showed keen interest in some topics such as how to look for a position in today's tough job market, how to maintain a position once it is acquired, and how higher education institutions function.

OUR GOAL

A well rounded doctoral graduate

Questionnaires were used at the beginning and end of the semester to assess the course. A participant information survey provided by the Preparation for College Teaching Project was used at the beginning, and a form designed locally to evaluate the activities of the course was used at the end. Further evaluation will be done by asking the participants to look back at the course after they have held faculty positions for a year.

While it is too early to make any judgment about the success of the course or its future in the department, it has provided information for designing an improved version for future semesters. It seems to have helped the participants with some aspects of their transitions to faculty status, and it certainly made them aware of some of the current teaching reform experiments and changes being advanced by the professional organizations.

The Department of Mathematics is committed to continue this program for the next year or two while assessing its value. As with any first attempt, there will be several changes in the program as we continue our efforts. Less time will be spent in the rehearsal talks and more in discussion sessions, the reading materials will be selected to provide more focus on issues of importance to the mathematics professor, and an attempt will be made to achieve more effective and systematic self-evaluation of teaching through the use of the teaching log books and classroom visits.

University of Tennessee
Department of Mathematics
Presentation Evaluation Form

Rate your reaction to the following statements:
1: Strongly disagree 2. Disagree 3. Agree 4. Strongly agree

I. Basic technique
1. The presenter structured the time properly and was well organized.
2. The presenter made appropriate use of metaphors, analogies, or stories.
3 The presenter made appropriate use of computers, calculators, concrete materials, and/or other equipment.
4. The presenter made appropriate use of pictures, diagrams, tables, and graphs.

II. Content
5. The presentation was based on sound and significant mathematics.
6. The presentation promoted problem-solving, mathematical reasoning, and increased understanding of mathematics.
7. The presentation helped the audience to make connections between several mathematical ideas and/or applications.
8. The presenter showed enthusiasm for the subject and fostered interest and enthusiasm among the audience.

III. Audience interaction
9. The presenter posed tasks that engaged the audience's intellect.
10. The presenter encouraged the audience to participate, raise questions, form conjectures, and become actively involved.
11. The presenter made appropriate decisions about how to respond to members of the audience and/or how to direct dialogue between members of the audience.
12. The presenter showed respect for the audience and enhanced the value of their ideas and modes of expression.

Presenter's Name:

Presentation topic:

General reaction to the presentation:

Additional comments or suggestions:

University of Tennessee
Senior PhD Course Evaluation

The objectives of this course included the following:

◇ To make participants aware of the role and responsibilities of mathematics faculty in activities other than mathematics research.

◇ To provide information about professional organizations such as AMS, SIAM, MAA, and NCTM and their literature.

◇ To provide knowledge about recent national movements in mathematics education from kindergarten through graduate school and their implication for the future of collegiate mathematics curriculum and instruction.

◇ To provide information and examples of teaching methods which differ from the standard lecture or lecture-discussion methods, and to create opportunities for participants to try those methods.

◇ To accomplish these objectives we created a program of readings, discussions, and participant presentations. We will also provide a series of talks next semester as a culminating experience (acid test) for the course.

Please help us evaluate the course by answering the following questions:

1. How well did the course activities help to meet the objectives listed above?
2. Which of the reading assignments or activities were not useful in meeting any of the listed objectives?
3. What additional reading or other activities would you recommend to improve the course?
4. What revision in the objectives would you recommend?
5. Assuming you are soon making the transition from graduate student to faculty member, do you think this course was helpful enough to merit the time and effort you were required to expend?
6. What is your overall evaluation of the course?
7. Please add any additional comment or suggestions that you may have.

Chapter 13

Washington University

Steven G. Krantz

Robert McDowell

An Overview. Washington University in St. Louis is a private university that was founded in 1853 by William Greenleaf Eliot (the grandfather of T. S. Eliot). Originally intended as an institution for educating the youth of St. Louis, the university now attracts students from all over the world. Besides the traditional academic departments, the university is noted for its medical school, its business school, and its law school.

The Mathematics Department at Washington University has about thirty faculty, counting both regular and temporary members. There are currently 48 students in the graduate program; of these, 45 are in the PhD program and 3 in the master's program in Statistics. For the past three years, the department has required that all PhD students take a three-credit course on the teaching of mathematics and the mathematics profession in general. This course involves both studying the principles of good teaching and engaging in a practicum. There are detailed discussions of tenure, of the role of academic scholarship in the life of a college teacher, of the departmental duties of a college professor, of math anxiety and other teaching roadblocks, and of collegiality. Normally, students assume various graduate-assistant teaching duties shortly after completing this course.

The 1991 Seminar, Robert McDowell. The Teaching Seminar was first offered in the Spring semester of 1991 by Robert McDowell. The Dean of the Graduate School had urged all departments to incorporate courses in teaching preparation into the graduate program. Professor McDowell, a member of the Committee on Preparation for College Teaching, had long been active in the areas of teaching, learning, and curriculum, and was eager to experiment with some of the Committee's ideas. The seminar is assigned to faculty on a rotating basis, and Stephen Krantz reports below on his 1993 version.

Structure and Participants. Not surprisingly, the goals of the seminar were precisely the twofold goals for graduate students of the College Teaching Committee's FIPSE project: help and improvement in teaching,

and mathematical broadening. The participants were in their second and third years (with one "volunteer" in his fourth year who asked to participate after hearing a few talks). The students had not had a previous TA training course. Our TAs typically conduct classroom sessions in which they go over assigned problems and administer a quiz prepared by the professor who presents the lectures. The seminar met for 90 minutes two times weekly; participants received three semester hours' credit. In a typical session, one of the participants presented an hour lecture on a topic of his or her choice. The topics illustrated concepts a good teacher of undergraduates should know.

Mathematical topics. They were asked to select topics that (1) were accessible to first-year graduate students;

(2) were not a formal part of the Washington University graduate curriculum, and (3) would be interesting to the other participants. The aim was to give the student-lecturer a genuine audience; the other participants were not in general familiar with the results presented, so the lecturer had to get and hold their interest, and teach them some interesting mathematics. Here are some topic examples:

A proof that π is transcendental

An introduction to linear programming with examples

Characteristics of the Cantor set

The Banach-Tarski paradox

Decomposition of rotations in 3-space

Orientable and non-orientable surfaces

An introduction to knot theory

Kepler's laws of motion

The five-color theorem

Evaluation. Each presentation was discussed immediately afterward. The participants made comments, and then the mentor summarized them and made additional points. Each talk had good features that could be praised, and each provided opportunities to discuss improvements. The elements of good teaching the mentor wished to emphasize arose naturally in the course of the discussion of the presentations. After the discussion, each participant filled out an evaluation form. Halfway through the term the group asked to rework the form, which led to an interesting discussion of student evaluations.

Each participant was given a videotape of the participant's own presentation, and also a compilation without attribution of all the participants' comments. The mentor and speaker used these as a basis for discussing ways in which the presentation could be improved.

Summary. At the end of the seminar, students were asked for comments. The consensus was that the format was interesting and helpful for the first half of the semester, but then began to pall. This was largely because this first group contained 21 participants, and that meant listening to a large number of lectures. Now that the doctoral student "backlog" has been worked through, we will have around ten students in the seminar, and will spend more time on teaching technique, educational literature, and related topics.

The 1993 Seminar, Steven G. Krantz. The aim of the course that I taught was to force students to consider for themselves the components and practices of good teaching, and to develop in their own minds a value system for quality teaching practices. Our byword, à la Walter Gropius, was that "God is in the details."

I began the course by having the students introduce themselves, describe their educational background and extemporize on what constitutes good teaching. Several of the students came from institutions where teaching, rather than research, is the primary faculty activity. As a result, they had been exposed to novel and intensive teaching experiences. We all learned from sharing these.

We read the book *How to Teach Mathematics* by Krantz and also *The Art and Craft of Teaching* by Gullette. We spent several class periods discussing issues that were raised in these books: how to prepare, how to field questions, blackboard technique, grading, discipline, racism and sexism in the classroom, pace, exams, and so forth. Students were anxious to filter the ideas in the books through their own sensibilities and to share their thoughts with the class as a whole. This was a productive portion of the course and one during which the members of the class became acquainted with each other.

The first key assignment in the class was that each student was to observe (with permission) an experienced professor teaching. For perhaps the first time, each student sat in a class for one hour without paying attention to the mathematics content but, instead, discerning teaching style and methodology. The students reported back to the seminar on what they had seen. The result was a lively discussion and interchange on what works in the classroom, how to gain and hold students' attention, how to bring a class to life, preparation, tone, and so forth. Throughout the semester students found themselves using the experience of observing an experienced professor as a touchstone for discussions.

I also arranged for each student to work as an apprentice to an experienced professor. Thus, students gained experience writing exams, running problem sessions, grading quizzes, and so forth. Again, this activity provided grist for our mill and gave students in the teaching seminar some hard information and experience on which to base their development of teaching style.

A significant component of the course consisted of having each student prepare two brief (15-20 minute) lectures on elementary mathematical topics, such as the chain rule or multiplication of negative numbers. The point here was to gain experience in (i) explaining and motivating fundamental ideas, (ii) providing good examples, and (iii) managing time in a lecture. The class participated enthusiastically in asking student-like questions and making comments. Each lecture was followed by detailed analysis of every aspect of the lecture: blackboard technique, notation, order of topics, graphics, and the like. Again, students learned first hand that good teaching is made up of many small details. Each student's two lectures were separated by several weeks, presumably so that the students could apply what they had been learning to give a better lecture the second time around.

Sample class projects. One day I brought to class a real examination, taken by a real student, from a course that I taught several years ago. We each graded it, and discussed why and how the grade was determined on each problem. Another class assignment consisted of writing a syllabus for a first-semester calculus course. There were several other assignments that constituted experience in "real life" teaching situations.

We had a spirited discussion, lasting several hours, of math anxiety and the circle of ideas pertaining thereto. We examined the documents *Everybody Counts* and *Moving Beyond Myths*. We considered and discussed the teaching materials and methods disseminated by the KUMON Institute of Chicago. We considered the Saxon method of teaching high-school mathematics.

We also considered at length the question of how much rigor should be used in presenting material to freshmen and sophomores. It tends to be the case that students who were trained in foreign countries, and those from more rigorous backgrounds, have a strong bent for the value of rigor. For others, this is less so. Students in the teaching course found it useful to address this issue both in the abstract and in the course of preparing their mini-lectures.

I maintain that a problem session, if taught properly, places greater demands on the instructor than does giving a lecture. The lecture can be delivered from prepared notes, with few interruptions. In a problem session, the instructor has to be prepared for almost anything, and the session can change course abruptly. I gave the students in the teaching seminar some experience at answering questions on the fly, working through problems blindly, and answering interruptions during the solution of a problem.

Professional responsibilities. An important component of the course, and one that generated considerable student interest, was a discussion of the infrastructure of the mathematics profession. We talked about what constitutes a PhD thesis, how one gets a job, teaching assignments, publishing, refereeing, departmental service, tenure, giving invited lectures, applying for grants, writing books, and so forth. Again, for many students this was the first time that any of these terms were fleshed out, and it provided their first opportunity to ask questions in a relaxed atmosphere. We compared life in an institution that is primarily devoted to undergraduate teaching to life in a research-oriented university. In this part of the course in particular, I was able to take advantage of the diversity of the students' backgrounds. One student was Chinese, and one was an American who had grown up in France. We were therefore able to learn about different approaches to the components of academic life.

During one class we had a guest lecture by Robert McDowell, who discussed his activities as director of the university's teaching center. He also discussed various modern theories of learning and confounded students with puzzles that exhibit the way our thinking processes operate.

Evaluation. I designed a course evaluation form, specifically for this course, to help me to learn how well it met the students' expectations and served their needs. While one student protested the freewheeling nature of the course, most felt that it gave them valuable experience and insights and that it dispelled some of their illusions.

In summary, the purpose of this course was to find activities that allowed students to confront, with as much direct experience as possible, the various components of the teaching profession. Part of the key to success was to conduct the class in a friendly, relaxed atmosphere. As a result, students were not uncomfortable discussing even delicate matters like political correctness, institutional racism and sexism, and the like. In the process, we gained new respect for each other and also respect for the complexity and value of the teaching profession.

Editor's Note: *How to Teach Mathematics* (AMS, 1993) was written by Steven Krantz as part of this project. The Preface is reprinted in Chapter 15.

You're the Professor, What Next?

Part IV

Invited Papers

Chapter 14 – Preparing New Faculty

Chapter 15 – Views and Experiences

These chapters include a variety of essays contributed for this book. Some were written by mathematicians about specific aspects of programs to broaden the training of graduate students or to orient new faculty; others consider the state of the profession more generally and give helpful information for new professors.

Preparing New Faculty

Clifton Corzatt & Paul D. Humke

Mary W. Gray

Diane Herrmann

Herbert E. Kasube

Edith H. Luchins

Richard D. Ringeisen

Pat Shure

Richard Tapia

Janice B. Walker

A Mentored-Teaching Postdoctoral Program in Mathematics *

Clifton Corzatt and Paul D. Humke

Abstract. In January, 1988, St. Olaf College established a mentored teaching postdoctoral program in mathematics. Over the four-year grant period, four interns were hired each for two academic years. Interns worked closely with senior faculty mentors with the aims of becoming master classroom teachers and establishing research programs.

Project Prehistory. In the October 1985 issue of *The Notices of the American Mathematics Society*, the Committee on the Status of the Profession stated:

---

* From the Proceedings of an AMS Special Session at Knoxville, in March 1993, "Interventions to Assure Success: Calculus Through Junior Faculty" organized by Bettye Anne Case.

The academic mathematics community must, in the coming years, maintain the atmosphere and conditions in which [the] dual teacher-scholar role will be nurtured.

Yet, in 1987 when we proposed this project we wrote:

Traditional training in the education of PhD mathematicians includes little if any formal work in learning how to teach. Indeed, the most talented graduate students of mathematics may be given fellowships or research grants which allow them to obtain the PhD without ever having taught a class. ... The result of this system is that new PhD's view teaching ability as marginally important to their profession.

Our purpose was to establish a postdoctoral program which addressed the dual needs of the teacher-scholar. The goals of our program were to:

◇ significantly enhance the teaching skills of new PhD mathematicians who were already considered excellent teachers.

◇ help each postdoc establish a firm foundation for a long and prosperous research career.

An Outline of the Project. With the help of funding from FIPSE, St. Olaf College established a mentored teaching postdoctoral program in January 1988. Over the grant period, four interns were hired; two for the academic period 1988-90, one for 1989-91, and one for 1990-1992. The final-year budget of the fourth intern will be entirely supported by St. Olaf College. The FIPSE Postdoctoral Position, included:

◇ teaching half-time ,

◇ providing sufficient funds for professional travel and library resources,

◇ creating a close professional relationship with a master classroom teacher,

◇ working in a department with an extraordinarily successful mathematics program,

◇ participating in and contributing to a rich professional environment.

We recruited the postdocs in much the same way we recruit any regular position. The response to our advertising was overwhelming and our postdoc positions proved exceptionally attractive.

An Outline of Activities. The postdocs taught half-time. (This means three three-hour courses in the course of the year at St. Olaf.) Of their remaining load, paid for with external funding, about two-thirds was devoted to research activity and about one-third to improvement of teaching. Each postdoc was assigned a mentor who in each instance was a senior professor.

Teaching Activities.

◇ Each postdoc was assigned a college paid student observer. Student observers are trained to observe what is happening in class and to meet with the faculty member weekly to share observations; they are not enrolled in the class.

◇ Videotapes of both postdocs and mentors were made and used as the basis for some lively discussions.

◇ Postdocs and mentors met as a group each week and visited each other's classes. Postdocs were also encouraged to visit each other and other instructor's classes.

◇ During the second year of the project, the postdocs and mentors organized a seminar which focused on educational issues and was open to all members of the department.

◇ The postdocs served on committees, did telegraphic reviews, worked on hiring, gave us advice, and helped organize social events.

◇ During the second year, the three postdocs and their mentors organized a national invitational conference entitled "The New Professor of Mathematics".

Research Activities.

◇ Working closely with their own mentor, the postdocs edited several publishable papers from their dissertations, examined which journals might be appropriate to send them to, and then submitted them. All of the postdocs published their thesis papers and one or two papers beyond their thesis work during their residence at St. Olaf.

◇ Each postdoc (as well as several department members) worked on the design and writing of at least one grant proposal.

◇ Two of the postdocs made good contacts at the University of Minnesota; one with the Geometry Center and the other as a weekly participant (and occasional speaker) in the *Real Analysis Seminar* in the Department of Mathematics.

◇ The postdocs could use their funds to bring research collaborators to campus or to visit other campuses. They did both.

◇ Funding from FIPSE enabled the postdocs to purchase books to establish the beginnings of a personal library.

Job Applications. During the last three years of the program the mentors and postdocs held regularly scheduled meetings focused on the *market*. We discussed what sort of mathematical environment each postdoc would prefer and at which schools one might find such an environment, and we examined several of the schools advertising jobs. All of the postdocs received multiple offers and was able to find a tenure track position at a school they were excited about. All felt that our program enhanced their *marketability*.

Conclusions.

◇ A teaching postdoctoral program in mathematics at a liberal arts institution with a high quality mathematics program is very attractive to a large proportion of the very best of this country's new PhD mathematicians.

◇ A teaching postdoctoral program in mathematics at an institution committed to good teaching is very effective in creating master classroom teachers. The primary vehicles for this are a vibrant mentor-intern relationship and a lively teaching seminar.

◇ Each intern learns a large amount of professional knowhow from a mentored teaching postdoctoral position in mathematics, and this process of learning both affirms and invigorates the existing faculty and staff.

◇ Effective classroom teaching of mathematics entails a complicated symbiosis of personality, technique, philosophy, and a deep understanding of mathematics.

Each new PhD in mathematics is a valuable resource for our mathematical community. The St. Olaf program was an attempt to protect and nurture this resource.

St. Olaf College, Northfield, MN

Producing PhDs *

Mary W. Gray

We frequently deplore that people of color are underrepresented in mathematics, but rarely does anyone make a serious attempt to do anything about it. Those in charge of hiring point to the small number of PhDs that are available; graduate schools note the fall-off from baccalaureate majors to those entering graduate school; those responsible for undergraduate education complain that they are sent ill-prepared students; the high schools say that they see the students too late to make up for years of neglect. In fact, there is plenty of blame to go around, but it is never too late to do what you can where you can. My focus is on how to increase the number of PhDs going to people of color.

There is no magic formula, no easy route to success; as is the case with any worthwhile goal, it takes a certain amount of effort. There are three parts to nurturing the nature that is there if one looks for it.

Recruitment. To recruit people of color it is necessary to go to where they are, in particular to Historically Black Colleges and Universities and to liberal arts colleges and state colleges and universities that have substantial numbers of students of color. These students may not have had as much mathematics as many Ivy League graduates have had, but many are very talented students. Funds for faculty to visit these institutions or for potential students to visit the graduate institutions are helpful but by no means essential. Using mail and e-mail and talking to faculty at meetings can be very effective if real commitment to having people of color in a program is conveyed. Regional MAA meetings are a good place to meet faculty and students from smaller schools. Having current students come to these meetings to give talks provides experience and exposure for the speakers and is a visible sign to those in the audience of the effectiveness of the program.

Another source of promising students is those who did not go straight to graduate school or who dropped out with only a master's degree. They may be found teaching in high schools, in community colleges or four-year colleges; they may be in government or industry. These older students sometimes need more support services, but they bring experience that can be profitably shared with other students. They can be sought out, made to feel wanted, and convinced that success is possible.

Recruitment may also have to include convincing the prospective students that they want a career in mathematics. Demonstrating that mathematicians are not necessarily isolated or arrogant and that our work can be cooperative, useful, and even socially relevant may per-

suade mathematically talented students that they might change the world — or at least have a personally satisfying life — as mathematicians just as effectively as they could as physicians or lawyers.

Above all, it is important to keep in mind that a student stands a better chance of success if that student is not the only person of color in the department. The commitment to recruitment needs to extend beyond mere tokenism, aiming, instead, at attracting a critical mass of students.

Support. Financial support is crucial, for students of color are less likely to be able to call on family resources to supplement their stipends or to cover emergencies. Even institutions with limited resources can often be persuaded to provide special funding. External sources of financial support also are possible. Nationally there are programs such as the US Department of Education Patricia Roberts Harris fellowships. Local businesses, industries, and foundations as well may be willing to provide some supplemental funding. A loan fund for crises; good summer jobs; opportunities to tutor, consult, or teach to increase income are all important — but it is essential to keep in mind that the students are primarily students and to help them limit their other responsibilities.

If there is a co-op education or internship program on campus, it can help with summer or academic year placements. The practical experience can be as valuable as the money. Other support services such as assisting students in forming groups to reduce housing costs or, in the case of single parents, in sharing babysitting chores or costs can help.

A particularly useful program that American University has devised is the creation of two-thirds-time instructorships, at two-thirds of a reasonable salary, for advanced PhD students. In addition to being helpful (financially), the instructorships look good on the students' resumes.

But support needs to take other forms as well. In particular, one can't let students jump in at too high a level. If what they need is some advanced undergraduate courses, they need to be assured that it is all right to start there, though some extra years of support might be necessary. However, much of what these students require is no more than most graduate students require, but, more important, making certain that they receive it.

A friendly department with a feeling of mathematical and social inclusiveness provides extra support for all students. Some care in matching students with effective advisors can help prevent dropouts among those students whom we have been at such pains to recruit. Something as simple as a quick turn-around time in responding to dissertation drafts may reinforce these students to stay with the program. If possible, one can steer the prople of color to those advisors with whom they have the best chance of success.

---

* From the Proceedings of an AMS Special Session at Knoxville, in March 1993, "Interventions to Assure Success: Calculus Through Junior Faculty" organized by Bettye Anne Case.

Persistence. Who has survived graduate school without the ups and downs of spirit? Who has not had moments of self doubt when the creative juices seemed unbearably sluggish? Faculty, but most of all other students or recent graduates, need to be there to share triumphs and setbacks. Getting students through a PhD program is often a cooperative effort.

But persistence is needed not only through receipt of the degree. Making sure the newly minted PhD has an appropriate job, has an idea for the next paper, has been clued in to what makes for success in the new department will assure that mathematics does not lose a carefully nurtured new doctorate. There comes the time when the graduates learn not only that they can make it on their own, but that with success goes the obligation to help along the next generation — to return the mentoring from which they have benefitted. A mark of success is not only that the graduates are doing good mathematics and effective teaching, but that they are themselves running programs to increase the participation of women and people of color in mathematics and science.

American University Washington, D.C.

The College Fellow Program as Professional Development

Diane Herrmann *

The College Fellow Program in the Department of Mathematics at the University of Chicago was initiated in Autumn, 1972. This program was designed as a formal method for preparing graduate students to become effective teachers of undergraduate mathematics, both as lecturers at the University of Chicago and as mathematics instructors in their post-PhD positions. Chicago is a major research institution whose mathematics faculty is primarily engaged in research. Traditionally, the PhD student from Chicago has a research position as a first job, and then a job which combines both research and teaching as a next position.

Each year, there are 25 college fellows and 30-35 lecturers involved in the teaching program at Chicago. The Program is designed as an apprenticeship model rather than as a special course in how to teach or in various teaching methods. The Program is required of all graduate students; while it begins in the student's second year, aspects of the Program continue through all the student's years at Chicago. Virtually all faculty at Chicago have been active participants in the College Fellow Program.

In the College Fellow Program, each second-year graduate student is assigned to a regular faculty member (faculty mentor) who is teaching in an upper-level undergraduate course. The assignment runs for three consecutive quarters. This gives the fellow a chance to relate to mathematics majors and to students interested in the sciences. The fellow is expected to attend class regularly and observe the pedagogical techniques of the mentor. At times this means watching what to do, and at times this means watching what not to do. Fellows hold regular office hours, usually three–four hours each week. They also run weekly problem sessions or hold review sessions before exams. In these informal encounters, the fellow begins to become comfortable with the students, learns a few names and a few faces. This also provides an opportunity for the fellow to hear complaints about the class and to understand exactly how to detect trouble in the classroom. The fellow also assists in constructing and grading exams. This helps the fellow learn about appropriate questions and how to use tests as an effective measure of student performance. By consulting with the mentor on individual student performance, the fellow begins to learn how to accomplish the difficult task of assigning grades to students.

During the course, the mentor and fellow consult frequently on mathematics content, student performance, and difficulties that have arisen in any aspect of the course. Eventually, after a sensible period of observation of the mentor, the fellow prepares and conducts a class of 50 minutes or less under the direct supervision of the mentor. This first lecture usually focuses on the mechanics of good lecture presentation. The mentor typically will remark on aspects of good teaching, like writing clearly, standing away from what's written, talking clearly and with appropriate expression, et cetera. Lectures by the fellow continue through the year. As the mentor and fellow work together, more substantive aspects of teaching are discussed. Attention is paid to effective organization, interaction with the students in the classroom, use of examples and applications. The aim is to help the fellow develop an expository style which promotes good learning in the mathematics classroom.

Throughout the year the college fellow is encouraged to develop a scholarly interest in what is being taught as well as how best to present it. Since all the students in a graduate class are participating in the College Fellow Program at the same time, a camaraderie develops around the teaching and learning of undergraduate mathematics. Several general meetings are held during the year to address these issues. Materials, especially those from the MAA, are made available to the fellows to help them with particular problems. Foreign college fellows are given particular attention so that they might be more comfortable with the American students they will be asked to teach.

Faculty at Chicago have been more than cooperative

* From the Proceedings of an AMS Special Session at the 1992 Joint Mathematics Meetings in Baltimore organized for the AMS-MAA-SIAM Joint Committee on Preparation for College Teaching by Bettye Anne Case.

in participating in the Program. In fact, faculty actually request to have a college fellow when they teach an undergraduate course. While teaching is not formally a part of tenure or promotion at Chicago, it is clear that the faculty take the teaching of undergraduate mathematics quite seriously. The climate is one where teaching is evidently an important part of what the faculty do. At the end of the year of participating in the College Fellow Program, the mentor is asked to give a recommendation about the advisability of appointing the fellow to the position of lecturer for the next academic year. Each college fellow must also provide the dean of the college and the provost with a curriculum vitae. Preparing a C.V. is a nontrivial task for some of the fellows, and this gives them a chance to work on such a document before they actually have to apply for a "real" job. In addition, the formality of the appointment procedure impresses upon the fellows the professional nature of what they are doing.

Lecturers at Chicago are "stand-alone" teachers in multi-sectioned elementary-level mathematics courses. At the beginning of the academic year, the lecturers attend two or three meetings in which the general policies of the Department of Mathematics are discussed. Content and syllabus of each lecturer's specific course are analyzed in some detail. An overview of the entire undergraduate curriculum is included. This gives the lecturers an idea of the interdependence of the curriculum and their responsibility in preparing their students to go on to the next level on mathematics.

During the year, meetings are held to address specific topics such as examinations and grading, supervising support staff like tutors or graders, and how to advise students about taking additional mathematics courses. These meetings are often attended by college fellows, lecturers, and junior faculty new at teaching in the college. The University is also a source of information and services about teaching. There are opportunities for graduate students to participate in campus-wide discussions of the nature of the College at Chicago, the mind of the learner, the virtues of the lecture versus the discussion method of teaching, and the academic job search process.

The importance of the evaluation process is emphasized in several ways. In the Department, each college fellow is evaluated each quarter, and a written progress report is submitted by the faculty mentor. Once each year, lecturers are visited by a senior faculty member, usually the student's thesis adviser, who consults with the lecturer concerning the classroom performance, and provides advice and guidance about possible improvements. Written reports from these faculty visits provide the basis for the letter of recommendation about the graduate student's teaching when it is time to apply for a job. There are also written course evaluations published by the college; if a graduate student is not performing up to standards, we know about it.

Teachers who do a good job are rewarded by the Department, the Division, and the college. There are monetary awards for excellence in effective and responsible teaching. Students nominate their good undergraduate teachers for these awards, and students from Mathematics are always included among the winners.

In addition to these activities which directly support the day-to-day teaching by the graduate students, there are several other activities in the Department that enforce the idea that teaching is important at Chicago. Some of the elementary courses include tutors as part of the regular staffing of the course. Lecturers who work in these courses have an opportunity to learn to supervise assistants. There is an active student chapter of the MAA at Chicago. The meetings of this group sometimes include expository lectures by graduate students. Here, some graduate students take advantage of the opportunity to interact with the undergraduate mathematics majors. There are also several campus-wide activities which focus on underrepresented groups in mathematics. Graduate students have been involved in teaching in these different programs, which range from a program for entering minority students in the college to a program for local community college students.

There is also active and visible support within the Department for the wider educational community. The NSF Young Scholars Program for mathematically talented seventh- through twelfth-graders that ordinarily runs in the summer has an academic year component that meets early on Saturday mornings. More than 25 graduate students participate as seminar leaders in this program each quarter. There are also seminars for teachers of secondary school mathematics, and graduate students participate actively in these.

Disciplinary issues like calculus reform provide an opportunity for discussions led by mathematicians and educators from inside and outside the department. Some topics for these discussions have been: using writing in the curriculum; using computer technology, like **Mathematica**, in courses; and the role of placement-testing in a mathematics department.

The extensive interaction around teaching in the department has prompted some innovative activities on the part of the graduate students themselves. One of the most popular weekly events is the "pizza" seminar. This is completely student-run, and the graduate students take turns presenting their work to one another. These talks are open to all levels of graduate students, so there is a demand for clear exposition. No faculty are allowed at these talks. Another idea that is very popular with our calculus students is the presentation each fall of the "Uncle Vanya Proof-in-a-Minute Method," which presents a simple way of approaching epsilon-delta limit proofs. Graduate students organize this presentation, which regularly draws crowds of 250 calculus students in a single lecture.

University of Chicago, Chicago, IL

The Preparation for College-Teaching of Mathematics:
A Personal Experience *

Herbert E. Kasube

My undergraduate training was at MacMurray College, a small liberal arts college in central Illinois. The emphasis of the faculty was on excellent teaching. It was here that I decided to pursue a career in college-teaching of mathematics. The faculty's commitment to quality undergraduate instruction had a profound effect on me.

While attending graduate school in the Mathematics Department at the University of Illinois at Urbana-Champaign, I received my master's degree. In December of 1974, an article appeared in the American Mathematical Monthly [1] that caught my attention. In it, Rick Billstein described a unique PhD program in mathematics at the University of Montana. This program had its initial funding from the National Science Foundation, and the focus was on the training of future college teachers of mathematics. I decided to leave the U of I and head west for Missoula, Montana. The remainder of this paper describes the special aspects of the program at the University of Montana and how I feel they have influenced me in my career.

In planning the PhD program in the mathematical sciences at the University of Montana, breadth of study was seen as a key element. In preparing a student to teach at a small college or university, this breadth proves essential. In such an environment we can not expect to teach only in our area of specialization. Versatility is a key. With this in mind, in addition to requirements in wide areas of mathematics, each student was required to have an minor outside of "traditional" mathematical subjects. I chose to have two minors — one in computer science and another in statistics. This component of my program served me well after I began my job at Bradley University. My department chair was looking for individuals within the department who could assist with our department's growing need to offer statistics courses. Even though my "specialty" had been algebraic number theory, my minor in statistics prepared me to teach our undergraduate statistics courses.

Students were also required to take a "current topics" course. The purpose of this requirement was to expose students to some contemporary topic in mathematics that may or may not fit into the traditional curriculum. I took a graduate-level course in topology that was taught with an emphasis on category theory. This

was not the typical approach to topology and offered students a chance to study a subject in a nontraditional way.

We were also required to take a seminar in the history of mathematics that was spread out over two quarters. The first quarter was spent researching a paper on some selected area of the history of mathematics. Topics had to be approved by the professor in charge of the course and had to be more than simply a chronology of mathematical events. For example, if we wanted to write about a particular mathematician, our paper had to say more than "They were born, did mathematics and died." The mathematician's influence on mathematics and other mathematicians had to be discussed. During the second quarter of this seminar students presented these papers in a colloquium setting to the entire department. In reading current recommendations [2] we see that the history of mathematics should be part of any successful graduate program in mathematics. I have continued this idea of a seminar in the history of mathematics in my own teaching. Our department does not have a formal course in the history of mathematics, but, when I have taught the mathematics seminar required of all mathematics majors, I have used a format similar to that from graduate school (scaled down to the undergraduate level). This approach has proved quite successful (see [4]) .

Most graduate teaching assistants never get an opportunity to teach a mathematics course beyond calculus (some never do more than recitation sessions). We were required to "intern" by teaching an upper-level mathematics course for one quarter.

This was under the supervision of a faculty member and proved quite enlightening. Dealing with post-calculus students can be quite different from dealing with pre-calculus students. While some [2] recommend "externships" for students going into industry, this internship served the same purpose for prospective college teachers of mathematics.

Another special requirement was enrollment in the "College Teaching Seminar". This one-quarter course had been the focus of Billstein's article [1]. This seminar was team-taught by two or three faculty members who shared their expertise with the graduate students in a way that was quite unique and somewhat informal. Time was spent reviewing and discussing exams that we had each designed. We would ask: how well were problems selected or worded? This peer review offered insights that proved very useful. One "assignment" was to develop multiple-choice questions for a calculus exam. In hindsight, this has proved to be quite helpful to me since I have been involved in giving a multiple-choice final exam in my business calculus course at Bradley. I feel that everyone involved (maybe even the faculty members) became better at making up an exam after this experience. Another eye-opening experience during this course was the videotaping and subsequent discus-

* From the Proceedings of an AMS Special Session at the 1992 Joint Mathematics Meetings in Baltimore organized for the AMS-MAA-SIAM Joint Committee on Preparation for College Teaching by Bettye Anne Case.

sion (critique) of one of the classes that one was teaching during that quarter. If one has never seen oneself in front of a class, it can be an interesting experience.

We also spent some time evaluating textbooks. In particular, we looked at calculus texts. We were reminded that, when we started our jobs, we would not always be handed the text to use by someone else. We would have to decide for ourselves. This discussion of textbook selection has certainly helped me. Not only have I been involved in textbook selection for my courses, I have also reviewed textbooks at various stages of production for numerous publishers. We were also prepared for the various professional demands outside of the classroom. We discussed advising of students, university and departmental committee responsibilities, research and creative production, and where to look for outside grants for research and/or educational purposes. I am currently very actively involved in the Illinois Council of Teachers of Mathematics, the Mathematical Association of America, and numerous University committees.

The final component of the program that should be mentioned is the dissertation. As stated in the Department's literature, this dissertation "must constitute an original contribution to mathematical scholarship."[3] . The exposition contained within this paper must be of the highest quality. Strictly pedagogical topics or undergraduate textbooks were not permitted. This was to be a PhD in mathematics . The paper might be an historical study or might present the mathematical content of some interdisciplinary subject. My particular choice was a study of the class number of an algebraic number field [5] . I felt that my advisor (Dr. Merle Manis) and my dissertation committee helped me become a better mathematical writer.

What about the success of this program? My situation was certainly not unique. Of the people who were graduated from the program at the University of Montana in the last 15 years or so (of whom I know), most are tenured or tenure-track faculty at numerous four-year institutions from Spokane, Washington, to Farmington, Maine, and Bogota, Columbia, to Baghdad, Iraq. At least four act as department chairs.

What else was so special about the program at the University of Montana? I think that there was a certain rapport between faculty and graduate students that made the graduate students feel like colleagues. As we moved on to our own jobs, the transition to a faculty position seemed less extreme. In January of 1991, at the Joint AMS/MAA Meetings in San Francisco, California, a session was held to discuss the preparation of college teachers of mathematics. In discussing rankings (ratings) of graduate department of mathematics, it was mentioned that the "traditional" criteria were based on research publication by the department. It was suggested by someone on the panel that the undergraduate textbook production within a department might be a good barometer of the quality of teaching done there and

consequently of the quality of teachers produced by the program. Of the approximately twenty mathematics faculty at the University of Montana during my stay, eight had written undergraduate textbooks, and five of these had written more than one book. This is a record to be envied. I am now co-author of two undergraduate mathematics textbooks, and it is fair to say that the urge to be involved in such an activity was nurtured at the University of Montana. The journal for the Rocky Mountain Mathematics Consortium was also based in the Department. It should also be pointed out that there was quite a bit of traditional mathematical research being done in the Department as well.

All in all, I would not trade the training that I received in the Mathematics Department at the University of Montana for anything. It has served me well, and I hope that I continue to serve it equally well.

Bibliography

[1] Billstein, Richard, A College Teaching Course for Future PhDs in Mathematics, *American Mathematical Monthly*, V. 82 , No. 10, December, 1974.

[2] Board on Mathematical Sciences, *Actions for Renewing U.S. Mathematical Sciences Departments*, National Research Council, Washington, DC , 1990.

[3] Department of Mathematics, University of Montana, *Guide to Graduate Programs in Mathematical Sciences at the University of Montana*, 1990.

[4] Kasube, Herbert E., The history of mathematics as taught in a seminar. *PRIMUS*, v. 1, No. 4, December, 1991.

[5] Kasube, Herbert E., The class number of an algebraic number field: its interpretation and use, PhD dissertation, University of Montana, 1979.

Bradley University, Peoria, IL

Preparation for College Mathematics Teaching:
The West Point Model from a Visiting Professor's Perspective *

Edith H. Luchins

I am enthusiastic about the Mathematics Faculty Development Program at the United States Military Academy (USMA) in West Point and about its implications for preparing college teachers. As the 1991-92 Visiting Professor of Mathematics, I participated in a six-week summer program for new staff. My subsequent

* From the Proceedings of an AMS Special Session at the 1992 Joint Mathematics Meetings in Baltimore organized for the AMS-MAA-SIAM Joint Committee on Preparation for College Teaching by Bettye Anne Case.

observations of these new instructors' teaching during the academic year pointed to the success of both the program and the goals for faculty development. My enthusiasm is shared by other observers; e.g., a panelist at the First USMA Mathematics Education Conference in September 1991 noted that West Point's Department of Mathematical Sciences has done more than any other math department in the nation in articulating faculty development goals.

The summer program is known as the Faculty Development Workshop I. It is numbered "I" because it is the first (and the longest) of four or five workshops for junior faculty in which senior faculty also participate. The workshops cover preparation for teaching the four core mathematics courses, which total 15 credits, and are required of all students in the Academy.

Technology has been in use for instruction since the Academy started : the first systematic use of chalk and blackboard in 1801, the first American math textbook in 1820, geometric string models and the slide rule introduced in 1845, the overhead projector and mechanical computer in 1947, an electronic computer in 1955, a calculator for every student in 1975, a personal computer for every student in 1985, and a computer algebra system introduced in 1989.

The Academy is in the forefront in the use of computational tools: 4,400 students each with an IBM 286 or 386, 700 faculty each with a similar PC, and 2,400 HP-28S calculators (for freshmen and sophomores — or, rather, plebes and yearlings). The Department of Mathematical Sciences has more than 50 mobile computers on carts with overhead screens for demonstrations in class, five lap-top computers for demonstrations, and computer labs, including a dedicated `Mathematica` lab. The whole campus is tied together by an electronic network, with e-mail regularly sent between teachers and students.

Rotational Faculty: Advantages and Disadvantages. The new junior staff members — about 16 to 19 each year — are part of what are known as the rotational (versus permanent) faculty, who come to the Academy for three years. Most of them are graduates of USMA, and following graduation, each has been an officer in the armed forces for eight to ten years. Most have spent the last two years working for their master's degree in an applied mathematical discipline at Rensselaer Polytechnic Institute, Georgia Institute of Technology, or the Naval Postgraduate School. These three are known as the Foundation Schools because they lay the graduate foundation for the core courses that will be taught by the rotational faculty. (Ulysses S. Grant was turned down as a rotational faculty member in the mathematics department but Omar Bradley served as one.)

Advantages of having rotational faculty are: they have current degrees; they have been successful as Army unit commanders, problem solvers, and decision makers; they are experienced in organization, execution and planning; they focus on teaching with enthusiasm, energy, and dedication; they identify with the cadets; and they serve as successful role models for these future Army officers. Moreover, their experiences provide applied mathematical problems for the classroom. Furthermore, their large numbers and presence make it possible to have small sections of no more than 18 students.

Disadvantages are the relatively low prior teaching experience, education usually at the master's rather than at the PhD level, and a potential for a lack of continuity as rotational faculty come and go. The department seeks to ensure continuity, by encouraging the interactive teaching made possible by small sections, by conducting faculty development workshops, and by continuing evaluation and support throughout the rotational period. Also, the most promising among them are invited to follow the rotational period with doctoral study and research and to return to USMA as permanent faculty members.

The Workshops: An Overview. Much planning goes into the workshops. A guiding objective is to make the new instructor and the instructor's family feel that they are part of the West Point mathematical community. In this highly technical, non-civilian environment, the personal touch and civility are central. The cordiality contrasts with the loneliness that some new TAs and new faculty members and their families complain about in civilian colleges. Each new instructor is assigned a sponsor, a second- or third-year instructor, who will make the transition smoother from graduate school to West Point. In April, the new instructors are sent letters of welcome with the names of their sponsors, a calendar of events, and a map to the home of the department head (Colonel Frank R. Giordano) for a kickoff picnic.

The social activities, all of them open to the families, were planned as early as February: in 1991 they included the picnic, a boatride, a cookout, bowling, a luncheon, and finally, a square dance after graduation from the summer workshop. The Superintendent's Change of Command, unplanned by the department, was very colorful.

The new instructors also were sent the course notes for the novel entry-level course Discrete Dynamical Systems (DDS), the textbook (by James Sandefur), as well as course notes for Calculus II. Most of the new instructors would teach DDS, but volunteers would be needed to teach the off-season Calculus II, and they were be selected from those with the most experience as TAs.

The six week summer workshop focuses on DDS introduced two years ago (and now more appropriately called Dynamical Systems, since the study of non-discrete systems has been added). The first workshop introduces the concepts of proportions, recurrences, dynamical systems, stability, difference equations, first-order and second order linear and non-linear equations, and systems of equations. Also considered are the necessary mathematical tools, e.g., matrix algebra, as well as the computational tools. The later part of the course introduces the limit concept and makes the transition from discrete to continuous dynamical systems and from difference to differential equations.

A two-week workshop during intersession prepares the instructors for the next course, Calculus I. Shorter workshops during intersession deal with Calculus II and Probability and Statistics. The programmable calculator HP-28S, the microcomputer, the spreadsheet QUATTRO PRO, and the symbolic computer algebra program DERIVE, are introduced in the first workshop and used thereafter in the core courses and other workshops. In addition, the statistical package MINITAB is introduced in the fourth workshop.

The summer workshop acquaints the new staff members with the mission of the department:

1. To prepare and offer mathematical sciences courses in support of the core curriculum and other programs.

2. To provide for the professional growth and development of its officers.

3. To facilitate the intellectual growth and development of the students.

The student growth model is interwoven with, and evaluated alongside, five educational threads that run through the core curriculum:

Scientific computing

History of mathematics

Writing in mathematics

Mathematical modeling

Mathematical reasoning.

The student goals are to learn the following processes:

◇ To solve problems with confidence and aggressiveness;

◇ To model mathematically within the discrete and continuous, linear and non-linear, and deterministic and stochastic realms;

◇ To use scholarly habits toward progressive student independence and life-long learning;

◇ To apply mathematics to real-world problems.

The student goals are also to acquire the following skills:

◇ Be able to communicate mathematics orally and in writing;

◇ Be able to use computers and calculators as tools for learning and problem-solving.

The role of the instructor is to be a motivator, a facilitator, and a developer, and to provide opportunities for student participation and intellectual growth.

Interactive teaching principles that are highlighted involve:

◇ Active classroom participation;

◇ Daily preparation by students;

◇ Frequent feedback to students;

◇ High performance required of students;

◇ Availability of additional instruction.

Illustrated are model lessons, with suggestions on what to do and what not to do. New staff members have the opportunities to work in teams with more experienced instructors, to prepare lessons, to teach, and to be critically evaluated by supervisors, peers, and others. The new staff members learn alternatives to the lecture method, including group problem-solving, conducting group and individual projects, how to encourage oral and written presentations, and how to assess and evaluate performance, e.g., individual and group board work and projects.

Projects play a very important role in all the math courses at West Point, e.g., the students in Calculus I do three major projects. Students recite frequently, make presentations to the class and to visitors, and have regular writing assignments (e.g., an essay every week) in every math core class, with an essay included on each test.

The Summer Workshop: Zooming In. Let me give a week-by-week account of the summer workshop. Following the previous day's kick-off picnic, the first official day began with a welcome in the department lounge in which all new staff members received a math department coffee cup, personalized with their own names. (This personal touch that permeates the program extends to the cadets. For example, each of the math majors has a personalized cup available in the lounge.) The new instructors briefly talked about themselves and their families. The department head gave an overview of math at USMA. Then the deputy department head spoke on the history of USMA and of the mathematics department as well as on how to integrate history into the teaching of mathematics.

The remainder of the first week was devoted to administrative matters, to a tour of the main library and the learning skills lab, and to a seminar on the department's teaching philosophy. There was a discussion of hardware-software issues, and of how to integrate the computer thread into the core curriculum. After the software was uploaded in the instructors' computers, computer training began with learning to use DOS, e-mail, and UNIX.

The second week was heavy on training in technology, in the context of the course, using the HP-28S calculator, MS/Word/Windows, and the QUATTRO PRO spreadsheet. DERIVE and the DERIVE study guide were studied in the computer lab. The technological tools were applied to calculator and computer problems and projects. An important topic was the nature of mathematical modeling and how to incorporate the modeling thread into the core curriculum.

The third week began with a two-and-a-half day seminar on the contents and techniques of MA103 held at a higher level than that on which it is taught. A second session on the teaching philosophy was followed by a seminar on how to incorporate the writing thread: how to write in core mathematics and how to evaluate the writing. A seminar on math reasoning processes showed the difficulties in incorporating this thread and in assessing and evaluating student growth in reasoning.

There were demonstration classes by second- and third-year instructors as well as discussions and illustrations of classroom procedures, e.g., hands-on projects. There were opportunities to learn how to use the mobile computers, the overhead screens, and other devices, and how to integrate the symbolic, numerical, and visual features into class and out-of-class activities. The new instructors then split into two groups. Working individually and in teams, they began preparations of lessons and gave presentations which were critiqued by their supervisors, peers, and visitors. Videotaping was available if requested by the new instructor.

The fourth week involved more demonstration and practice classes and critiques. There were briefings of Calculus I and Calculus II. A small group of five volunteers split off to focus on the latter with lesson plans, practice classes, and critiques, because they would be teaching this course in the fall. The whole group discussed problems in and methods of evaluation and assessment and submitted written critiques of the Faculty Development Workshop I.

In the fifth week, several hours were spent discussing the workshop critiques. A seminar was led by the department's Director of Faculty Development and Research on opportunities in these areas. (It is significant that such a position exists.) By the end of the fifth week, computer and calculator projects were completed and discussed. Teams prepared block reviews (which are held biweekly during the academic year). Practice classes continued with critiques during this week and the final week.

The sixth and final week included more time on demonstration of classroom procedures. The requirements for mathematics as a major and as a field of study (or minor) were discussed. To dramatize what was involved in preparing and evaluating a one-hour test, the participants were given such a test in DDS and they discussed it. An overall critique of the workshop concluded the session. Its symbolic conclusion occurred at a graduation marked by the award of certificates of accomplishment and celebrated with a square dance.

Among the critiques was the complaint that during the demonstration and practice classes new instructors took on the roles of cadets, going to the boards and asking questions that they thought cadets might ask. Although intriguing at first, it became irksome. Accordingly, the possibility is being explored of having cadets or high school or college teachers or students participate. Perhaps because of the "play acting" feature, some instructors asked for fewer practice sessions, or, at least, for not being regarded as cadets during the sessions. They asked for more demonstrations and more block reviews of the materials. Another request was for earlier and more frequent use of teams for preparation of actual lesson plans to be used in the fall. They also suggested more self-paced training, e.g., for computers. These requests have been incorporated into the workshop scheduled for Summer 1992.

Other summers also are devoted to instructor development. Following the Faculty Development Workshop I in Summer 1, Summer 2 offers eight weeks of research opportunities, usually in army agencies or labs. Summer 3 is spent in six weeks of military training of cadets and field work. In Summer 4 six weeks is spent teaching in the Summer Term Academic Program, primarily for cadets who retake courses, but also for enrichment experiences.

Summary. The strengths of the USMA Faculty Development Program include the following: it is an on-going program; it is characterized by a personal touch and concern for families; it contributed to articulation of faculty development and student growth goals and it made instructors aware of them; it contributed to formulation of educational threads running through the curriculum and made instructors aware of them; it familiarized instructors with the Academy and with the department's educational philosophy; it brought pedagogical issues to the instructors' attention; and it immersed instructors in course materials and technologies their students would be using.

Concluding Remarks. Do West Point's faculty development and student growth programs have implications for preparing faculty and TAs in non-military schools that lack the uniformity and uniforms, the computer for each student and faculty member, or the time and money for six weeks of preparation? Many of the features and guiding principles of the USMA model can be incorporated into shorter courses, e.g., during the summer or in intersessions. We can no longer accept the myth that, although pre-college teachers need extensive training and supervised practice teaching, college teachers are born and need no preparation for teaching.

Acknowledgement: Thanks are offered to Colonel Frank R. Giordano and to Lieutenant Colonel Richard D. West for reading a draft of this paper and for their helpful comments.

Rensselaer Polytechnic Institute, Troy, NY

Support for
PhD Teaching Projects *

Richard D. Ringeisen

Introduction. As both a charter member of the joint committee on preparation for college mathematics teaching and the department chair where there is a trial program, I have a keen interest in seeing trial programs succeed as well as a perspective on how we all hoped that might occur. I intend here to share some thoughts about successfully implementing such programs, particularly from the point of view of obtaining support for them at your university.

I will discuss two different, but equally important, kinds of support for these programs and will try to give the reader some hints as to how to be successful in both categories. First, of course, is financial support, for without that there could be no such program. On the other hand, spending the entire department budget will not make the programs successful in the way the committee (and through it the mathematical community) has always hoped, unless it is supported by the department's faculty and PhD students. So I will offer advice for obtaining each kind of support.

It may help the reader to know that we have an experimental program underway at Clemson and that we hope to make it an integral part of our PhD program. It is very nicely and carefully described in a paper by our site mentor, Joel Brawley, in this volume. You are encouraged to give that a look because particulars on what happens in our professional seminar will not be given here, except where necessary for the discussion. One also needs to know that I am department head at Clemson, and that I spend much of my time seeking support for various things.

Financial Support. Financial support at any university is difficult to obtain for nearly anything these days, and thus the department itself will almost certainly need to make some contributions. Some of those contributions are real money, and others are not quite so real. The first and perhaps most important thing to do for a successful seminar experience is to make it a "real course," wherein the students receive academic credit and the professor receives teaching load credit. This is very important not only because it is fair and correct, but because of the message it sends to students, faculty, and the upper administration. You are saying, in a manner, "This is an important activity for this department and this university." Furthermore, without credit for the students, one should give up now on any hope of making it a regular

---

* From the Proceedings of an AMS Special Session at the 1992 Joint Mathematics Meetings in Baltimore organized for the AMS-MAA-SIAM Joint Committee on Preparation for College Teaching by Bettye Anne Case.

part of the PhD program. Now load credit, of course, is not free. On the other hand it is not that expensive either, if handled carefully.

How does one go about covering this extra course? The most inexpensive way is to offer it in place of another graduate course once a year. However, this could lead to faculty resentment in one form or another (i.e., whose favorite course is not offered to make room for the extra course?). Replacing an undergraduate course or offering one less undergraduate course isn't going to be popular with the upper administration, so that the idea has some weaknesses there as well. Most departments have at their disposal some discretionary money, often in the form of lapsed salary funds for leaves or grant release funds, or perhaps a relatively flexible TA budget. In the former case, covering the courses released can almost always be done with less expense than the salary vacated and thus "squeezing" an extra course out isn't too difficult. In the latter case, upper level PhD students (such as those exiting the professional seminar!) can be assigned an extra course for some extra pay. At any rate, the important thing is to realize that real load credit does cost something. This is a good point to make when explaining to the dean how important you consider the project to be to the university and the profession!

Another minor but important expense comes by way of travel and equipment/supply cost. The amount of national attention drawn to the teaching aspects of our professorial positions has increased the number of forums for discussion and presentation. It is a good idea to have your proseminar person attend as many of these as possible, and that may cost some extra travel funds. (Later in this article I describe an ideal professor for such a course, and such persons are likely already to be traveling relative to their own mathematical research.) As described in many other papers in this volume, there is a growing literature on the subject of college mathematics teaching, and you are likely to need to purchase some books to be available to the student participants. Many of the pilot sites, including Clemson, use some videotaping as part of the process, and so one might need to make a departmental investment in such equipment as well.

Obtaining university support from outside the department involves looking carefully at what kinds of funds are sometimes made available as well as knowing what impresses the upper administration. There is a definite local advantage to such projects, and that is the strongest point which needs to be made when seeking funds. Although these professional seminars are usually aimed at upper level (say, post-preliminary-exams) graduate students, it is nevertheless true that many of these students will continue as teaching assistants at their doctoral universities for a few more years. Thus you are improving the quality of instruction for the university's undergraduates. Nothing sells these days at any university as well as the words just written. Besides, they are true! One should also note the national attention that can be drawn to your university for being involved in such a project. Oh, and for those of us at state-assisted universities (which often were state "supported" until recently), the legislators simply love to hear about innovations at the undergraduate level, and are equally fond of national attention.

So, assuming that one has made all of these salient arguments, exactly where does the money come from? Many universities have special funds available for innovative programs relative to teaching. It was precisely such funds which we were able to tap at Clemson, through a proposal by our site mentor which survived an internal review process. Remember we are never talking about a great deal of money, so that the projects are ideal for these sort of small internal university grants. In addition, deans and provosts sometimes have discretionary funds at their disposal, and they often like to spend them on very visible projects. In short, obtaining funds external to the department depends on a good job of letting folks know what you are doing and allowing (in fact, encouraging) other segments of the university to share in the credit. It can be done and is being done at several of the project pilot sites.

Moral Support. There is probably nothing more important to the success of such a project than to have exactly the proper person running it. I believe it to be absolutely essential that the person chosen to lead the pilot project be one of the department's very best all-around faculty members. Joel Brawley at Clemson probably still suffers some pain from the severe twisting his arm endured when this department head discovered we might have an opportunity to try out such a project. Nonetheless, the expectations for success which our project enjoyed from the beginning and the very positive manner in which it continues to be viewed by our faculty and graduate students alike is in part due to the reputation of this faculty member. If at all possible this project person should be a quality researcher; for, after all, we are talking about advanced PhD students enrolling in the course and about convincing a department that it is a suitable part of a "real" PhD program. Thus, the project person's research reputation is important. However, it goes without saying that the person in charge must be extremely well respected for that individual's teaching abilities. In our case we have an alumni professor (chosen by the university, with special concern for high quality teaching) who also is very highly regarded in his research field, and is well respected as a colleague. If you start with such a faculty member, then the likelihood for departmental support is greatly enhanced.

To obtain general departmental faculty support it is very important to have everyone informed about the project from the beginning. Perhaps a colloquium could be devoted to a visitor who is involved in a teaching project elsewhere. That could be used to spur conversations around the coffee lounge and to begin building interest. Nearly every department has some sort of graduate curriculum committee, and it is important that this group be consulted often and early about the seminar course, particularly its goals and successes. If your end result is to have the course become a regular part of the curriculum, perhaps even required, then such departmental communication is critical while the course is still an experiment. It is interesting to note that Clemson had several faculty members, including journal editors, outstanding teachers, and even lowly department heads actually meeting with the students at times during the course. We now have considerable interest from other faculty members about involvement next fall when the course is offered again.

Another constituency one dare not forget to court is the graduate-student population itself. These are very busy young people, concerned about their graduate study, their current teaching responsibilities, and just keeping up from day to day. Ideally, they would want to take this course, and we would like them to want to do so. If you have strong faculty support, then part of this latter battle is already won. When your PhD advisor tells you it looks like a good thing to do, then you are very likely to think so yourself! Those students who are headed for academic careers should know that such a course "looks good" to potential employers. (In point of fact, every industrial employer with whom I have spoken has heartily endorsed the idea as well.) All of our students who took the course last fall had a special sheet to insert with their own vita wherein the department head described the course and what it has done for the student's teaching experience. This seems to have worked quite well. It is fair to say that our students found the seminar exciting. In fact, we have no problems with enrollment for the next offering at all, probably very closely correlated with what this year's students had to say about it. It could be that the course is especially helpful and attractive to foreign graduate students, both because they learn a great deal and because it does look especially good to potential employers. We have very few foreign students here, and so that discussion would best be carried on by people at some of the other sites.

Summary. In summary, it should be said that these comments were certainly not made with the feeling that I know something that no one else does. Rather, these are just helpful hints that seem to be transferable to any institution. In short, the program is relatively inexpensive, the students and faculty genuinely enjoy the experience, and it is viewed as an important part of the PhD program. Putting the course together is real work and should be regarded as such, and that is an important point. However, gaining support isn't necessarily difficult, and the payoff to the mathematical profession could be substantial.

Old Dominion University, Norfolk, VA
(formerly at Clemson University, Clemson, SC)

Newly-Hired Faculty at a Large Research University:
Setting the Stage for Success *

Pat Shure

At the University of Michigan Math Department we recently designed a professional development program for newly-hired faculty. Our Department wanted to emphasize the importance of successful undergraduate teaching and needed, at the same time, to retrain our teachers to implement calculus reform. Thus we were led to examine the difficulties faced by new PhDs who are trying to establish themselves both as researchers and as teachers. In a large department, junior faculty sometimes find themselves feeling isolated and uncertain. They worry about finding enough time to do all the things they have to do. We set up a program to give our newly-hired faculty immediate companionship and to help them become as comfortable as possible with teaching.

All Math Department newcomers, both newly-hired faculty and new teaching assistants, were required to take the same one-week workshop just prior to the fall term. The official topics were generally instructional: interactive lecturing, cooperative learning, team homework groups, graphing calculators, et cetera. However, we also had other more personal goals; we hoped to:

◇ get them to like the students and get them to focus on the students' feelings, rather than their own.

◇ get them in the habit of asking questions instead of pretending that they know everything.

◇ start friendships among the instructors.

◇ demonstrate that the senior faculty is committed to good teaching.

We wanted to change their view of effective teaching. At the beginning of the training program the new instructors believed that the primary element of good teaching was didactic — to present the theory in logical progression, thoroughly covering all details. In order to develop as instructors, they needed to shift their attention to the needs of the students. The students wanted instructors who explained things clearly and whose approach was supportive and inspiring.

When classes began, the new teachers were each assigned to one of the elementary courses. The courses were deliberately organized to simplify the teachers' job. Teachers had access to course-wide assignments, teaching notes, and sample exams. Their students took Departmental exams and were graded along course guidelines. There were weekly course meetings to discuss what to teach and how to teach it. This amount of structure

and feedback was useful, since many of the new PhDs had little experience being fully responsible for a class.

Apart from their role as teachers, it was also crucial to ensure that the new people were absorbed into the research life of the department. Before any newcomer was hired, a senior faculty member agreed to spend time with that person. These senior mentors helped the new PhDs establish research contacts, visited the classrooms, and gave additional teaching advice.

Our overall goal was to get the newly-hired faculty established and comfortable both as teachers and as researchers.

University of Michigan, Ann Arbor, MI

Nurturing Women and Minority Graduate Students*

Richard Tapia[†]

The Rice Department of Computational and Applied Mathematics is recognized for enrolling and graduating a significant number of women and underrepresented minorities in its PhD programs.

The history of this involvement traces interventions and retention issues that have played a critical role in this success.

Among these are:

◇ adequate institutional commitment signified by funding;

◇ implementation by a faculty member who has both adequate program authority for effectiveness and real empathy with the students;

◇ direct and personal involvement of most faculty members with graduate students;

◇ accurate perception of the special needs of students from underrepresented minority groups (e.g., students with responsibilities and family commitments; students with less confidence and experience with unwritten academic requirements and conventions);

◇ sufficient flexibility in departmental structure to address a broad spectrum of needs;

◇ recruiting a critical mass of women students and students from underrepresented groups.

Rice University, Houston, TX

---

* From the Proceedings of an AMS Special Session at Knoxville, in March 1993, "Interventions to Assure Success: Calculus Through Junior Faculty" organized by Bettye Anne Case.

[†] An article in *SIAM News* 24:3 (May 1991) summarizes some of Richard Tapia's work with intervention programs, and notes: "He rarely stops to write about his work. . ." The short abstract of his Special Session talk is followed by this article.

Success at Michigan for African-Americans in Graduate Mathematics Programs *

Janice B. Walker

I became interested in the master's program in mathematics at the University of Michigan during my junior year at Tuskegee Institute (now Tuskegee University) in Alabama. This interest was sparked by two doctoral candidates from the University of Michigan who were teaching mathematics part-time at Tuskegee in the Michigan-Tuskegee Exchange Program. They were Bruce Palka and John Quine, who are professors of mathematics at the University of Texas at Austin and at Florida State University respectively. They encouraged me to apply to the graduate school at Michigan and helped me gain the confidence needed to pursue a graduate degree. At the end of the school year, each sent a letter of recommendation to the University of Michigan on my behalf.

During the fall of my senior year, I applied to several prestigious graduate mathematics programs, such as the University of California at Berkeley, the University of Chicago, and the University of Michigan. I was accepted and awarded financial support at each of these schools. The chair of the mathematics department at Tuskegee had attended Michigan for graduate work, and several former Tuskegee students were in the master's program in mathematics there. However, I.N. Herstein had recruited me for the University of Chicago, and another Tuskegee graduate in mathematics was doing well at Berkeley. After a lot of deliberation, I chose to attend the University of Michigan.

I arrived at Michigan during the fall of 1971 as a student in the master's program in mathematics. I was relieved and excited to see more than six other new African-American graduate students there. There were at least eight others in mathematics (or mathematics education) who had been admitted prior to 1971. For the next three years, each new class of graduate students averaged at least six African-Americans. One of our first experiences was making a schedule for taking courses. Soon afterwards, we found ourselves immersed in the challenges of the courses. Most of us did at least satisfactory work, but some of us found certain courses and professors difficult. Some professors were rumored to be suspicious of our capabilities. However, we made few complaints of bias. We were generally treated fairly, evaluated solely on the quality of our work. Many of us wanted to prove that we could compete, but some of us did not have enough experience writing proofs and thinking abstractly and thus experienced some disheartening

moments. However, after completing several graduate courses, these essential activities became much more natural to all of us.

Who were we? Many of us grew up in small southern cities and went to historically black colleges such as Texas Southern, Southern, Hampton, Howard, Fisk, Jackson State, Morgan State, Morehouse, North Carolina Central, North Carolina A & T, Talladega, Tougaloo, and Tuskegee. (However, there were a few students from other universities like Lehigh, Hampshire, and the University of New Orleans.) Most of us came from families of little or modest means.

Why did we choose Michigan? One very important reason was that Michigan had and still has a reputation as an institution of the highest academic quality. Perhaps a more compelling reason was that Michigan made enticing offers for financial support. Most of us were awarded Opportunity Fellowships. (At our request these were later renamed.) A few had NSF or Ford Foundation fellowships. Michigan even supplemented one NSF fellowship. Some were attracted to Michigan because the South offered African-Americans few opportunities for graduate study in mathematics at that time, while many of us knew someone who had attended or was attending Michigan. In fact, I have learned that Michigan has had a long history of attracting African-American students and leads the nation in the number of doctorates awarded to African-Americans in mathematics. Finally, we cannot overstate the recruitment efforts of people like Maxwell Reade, who was an associate chair of the Department of Mathematics at Michigan.

While Michigan's reputation and fellowships were very attractive, the number of African-Americans who enrolled in the graduate mathematics programs would have been dramatically smaller without the work of Maxwell Reade. With the exception of a sabbatical year, he diligently recruited from 1969 to about 1977 at historically black colleges in the South. Reade was given authority by the graduate studies division to offer fellowships to African-American students in mathematics. In particular, he could guarantee fellowship funding to a potential applicant until the completion of the master's degree. Moreover, a student could begin at a level which was comfortable, even if it meant enrolling in a course that was traditionally taken at the undergraduate level or in one which appeared to be a course already taken. Reade was able to recruit extensively in the South because he had an agreement with the mathematics department whereby he would teach only in the fall and would recruit African-Americans during the winter semester. His warmth, humor, and passion for Michigan became the deciding factor for many to choose Michigan for graduate work.

Why did we succeed? Probably the most important reason was that we were talented. As undergraduates, we had all been good students and some had been exceptionally good. Although the University of Michi-

* From the Proceedings of an AMS Special Session at Knoxville, in March 1993, "Interventions to Assure Success: Calculus Through Junior Faculty" organized by Bettye Anne Case.

gan wanted to raise the number of black students as demanded by the Black Action Movement in 1970, we were not admitted solely for that goal. An equally important reason was that we were extremely supportive of each other, both professionally and personally. The sheer number of us attending made it easier to develop a group sense of power, courage, and self-esteem. Also we were warmly accepted and supported by a number of faculty members. The cultural and ethnic diversity among students and the liberal atmosphere in the general community were important factors for our general comfort. Racial tension was not common. Finally, there were sufficient avenues for escape from the stress and pressures of academic life, like the insanity of Michigan football. (GO BLUE!)

The support of the mathematics faculty was best illustrated in the actions of Maxwell Reade. I remember his stopping me in the corridors of Angell Hall to inquire about how well things were going. He was the only person who was consistently mentioned by black students as being genuinely interested in our well-being and actively helpful whenever possible. Reade did whatever he could to ease our transition into graduate school, such as finding doctoral students to serve as tutors for some until they felt more confident. He did not feel that his task was done when someone accepted an offer to attend Michigan. He knew that frustrations and disappointments were not uncommon for beginning graduate students and painstakingly tried to prevent or alleviate them.

There were other faculty members who were unmistakably supportive to us. Some eventually became dissertation advisors. Some became unofficial mentors who were simply interested in our progress and had open doors for advice and assistance. Unfortunately, most of the professors were focused on their research and formed few alliances especially with students in the master's program.

However, the African-American graduate students in the doctoral program formed a closely knit group that still exists. We were a family. We celebrated successes and shared failures. After the Qualifying Review was passed, you became a member of a team for which quitting was no longer an option. The African-American doctoral students also formed the core of a mathematics society which was organized as a forum for providing support and information to each other, presenting mathematical talks to each other, and interacting socially.

This society was named the Ishango Mathematics Society. (The name "Ishango" was taken from the Ishango culture that existed in the Congo basin of Central Africa between 9000 and 6500 B.C. A bone tool handle was found there in the 1950s with markings that suggested a knowledge of doubling, base 10, and prime numbers. Some believe it represents a lunar calendar of even greater antiquity.) The concept and formation for this group are due primarily to Curtis Clark and William

Hawkins, who came to Michigan as doctoral students in 1973 and 1974 respectively. We usually met bimonthly from February, 1975, to April, 1978, during the school year and had a membership of 18 or more, but only half of us were very active. The Society encouraged cooperation among African-American graduate (and undergraduate) students on issues such as lending books, tutoring, and collaborative studying for courses. Tutoring was especially helpful to the beginning graduate students. When I arrived, I went to Bruce Palka for help in an analysis course. A few years later, the new black students could and did come to the more senior black students for assistance. These arrangements were very likely as valuable to those giving assistance as to those receiving it. We also studied together in many courses. The Society also provided a means for us to bond in personal ways that were extremely instrumental to our success via various social events such as picnics, potluck dinners, and pizza-and-beer nights at various restaurants.

How many of us succeeded? Using data from the mathematics department at Michigan, a failing memory, and conversations with other graduates, I have been able to count well over 40 African-Americans who were admitted to the master's and doctoral programs at Michigan between 1970-1976. Personally, I am unaware that any of these students failed to complete the requirements for the master's degree. Moreover, I can recall 10 African-Americans who pursued a doctorate in mathematics, and six of them eventually received doctorates from Michigan between 1976-1984.

Where are the "Michigan Six"? Lawrence R. Williams was the first of the six to graduate with a PhD in 1976. He studied operator theory under Carl Pearcy, and he is now an associate professor in the Department of Mathematics, Computer Science, and Statistics at the University of Texas at San Antonio. Lawrence also serves as Associate Dean of the College of Sciences and Engineering. Johnny E. Brown and Robert Charles Hagwood graduated in 1979. Both started in the master's program in 1973. Johnny, who worked under Peter Duren in geometric function theory, is a professor at Purdue University and Director of the Graduate Program in mathematics. Charles, who completed his dissertation in mathematical statistics under Michael Woodruffe, is employed at the National Institute of Standards & Technology. William A. Hawkins entered the doctoral program in 1974 and graduated in 1982 in algebraic geometry under James Milne. He is on leave from the mathematics department at the University of the District of Columbia where he is an associate professor and was chair for five years. He is currently at the MAA Headquarters where he has served as the head of SUMMA (Strengthening Underrepresented Minority Mathematics Achievement) since August of 1990. In 1973 Curtis Clark also came to Michigan to pursue a doctorate. He graduated in 1984 and worked as an assistant professor in mathematics at Georgia State University. Several years ago, he returned to teach mathematics at Morehouse College, his alma mater. Curtis studied under Thomas Storer, perhaps the first Native American to receive a doctorate in mathematics in the United States.[1] Finally, I left Ann Arbor with a five-month old daughter in 1980 and completed my doctorate in 1982 under the direction of Douglas G. Dickson. I am an associate professor and chair of the Department of Mathematics and Computer Science at Xavier University in Cincinnati, Ohio.

Xavier University, Cincinnati, OH

---

[1] In compiling the *Archival Record of Minority Mathematicians with PhDs in Mathematics or Mathematics Education*, SUMMA (Strengthening Underrepresented Minority Mathematics Achievement) and NAM (National Association of Mathematicians) found no Native American awarded a PhD in mathematics before Thomas Storer, USC, 1964. Updated information about the minority mathematicians is welcome; see SUMMA, Chapter 16.

Chapter 15

Views and Experiences

Manuel P. Berriozábal	Brian Lekander
Donald W. Bushaw	Eleanor G. Palais
Donald R. Cole	Wanda M. Patterson
Stephen A. Doblin	T.G. Proctor, *et.al.*
Ed Dubinsky	A. Selden & J. Selden
Michael Freeman	Barbara E. Walvoord
Leonard Gillman	Ann E. Watkins
Bruno Guerrieri	Roselyn E. Williams
Steven G. Krantz	

The San Antonio Prefreshman Engineering Program:
A Mathematics-Based Minority Intervention Program *

Manuel P. Berriozábal

The Greater San Antonio Area has a population of approximately 1.2 million people, at least 55% of whom are minorities. According to a 1986 study and subsequent updates conducted by the Texas Higher Education Coordinating Board, less than 12% of the annual Texas college graduates who earn baccalaureate degrees in science or engineering from the state colleges and universities are minorities; yet minorities constitute 35% of the state population.

---

* From the Proceedings of an AMS Special Session at Knoxville, in March 1993, "Interventions to Assure Success: Calculus Through Junior Faculty" organized by Bettye Anne Case.

Thus, a serious underrepresentation of minority scientists and engineers exists in Texas. A major goal of the San Antonio Prefreshman Engineering Program (PREP) is to increase significantly the number of minority students who pursue engineering or science studies at the college level.

Interim goals of San Antonio PREP are the following:

1. To acquaint program participants with professional opportunities in engineering and science;

2. To reinforce the mathematics preparation of these students in the pursuit of mathematics and science/engineering studies at the precollege and college levels; and

3. To increase the retention rate of these students in college.

San Antonio PREP began in 1979 with 50 high-school students. In subsequent years, lower grades were added to the program so that by 1984, students from all middle school grades participated in San Antonio PREP.

A significant reason for reaching into the middle

schools was to get middle-school students to start thinking about career choices. Hopefully, after at least one summer of experience in PREP, many will consider science or engineering as career choices. In the immediate past years at least 85% of the participants have been middle-school students with 1400 participants in 1992.

Because of the large size of San Antonio PREP and space limitations on the college campuses during the summer, local school districts have offered high-school facilities for some first-year middle-school participants, specifically sixth-grade participants. All other participants pursue PREP at one of the following eight college campuses: The University of Texas at San Antonio, St. Philip's College, San Antonio College, Palo Alto College, Incarnate Word College, Our Lady of the Lake University, Trinity University, and St. Mary's University. The PREP campuses contribute all necessary classroom, laboratory, and office facilities.

Traditionally, the principal academic obstacles for students in engineering and science include mathematics and its problem-solving applications. A major objective of the PREP program is to strengthen the participants' abstract reasoning and problem-solving skills. Part of the program efforts are directed toward the reinforcement of the normal precollege mathematics instructional program by discussing college-level topics not offered in high school or middle school. A science and engineering awareness thrust is given in the physics and engineering courses, research and study sessions, and career opportunities components.

The program spans eight weeks during the summer with participants attending five hours daily. Participants may return for a second and third summer.

First-year participants take eight weeks of Logic and Its Applications to Mathematics, eight weeks of Problem Solving, four weeks of Engineering and four weeks of Computer Science.

Second-year participants take eight weeks of Algebraic Structures, Physics, and Problem Solving.

Third-year participants take eight weeks of Probability and Statistics, Technical Writing, and Problem Solving. Each class is offered one hour each day.

The program schedules daily presentations from a variety of speakers, including women and minority scientists and engineers. Each Friday, activities such as field trips or practice SAT examinations may replace the regular academic components.

Participants are given homework assignments, group projects, and examinations. Participants are expected to maintain a 75%+ average as a condition for program retention.

On the first day of the program, participants take a mathematics placement examination. They are subsequently assigned to groups of twenty participants called seminar groups. Participants in each group are usually in the same academic grade level and mathematics achievement level. They participate in all program components as a group. At the end of the program each participant is assigned a final grade which may be reported to that student's school.

Instructional staff consist of college professors, high-school mathematics and science teachers, Air Force and Naval officers, industrial engineers, undergraduate engineering or science majors (who serve as program assistant mentors), and high-school students who have completed three years of PREP (and who serve as junior program assistants).

In December, the PREP office sends brochures and applications to all middle schools and high schools within the region. Selected applicants are expected to have a B or better average and to have attained a level of precollege mathematics achievement appropriate to their grade level.

Various types of support services may be found at the San Antonio PREP sites, for example: professional counseling, free bus passes, free breakfast and lunch for low-income family students, stipend/wages of $600 or more for subpoverty level participants, and modest stipends for other low-income family students.

Each year San Antonio PREP conducts an annual survey of former participants. The main purpose of this survey is to track the number of high-school graduates, college attendees, college graduates, and science and engineering graduates. An annual report on the survey results is sent to the program sponsors and benefactors, thus providing accountability for the program costs and sponsors' contributions.

Since 1979, nearly 4,100 students have completed at least one summer in PREP. At least 79% have been minority, and 53% have been female; 53% of all participants have come from low-income families as defined by the Texas School Lunch Program. Through a major contact effort of the PREP offices, 1,342 of the 1,871 college-age former participants responded to the 1992 annual survey. The high-school graduation rate is 100%. The college attendance rate is 87%. The college graduation rate is 80%, from early PREP years. The rate of engineering and science graduates is 55%.

PREP charges no tuition or fees. This fundamental decision was made in 1979 so that students from low-income families would be encouraged to participate.

The maintenance and growth of PREP has been sustained over the years through its partnership with local, state, and national public and private-sectors agencies and industry. New partners are added each year, and old partners stay with the program. Our success in keeping partners is due to the achievement record of our PREP graduates as previously cited.

Our partners include colleges and universities, in addition to the host institutions, local school districts, State of Texas, Texas Higher Education Coordinating Board Eisenhower Program, National Science Foundation, Department of Energy, US Air Force, US Navy, Defense Mapping Agency, local utility companies, Summer Child Nutrition Program, private industry, local Private Industry Council Summer Youth Employment and Training Program service providers, and local volunteers.

These partners provide financial and in-kind support. For example: some school districts provide mathematics and science teachers at school district expense, and the Air Force and Navy assign many officers as instructors on a full-time basis. Also, some industries recruit engineering and science undergraduates for summer work and contribute their services to PREP on a full-time basis.

Beginning with the Summer of 1986, PREP was replicated throughout Texas. The statewide program is called the Texas Prefreshman Engineering Program, or TexPREP. Over 2,200 students completed 1992 TexPREP; 1,168 of these participants were in San Antonio PREP. TexPREP currently operates in 11 cities on 18 college or university campuses. Results are similar to those in San Antonio.

San Antonio PREP and TexPREP have both a documented record of achievement and national recognition with numerous citations and awards from the US Department of Education, Texas Senate, The United Negro College Fund, Inc., Change Magazine, the 1989 Hispanic Engineer National Achievement Awards Conference, San Antonio Mind Science Foundation, National Research Council Mathematical Sciences Education Board, Mathematical Association of America SUMMA Project, US Department of Energy Mathematics/Science Leadership Development and Recognition Program, The Society of Mexican American Engineers and Scientists Bravo Award, Ford Motor Company Hispanic Salute, Lone Star Showcase and Salute to Community Service, and most recently an Anderson Medal Competition Certificate of Merit.

In 1991, the National Science Foundation Instructional Materials Development Program funded a San Antonio PREP proposal to complete our academic writing project, designed to produce a kit consisting of an operational manual and a complete set of curricular materials for all PREP academic components. The writing team members were San Antonio PREP faculty and students. These materials were tested in the San Antonio sites during the 1991 and 1992 summers, and are now available to any institution interested in starting a similar intervention program. Indeed, as part of our dissemination component in the NSF proposal, a national San Antonio PREP Workshop/Conference was held in San Antonio in February of 1992. Participants included representatives from 70 college and universities interested in starting an intervention program similar to PREP. All participants received a complete kit of materials and were authorized to copy freely the materials for non-commercial purposes. If adequate funding could be found, almost all of the represented institutions would be interested in starting PREPs.

The National Science Foundation has proposed as a goal for the year 2000 and beyond an annual output of 50,000 minority college science and engineering graduates. The current annual output is 17,000. If San Antonio PREP, or similar intervention programs, could reach 250,000 minority middle school or high school students each summer, 125,000 of whom are first year participants, then a steady state annual output of 50,000 minority science or engineering college graduates would eventually be achieved. The annual direct cost in today's dollars in program operation and participant support would be approximately $500,000,000. Not only would equity for minorities in science and engineering be achieved, but also a future technical manpower shortage would be averted.

University of Texas at San Antonio, San Antonio, TX

Aspects of Assessment of Future College Mathematics Teachers *

Donald W. Bushaw

Assessment of student outcomes in general. In the past seven years or so, more and more people, including some mathematicians, have become more and more concerned with something called "the assessment of educational outcomes." This movement has been supported by many national, regional, and local bodies; for example, the American Association for Higher Education, the Educational Commission of the States (a governors' organization), the Fund for the Improvement of Postsecondary Education (FIPSE), specialized and regional accrediting agencies, state governments, individual educational institutions, departments, and academic people.

Many general in-print discussions of assessment of educational outcomes are available. See, for instance, [3], [4], and [9].

The underlying premise is that higher education efforts require various types of thoughtful and effective evaluation based not merely on the traditional "input" or "process" variables (quality of faculty and facilities, size of library holdings, et cetera) but also on "output" or "product" variables (representing the quality of the outcomes of those efforts). Given the circumstances, one asks not only whether the objectives of an educational

---

* From the Proceedings of an AMS Special Session at the 1992 Joint Mathematics Meetings in Baltimore organized for the AMS-MAA-SIAM Joint Committee on Preparation for College Teaching by Bettye Anne Case.

program *should be expected* to be attained but whether they *are* being attained. This is a natural and supremely appropriate question, and it is remarkable that anyone should think of it as being "new," "radical," or the slogan of a fad.

But *why* assessment? I will pass over in silence the historical answer and move at once to an educational answer.

As mathematical educators we are interested in the quality of our programs (courses, curricula, departments). Since the objective of the programs is to produce students who have certain characteristics, the programs' quality can be judged by determining to what degree the students who are produced by those programs demonstrate those characteristics. Those determinations give us information about individual students which may be useful in various ways; e.g., to diagnose the students' problems and thereby to assist them in benefiting from subsequent educational experiences, or to assist prospective employers in evaluating candidates for employment. Collectively they may provide bases for identifying desirable program improvements or simply evidence for our publics, whatever they may be, of the quality of the programs.

This somewhat complex situation is conveniently summed up in a little table:

Basic types of outcomes assessment in higher education

unit of analysis	improvement	purpose demonstration
individual	diagnosis	certification
group or program	planning	accounting

(adapted from [5, page 17]).

I hope that the terms are sufficiently self-explanatory. The "individual" might be a student or teacher; the "group or program" might be a class, a curriculum, a department, or a whole university. The only term in this table that might fairly be called jargon is "certification," which in this context means announcing, by means perhaps of a grade, a degree, or a license, that the student has achieved a certain level of accomplishment of a certain kind. Of course, assessment of one kind might depend on the results of other assessments; for example, the assessment of a curriculum might be based, at least in part, on the assessed achievements of students who have gone through it.

We know how to use the results of many kinds of academic assessment, but, curiously, serious attention to the question of how *program* assessment results should be used is a relatively recent phenomenon. Certainly it is generally agreed that they *should* be used. It may be that *general* statements on this subject are simply difficult, and that it may be sufficient that well-trained and dedicated professionals can be counted on to respond appropriately to specific assessment results. For example,

if the assessment of a particular college teacher preparation program reveals that the participants are left with the feeling that they do not know enough about the literature on college-teaching, the program's emphasis on that topic can be increased. The *solution* is trivial but identifying the *problem* may require assessment!

Let us look briefly at some types of assessment of educational outcomes.

Assessment of the individual student. This is the whole matter of student evaluation, usually leading to grades or to qualitative statements (as in recommendation letters) about students' qualities and achievements. Most of the mechanisms are well known: tests, term papers, laboratory reports, special projects, one-on-one conversations, and the like. Assessments of this kind are not always related to particular courses, however, but may attempt to measure, in an individual student, the overall effects of an educational program. Comprehensive examinations such as placement tests, the GRE or the ETS Major Field Achievement Tests, or activities such as the increasingly common writing proficiency examinations, are examples.

Assessment of a group or program. This is the real focus of the "assessment movement." Most often, assessments of groups or programs are based on suitably aggregated assessment results for individual students or teachers. But they may also involve such things as examination of employment records of graduates, satisfaction surveys of past students or of employers of past students, subsequent performance in graduate or professional programs, and so on.

Assessment in programs of college teacher preparation. Assessment in some forms (e.g., grading students) has been with us for a long time. It is widely, though not unanimously, agreed that the assessment movement is well established, and is not likely to go away. (For a cautious view, see [8].) Surely any initiation into the mysteries of college teaching should acknowledge these facts. Graduate students who are being prepared for a career in college-teaching should be introduced, by precept or example (preferably both), to the theory and practice of assessment in a multiplicity of forms. The students themselves, should be prepared to practice sound assessment, for example in evaluating individual students and courses. They will be assessed as individual participants in the college-teacher preparation program, so this assessment also should be done well. (A mere grade in a seminar is not likely to be adequate.) Faculty contributors to the program will themselves be assessed. And, finally, the very college-teacher preparation program should be regularly assessed to provide guidance for efforts to improve it, and to demonstrate to persons in positions of authority and other skeptics how well it is working.

Thus almost everything that has been said or done about assessment in higher education potentially applies to college teacher preparation programs and their participants. One thing that is special about this application, and about assessment for prospective teachers in general, is that those who are subject to assessment as participants in such a program need also to learn something about doing assessment themselves.

In general, participants in college-teacher preparation programs may be expected to be relatively mature and dedicated. Thus, they may make better use of assessment mechanisms than less motivated students might, particularly in the areas of self-evaluation and evaluation by portfolio.

These forms of assessment are notable for not necessarily relying on multiple-choice tests and other similarly rigid formats, although they may use them. I will conclude by saying something about each.

Self-evaluation. In a talk given at a 1989 conference concerned with the nurture of teaching-assistants, Patricia Cross, now of the University of California-Berkeley, said:

> *If teaching is to become a satisfying and rewarding career, lifelong self-assessment is the most important skill that you can possess.*

This very strong statement implies that satisfied and rewarded teachers must be *capable* of self-assessment, presumably realistic and useful. Surely any "reflective practitioner" of instruction in mathematics, or indeed of any other calling, must think often about what one is doing, in particular how well it is being done.

Although a significant number of institutions (about 40% of liberal arts colleges in 1983; see [12], p.45) "always use" a "self-evaluation or report" in evaluating teaching performance, these have little credibility as means for evaluation. One reason is that they often do not accord well with evaluations obtained in other ways; another reason is that there is a certain natural skepticism about their objectivity. The following statement from a University of California document puts the matter one way:

> *Although information from the individual faculty member can contribute to a comprehensive evaluation of teaching, self-evaluation cannot suffice as the sole documentation of teaching evaluation or stand as an unbiased or objective source of information ([11], p.41).*

Nevertheless, a careful, somewhat formal analysis of one's own teaching can be a valuable synthesis of a faculty member's credentials, achievements, and ideas as a teacher and can be a useful guide to improvement efforts. Self-evaluation is frequently recommended for these purposes in the literature (see [1], Chap.3; [12], passim). For apprentice teachers, composing such an analysis can also foster a reflective attitude about one's teaching. Participants in a college-teacher preparation program who are also currently teaching, probably as teaching assistants, should be expected, or at least encouraged, routinely to evaluate their own teaching in this manner.

Several publications offer extensive advice on the construction of "forms and activities" for self-evaluation (for example, see [1], [7], and [18]).

There is some current interest in *student* self-evaluation (see [16]). Participants in a college teacher preparation program are, among other things, students, and so can and usually will be assessed by various methods, possibly including this one. In this presentation, however, the emphasis is on their role as a developing teacher.

Portfolios. Teaching portfolios, also known as teaching dossiers, are the subject of several recent free-standing publications. (Examples are [6], [13], [14], and [15].) These publications typically define portfolios, describe rationales for their use, offer advice on their preparation and use by administrators, and present samples. A brief report on the mounting use of portfolios is given in [17], a more extensive account in [10].

"The teaching dossier is intended to provide selected short descriptions that will accurately convey the scope and quality of the professor's teaching.... A teaching dossier would not normally be more than about three pages long, a reasonable amount to ask someone to read" ([15], p.1). (In fact, they tend to be about twice as long, not counting appendices, as this passage suggests.) The "selected short descriptions" can contain statements of responsibilities and objectives and such information as:

  courses taught; other teaching responsibilities

  syllabi, assignments, exams, other handouts from courses taught

  summaries of student evaluations

  summaries of peer evaluations

  statements from students, past and present

  records of former students in subsequent courses, graduate or professional study, careers

  descriptions of efforts to improve teaching

  graduate students supervised

  mentoring of colleagues

  publications related to teaching

  professional activities outside the department or off campus related to teaching

  participation in faculty development activities

  awards

  goals for the future

Portfolios can be used only for self-evaluation or can be evaluated by small teams of peers, faculty members, or external examiners. In the latter case, conclusions should be drawn from the evaluation of portfolios — whoever does it — about activities that might lead to improvements in the subject's teaching.

The portfolio should take not more than a few hours to prepare; but it should be treated as an organic entity, subject to frequent updates.

The subjects of self-evaluation and portfolios are not unrelated. A serious self-evaluation can be an important part of a teaching portfolio, and a teaching portfolio can act as a springboard and documentation for self-evaluation.

It should also be said that although portfolios have been around for a long time, the research literature of their use is not so rich as that for many other types of assessment.

Conclusion. I have tried to sketch the scope of assessment in higher education, with emphasis on a few points particularly pertinent to the preparation of college teachers of mathematics. I may have impressed you with the complexity of the assessment scene, and left you with the impression that so much has been done and so much is going on that a choice must be made between neglecting assessment or neglecting other things that are probably more important.

I would be sorry to have left you with that idea. I *would* like you to believe that assessment of educational outcomes has become a major phenomenon in higher education these days, and — I would say — rightly so. I would like you to believe, too, that assessment has a rightful place in college teacher preparation programs both as part of the syllabus and as a dimension of the program itself. The participants should learn something about assessment, and their learning should, like the program itself, be assessed in reasonable ways. Self-evaluation and the use of portfolios are two forms of assessment that may be especially useful along the way. The scope and complexity of the literature on assessment makes it clear that programs charged with the training of future college mathematics teachers must include their learning something about assessment, and their learning itself should be assessed in reasonable ways.

References

[1] John A Centra, *Determining Faculty Effectiveness: Assessing Teaching, Research, and Service for Personnel Decisions and Improvement* (San Francisco, etc.: Jossey-Bass Publishers, 1982).

[2] Robert M. Diamond and Franklin P. Wilber, "Developing Teaching Skills During Graduate Education." *To Improve the Academy*, 9 (Stillwater, OK: New Forums Press, 1990) 199-216.

[3] Peter T. Ewell, ed. *Assessing Educational Outcomes* (New Directions for Institutional Research, 47) (San Francisco, etc.: Jossey-Bass Publishers, 1985).

[4] Peter T. Ewell, "Outcomes, Assessment, and Academic Improvement: In Search of Usable Knowledge." In John C. Smart, ed., *Higher Education: Handbook of Theory and Research*, 3 (New York: Agathon Press, Inc., 1988).

[5] Peter T. Ewell, "To Capture the Ineffable: New Forums of Assessment in Higher Education." *Review of Research in Higher Education* 17 (1991) 75-125.

[6] Aubrey Forrest *et.al.*, *Time Will Tell: Portfolio-assisted assessment of general education* (Washington: The AAHE Assessment Forum, American Association for Higher Education, 1990).

[7] Harold I. Goodwin and Edwin R. Smith, *Faculty and Administrator Evaluation: Constructing the Instruments*, 3rd ed. ([Morgantown, WV]: Department of Education Administration, West Virginia University, 1983).

[8] Pat Hutchings and Ted Marchese, "Watching Assessment: Questions, Stories, Prospects." *Change* (September/October 1990) 13-38.

[9] Reid Johnson *et.al.*, *Assessing Assessment: As In-Depth Status Report on the Higher Education Assessment Movement in 1990* (Higher Education Panel Report Number 79)(Washington: American Council on Education, May 1991).

[10] Barbara J. Millis, "Putting the Teaching Portfolio in Context." *To Improve the Academy*, 10 (Stillwater, OK: New Forums Press, 1991) 215-229.

[11] David Outcalt *et.al.*, *Report of the Task Force on Teaching Evaluation* (Santa Barbara: The University of California, 1980).

[12] Peter Seldin, *Changing Practices in Faculty Evaluation* (San Francisco, etc.: Jossey-Bass Publishers, 1984).

[13] Peter Seldin, *The Teaching Portfolio: A Practical Guide to Improved Performance and Promotion/Tenure Decisions* (Bolton, MA: Anker Publishing Company, Inc., 1991).

[14] Peter Seldin and Linda Annis, "The Teaching Portfolio." *Teaching at UNL* [Teaching and Learning Center, University of Nebraska-Lincoln] vol. 13, no. 2 (September, 1991) 1-2, 4.

[15] Bruce M. Shore, *et.al.*, *The Teaching Dossier: A Guide to its preparation and use*, revised (Montreal: Canadian Association of University Teachers, 1966).

[16] Carl Waluconis, "Student Self-Assessment: Students Making Connections." *Assessment Update* 3:4 (July-August, 1991) 1-2, 6.

[17] Beverly T. Watkins, "New Technique to Evaluate College Teaching: Effort uses portfolios to document professors' in-class performance." *The Chronicle of Higher Education* (May 16, 1990) A15f.

[18] Maryellen Weimer *et.al.*, *How Am I Teaching? Forms and Activities for Acquiring Instructional Input* (Madison, WI: Magna Publications, Inc., 1988).

Washington State University, Pullman, WA

Evolutionary Intervention Programs:
The FAMU Method *

Donald R. Cole

Florida A & M University (FAMU) is an historically black public university with years of successful experience in producing many of the nation's leading black scientists, engineers, and mathematicians. Although FAMU currently leads the nation in the recruitment and enrollment of National Achievement Scholars, 10% of the first-time-in-college class comprises promising students whose academic admission standards are below normal acceptance standards. The historical commitment to work with students throughout the academic spectrum has led to years of developing what is termed a "holistic" approach to student success which centers on recruitment, sustainability, and post-baccalaureate activities and programs. This paper discusses a few of the more successful FAMU intervention programs and analyzes their major common components which assure success in these programs.

Table 1 below lists some of the more successful technical intervention programs at Florida A & M University, their funding agencies, and key features.

Table 1. "The FAMU Method"

Program; *Funding Agency*; *Program Features*

Minority Access to Research Centers; NIH; Entry into PhD Program, Curriculum Enrichment, Summer

BIONR; DOD; Entry into PhD program, Pre-College work, Institute status

Florida - Georgia Alliance; NSF; Alliance with 9 Senior Colleges and 3 Junior Colleges

Florida Comprehensive State Center for Minorities; NSF; Saturday Academy

Packard; Packard Foundation; Summer, Tutorial, Colloquium

Ronald McNair; NASA; Mentor, Summer Research Paper

Life Gets Better; NASA/NSA; Scholarship.

The interdependence and interdepartmental relationship among these programs form a holistic approach to student success termed "The FAMU Method". Over the last year this approach has attracted over 80 National Achievement Scholars, generated over 30 students entering PhD programs and created a pipeline of over 50 honor-roll students with PhD aspirations in the Natural Sciences and Mathematics.

Analysis of the Programs. In analyzing each program listed in Table 1 in terms of intervention through surveys of faculty, staff and particularly students, we found certain common features immediately.

A Program Director – In each case this crucial individual could be described as one who "saw the need, and proceeded almost independently to establish the program". The sincerity of the commitment to the program established by this natural chain is easily passed on to the students.

Mentors/Academic Advisors – These staff members provide a one-on-one bonding to the program. Achievements of mentors' expectations quickly become a goal of students and the I-am-a-team-member attitude is soon adopted.

Seminars, Presentations, and/or Colloquium – When associated with a particular program, these features distinguish the importance of that program to the participating students. While promoting professionalism, these components also promote group-study and preparation.

Summer Internships – These provide exposure and practical application to the discipline and enlighten the "theory-practice" component.

Financial Support – This provides a means by which total effort may be spent within the subject matter area.

Academic Support – This is especially important in the early years. Activities in this component range from faculty-student to student-student arrangements. Test preparation, time management, and mandatory study hours were all observed under this component.

Tracking/Evaluation – Such follow-up of students and program should be a continual entity of each component. Many programs had outside personnel evaluate their programs.

Conclusions. Successful intervention programs for underrepresented students must begin by abandoning the older perceptions that such students are incapable at a certain technical level, lack motivation, and are inadequately prepared — perceptions the FAMU experience has invalidated. Through the intervention programs conducted at Florida A & M University, it is clear that group-learning, student-teaching-students, and faculty involvement are key factors in these successful programs.

Older theories which relied mostly on recruitment are being modified by "Recruitment-Intervention" as we discover that "In Life, as in Golf, it's the 'follow through' that makes the difference!"

University of Mississippi, University, MS

(formerly at Florida A & M University, Tallahassee, FL)

* From the Proceedings of an AMS Special Session at Knoxville, in March 1993, "Interventions to Assure Success: Calculus Through Junior Faculty" organized by Bettye Anne Case.

A Comprehensive Approach Can Make the Difference *

Stephen A. Doblin

The University of Southern Mississippi is a relatively small comprehensive university located in a portion of the country with little high-tech industry. It attracts many first-generation college students and has traditionally been provided few state resources. Nevertheless, during the decade of the 1980s the Department of Mathematics was able to maintain a stable faculty, more than triple (to 200+) the number of majors, achieve notable research productivity, and place its graduates in advanced degree programs and mathematics-related employment throughout the country. The major reasons for this success were a comprehensive approach to recruitment, retention, student and faculty development, and internal and external interactions; commitment; and teamwork.

The purpose of this paper is to share some of the approaches that worked for us, along with a brief description and rationale for their inclusion. While certainly not exhaustive, this discussion does touch upon those activities that some consider to have been the most important in helping us achieve our primary goal of producing intelligent, thoughtful, highly motivated graduates who have the ability to succeed in advanced degree programs, contribute to mathematics-related professions, continue learning, solve problems, and make informed choices. Activities fell into three broad categories: the major program; external communication and interaction; and faculty support and development. As will become evident, several of the activities cross categorical lines.

The Major Program. It was important to provide a structured yet flexible curriculum for the mathematics major, one that could be somewhat tailored to career intentions and completed within the normal four-year time frame. Students were to be nurtured from the onset, effectively informed of various rules, regulations, and opportunities, and always challenged.

1. *Academic Advisement.* As soon as a student declared mathematics as a major, that student was assigned a tenure track faculty member to serve as an advisor for the duration of the degree program. The student met with the advisor several times each semester to select courses, construct a comprehensive program, and discuss post-baccalaureate options. Having a personal friend with whom to discuss university and departmental requirements, academic problems, campus assistance facilities, and graduate school and career opportunities proved to be an

important support mechanism, particularly for first-generation college students.

2. *Guide for the Mathematics Major.* A pamphlet containing information about the department as well as the discipline was given to each new major. It contained sections on such topics as undergraduate mathematics major requirements; activities, honors and facilities; financial assistance; post-graduate opportunities; contact after graduation; current mathematics faculty and staff; job titles of graduates; recent faculty research/scholarly activity; five-year projected advanced mathematics course offerings; and advisement checklists. Experience had shown that the student would read this guide in preference to university publications such as the undergraduate catalog and student handbook.

3. *Undergraduate Research Program.* Our most promising undergraduate scholars were encouraged to work on mathematics problems under the direction of faculty mentors. Participants presented the results of their work at local seminars, nearby universities, and regional and national meetings. Several of their papers were published in the *Journal of Undergraduate Mathematics* and in *Mathematics and Computer Education.*

Other important activities in this category included the formation of a mathematics club, active encouragement of students to participate in the cooperative education program and to seek summer employment in math-related positions, the sponsoring of several departmental lecture series, and the broad dissemination of career information in the mathematical sciences.

External Communication and Interaction. This category addresses ways in which we could "tell our story to" and assist students in our service courses, our colleagues in other academic areas, our alumni, potential students, and the community at large.

1. *Developmental Mathematics and the Mathematics Learning Center* (MLC). Our developmental mathematics course provided the underprepared student with the math background requisite for success in college algebra. Surprisingly, a few students began in this course and completed a baccalaureate degree program in mathematics. The MLC, staffed by undergraduate mathematics majors and graduate students, offered tutorial assistance in precalculus courses along with several other services. These initiatives, geared toward assisting the students who took our service courses to succeed, helped enhance our image among the students as well as among colleagues in other departments. In addition, the experience gained by our majors who worked in the MLC made them feel more a part of the department and proved to be valuable when they sought teaching assistantships and employment after graduation.

2. *Instructor Corps.* Several of the department's faculty were non-tenure track, MS-level mathe-

* From the Proceedings of an AMS Special Session at Knoxville, in March 1993, "Interventions to Assure Success: Calculus Through Junior Faculty" organized by Bettye Anne Case.

maticians whose principal responsibility was lower-division instruction. Their experience, dedication to teaching, and concern for the student provided an effective transition from high school to college for students whose academic major, at least initially, lay outside of mathematics. They were demanding teachers, and all participated in scholarly activity related to collegiate mathematics. The senior instructor also directed the department's training program for graduate teaching assistants, which had a significant and positive impact on our new graduate students as well as upon the quality of teaching in college algebra.

In this category were also included the establishment of interdepartmental and intercollege liaison committees which often developed syllabi and selected texts, the production of a departmental newsletter for alumni and friends, sustained interaction with prospective majors and their parents, and the development of a placement testing program for students majoring in areas other than mathematics.

Faculty Support and Development. The success of a department and its programs depends upon the hard work and support of its faculty. Initiatives for which a majority do not feel ownership as well as those which are not recognized within the departmental reward structure are doomed to either short-term existence or failure.

1. *Scholarly Activities Bulletin Board*. Reprints of publications, the cover page of successful grant proposals, and presentation descriptions (meeting, dates, title) were displayed on a bulletin board mounted in a prominent location near the department office. Visitors to the department invariably stopped to observe the bulletin board and inquire about its contents.

2. *Evaluation and Reward*. The department created its own evaluation instrument, approved by the dean and used during the annual performance review each spring, which provided a mechanism to recognize an individual's activities that contributed directly to the overall department mission. Hence, it was perfectly clear to each faculty member how teaching, committee service, student advisement, recruitment, community service, professional presentations, et cetera, "counted" in the allocation of merit raises.

Additional thrusts included: the appointment of a departmental public relations coordinator; establishment of a restricted fund account within the university foundation so that alumni and friends could contribute directly to the department's programs; and a broad-based approach to faculty recruitment which involved attempts to hire mathematicians with research interests that complemented existing expertise, who broadened the department's diversity, and whose concerns and philosophy would make them good teachers and colleagues.

During the past few years, enrollments have remained relatively stable, and most of the initiatives described

above are still in place and are effective. A new computerized degree audit program has been implemented, which should enhance the advisement process and reduce the paperwork burden on the faculty. An interdisciplinary doctoral program in scientific computing has begun, and the department's computing resources have improved dramatically. In addition, participation in the Core Calculus Consortium and Mississippi's Alliance for Minority Participation has had a positive impact on student learning.

University of Southern Mississippi, Hattiesburg, MS

Editor's Note: *The essays by Ed Dubinsky and Steven Krantz present an example of the differences of opinion in the mathematical community on how to accomplish the same goal—better teaching. The circumstances of this inclusion may be of interest: Steve Krantz wrote a small book as part of his efforts directing the program at Washington University which was associated with the FIPSE assisted project of the Committee on Preparation for College Teaching. The preface of that book is reprinted later in this chapter. It sets the stage of opinion for Krantz's book. His early drafts were read by a number of mathematicians. Krantz and Dubinsky corresponded about some of the points. With knowledge that Dubinsky's opinions might be considered a counterpoint to Krantz's, this editor invited Dubinsky to write about what he thinks most important in the current generation of changes in mathematics teaching.*

Pedagogical Change in Undergraduate Mathematics Education

Ed Dubinsky

If you read the early reports such as *Toward a Lean and Lively Calculus* [D] or *Calculus for a New Century* [ST], it will be clear to you what the goal of calculus reform was. The original intention was to revise the *content* of calculus—to make it lean and lively, to make it more relevant to applications, to include some contemporary mathematics, and to reflect the new technology. All of the early talk was about changing the topics that were covered in a calculus course.

This is not what happened. The calculus reform syllabi are beginning to appear, and their lists of topics invariably include: polynomial, rational, exponential and trigonometric functions; differentiating by using product, quotient and chain rules; applying derivatives to motion, curve graphing and economics; integrating by techniques and by numerical approximations; applying integration to physics, economics, and statistics; and, of course, the fundamental theorem of calculus. The textbooks are still hefty, and many of them are still in a format of discussion sprinkled with worked examples followed by exercises to practice.

It is my opinion that, after a fairly short period of relatively superficial manipulations, the set of topics in the calculus will settle down to pretty much what it was. Indeed, it is hard to see how it could be otherwise. There is certainly some evidence that the topics have not changed for a very long time [DR]. This could be due to inertia, but it could also be that it is not the content that we want to change.

Today, many people are coming around to the point of view that what needs to be changed is not the content, but the pedagogy—not *what* we teach, but *how* we teach. This is true, not only for calculus, but throughout undergraduate mathematics.

True, this is not yet a majority view. Many people appear to believe that effective teaching is actually quite easy to achieve, if only you care enough to give it a certain amount of attention and energy. The suggestions offered by proponents of this view are, in my opinion, little more than common-sense, well understood by a very high percentage of members of our profession. The analysis ignores the fact that a really large number of mathematicians *are* conscientious and dedicated in their teaching. Very many of us *have* used these suggestions in our teaching and have been doing so for many years. The important point is that, in spite of all this, our students are still not learning mathematics. For me, the inescapable conclusion is that much more is needed than common sense suggestions gleaned informally from experience. I am convinced that we need to reconsider and revise our pedagogy—and we need to do it in conjunction with research into what it means for a student to learn a mathematical concept.

I would like to touch on some questions about this pedagogy issue. What are the new pedagogies that people are talking about? Why do their proponents expect them to lead to improvement? To what extent is pedagogy actually changing? And, do we see any results that suggest things are getting any better?

I think there may be something to say about the first two questions, but it is much too soon to expect anything definitive regarding the last two.

What are some pedagogical innovations in collegiate mathematics?

There are three major teaching changes to which a growing number of college faculty are giving at least lip service. The first is fundamental in our conception of the very nature of teaching and learning. We are beginning to move away from a feeling that mathematical knowledge is a kind of mental commodity that can be delivered by a teacher to a collection of students using various media such as speaking, writing, and even demonstrating on computers. We are beginning to understand that one must construct knowledge for oneself and that the role of the teacher is not to explain mathematics in a classroom, but to induce students to construct it in their minds.

The second change we are talking about has to do with using computers to support the first change—that is, to foster students' constructing mathematical knowledge. This can be done in many ways. A small but growing number of people, including this writer, are convinced that the best way, by a long shot, is to have students implement mathematical ideas on computers — that is, write appropriate programs that implement mathematical processes and objects.

The third change that is fairly sweeping the country—if not in our classrooms, at least in our conference reports—is called cooperative learning. Students work in teams to learn mathematics cooperatively, not competitively and not as isolated individuals.

There are many other teaching innovations that are discussed in *UME Trends*, in the publication *A Source Book for College Mathematics Teaching* [SC] and many other places. They include such methods as writing, self-paced instruction, student projects, and the Moore Method. But I believe that the three I mentioned are the most important today, and I would like to concentrate on them for the remainder of this paper.

Replacing lectures. So what *do* you do if you are not standing in front of the class explaining mathematics? Some people talk for a while, ask a question, try to get a discussion going, and go on from there. Some people send the students to the blackboard to work on problems.

In Our Project. As one alternative, here is what we do. It is all part of a general scheme for getting students to construct mathematical concepts for themselves. The students work cooperatively in teams (see below) on tasks that, from their point of view, are related to work they have done in the computer lab (see below). From my point of view, the tasks are related to the mathematics that we would like them to construct but broken down according to the particular mental constructions that my research has suggested they can make in order to learn the mathematics.

More specifically, we will give the students a (paper-and-pencil) task to work on in their teams and then try to get them to discuss what they have done. At the end we will summarize and emphasize the mathematical ideas that we think they have constructed.

Example. Suppose, for example, I would like them to understand the mathematical construction of a function E which assigns to each real number M the sequence $E(M)$ whose j^{th} term is $Mj(j + 1)$, $j = 1, 2, \ldots$.

This first requires that they understand a sequence of numbers as a function which consists of a process of transforming positive integers (domain) into real numbers (range). Then they must understand that a sequence is itself a mathematical object, a notion that permits the construction of a set of sequences or a parametrized family of sequences, or a function which maps numbers to sequences. I will discuss in the next section a precursor to the following program which I would have the students work with in the computer lab. We are using the mathematical programming language ISETL here. It is a high-level language whose syntax is very close to standard mathematical notation (see [DS] for more details). We find that students begin to feel comfortable with it after a week or two.

```
E := func(M);
        return func(j);
                return M*j*(j+1);
        end;
end;
```

Now, the task in class might be to explain what the computer does if you enter E(32.78)(6). (It returns the value at 6 of the function E(32.78).) Or I might ask for an explanation of the computer's response if you say E(32.78) (it returns an internal designation for this function) and compare that with the previous response. Or, I might ask them to explain why E(32.78,6) would make no sense to the computer. It turns out that most of the conceptual issues — such as those concerning a function whose value is a function (sequence), a function as a process versus a function as an object, the meaning of the graph of a family of functions and so on — will tend to arise in such a situation. Not only do the conceptual issues arise, but you can actually see the students constructing appropriate and powerful mathematical ideas as a result of their experiences with the computer and the opportunity for reflection that is provided by this kind of classroom activity.

Using Computers. As with trying alternatives to lectures, there are a multitude of ways in which people are using computers in college-level mathematics courses. The role of the computer in these approaches ranges from the easy-to-implement but fairly barren "electronic blackboard" to a tool for investigation and discovery or an environment in which it is natural and helpful to construct sophisticated mathematical notions (such as function as a total entity).

The Source Book mentioned above [SC] delineates some of these computer uses. I would like here to expand only slightly on the comments I made previously regarding the use of computers in our project.

Students can use the computer to perform calculations, discern regularities and make discoveries about patterns. They can use it as a powerful tool to solve problems and to apply various mathematical concepts to problems in other fields such as management, engineering, and the physical sciences. Details about such uses in our project and others for Calculus courses appear in [T,L].

Example. In our project, we use computers in an unusual way. Our theoretical and empirical investigations

of how people learn help us to lay out a series of mental constructions that students can make in order to understand a mathematical concept. Often, however, they do not find it easy to make such constructions. For example, it turns out that constructing the idea of a sequence as a process which converts a positive integer to a real number is both important and difficult for students. We help them perform this mental construction by asking them to write and work with code such as the following (which explicitly constructs the function E(32.78) discussed above and evaluates it at three points.)

```
M := 32.78;
s := func(j);
        return M*j*(j+1);
    end;
s(3); s(17); s(23167);
```

Working with sequences requires the student to see them not only as processes, but also as objects which can be manipulated. For example, the concept of sequence of partial sums is an operation on a sequence that transforms it to a different sequence. We help students construct this extremely difficult (for them) notion by asking them to write a program such as

```
PS := func(s);
        return func(j);
                return%+{s(k) :  k in [1..j]
        end;
end;
```

Working with this code helps students understand that PS(s) represents the sequence of partial sums of the sequence s and P(s)(17) represents its 17th term. It is not that programming helps them directly with notation but rather that it helps them understand the underlying concepts, and *that* makes the notation less problematic.

Working cooperatively. As I have indicated in discussing alternatives to lecturing and the use of computers in our project, we have students working cooperatively in teams. This has recently become a very popular idea in undergraduate mathematics teaching, and, once again, there is a multitude of ways in which students can work cooperatively. These range from the informal suggestion, made from time to time by the instructor, that students should get together to work on some problems, to a more formal, permanent arrangement.

In our project, we work at the latter extreme. Students are assigned at the very beginning of the course to teams of three or four. Unless special difficulties arise, the assignments remain intact throughout the course, and students do all of their work in these teams. This includes lab assignments, homework, classroom interaction, and even some examinations. We work hard to help the students develop a team spirit and a commitment to the progress of all members of their team.

Many questions arise when you begin to think about

cooperative learning. There are practical questions. How are the teams selected? What are productive and non-productive ways in which a team can function, and how can the teacher influence this? How do you translate the work of a team into an individual grade? There are also more foundational questions. What is the relation between learning and problem-solving in teams? How are evaluation and assessment affected? What role does language play? How can technology be incorporated into cooperative learning?

Although a great deal has been written about cooperative learning in grade-school, almost nothing exists about using it in collegiate mathematics classes. Based on our project, we are writing a report which will be a summary of our experiences and a practical guide for using this approach. It will also include an annotated bibliography. Hopefully, it will be available by the time this article appears.

Why do we think these new pedagogies are worth trying? I would like to try to explain my contention that research into how people can learn mathematical concepts is, or should be, the main answer to this question. That is, the interaction among our theoretical analyses, implementations, and observations provides specific pedagogical approaches that are based on more solid ground than just the personal impression that comes from informal reflections on experience which, I submit, have proven inadequate for the educational problems our profession is facing. I have not always been committed to this point of view, and it may be useful to share my view of my own development.

I come to the question of how to find effective instructional strategies as someone who has been teaching college mathematics courses for almost 40 years, since 1956. For all of that time I have felt unsatisfied with the results — in terms of student-learning. Most of the time I was trying one or another kind of teaching innovation, either my own idea or a technique suggested by someone else. Where did these ideas come from, and why did I think they were worth trying? They came, for the most part, from honest, sincere, and dedicated people trying to use their experience to think of the best ways of helping their students learn. I believe that always, especially today, there are hundreds if not thousands, of people who are doing this, and until recently I considered myself to be one of them. I think we should applaud their intentions but at the same time face the hard reality that it is not working. That is, it is in the face of all this experience, sincerity, dedication, and hard, creative thinking about specific innovative teaching methods, that we are still having the same results. Our students are not learning mathematics as well as we would like them to—as well as we believe it is possible for them to do.

If we are to succeed in materially improving undergraduate mathematics education, we have to find a better way of coming up with new and different things to do. Moreover, after we try our new methods, we also have to

determine how well they worked. This is the assessment issue and that is a matter for another discussion. Here, I am discussing sources for innovative pedagogy — how we come up with it and what might be useful criteria for deciding what to try. I think that one lesson we must accept from our experiences is that it is not enough just to think of ideas in any way that we can and try everything that occurs to us. We must look for guides to which pedagogical approaches are promising and worth the tremendous investment of time and money that it takes to try them.

I would like to suggest that one guide should be the theoretical analyses that are part of research in how people learn. Indeed, the three pedagogical innovations that I laid out in the previous section arose as applications of the theoretical perspective with which I am working. I would like to spend the rest of this section indicating very briefly how our ways of using computers, what we do in the classroom, and our approach to cooperative learning are all guided by this theoretical perspective.

What is our theoretical perspective? Let me begin this very brief sketch of my theoretical perspective with the following general statement which expresses my constructivist point of view.

> *An individual's mathematical knowledge consists of a tendency to respond, in a social context, to a perceived problem situation by constructing, re-constructing, and organizing in that individual's mind, processes, objects, and schemas with which to deal with the situation.*

To this I would add two things. First, constructing and organizing takes place by performing mathematical procedures in these problem situations and *reflecting* on what you are doing. Second, the statement alone is fairly barren and comes to life only when the theoretical analysis proceeds to specifications of the nature of specific constructions that are to be made. This latter activity is the heart of our project's research. The present short essay is not the place to go into details, but the interested reader can consult our papers in the literature.

What I *can* do here is begin to describe how these points relate to the three pedagogical innovations of the previous section.

How does our theoretical perspective guide computer activities? The theoretical perspective provides a framework for what follows, and that has to do, in terms of the constructivist point of view expressed above, with the construction of specific processes and objects, organizing them into schema, and reflecting on all of this mental activity.

Here is how the theoretical perspective guides the design of computer activities. The theoretical analysis (buttressed with the results of empirical investigations) tells us, roughly, what processes and objects need to be constructed. We then design computer activities to make these constructions on the computer. We have found that this very often results in the students making analogous constructions in their own minds.

We have given some examples above of how this works in calculus, and we have described elsewhere some examples of how it can work in abstract algebra [LD].

How does our theoretical perspective guide classroom work? Classroom work consists mainly of students working in teams to perform tasks using only paper, pencil, textual material, and their minds. A task might ask the students to solve a short mathematical problem, or explain some phenomena that occurred in their computer work, or look for an example/counterexample (often without knowing beforehand which is the case). A good example is connected with the computer program for the functions E and PS together with the classroom tasks and questions discussed earlier in this paper.

After the teams have worked on the task, the instructor tries to get a discussion going. Sometimes, when the instructor is convinced that a fair number of students have made appropriate mental constructions, that instructor will summarize the mathematics, taking the opportunity to point out standard terminology and notation.

There are a number of pedagogical goals for this kind of classroom work that relate to our theoretical perspective. In some cases, the tasks are chosen as additional stimuli to have students make the same kind of mental constructions that were expected from the computer work. Using recently constructed objects and processes to solve an unfamiliar problem can induce students to form these mental constructs into coherent schemata that can be used for solving similar problems. Also, this work can get students to reflect on their mental constructions, and how they may be used.

How does our theoretical perspective guide our approach to cooperative learning? The most important theoretical support for our use of teams lies in the need, according to our theoretical perspective, for students to reflect on the calculations that they perform. Working in a group, it is natural for students to ask each other questions, explain what they are doing and, in general, to discuss the mathematics they are working on. We see this in the classrooms, in the labs, and during informal interaction among students. This sharing of ideas gets students used to reflecting on what they are doing. It helps mathematics become, for them, not so much a mindless "repertoire of imitative behavior patterns" [M], but a form of thinking in which they are consciously engaged.

There are several other, mainly affective, contributions that we see coming out of cooperative work. The support that students give each other, both emotionally and just in terms of getting the work done, is invaluable. Setting up a permanent (for the semester) team and suggesting to students that each member of their team is responsible for the learning of all members of the team, tends to make their experience with mathematics more joyful. If nothing else in our approach works, the more positive attitude towards the course that students seem to get from working in teams is a big gain that alone can justify our efforts.

Conclusion. In conclusion, let me summarize what I have tried to say here.

My main point is that real improvement in undergraduate mathematics education cannot and will not occur in the absence of substantial pedagogical change. Moreover, one activity that can make a major contribution to this change is research in how people learn mathematical concepts.

I have tried to describe the major pedagogical changes that people are thinking about: alternatives to lecturing; technology; and cooperative learning. I have discussed ways in which these three instructional strategies play a role in the curriculum development project with which I am involved. In particular, I have pointed out the relationship between our use of these strategies and our research in learning, especially from the point of view of theoretical analyses.

In closing, I would suggest that major pedagogical changes are today more honored in the conference report than in the classroom, and the jury will still be out for a long time on deciding about their long term value. But it would be a mistake to end this article on such a negative note. As I tell my students struggling to understand mathematics, one should not always concentrate on how far there is to go; sometimes it is helpful to look back and see how far one has come. In the case of pedagogical change in undergraduate mathematics education, it is possible to hope that our dismay at the daunting length of the former may be overcome by the awe inspired by the substance of the latter.

References

[D] Douglas, R. G. (Ed.) (1986), Toward a lean and lively calculus, Report of the Conference/Workshop to develop Curriculum and Teaching Methods, *MAA Notes*, **6**.

[DR] Dubinsky, E. and A. Ralston Undergraduate mathematics—now and then, unpublished report.

[DS] Dubinsky, E. & K. Schwingendorf (1992), *Calculus, Concepts, and Computers, Preliminary Version*, West: St. Paul.

[L] Leinbach, C. (Ed.) (1991), "The laboratory approach to teaching calculus, *MAA Notes*, **20**.

[LD] Leron, U. and E. Dubinsky, An abstract algebra story, *American Mathematical Monthly*, to appear.

[M] Moise, E.E. (1984), "Mathematics, computation, and psychic intelligence" in "V.P. Hansen & M.J. Zweng (Eds.), *Computers in mathematics education* (1984 Yearbook of the National Council of Teachers of Mathematics), Reston, VA, pp. 35-42.

[SC] Schoenfeld, A.H. (Ed.) (1991), *A Source Book for*

College Mathematics teaching, MAA Reports, **2**.

[SC] L. Steen (Ed.) (1989), *Calculus for a New Century, MAA Notes*, **8**.

[T] Tucker, A. (Ed.) (1990), *Priming the Calculus Pump: Innovations and Resources, MAA Notes*, **17**.

Purdue University, West Lafayette, IN

MathExcel:

A Collaborative Calculus Workshop at the University of Kentucky *

Michael Freeman

MathExcel is an intervention in first-year calculus at the University of Kentucky on behalf of women, minorities, and students from rural and smaller school districts. Modeled on the Emerging Scholars Program of the University of Texas at Austin, MathExcel is designed to give good students from these groups the best possible chance to excel in this crucial course. Here are results of the Fall 1992 MathExcel class of 38 women, 9 African-Americans, and 51 rural students[1] in first-semester calculus:

Results: Fall 1992	MathExcel (77 students)	non-MathExcel (459 students)
Average Grade	2.97	1.86
% A's	44.4	17.6
% Withdrawing	0.0	9.6
% Failing	6.5	26.6

Note the increase of 1.11 points in average grade (on the usual four-point scale). None of the 77 MathExcel students withdrew, and their failure rate was relatively small. Five of the top six students out of the 536 in first-semester calculus were in MathExcel. Four of these were women, and the man was from a rural district. The top two overall were MathExcel women from rural districts.

MathExcel students are very well motivated, but they do not have extraordinary natural ability: their average math ACT score was only about two points higher than that of the rest of the class. MathExcel is now in its third year, and these results are our best yet. Each semester since the Program began in Fall 1990 the comparative results have been extraordinary.

Calculus and MathExcel. The first year of calculus is required of majors in engineering, all the physical sciences, mathematics, secondary-math education, and many biology majors and pre-meds. The dropout and failure rate of 36% for regular students is unfortunately normal, but it doesn't tell the whole story. Getting a D or even a C can also discourage a student from continuing, with some reason. Our experience shows that fewer than half of students receiving a C or lower in calculus will be able to complete a major requiring it. Since a regular student has about only one chance in four of doing better than a C, the calculus sequence is a serious barrier to success.

MathExcel is *not* remedial. It is a different kind of honors class, celebrating achievement attained not just by raw talent, but also by cooperation, hard work, and perseverance. We want to encourage good students from these groups to enter the professions requiring calculus, and we are after the best ones we can find.

MathExcel addresses the underrepresentation of women, minorities, and students from smaller and rural school districts in majors requiring calculus. MathExcel aims to help capable students from these groups achieve success in their majors by:

◇ giving them the best chance to master this subject,

◇ providing them with the powerful technique of collaborative study,

◇ easing their transition to the University, and

◇ providing a friendly cohort to support them throughout their careers at UK.

Organization and Resources. MathExcel is a one-year program beginning each Fall and consisting of special sections of Calculus I in the Fall and Calculus II in the Spring. These are four-credit courses of about 500 students and around 20 sections. Regular students attend three lectures by a Faculty member and two recitations led by a teaching assistant.

MathExcel students, who currently form three of the sections of Calculus I/II, attend the same lectures (1 MathExcel section in the same room with two regular sections), do the same homework, take the same tests at the same times, and are subject to the same grading standards as regular students[2]. However, instead of the one-hour recitations, MathExcel students attend special two-hour MathExcel Workshops three times each week.

MathExcel requires four more hours of class per week. To recognize this extra time, a MathExcel student is also enrolled in the special course Calculus Workshop which carries two hours of Pass/Fail credit. Thus a MathExcel student earns the four hours of normally graded credit received by every Calculus I/II student plus two hours of Pass/Fail credit.

The workshops are organized around cooperative

---

* From the Proceedings of an AMS Special Session at Knoxville, in March 1993, "Interventions to Assure Success: Calculus Through Junior Faculty" organized by Bettye Anne Case.

[1] Many of the women were from rural areas. Only 3 of the 77 MathExcel students were white males from urban schools.

---

[2] Thus the above comparison of grades is fair. Moreover, the Director of MathExcel does not teach Calculus I or Calculus II concurrently with MathExcel, and has nothing to do with the administration of either course. MathExcel grades are not inflated.

study, and are substantially more stimulating and challenging than regular course activities. Each workshop is led by one of our best teaching assistants, assisted by two paid undergraduate assistants chosen from among the top students who have recently completed the first year calculus sequence.

Since its inception in Fall 1990 MathExcel has been supported entirely by institutional resources. The yearly incremental cost of offering MathExcel to three sections of 77 students this year is around $57,000. This includes a portion of the Director's salary, stipends and training for the teaching assistants serving as leaders, and salary for undergraduate assistants (At $8/hour it costs about $800 per assistant per semester, and we use two per section). To add another MathExcel section of 25 students would cost an additional $9,000.

At the beginning of each workshop session the students are given a Worksheet of calculus problems. MathExcel students know they are expected to work together in small groups, and to get busy right away, with little preparation by the leader. The problems are more challenging than assigned homework. It is done prior to the workshop, and there are more problems than anyone could reasonably expect to complete in two hours. When a group has difficulties, leaders and assistants do not simply give a solution. They serve more as guides, in a Socratic role. In this way the students learn mainly from each other.

Leaders and their assistants constantly seek to provide a congenial, supportive, and above all personal atmosphere. In a nutshell, MathExcel is a much more personal experience. Probably no other undergraduates at the University of Kentucky receive as much personal attention. We see many friendships develop. The importance of these friendships is a recurring theme in the several surveys of MathExcel students we have conducted.

Our results are due to the MathExcel Workshops and the motivation of MathExcel students. Attending workshop is a much more effective and efficient way to learn than the same time spent studying alone or watching someone else work problems. Workshops are self contained and non-judgmental. Thus students usually do not take home unfinished Worksheets, and their Worksheets are not graded. The atmosphere is informal, intense, and very constructive. Any competition is purely recreational. MathExcel students help each other compete with the rest.

Personnel. Workshop leaders are selected for proven ability as teachers, interest in collaborative learning, and their personal qualities. We expect and get extraordinary effort from them. A MathExcel class and its leaders and assistants develop bonds which promote a special level of engagement from all concerned. Most leaders have experience as high-school teachers, and all have at least one year of teaching at Kentucky. One of this year's leaders, whose section is consistently at the top in exam scores, says: "I am having the time of my life. My students are not just succeeding, they are excelling."

Admission to MathExcel is selective, and by application. We look for ability, motivation, and affinity for cooperative effort. We have accepted students with unimpressive standard test scores who have otherwise distinguished themselves, for example, as valedictorians. We have not accepted a few with high standard test scores who seemed to lack commitment. We interview applicants personally in visits to high schools in Kentucky and at the Freshman Advising Conferences on campus in April, June, and July.

The undergraduate assistants are a great asset to MathExcel. Several of the twelve current assistants were Governor's Scholars in high school, and several hold scholarships at UK. Two hold four-year full-support Scholarships, the best UK has to offer. Nine are women, and two are African-American. The two men have at least an avocation for the ministry.

The assistants are wonderful role models. Moreover, the experience is very good for them. They feel honored by their appointment, and their role as teachers of calculus leads them to a real mastery of it. Last year one said, "I was thrilled to be asked to be a part of such an important and worthwhile program and felt more a part of the University."

We are continually struck by how MathExcel seems to bring out the best in everyone involved. Students like these respond to the constructive atmosphere, the presence of so many others like them, and our explicit and implied expectations. The workshop leaders, their undergraduate assistants, and MathExcel students are all energized and inspired by each other and the workshop setting. Perfect attendance at a workshop is not unusual, and the norm is above 90%.

Improvements. The new "Math House" is a valuable addition to MathExcel. The Math House is a single-family house owned by the University and turned over to the Math Department at our request in Fall 1992 as a place for selected students and faculty to gather for study and socializing. It has proved to be the best location for MathExcel Workshops. Its intimacy and informality are especially conducive to the atmosphere we seek, and the sense of community among our students is increased by meeting in "their" house. MathExcel students have free access to the House and make additional unscheduled use of it.

We have scheduled two of our current workshop leaders to teach sections of Calculus III and IV next year in which current MathExcel students will be encouraged to enroll. This will keep MathExcel students and their workshop leaders together for another year of cooperative study.

University of Kentucky, Lexington, KY

Teaching Programs that Work*

Leonard Gillman

During the past several years, a number of innovative programs have been established for helping students succeed in mathematics, particularly minority students. I have picked out a few to talk about, but you should be aware that there are many others and their number continues to increase. The ones I have picked represent a fair mix of type of student and type of school.

San Antonio, PREP *(Prefreshman Engineering Program)*. This program was founded in 1979 by Professor Manuel P. Berriozábal of the University of Texas at San Antonio. It is designed to encourage Hispanic students who are talented in mathematics and science to continue through college with majors in those fields, rather than switch in high school or college to easier courses or drop out altogether. (I suspect he calls in an engineering rather than a mathematics program because it is easier to get money that way.) What he does is to get students from grades 6-11 and drench them with mathematics and science to keep them interested. PREP has been expanding across the state, but I'll talk about the main program in San Antonio.

It is an academically intensive, eight-week summer program; and students may return for a second or third summer. Admission and retention standards are strict. There is an orientation session for parents, most of whom come from backgrounds without traditions in higher education. Here's the daily program: *first summer*: logic, engineering and computer science; *second summer*: algebraic structures, physics,; *third summer*: probability and statistics, technical writing; *every summer*: problem solving, study sessions, guest speakers. There are also occasional field trips and some practice with SAT tests.

You may wonder what is taught in logic and in algebraic structures. The logic course discusses propositional calculus — including de Morgan's laws, implication, *modus ponens*, and quantifiers — and elementary switching networks; the older students are also introduced to the algebra of sets (finitary operations) and abstract Boolean algebras. Algebraic structures, for the younger children, are limited to a careful discussion of the algebra of real numbers; the older ones also meet abstract groups and other structures.

The guest speakers, who come from universities, and

industry, and government, give pep talks and describe career opportunities for math majors; this is of great importance, as students are usually unaware of such opportunities-part of the reason they switch to easier programs or drop out of school. Class size is in the low twenties except for logic, which is doubled up because of a shortage of faculty and classrooms. Faculty teach three courses each day and are on campus from 9:00 to 5:00; students, too, are on campus all day. Over 600 students completed the 1989 program; three-quarters were Hispanic and one in fifteen were Black; half were from low-income families and received daily lunches, bus transportation, or stipends. I visited six classes, attended a guest lecture, and gave one.

The talk I heard was by Dr. Susan P. Stattmiller of San Antonio. She is an unassuming but determined young woman, white, from a family culture in Ohio where no one had gone to college and where women were thought of only as homemakers — a culture similar to that of many of the students in the audience. She liked science and math, pushed on into college, graduated with a joint major in math and chemistry, continued to a master's degree and a teaching certificate in mathematics, and got a job teaching high school. Then one day she decided to become — guess what — a dentist. When she was accepted at dental school and phoned to tell her mother about it, her mother said, "That's a very nice job — a dental hygienist." "No, mother, I'm going to be the dentist, the doctor." "Yes, I'm sure you will have a very nice doctor to work with." They straightened it out after another cycle or two. Today she is a practicing dentist — though she hopes to return to teaching mathematics, if only part-time. Now that's what I call a role model.

Many academic people derided PREP when it was first announced, maintaining that a mathematics professor would degrade the profession by being associated with it, that you can't get young kids to study logic for eight weeks, that minority students never do well in a traditional academic setting. In fact, the 1989 class (presumably typical) voted logic the overwhelming favorite, with an interest rating of 98%. In the 1989 survey of the 1,115 PREP graduates who are now of college age, 759, or 68% responded. All of these completed high school (versus 55% of Hispanic students nationwide), and 89% of them are now in college or have graduated. While only 50% of Texas freshman go on to graduate, with a mere 20% in science or engineering, the figures for PREP graduates are 75% and 67%. (This last figure will drop some in the next few years because of a wider base of PREP admissions.)

Escalante. Everyone in this room has heard of Jaime Escalante and his incredible successes with AP calculus at Garfield High School in East Los Angeles. Probably most of you have seen the movie *Stand and Deliver*. You should also know that there is a book, *Escalante* [1]. It too is somewhat romanticized, but it includes gripping details not given in the movie. Of course there are math-

* On 19 January 1990, Professor Leonard Gillman delivered his Retiring Presidential Address at the Seventy-Third Annual Meeting of the Mathematical Association of America, in Louisville, Kentucky. An earlier version of Professor Gillman's address first appeared in the January-February 1990 issue of *FOCUS*, the Association's newsletter.

ematical gaffes, such as the reference to "a new book on the integration of π." A blooper readers are likely to miss is the reference to Richard Strauss's Opus 314: his highest opus number was 88. (There is a 314 in another list, but it's a collection of essays.)

Garfield High School has 3,500 students, 95% Hispanic, predominantly from middle or lower income families in which neither parent has completed high school. Eighty percent qualify for the federal free or reduced-cost lunch program. Escalante arrived in 1974. He had taught math and physics in his native Bolivia, and he had ideals and standards. What he found at Garfield was debris, graffiti, and gang fights; students who were surly, bored, unruly, and hostile, wedded to a life of academic failure; and teachers who didn't seem to care whether or what the kids learned. The basic mathematics text was fifth-grade level by Bolivian standards. To win students over, Escalante devised gadgets, gimmicks, and a special vocabulary — like Paul Erdős — all of which he still uses as he entertains, challenges, cajoles, encourages, praises, warns, scolds, and threatens in English and Spanish and two Bolivian dialects. He plays music, sometimes soft, sometimes loud. He hands out candy, has the class chant and clap, and tell jokes. "Red light" means stop and think; "green light" means smooth sailing ahead. He squeals in ridicule of the marching band — a waste of time away from mathematics — and barks on behalf of his toy bulldog, which he asserts is "45-carat" gold. (The bulldog is the school mascot.) A poster warns, "You don't do your homework, you gonna be working the rest of your life at Jack-in-the-Box."

Escalante also wins students by example. They respect a teacher who is at the school from 7:00 a.m. to 7:00 p.m. Even students not much interested in math work hard for him when they see him working so hard for them. He also challenges student athletes to handball: "You choose. I use left hand or right hand. You beat me, you get an A. I beat you, you do this homework" [1]. The student does the homework. Finally, Escalante provides personal support by accepting telephone calls at home, interceding with parents, arranging field trips, and organizing expeditions to McDonald's; even a clean-up-and-paint squad for the classroom or a money-raising car wash enhances a student's feeling of participation and self-esteem.

Escalante rejects placement tests as just one more bureaucratic obstacle. A student who wants to study calculus is admitted to the class — though is admonished that once in you are expected to stay in. Many students underestimate what is in store for them, and Escalante has to work to persuade them to stick it out by encouraging them and sometimes bamboozling them; and he usually succeeds in converting the hesitant student into a citizen of the world of hard work and accomplishment. Calculus students acquire a community spirit. The better ones help out as tutors for the others. All tend to be looked upon by the rest of the school with respect and as role models.

Now that Escalante is established, his showmanship is needed less if at all. Still, many students go for it. As one of his recent students, whom I'll call Jeff, told me, "He's funny. He's very funny. He's creative when he teaches. He'll make an opening monologue, very fast, and he'll start telling jokes, and then all of a sudden he's hit you with a quiz — and you want to do it because you're in a good mood. He comes up with some different stuff every time you go to class: we go to class and there's some apparatus there. He gets you interested." I visited two of his classes, an algebra and a calculus. (He told the algebra class the bulldog cost him $1,000; for the calculus students he revalued it at $2,000.)

I also visited two classes of Ben Jiménez, his colleague: a trig and a calculus. The four classes averaged about twenty-three students. Jiménez is the exact opposite in character: quiet, undramatic, and no-nonsense. He doesn't use gadgets, nor even "red light" or "green light," but he achieves the same results. Like Escalante, he emphasizes on day one that he is a serious teacher teaching a serious course; that the students are going to have to attend class, study hard, and do a great deal of homework; and that the standards will be uncompromising. At the same time he makes clear that he is the students' friend; that the believes in them; that he is available in his classroom or adjoining office from 8:30 on, in particular during the noon hour and after school; and that he will work hard to help any student who is working hard.

Garfield students tackled AP calculus for the first time in 1979 and gained national fame in 1982. The AP was not an end in itself but a lofty, almost impossible challenge. The mere act of studying for it, whatever the result, was itself a worthwhile discipline. Those who passed acquired a strong sense of achievement for having performed something difficult as measured by a national standard. The results at Garfield started out modestly but picked up to an astounding record. In one 6-year stretch involving 250 students, the passing rate was 89% versus a national average of 70%. (Then it dropped when a number of marginal students swelled one of Escalante's classes to 40 students.) In the past few years, one quarter of all Hispanic students in the country who passed calculus AP have come from Garfield High. Did you hear that right? *One quarter in the country came from Garfield.* (One may wonder what this says about the rest of the country.)

What do AP students do for an encore? Of the eighteen students in the famous class of 1982, fifteen entered college, and by 1987 nine had graduated — well above average for Hispanics. Many of their degrees were non-trivial: Columbia University, UC Santa Cruz, UC Berkeley in microbiology and immunology, Cal Poly (Pomona) in aerospace engineering and computer science. Several of the group help professional or technical jobs. One had passed her CPA exams and another had completed an MBA. The spirit of success carried into the rest of the school. Grades rose in *all* math courses. The percentage of *all* Garfield students admitted to college increased during those five years from 60% to 70%.

Recitations like this do not tell what goes on behind the scenes. Jeff (whom I quoted earlier), now a sophomore math major at Berkeley, made that clear to me. Neither of his parents finished high school. His father drove a truck until he hurt his back and had to quit, then went into business for himself, buying property, fixing it up, and selling it. He now makes more than his relatives who went to college, so he sees no need for a college education and informed Jeff he would not give him money to help get one. Jeff went on to Berkeley anyhow. He was getting by on Financial Aid until one day irony and bureaucracy struck: his father made a little extra money on some property and the financial office informed Jeff he was no longer eligible. Now he is managing with loans and by teaching at a high school in Oakland, commuting by bicycle. His tone when reporting all this portrays his father as merely ignorant, not evil. "My dad's mad at me now. But Mom's coming around; she's beginning to understand it's important to have an education." I think he loves them both.

UC Berkeley, PDP (*Professional Development Program*). PDP is Berkeley's universally acclaimed intervention program for minority freshmen. It is an honors program, created by Uri Treisman and inaugurated in 1978. The freshman class at Berkeley was 20% Hispanic and 10% Black. Treisman was concerned about the high rate at which black students with strong high school records in math were failing freshman calculus; moreover, only one in eight of those who did pass got through sophomore differential equations or premed organic chemistry, the prerequisites for careers in engineering and the sciences. (In a tragic twist of irony, when many of the students switched to other majors and graduated, the university counted them as successes, whereas they thought themselves as failures.)

Uri set out to discover the roots of the problem. Rejecting the pat answers of armchair sociologists, he made a detailed study of the backgrounds and study habits of the Black students, in true anthropological fashion, going so far as to observe them in their dormitories — certainly the first such study by a mathematician if not the first by anyone. Those from predominantly Black high schools typically had less exposure to mathematics than other Berkeley students, causing them to overestimate their understanding of course concepts; those from predominantly white high schools also had trouble adjusting as they discovered to their dismay that they were not welcome in Berkeley's white community. Uri observed that the Black students tended to keep their academic and social lives separate. They almost invariably worked alone, with all the attendant frustrations that come from having no frame of reference; in one group he observed, only two of the students regularly studied with others: with each other — and at the end of the year they married and quit college. Chinese students, in contrast, studied in groups, exchanging hints and tips and offering constructive criticism of one another's work. Uri was also

quick to reject ordinary tutoring or other remedial programs, which focus on failure, dealing with all the stuff the kids don't know but never actually doing the stuff they're supposed to be preparing them for.

Accordingly, Uri established PDP as an intensive, demanding program for talented students, particularly minority students, who are planning a career in a mathematics-based profession. It is designed to help them excel at the university. Candidates are selected by an elaborate process that ensures an ethnic mixture, diversity of high school background, and gender balance. They are assured that they are among the most promising freshman and that the program is seeking students with a deep commitment to excellence and the desire to continue to graduate school and become leaders in their profession and society. The program itself consists of enriched, intensive work sessions to replace the regular calculus recitations. The emphasis is on students' strengths rather than weaknesses: a student with an identified deficiency will be handed difficult problems where it will have to met head on — the direct opposite of remedial programs. There are two 2-hour intensive sections per week, 15-20 students to a section. Students come in having already done the regular class homework and are handed worksheets containing challenging problems of the sort that separate A-students from B-students. They begin working the problems individually, then, when things get tough, in collaboration with another. These experiences lead to a strong sense of community and the forging of lasting friendships. Several of Escalante's graduates have entered Berkeley and participated in the program.

The session is conducted by a "facilitator," a teaching assistant (TA) who guides the work but does not give out answers. An important goal is to ensure that the students will go on to excel in their sophomore and later courses without the program there to help them — to become "independent but not isolated learners," as the PDP people put it. The facilitator is a role model and peer counselor, and together with other program staff is sensitive to warning signs such as a distracted appearance or nonattendance; they act as a support group, intervening to help solve outside problems (housing, for instance) before they have become crises. (In contrast, university counseling offices usually see students only after they are already in trouble.) Activities such as pizzaparties and volleyball, as well as the informal atmosphere of the sections, provide additional social contacts that help counter possible feelings of isolation. This combining of the academic and social functions is regarded as an important feature of the program.

The proportion of Black students who graduated or are still in school after six years or more is only 39% for non-PDP students but 65% for PDP graduates. Two years ago, a Black female student from PDP became the first [black] Rhodes Scholar at Berkeley in twenty-four years. There are now more than thirty satellite programs across the country. I visited the new ones at Cal

Poly (San Luis Obispo) and UT Austin, as well as the main one at Berkeley, speaking with students, facilitators, administrators, and faculty, and sitting in on work sessions. The excitement of the students as they shout and argue about mathematics is a joy to behold. At Austin the Calculus 1 students in the program averaged 21 points higher than the rest of the class.

New Mexico Calculus Project. This is an intriguing program at New Mexico State University at Las Cruces. Its goal is to improve calculus teaching (for all students) by means of student research projects. About 30% of the students at the university are Hispanic, and there are satellite programs at several nearby schools, of which one is predominantly Hispanic. The project stems from a 1987 experiment by two young faculty members, Marcus S. Cohen and David J. Pengelley, whose aim was to get students "to discover the excitement of calculus, build their self-confidence in theoretical thinking, and thereby fundamentally alter their perception of what doing mathematics is really all about." For the formal project, they were joined by colleagues Edward D. Gaughan, Arthur Knoebel, and Douglas S. Kurtz.

A research project is a calculus problem, more elaborate and challenging than a standard problem. Ten calculus sections (of the twenty in the college) use these projects, which replace the three one-hour exams. The instructors are the above five and five TAs; each TA is assigned to one of the professors and teaches a section of the same course (first, second, or third term). Help labs staffed by the TAs are open two to four hours daily during the two weeks of a project. Students are expected to turn in polished solutions, and, later, explain them in a private 15-minute interview with the professor. With two sections of forty students, an instructor spends a lot of time reading papers, holding office hours, and interviewing. (Creating the problems in the first place is also time-consuming, but once a data bank has been established, that may no longer be a big consideration.)

A problem often involves simply fitting several concepts together, as when computing the volume cut from the first octant by a tangent plane to a given surface. Sometimes students have to consult a section of the text that the course did not cover. They may be asked to develop a topic on their own, such as integration by hyperbolic substitution. Some problems are stated without numbers or variables, such as the following greenhouse problem:

"Your parents are going to knock out the bottom of the entire length of the south wall of their house and turn it into a greenhouse by replacing some bottom portion of the wall by a huge sloped piece of glass (which is expensive). They have already decided they are going to spend a certain fixed amount. The triangular ends of the greenhouse will be made of various materials they already have lying around. The floor space in the greenhouse is only considered usable if they can both stand up in it, so part of it will be unusable, but they don't

know how much. Of course this depends on how they configure the greenhouse. They want to choose the dimensions of the greenhouse to get the most usable floor space in it, but they are at a real loss to know what the dimensions should be and how much usable space they will get. Fortunately they know you are taking calculus. Amaze them!"

The students' jaws drop when they encounter problems of this type, and only the better students can manage them; so most problems proceed by a series of hints. Whatever the form, the work is intended to require reasoning over a period of days and so reward good thinkers rather than god test takers. A problem that wows the students is to show that the series $1 - \frac{1}{2} + \frac{1}{3} + \frac{1}{4} + \ldots$ can be rearranged to converge to any number: their invariable reaction is that it is not possible, and, the faculty report, deriving the result "revolutionizes their view of mathematics."

The program has been underway for only a year and a half, but the experimenters are highly encouraged by the preliminary data and in fact amazed by what some of the students can do. In our conversations, they had nothing but praise for their students — in sharp contrast to the universal pastime of ridiculing or grousing. (I've been a champion grouser for over forty years: they think every function is linear, they don't know the chain rule, they can't remove parentheses, all they care about is whether it will be on the test.) Many students find the projects exciting, but some of the weaker ones do not enjoy the challenges, and require lots of help. In my view, if a lot of students are learning to think more mathematically and enjoy it at the same time, the program is worthwhile; the weaker ones can always opt for the traditional course.

SUNY Potsdam Mathematics Program. This program was started in 1969 by Clarence F. Stephens, who had just joined the department as chairman. His determination to develop a flexible mathematics program that emphasizes the "human factor" has fostered marked success — a substantial number of students major in mathematics and those that concentrate in the discipline constitute over 40% of the College's honors students.

Potsdam is a liberal arts college of 4,000 students, located in the upper reaches of New York state. Ninety-five percent are from New York high schools, primarily from lower middle class backgrounds, often from farming communities and small villages. The mean combined SAT for mathematics majors is about 1100; for the college as a whole, slightly lower. Practically all the students are white, but Steve (as Dr. Stephens is known to his friends) had tested and honed his ideas during twenty-two years at Prairie View A & M in Texas and Morgan State University in Baltimore, where the students were predominantly Black. The program at Potsdam soon started bearing fruit but did not become widely known until 1987, when John Poland's article appeared in the *American Mathematical Monthly* [2]. Steve retired the same year, and Vasily C. Cateforis, his colleague and former student, became chairman. Perhaps the best way to describe the program is to read from a letter Steve wrote me last spring:

"You may be disappointed with the description of our program. We have not used grant money to develop it, and the only innovation in the curriculum is our BA-MA program established in 1970, which provides an opportunity for able mathematics majors to earn both degrees in four years without attending summer school. Most efforts to improve undergraduate mathematics focus on curriculum and educational technology. While we acknowledge the importance of these two factors in the improvement of mathematics education, we focus on the human factor of changing students' perceptions that mathematics is an almost impossible subject for students to learn and only the most gifted can be expected to achieve any degree of success."

"We simply established a humanistic academic environment for learning mathematics in which students in mathematics courses feel good about themselves and find enjoyment in the study of mathematics as a result of proper teaching strategies and supportive environment which promote student success and academic excellence. We help our students to understand the meaning of a mathematical proof and have respect for it, to learn how to learn mathematics, to read a mathematics textbook with understanding and pure enjoyment, to study independently and as a member of a group. We use many different methods of teaching undergraduate mathematics. We teach in the spirit that *Everybody Counts*."

The major at Potsdam consists of 30 to 40 hours of mathematics (out of 120): calculus (12 hours) , set theory and logic, linear algebra, modern algebra, advanced calculus, a problem seminar, and electives. Class size ranges from 40 at the lower level to 15 in the problem seminar. The predominant spirit is the culture of success: continual encouragement, recognition of every accomplishment, successful role models — honors students, BA-MA students, graduates — enough success to develop self-esteem, enough time to develop intellectually, recognition of one's achievement, and the belief that the study is worth while.

Instead of racing through a long syllabus that students are largely not going to absorb anyway, the faculty want their students to learn enough of the subject well enough to understand the essential idea and general strategy. The students solve very hard problems in small groups, teaching one another as the professor guides the effort with helpful questions. Tests are regarded as articles of learning rather than measures of ability. Grading is flexible, to allow for late bloomers. Teachers focus on developing the students' skills rather than on the transmission of knowledge. They challenge the students — but within reason, consistent with Steve's maxim: "Teach the students you have, not the students you wish you had." No one motivates by threat. No one says "I taught them, but they didn't learn it" — that would be likened to a

salesperson saying, "I sold it to them, but they didn't buy it."

The faculty believe that the best basis for understanding mathematics and its wide applications is experience in classical mathematics with its emphasis on logical structure, precision, careful analysis, and clarity of expression. Consequently, there are no service courses; nor are there remedial courses, nor even placement tests. (Have faith in them: throw them in and they'll probably swim.) There are no special mathematics courses for math education majors. There is no course titled "Calculus for two-headed football players." The only special course is honors calculus.

As the mathematics program has prospered, it has attracted more and better students. There is no mathematics requirement for graduation — but half the freshman class have been known to take calculus as an elective. There is a large and flourishing Pi Mu Epsilon chapter. Although nationally, post-calculus courses account for only 10% of mathematics credits, at Potsdam the figure is 50%. While nationwide only 1% of graduating seniors are math majors, at Potsdam it has been running about 20%. While the proportion of women among all graduating seniors is 51% nationally and just a bit higher, 54%, at Potsdam, the proportion among graduating mathematics majors is only 46% nationally but 55% at Potsdam — although the department makes no particular effort to attract women into mathematics. Over 40% of honors graduates are math majors. Thirteen of the last 16 valedictorians were mathematics majors; 8 of the 13 were women. Did you get that? *Half the valedictorians were women mathematics majors.* I spent three days on campus, where I visited five classes and conferred at length with students and faculty and with the top administration — who, I am happy to report, support the mathematics program unequivocally. Again the faculty have nothing but praise for their students.

Most of the mathematics graduates go into industry, many to places like Kodak and IBM, or various insurance companies, where they prove to be able to think independently, read and write technical reports, work cooperatively with others, present and defend their work, and offer criticism without devastating the recipient; many rise to high managerial positions. Others go on to do graduate work in mathematics-related fields, often at Cornell or Big Ten schools, where they find themselves well prepared for independent work despite a mathematical education that may be less broad than that of other students.

All the departmental faculty have engaged in research, all teach courses at every level, all are dedicated to the program. There are no TAs; instead, the better or more advanced students help the others in study sessions. The faculty work long hours, subordinating any research ambitions to the success of their students. The dean told me of the time he was walking across the campus late one Friday afternoon when spring had just broken out and everyone had fled home early — except that as he passed the mathematics building he happened to look up and there were the professors still working at their desks.

Some mathematicians believe only potential researchers should major in mathematics. But this would write off a mathematically educated citizenry. I for one hope to see many more Potsdam graduates sitting in our state legislatures when university budgets next come up.

Conclusions. The five programs I have dealt with represent various combinations of school level, student achievement, faculty and student ethnic group and sex, and mathematics courses. All these programs are based on teaching strategy or attitudes rather than curricular details. Most of them emphasize sympathetic attitudes toward students that mathematics faculty do not generally adopt in any systematic way. Indeed, there is no reason to believe that mathematicians possess any special skill at understanding human sensitivities so that such attitudes will come naturally. On the contrary, given the character of our training in abstract reasoning and logical precision, devoid of any reference to human frailty — or perhaps because of our inner nature that impels us to such study in the first place — it is likely that most of us have very little such skill.

The fundamental precepts of those programs are: challenge your students with difficult problems, demand hard work, and adhere to uncompromising standards; at the same time, constantly promote and bolster their confidence and self-esteem, assuring them that they can succeed and praising each accomplishment, and make yourself available outside of class for sympathetic help and encouragement. "Remediation" is a dirty word. Classes are kept small. Where the program is part of the regular teaching assignment (Garfield, New Mexico, Potsdam), the faculty generally work long hours at their teaching.

The principle of having students work in groups is basic to the Berkeley and Potsdam programs. (I permit myself the irreverent observation that when students learn from one another they take up less of *your* time.) Berkeley maintains a support structure to help entering students adjust to the new environment and to help keep nonacademic problems out of the way — important for minority students just entering the university. Potsdam tenders support via its Pi Mu Epsilon chapter and other activities typical of a big happy family. Escalante provides his personal support. All these support systems are closely associated with the actual program faculty and staff — in contrast to the typical counseling group in an office in the basement of the administration building with its name on the door, two telephones, and a xerox machine, but without academic expertise and isolated from the faculty.

I learned something about role models. I asked Escalante, "Suppose there was a teacher who was your clone except that he was anglo; would he have the same success with these students?" I expected him to say,

"No, he has to be Hispanic." Instead, he said, "No. I pronounce their names right: I say 'Hulia' instead of 'Djulia,'" adding, in explanation, "It comes from within." He went on to say that an anglo who lived for several years in Mexico, say, and absorbed its culture, could also succeed. Then he added, " I could teach an anglo class."

Even more interesting is the story at Potsdam, with its incredible record of female mathematics majors. Professor Patricia Rogers of York University recently concluded a detailed study of the department, in which she states: "Given the importance placed by some writers on providing female students in male-dominated fields with female role models, it surprised me to find that in a department of fifteen faculty, only one is female [3]." The women students spoke of their male teachers as father-figures and, in fact, role models, and were unaware that females are often discouraged from taking math. The article presents an interesting discussion of power and other social or psychological factors and their relation to the nature of mathematics and concludes that apparently "in an environment which is genuinely open to and supportive of all students and in which the style of teaching is true to the nature of mathematical inquiry, women are attracted to mathematics and are just as successful as men."

Here then are five programs with success records ranging from substantial to unbelievable. I'll temper that. I have not defined what I mean by success. I do not agree with every detail of every program. Some reports are skewed, as when they are limited to those who responded to a survey, or when they do not separate the effect of the program itself from the fact that its students were preselected high achievers. I picked up occasional infelicities on the part of lecturers. Some wrote faintly on the board or spoke too softly. One calculus instructor, in a drill on derivatives, wrote $f(x) = x^2 - 3x = 2x - 3$, which the students dutifully copied. (What about $3x = 21 = 7$? Equivalently, $3x = 21 = 8$?) But my statement stands. The problems we face in educating our youth, especially minorities, are daunting, and it behooves us to study these programs and learn from them.

There is one common feature of these programs I have not yet mentioned, perhaps the most outstanding one of all. Can you guess what it is? They were all created by dedicated, imaginative individuals, working by themselves. *MAA had nothing to do with it.* (I refer to promulgating an official philosophy that led to the creation of the programs in the first place, not just providing fortuitous support or cooperating after the fact.) Instead, we were sitting around in committees talking about — oh, never mind!

Finally, it appears that good teaching is good teaching, whether your students are above or below average, live in poverty or in affluence, or represent any particular color or sex.

I wish to express my thanks to Lewis W. Wright (Assistant Vice-President, Administration) Sarita E. Brown (Assistant Dean, Graduate Studies), and Robert E. Boyer (Dean, College of Natural Sciences) at UT Austin for providing a travel grant in support of my study; to Reba Gillman, who is my wife, and Jackie McCaffrey, the coordinator of UT Austin's "Berkeley" program, for many insightful comments on earlier drafts of this talk; and to Richard D. Anderson (former MAA President) for picking up a statistical misquotation.

References

[1] Mathews, Jay. *Escalante, The Best Teacher in America.* New York: Henry Holt, 1988.

[2] Poland, John. A modern fairy tale? *American Mathematical Monthly* 94 (1987): 291-294.

[3] Rogers, Pat. Gender differences in mathematical ability — perceptions vs. performance. *Association for Women in Mathematics (AWM) Newsletter*, Vol. 19, No. 4 (1989): 6-10 (Based on a forthcoming article in *Gender and Mathematics: An International Perspective*, UNESCO, 1990.)

University of Texas, Austin, TX

The Calculus Calculator:
Or, the Poor Man's Version of Mathematica *

Bruno Guerrieri

The following statements are perhaps most applicable to faculty members who teach mathematics courses at the undergraduate level and have a desire to use computer technology (IBM-compatible based) to supplement their instruction but who may have limited resources. In my particular case a given semester may involve four classes (say two Algebra and Trigonometry, one Calculus II, and a Math Modeling course) with an average of 35-40 students in each and a rather limited Computer Laboratory containing a network of 18 286-IBM machines and four printers. The Calculus Calculator resides on a file server and is easily accessible from each of the terminals. Despite its name, the Calculus Calculator is a computer algebra system (CAS) perhaps not as powerful as better known CASs such as **Derive**, **Mathematica**, or **Maple** but which has the advantage (so far) of being given away, in the form of a site license, to anyone who recommends the accompanying manual to their students. The manual, "The Calculus Calculator", and any other information can be obtained from Prentice-Hall for about $20.00. While programmable graphics calculators are getting better, I still believe that graphics can

be better seen on a computer screen and the ability to obtain hard copies makes the Calculus Calculator (on a 286 machine) an ideal system in between the two above extremes.

While I am not ready to change completely the way in which mathematics is delivered, meaning that I do not subscribe for the moment to the idea of teaching a series of cases in an electronic classroom, I believe that the computer, with its computational capabilities, enables one to explore more realistic and meaningful problems. At the same time my audience is usually made up of average students who need to be drilled to a certain extent rather than being given complete license to explore. Consequently, so as to obtain the best of both worlds, I will usually teach my classes in a traditional manner and will augment the course with take-home projects that will necessitate the use of the computer. In these projects students are expected to read sections in the book (perhaps not covered in class) and somehow synthesize what their calculations show with concepts developed in the book. I must say, before proceeding, that this does NOT seem necessarily to bring the level up a notch. Some students do take well to the experiment and in fact show initiative and curiosity, while many view the process as one more imposition. But I seem to have perceived that I was able to talk, at times, at a more "conceptual" level than I would have been otherwise with the class seemingly following, as opposed to resisting.

The Calculus Calculator (CC) program basically has three windows. The first is an input window where parameters of the problem under investigation are entered or calls are made to built-in functions such as graphing or performing linear regressions or integrating (numerically) or differentiating (perhaps symbolically). Another window is the graphics one which supports many graphs at once and can easily be printed. Finally, the last one is a function window in which the user, relying on a simple programming language (along the lines of Pascal and without all the required declarative statements), can build a library of user-defined functions that will perform for an applied mathematician such as myself numerical solutions of most applied mathematics one is likely to see in an undergraduate course. While CC can be used with very little explanation in an Algebra and Trigonometry course to help students compare graphs and grasp concepts such as symmetry, shift, period, inverse function, and, while CC can be used intelligently in a course such as any of the calculi (please refer to the plethora of calculus books with computer applications flooding the market these days), it will really shine in a course such as mathematical modeling. As mentioned before, the learning curve (in terms of writing simple user-defined functions) is not that hard for a junior or senior. Standard problems one sees in math modeling, such as the predator-prey problem, repeated drug dosages, study of defective genes, or pollution in lakes can *easily* be investigated by relying on simple functions which students

---

* From the Proceedings of an AMS Special Session at Knoxville, in March 1993, "Interventions to Assure Success: Calculus Through Junior Faculty" organized by Bettye Anne Case.

can be asked to generate or which can be provided by the instructor.

In conclusion, I must say that CC (or any other such CAS) is a worthwhile tool that enables one, I think, to deliver mathematics in a more interesting way. Not all students, of course, see this. But, more than ever, this tool forces me to reconsider many of the methods that were passed on to me and that I will seemingly pass on to others if I am not careful — methods which have more to do with handing down recipes because of time constraints and which are a clear disservice to mathematics. While it is important that the so-called "basics" be drilled in the students so that they can be ready for the next course, it is paramount that this approach be softened and that students be given a chance to ponder and synthesize as well. To achieve this, a friendly tool that reins in some of the mundane tasks of mathematics and allows a student to focus further out can be helpful if integrated in the curriculum by CAREFUL instructors.

Florida A & M University, Tallahassee, FL

Preface *

Steven G. Krantz

While most mathematics instructors prepare their lectures with care, and endeavor to do a creditable job at teaching, their ultimate effectiveness is shaped by their attitudes. As an instructor ages (and I speak here of myself as much as anyone), he finds that he is less in touch with his students, that a certain ennui has set in, and (alas) perhaps that teaching does not hold the allure and sparkle that it once had. Depending on the sort of department in which he works, he may also feel that hotshot researchers and book writers get all the perks and that "mere teachers" are viewed as drones.

As a result of this fatigue of enthusiasm, a professor will sometimes prepare for a lecture *not* by writing some notes or by browsing through the book but by lounging in the coffee room with his colleagues and bemoaning (a) the shortcomings of the students, (b) the shortcomings of the text, and (c) that professors are overqualified to teach calculus. Fortified by this yoga, the professor will then proceed to his class and give a lecture ranging from dreary to arrogant to boring to calamitous. The self-fulfilling prophecy having been fulfilled, the professor will finally join his cronies for lunch, confirming (a) the shortcomings of the students, (b) the shortcomings of the text, and (c) that professors are overqualified to

teach calculus.

There is nothing new in this. The aging process seems to include a growing feeling that the world is going to hell on a Harley. A college teacher is in continual contact with young people; if one feels ineffectual or alienated as a teacher, then the unhappiness can snowball.

Unfortunately, the sort of tired, disillusioned instructors that I have just described exist in virtually every mathematics department. A college teacher who just doesn't care anymore is a poor role model for the novice instructor. Yet that novice must turn somewhere to learn how to teach. You cannot learn to play the piano or to ski by watching someone else do it. And the fact of having sat in a classroom for most of your life does not mean that you know how to teach.

The purpose of this booklet is to set down the traditional principles of good teaching in mathematics — as viewed by this author. While perhaps most experienced mathematics instructors would agree with much of what is in this booklet, in the final analysis this tract must be viewed as a personal polemic on how to teach.

Teaching is important. University administrations, from the top down, are today holding professors accountable for their teaching. Both in tenure and promotion decisions and in the hiring of new faculty, mathematics (and other) departments must make a case that the candidate is a capable and talented teacher. In some departments at Harvard, a job candidate must now present a "teaching dossier" as well as an academic dossier. It actually happens that good mathematicians who are really rotten teachers do not get that promotion or do not get tenure or do not get the job that they seek.

The good news is that it requires no more effort, no more preparation, and no more time to be a good teacher than to be a bad teacher. The proof is in this booklet. Put in other words, this booklet is not written by a true believer who is going to exhort you to dedicate every waking hour to learning your students' names and designing seating charts. On the contrary, this booklet is written by a pragmatist who values his time and his professional reputation, but is also considered to be rather a good teacher.

I intend this booklet primarily for the graduate student or novice instructor preparing to sally forth into the teaching world; but it also may be of some interest to those who have been teaching for a few or even for several years. As with any endeavor that is worth doing well, teaching is one that will improve if it is subjected to periodic re-examination.

Let me begin by drawing a simple analogy: By the time you are a functioning adult in society, the basic rules of etiquette are second nature to you. You know instinctively that to slam a door in someone's face is (i) rude, (ii) liable to invoke reprisals, and (iii) not likely to lead to the making of friends and the influencing of people. The keys to good teaching are at approximately the same level of obviousness and simplicity. But here is

---

* See **Editor's Note** preceding the article in this chapter by Ed Dubinsky. Written as part of the FIPSE-assisted College Teaching Project. Reprinted from *How to Teach Mathematics: A personal perspective*, by Steven G. Krantz, by permission of the American Mathematical Society, 1993.

where the parallel stops. We are all *taught* (by our parents) the rules of behavior when we are children. Traditionally, we (as mathematicians) are not taught anything, when we are undergraduate or graduate students, about what comprises effective teaching.

In the past we have assumed that either

(i) Teaching is unimportant, or

(ii) The components of good teaching are obvious, or

(iii) The budding professor has spent a lifetime sitting in front of professors and observing teaching, both good and bad; surely, therefore, this person has made inferences about what traits define an effective teacher.

I have already made a case that (i) is false. I agree wholeheartedly with (ii). The rub is (iii). If proof is required that at least some mathematicians have given little thought to exposition and to teaching, then think of the last several colloquia that you have heard. How many were good? How many were inspiring? This is supposed to be the stuff that matters — getting up in front of our peers and touting our theorems. Why is it that people who have been doing it for twenty or thirty years still cannot get it right? Again, the crux is item (iii) above. There are some things that we do not learn by osmosis. How to lecture and how to teach are among these.

Of course the issue that I am describing is not black and white. If there were tremendous peer support in graduate school and in the professorial ranks for great teaching, then we would force ourselves to figure out how to teach well. But often there is not. The way to make points in graduate school is to ace the qualifying exams and then to write an excellent thesis. It is unlikely that your thesis advisor wants to spend a lot of time with you chatting about how to teach the chain rule. After all, your advisor has tenure and is probably more worried about where the next theorem or next grant or next raise is coming from than about such prosaic matters as calculus.

The purpose of this booklet is to prove that good teaching requires relatively little effort (when compared with the alternative), it will make the teaching process a positive part of your life, and it can earn you the respect of your colleagues. In large part I will be stating the obvious to people who, in theory, already know what I am about to say.

It is possible to argue that we are all wonderful teachers, simply by *fiat*, but that the students are too dumb to appreciate us. Saying this, or thinking it, is analogous to proposing to reduce crime in the streets by widening the sidewalks. It is double-talk. If you are not transmitting knowledge, then you are not teaching. We are not hired to train the ideal platonic student. We are hired to train the particular students who attend our particular universities. It is our duty to learn how to do so.

This is a rather personal document. After all, teaching is a rather personal activity. But I am not going to advise you to tell jokes in your classes, or to tell anecdotes about mathematicians, or to dress like Gottfried Wilhelm von Leibniz when you teach the product rule. Many of these techniques only work for certain individuals, and only in a form suited to those individuals. Instead I wish to distill out, in this booklet, some universal truths about the teaching of mathematics. I also want to go beyond the platitudes that you will find in books about teaching *all* subjects (such as "type all your exams", "grade on a bell-shaped curve") and talk about issues that arise specifically in the teaching of mathematics. I want to talk about principles of teaching that will be valid for all of us.

My examples are drawn from the teaching of courses ranging from calculus to real analysis and beyond. Lower-division courses seem to be an ideal crucible in which to forge teaching skills, and I will spend most of my time commenting on those. Upper-division courses offer problems of their own, and I will say a few words about those. Graduate courses are dessert. You figure out how you want to teach your graduate courses.

There are certainly differences, and different issues, involved in teaching every different course; the points to be made in this booklet will tend to transcend the seams and variations among different courses. If you do not agree in every detail with what I say, then I hope that at least my remarks will give you pause for thought. In the end, you must decide for yourself what will take place in your classroom.

There is a great deal of discussion these days about developing new ways to teach mathematics. I'm all for it. So is our government, which is generously funding many "teaching reform" projects. However, the jury is still out regarding which of these new methods will prove to be of lasting value. It is not clear yet exactly how **MATHEMATICA** notebooks or computer algebra systems or interactive computer simulations should be used in the lower-division mathematics classroom. Given that a large number of students need to master a substantial amount of calculus during the freshman year, and given the limitations on our resources, I wonder whether alternatives to the traditional lecture system — such as Socratic dialogue — are the correct method for getting the material across. Every good new teaching idea should be tried. Perhaps in twenty years some really valuable new techniques will have evolved. They do not seem to have evolved yet.

In 1993 I must write about methods that I know and that I have found to be effective. Bear this in mind: experimental classes are experimental. They usually lie outside the regular curriculum. It will be years before we know for sure whether students taught with the new techniques are understanding and retaining the material satisfactorily and are going on to complete their training successfully. Were I to write about some of the experimentation currently being performed, then this book would of necessity be tentative and inconclusive.

There are those who will criticize this book for being reactionary. I welcome their remarks. I have taught successfully, using these methods, for twenty years. Using critical self-examination, I find that my teaching gets better and better, my students appreciate it more, and (most importantly) it is more and more effective. I cannot in good conscience write of unproven methods that are still being developed and that have not stood the test of time. I leave that task for the advocates of those methods.

In fact I intend this book to be rather prescriptive. The techniques that I discuss here are ones that have been used for a long time. They work. Picasso's revolutionary techniques in painting were based on a solid classical foundation. By analogy, I think that before you consider new teaching techniques you should acquaint yourself with the traditional ones. Spending an hour or two with this booklet will enable you to do so.

I am grateful to the Fund for the Improvement of Post-Secondary Education for support during a part of the writing of this book. Randi D. Ruden read much of the manuscript critically and made decisive contributions to the clarity and precision of many passages. Josephine S. Krantz served as a valuable assistant in this process. Bruce Reznick generously allowed me to borrow some of the ideas from his booklet *Chalking It Up*. I also thank Dick Askey, Brian Blank, Bettye Anne Case, Joe Cima, John Ewing, Mark Feldman, Jerry Folland, Ron Freiwald, Paul Halmos, Gary Jensen, John McCarthy, Alec Norton, Mark Pinsky, Bruce Reznick, Richard Rochberg, Bill Thurston, and the students in our teaching seminar at Washington University for many incisive remarks on different versions of the manuscript. The publications committees of the Mathematics Association of America and of the American Mathematical Society have provided me with detailed reviews and valuable advice for the preparation of the final version of this book.

Washington University, St. Louis, MO

The Importance of Preparing Future College Faculty:
What Mathematics Can Learn from National Projects Outside the Discipline *

Brian Lekander

Recent efforts to improve undergraduate education, important as they have been, have largely neglected a promising strategy: that is, to improve the teaching abilities of those entering the professoriate. New attention to outcomes assessment, school-college articulation or curricula — such as the recent reforms in calculus — will all be empty if those in the classroom remain wedded to outmoded pedagogies, subordinate their instructional responsibilities to other interests, or, because of inexperience or insufficient training, cannot meet the demands of today's students. Despite the criticisms that have been leveled at nearly every other sector of American education, we generally believe that our graduate universities are unsurpassed in quality. That they produce excellent researchers is certainly true. But for too long we have failed to question seriously whether these universities are doing their best to prepare doctoral candidates for another of their major, if not primary, responsibilities: teaching. We have roundly criticized the excesses of school teacher preparation, often justly, but we have quite comfortably accommodated ourselves to the notion that doctoral programs are doing just fine at producing college instructors.

Now, however, there are a number of projects sprouting up across the country that are beginning to challenge the way that doctoral students are initiated into the teaching profession. These projects are testing the idea that formal attention to teaching issues during doctoral education can hasten the development of good teachers. These projects usually combine several modest but sensible strategies: seminars devoted to successful teaching practices, mentor relationships between senior faculty and graduate students, and supervised classroom teaching experiences. The projects do not challenge the importance of disciplinary research; they seek instead to restore an appropriate balance between teaching and research. The Fund for the Improvement of Postsecondary Education (FIPSE), a federal agency committed to funding innovative instructional models, has supported many of them. What I intend here is first to describe more fully why FIPSE has found it important to fund these projects to prepare future college faculty, not just in math but across the disciplines. Then I wish to describe some of the different experiments now underway, citing the specific institutions and associations that have initiated them, and suggesting some ways that mathematics departments might adapt their innovations.

What's Wrong with the Graduate Preparation of College Teachers? By way of introduction to the problem, let me begin by describing generally (in an admittedly slanted way) a few features about the process by which graduate students become higher education faculty, whether in math or in other disciplines. Nationally, the environment is, of course, shaped by the major research universities. They confer the most doctoral degrees, and their graduates tend to be the most attractive to hiring institutions. Significantly, though, it

From the Proceedings of an AMS Special Session at the 1992 Joint Mathematics Meetings in Baltimore organized for the AMS-MAA-SIAM Joint Committee on Preparation for College Teaching by Bettye Anne Case. Mr. Lekander is a Program Officer with the Fund for the Improvement of Postsecondary Education. This article represents the personal views of the author, and is not a policy statement of the U.S. Department of Education.

is the individual graduate departments in these universities that have become the gatekeepers for the college-teaching profession.

In doctoral departments, students are trained to become researchers in a discipline. Although students may financially support themselves as teaching assistants, teaching may delay completion of the degree. Frequently, it takes place in the most difficult large introductory sections, an environment made necessary by fiscal pressures, not by the needs of undergraduates or of new teachers.

The best graduate students, especially in the sciences, may not teach at all. They support themselves via research assistantships or fellowships. Likewise, the most highly regarded faculty may not teach at all, though their salaries and status indicate tangibly that they have reached the top of their profession.

At the end of graduate training, when students have completed their doctorates, they enter a job market that often most highly values the best researchers, and their new jobs challenge them with the hurdle of tenure, for which they are often judged principally, if not exclusively, on the quality of their publication records.

Allow me to make a few generalizations about this process, which are largely accurate, I think, despite the inevitability of many exceptions across the country:

1) The system of graduate education that's now in place exerts heavy socializing pressure on graduate students to devalue teaching, not only relative to research, but generally. If teaching is not always explicitly considered a waste of time, it is frequently considered of secondary importance and a distraction from graduate study. Furthermore, students develop unhealthy hierarchical notions about the relative merits of research universities versus other institutions, such as liberal arts colleges, comprehensive universities, and community colleges, where teaching is the chief priority. Graduate students quite naturally develop the values of the environment in which they are trained, despite the fact that most will earn their living as classroom teachers at schools other than research universities.

2) The system that's now in place does not serve the needs of undergraduates, who are suffering from a lot of bad teaching. For example, droves of students are failing calculus, the gateway to math and science careers. Some large universities have attrition rates in calculus hovering around 50–60%. It is reasonable to assume that this is happening at least partly because the most common models for teaching are not working. Of course, graduate education is by no means solely responsible either for bad teaching or for the devaluing of teaching. To a large extent faculty reward systems, the quest for grant dollars, and even the excitement of new and intriguing research questions may be held partly responsible. The important point, however, is that graduate education, because it is training college teachers, has a responsibility to undergraduate education, a responsibility that should be shared by, but not wholly relegated to, hiring institutions.

3) In many important ways, graduate schools offer the same training to everybody. The system that's in place is by and large the only one. There are no real alternatives. (The doctor of arts, for example, has been tried for those intending to teach, but it creates graduates who are perceived as second-class citizens, and indeed they are in terms of research preparation.) Among the nation's doctoral programs, great diversity can be found in research methodology, or schools of thought, but there is not generally much diversity of mission for preparing students. Doctoral programs of the second or third tier usually try their best to become research universities of the first tier, imitating their better cousins rather than staking out turf of their own. The differences that exist are too often those of size and hierarchy rather than character and mission.

What Features Should Be Characteristic of the Graduate Preparation of College Teachers? The development of teachers takes years, and certainly doesn't happen via a semester-long seminar or a series of workshops. But graduate students' development as teachers should be accelerated. They should develop aspirations of classroom creativity and productivity that will serve them well for a long time. Early in their careers, they should learn strategies that work so that fewer students will be guinea pigs for mistakes of inexperience or trial and error.

The development of research skills takes a great deal of energy, too. We don't want to cut into that. We want teachers who are good researchers. But faculty have other responsibilities as well. They serve on committees, they advise students, they serve the community. They serve as peer reviewers for grant applications and journal submissions.

So, in supporting graduate education projects to prepare future college faculty, the Fund for the Improvement of Postsecondary Education (FIPSE) has sought to identify those with the following characteristics:

1) Good projects socialize graduate students in such a way that they learn to value teaching. Graduate students are at an important formative stage. They're just learning what will be expected of them in their professional lives, just assimilating the values and behavior that characterize academics. As a result, FIPSE believes it's important to target graduate students, rather than exclusively supporting new faculty training and other faculty development programs to improve college teaching. For brand new teachers, it's especially important to "get their feet wet" in a supervised setting, where they have the freedom to fail and ask questions. Graduate students have this freedom. Newly hired faculty, however, are burdened with the pressures of a full teaching load, tenure review, and departmental politics. How many of them have been armed with the conviction that development as a teacher is a priority that should not be compromised?

2) Good projects train graduate students in the fundamentals of good teaching — not just bare-bones survival skills, such as learning to lecture without nervous tics, to lead a discussion, or to grade papers — but in skills and topics that are genuinely valuable to all teachers, regardless of their level of experience. A basic familiarity with the literature on how to teach mathematics, for example, should be expected of everyone teaching in the discipline. Generally (but not always) FIPSE-funded projects have been targeted at dissertation-stage graduate students and presume some prior teaching experience. They directly address professional preparation in a way that TA-training programs do not.

3) Good projects help doctoral students adjust to the full range of their responsibilities as faculty including committee participation, advising students, proposal writing, and even research. At St. Olaf College, where FIPSE sponsored a teaching postdoc program in mathematics, senior faculty were surprised to find that several newly minted doctorates did not come to them familiar with the basics of how to write a paper and select a journal for publication.

4) Good projects tend to engage the doctoral institution, or department, in a reflection about its mission. What kinds of students typically graduate from the institution? Where are these students most typically hired? In what ways can institutional resources such as endowed fellowships be used to encourage better teaching? These projects clearly define what is expected of individual graduate students as teachers, making the educational experiences distinctive and well coordinated.

What Specific Strategies has FIPSE Supported?

Finally, let me get down to some specifics: These are some of the strategies FIPSE has supported, along with a few comments on each, and some suggestions for how math departments might think about improving the preparation of their own graduate students for the teaching profession:

1) *Collaborations Between Graduate Schools and Liberal Arts Colleges.* The Association of American Colleges and the Council of Independent Colleges have each created formal partnerships between some of their member liberal arts colleges and nearby graduate schools. In these projects, graduate students not only attend teaching seminars, but they also gain experience teaching on liberal arts campuses under the supervision of liberal arts faculty. The great advantage of this approach is that students gain exposure to a different learning environment and can learn from faculty who are not overseeing their doctoral study. (Graduate students in each of these projects have expressed surprise at the vitality of the liberal arts faculty, about whom they had many misconceptions.) This type of project is probably limited to schools that are close to each other geographically, and, depending on the breadth of the activities, sharing program expenses can be a difficult proposition. Nevertheless, there's no reason math departments couldn't

initiate some similar experiments of their own.

2) *Programs Geared to the Needs of Hiring Institutions.* The CUNY Graduate Center has instituted a program in which graduate students attend a formal seminar and then participate in a field-teaching experience specifically intended to prepare them for teaching positions at two- and four-year colleges in New York City. The University of North Texas, on the other hand, has created a program in which doctoral students are prepared to teach undergraduate survey courses, specifically those that are part of the University's Classic Learning Core program. In each of these cases, the graduate students are perhaps not so well trained as they might be for some learning situations, but they're better prepared for their probable destination: if most CUNY graduate students teach in urban colleges within the large citywide system, why not prepare them for it? Math departments might ask themselves whether their students typically head for a particular kind of teaching situation; or whether their students demand a particular kind of training. If, for example, graduate students will be expected to understand and use mathematical computing, to teach service courses to underprepared undergraduates, or to teach minority students at urban commuter campuses, then perhaps they should be trained for it.

3) *Programs that Recognize Different Stages in a Teacher's Development.* Syracuse University, American University, and Emory University have quite different projects, but each has created a program in which graduate students are not expected to master all teaching responsibilities at once. Instead, they recognize that a graduate student is developing as a teacher as well as a scholar. Over the course of a few years, teaching responsibilities gradually increase; a student may begin leading a discussion section, then teach under supervision, and then finally, lead one's own course. Along with the increase in responsibilities come different kinds of training, increases in pay, and changes in title. Math departments in particular might think about how they can allocate course assignments in creative ways so as to avoid graduate TAs' perpetual assignments to the same service courses.

4) *Teaching Postdoc Programs.* Although not aimed at graduate students, postdoc positions are a common entry point to the teaching profession for newly minted PhDs. St. Olaf College's math department experimented a couple of years ago with the idea of specially funded teaching postdocs. Although funding wasn't available to continue them, they proved to be valuable to both faculty and to students. If someone in your department has a postdoc position funded through NSF or a foundation research grant, why not explore ways in which teaching experiences can be used to complement a postdoc's research responsibilities?

5) *Centrally Administered Interdisciplinary Programs.* The University of Oklahoma and the University of California at Riverside are among the many schools that

have organized special programs aimed at doctoral candidates in the late stages of their programs. Often these programs are cross-disciplinary and are run by the graduate school, an education department, or a faculty development center. Sometimes these programs have difficulty gaining participants, largely because they operate outside of graduate departments, and consequently they have little support and funding. If a program like this is mandated, the best way to ensure that good things happen is to participate, especially as faculty mentors; another is to volunteer to run your own teaching program in lieu of that offered by the central administration. If the program is run by Education, English, or other faculty, it might not seem relevant, but math graduate students still might gain valuable lessons about teaching by participating.

6) *Centralized Fellowship Programs.* Loyola University of Chicago and American University have shown that cross-disciplinary projects similar to those described above may work quite well, especially at smaller campuses where there are teaching-fellowship programs. There may be an informal community on a campus that is ripe for some kind of planned teaching-oriented activities, and the fellowships can help to overcome departmental boundaries in soliciting participation.

7) *Programs Led by Professional Associations.* One obstacle to greater emphasis on teaching in a graduate program may be the peer pressure that exists within a discipline. But if the discipline as a whole demonstrates that teaching is a priority, then graduate departments will follow. Fortunately, the mathematics professional associations — the MAA, AMS, and SIAM — have been exerting their leadership among math departments. A FIPSE grant to the MAA has sponsored a series of pilot projects in math departments: Cincinnati, Clemson, Dartmouth, Delaware, Oregon State, and Washington (St. Louis) have all participated. Of course, even if your department is not funded under such an effort, you can benefit directly from the lessons and materials that come out of these pilot efforts. Expose your students to those senior faculty who are not only excellent researchers but care about their students. Engage your doctoral candidates in teaching-oriented activities. Take them to professional math education meetings, and encourage them to write and present papers on teaching topics.

A Final Note. If you are in position to hire recent graduates of PhD programs, consider seriously those who have worked to develop their teaching skills. Don't ignore the need for these new hires to continue this development in their initial years of teaching, and give them a reward system that will encourage them to do so for the life of their careers. Changing graduate preparation alone is not going to be sufficient to improve college teachers, but it is a good first step.

Department of Education, Fund for the Improvement of Postsecondary Education, Washington, D.C.

Math Education Reform:
How Is It Proceeding — Are We On the Right Track? Update on a Differing View

Eleanor G. Palais

Three years ago a paper which I published in the *Notices* of the American Mathematical Society, "A Differing View on Mathematics Education Reform", started with the sentence; "I have an uneasy feeling of *deja vu* as once again there is talk of what's wrong with our math curriculum and why aren't our students doing better in their scores and in their general math achievement and understanding". Now three years later, although some of my fears have been calmed, there is still a serious concern on the part of at least this teacher that we may again be pursuing a costly "Reform" without learning lessons from the past. A band-wagon atmosphere exists as teachers, college professors, and administrators all claim to have found a cure for students' inability to learn mathematics by changing curriculum, adding technology, and restructuring our traditional ability grouping. The implication is that such changes will improve the nation's math ability. The enthusiasm, time, and effort are commendable, but gnawing questions are still there. Will change work? Will scores go up? Will more students go into pure math? Will students find math more meaningful? Will industry find a more competent work force?

The new texts being prepared today are sensitive to the changes which are occurring. Transformation theory and symmetry are being emphasized while the role of proof in Euclidian geometry is lessened. The changes are those suggested in *Standards*, a document written by the National Council of Teachers of Mathematics, the Bible of the reform movement. Iteration of functions and chaos and fractal theory are subjects which are now included in many new high-school textbook series in order to introduce students to the "cutting edge" of mathematical research and make math more meaningful and interesting. Wonderful software is available for schools which can afford it. To use the software, students need a computer laboratory with updated machines, LCD displays, and talented teachers who are familiar with teaching using the new technology. At present there is money available as the government and private sector join the movement to improve the nation's math ability. Grant-writing has become the norm as school systems on tight budgets vie to get a small piece of the pie.

It is an exciting time to be a teacher, and yet I continue to be very uneasy about aspects of the reform movement, perhaps because I want so much for it to succeed this time around. My major sources of concern focus around several issues. One is the movement toward restructuring. In many schools this means dropping or reducing the number of heterogeneous ability groups, even in the high-school grades. Homogeneous grouping sounds very attractive and democratic, and it certainly may be so in the earlier grades. As youngsters develop at different speeds, homogeneous grouping in early middle school allows all students to begin on a level playing field, while not relegating some to honors and some to other levels. In mixed-ability classrooms the stronger students can encourage and help the slower students improve by working together. Even though I myself have concerns about whether this method is fair to the more adept student, I can be persuaded that in the earlier years it may have some merit. By ninth grade, however, ability grouping, which has worked well in the past in giving many students a feeling of success while learning, should remain in place. Of course, teachers should always encourage students to move up a level, if appropriate, and there should be flexibility between levels. In our school we have four ability tracks, each of which serves a very different student population. Though we address the same curriculum in both middle levels, it is the way we teach that is quite different. At the start we go more slowly in the lower level while giving much support and encouragement. There is more explanation of homework and less time-pressure in tests. It is not uncommon for students who are in the top of a lower-track class to move up to the next level, usually at the start of the next year, as they have gained in both confidence and skill. Some of these students actually double up and take two math courses during the tenth grade as they feel competent and able in the subject. We must then take care that we do not hold back any of our best students by homogeneous grouping, and that we allow our slower students a measure of success in a lower group. It is often the taste of success which is the catalyst that encourages a student's marked improvement in mathematical ability. That success might never be achieved in a homogeneous ability-grouped class where a student is only an average member of the group.

A second concern, one I have mentioned in my previous article, is that in order to truly understand why our students' math ability is poor, we must examine the students themselves. One reason I believe our young people today are doing poorly in their school work is because they have been brought up as passive listeners in a TV generation. When I am teaching, my students are often glassy-eyed and watching what is happening as if I were a TV performer. *The students do not receive what is being said!* Students must be taught about active listening and how to distinguish it from the passive listening they do while watching entertainment. In addition to developing poor listening skills, watching television has made this generation forget how to visualize since this is always done for them. For example, if we say "imagine a trapezoid", students today do not see the picture the way we do. The same applies also to thinking through a problem. Students have difficulty in visualizing the steps ahead, as they are not accustomed to imagining or visualizing anything. Our major efforts must be placed on

helping students overcome these deficits if we are really going to improve their ability to learn. This is where our creative education and monies should go. For the first time in history, we are teaching a generation of students who for the most part have been brought up to tune out whatever noise is going on around them—even if it's teaching. There are some researchers who believe that there have been actual changes in the brain's neural wiring during this TV-generation's early childhood. These changes may seriously affect the attention span of young students today. An interesting discussion of this is given in a book by Jane M. Healy: *Endangered Minds: Why Children Don't Think and What We Can Do About It.* It is clear we must pay further attention to the students themselves.

For the first time we have also a generation in school which has not *had* to learn to survive. Many of our students don't have the competitive worries people their age used to have. They know they will be able to make it because they come from well-off families, while, on the sadder side, others feel that the world has become so upscale that they have little chance of making it no matter how hard they work. In short, many feel that education is no longer the guarantor of making it in the world. It is a sad commentary on our society today that it also does not value education as it has in past generations. In many school programs, academics take second place to a variety of extra-curricular activities including sports, band, et cetera, and teachers sit helplessly by as numerous field trips disrupt the continuity of class attendance and lessons, while family extended vacations often take priority over school presence.

As a band-aid for the difficulties in education the politically popular cry is for longer school days, longer school years, and more accountability by teachers. In fact none of these will improve the situation, for more of the same will bring only the same poor results. The problem is not the length of a student's time in class, or the grade that a teacher gets in taking some ability test devised by an education board. We should be looking at the students themselves and also at societal values, as possible reasons for low math achievement. While we should look for ways of improving our teaching styles and updating our curriculum as well making the subject matter meaningful and dynamic, we must also be cautious in making sweeping revisions of those styles and methods of teaching math. For example, I enthusiastically endorse the use of technology in the classroom as a vital tool. But we must be wary that as we encourage greater interaction with the computer, which is clearly less passive than a student's watching a TV screen, we are nevertheless teaching by having students watch a TV screen. Are we sure this is good? Will it work? Are we evaluating our results as we proceed and maintaining the flexibility to stop technology overload if the results don't noticeably improve within a reasonable period of time? Please educators, keep the pencil, the paper, and the mind still active. Mathematics is learned only by doing, not by watching!

The economic costs involved in this reform movement are staggering! The National Science Foundation has joined many private organizations in funding grants for schools, colleges, and universities. The economic burden of the computerized classroom with its hardware and software being frequently upgraded will be an ongoing problem for schools already feeling the strain of overloaded budgets. Improved teacher salaries as well as incentives for talented young college graduates to pick careers in education must be a top priority for our nation, or we will face a serious shortage of quality teachers as the enrollments grow in our lower grades.

Here is one teacher who desperately wants this reform movement to succeed. As we change curricula and teaching styles, and as we introduce new technology into our classrooms, let us not forget the students themselves. We must make certain that we are sensitive to their needs as well as to those of our society today. We must be sure that students can still read the printed word and express in clear writing their questions and their ideas. They must be taught to listen, to learn, to question, and to probe. Society must give value to their successes along with society's demands for excellence. The solution to the problem may be more complex than merely introducing technology and updating curriculum. Technology is a powerful tool, and it must be used as such. It is not a substitute for the mind. Finally, we must implement ongoing appraisal and evaluation of our changes as we proceed with this reform.

Above all, society must decide by its actions, and not by words alone, whether or not it values quality education for its youth. If it does, it must make teaching a profession which attracts our best college graduates. Our country has shown its concern over our students' poor math achievement. It must now show that it supports the reform movement and excellence in education with deeds as well as with words. Academic excellence must be a top priority in family life, and the education of our young a top priority on local, state, and federal agendas.

Belmont High School, Belmont, MA

Scholars in Mathematics at Spelman *

Wanda M. Patterson

There was standing room only, thunderous applause and overwhelming enthusiasm, awards to top achievers and a festival atmosphere at the culminating activity, "A Scholarly Affair," of Scholars in Mathematics at Spelman College. This activity grew out of the enthusiasm and excellence of the program's weekly forum at which program participants gave their first research presentations,

discussed interesting problems from their fields, gave in-progress reports on their own research and shared with professionals from the immediate geographical area the nature and preparation of their research careers. An analogous weekly meeting for freshman has been added as a new course entitled "Research Issues in Mathematics." This course introduces students to mathematical literature, great mathematical ideas of the past and present, and an axiomatic approach to a problem; that is, given a set of definitions or axioms, how do you use them to approach a given problem? Included in this course also is much reading and discussion on mathematical organizations, twentieth-century mathematicians and ethical issues facing mathematicians today.

The project Scholars in Mathematics at Spelman (SIMS) aims to identify, develop, and encourage Spelman students to pursue graduate degrees in mathematics. The program provides talented students with a strong academic background, strengthened further by very rigorous electives and research experiences, to assure their preparedness as they continue on to pursue degrees in mathematics at the best graduate schools in the country. Mentoring, summer research experiences, and attendance and research presentations at selected regional and national conferences are also a part of the program.

The SIMS program was funded in the Fall of 1991 by the Office of Research Careers for Minority Scholars of the National Science Foundation. That first year of the program seven scholars were selected: one senior, three juniors, and three sophomores. In the 1992-93 academic year the SIMS program consisted of nine (six continuing from the previous year) upperclass scholars and seven freshmen. The freshmen were dubbed pre-SIMS scholars and received a partial scholarship, mentoring, and tutorial support. Two of these are expected to continue next year as SIMS scholars. In addition to these, in the 1993-94 academic year there will be two or three freshmen pre-SIMS scholars and the continuing scholars (three rising juniors and three rising seniors), making a total of ten.

Thus far the overall progress of the SIMS program as measured by the activity and success of its scholars is excellent. All four scholars who have graduated from the program will be enrolled in graduate mathematics programs in the Fall (three in PhD programs). All graduates have given research presentations at either national or regional conferences. All scholars (including four of this year's freshmen) have found meaningful summer research internships each year since being selected for the SIMS program. Spelman expects great things from their scholars in the future and shall continue to equip them to "Go for the Gold."

Spelman College, Atlanta, GA

From the Proceedings of an AMS Special Session at Knoxville, in March 1993, "Interventions to Assure Success: Calculus Through Junior Faculty" organized by Bettye Anne Case.

Calculus and Pre-Calculus Laboratories:

SC–AMP at Clemson University and the University of South Carolina *

T. G. Proctor, Susan J.S. Lasser,

Mary Ellen O'Leary, Robert W. Snelsire

A consortium of institutions in South Carolina was funded by the National Science Foundation in the fall of 1992 as part of the Alliance for Minority Participation in Science, Engineering and Mathematics Program. Components of the consortium project were initiated at that time at Clemson University and at the University of South Carolina. Other institutions in the Alliance will have the academic elements of the program underway by summer 1993. This report describes activities at Clemson and the University of South Carolina during the academic year of 1992-93.

There are two programs for minority students at Clemson University. The Program for Engineering Enhancement and Retention (PEER) and the Math Excellence Workshop, both components of the Minority Engineering Program, are well established in the College of Engineering,. Another program, Tools to Enrich and Advance in Mathematics and the Sciences (TEAMS) is being developed in the College of Sciences. At the University of South Carolina, the Emerging Scholars Program involves students from both the College of Science and Mathematics and the College of Engineering. This preliminary report will give abbreviated descriptions of (1) PEER — the most established element of the project, (2) the Math Excellence Workshop (MEW), (3) the activity in freshman mathematics courses, and (4) a course designed for freshmen students — "University Survival Skills".

PEER. This program at Clemson University began in 1987 using funds provided by the College of Engineering; supplementary funds were soon obtained from a research foundation. Development of the program was based on information obtained by reviewing existing research on African American students at predominately white universities and by interviewing successful junior and senior African American students. The initial element was a mentoring program. This proved to be an essential ingredient. At present there are 15 African American seniors with good academic records each mentoring eight to 10 new (freshman and transfer) students. The mentors themselves are given a six-hour training session and meet with the minority engineering program director once a week. The purpose of the mentoring program is

to: (1) treat isolation problems; (2) close an information gap which is typical of first generation college students; and (3) encourage interaction with University support services and teaching staff.

A Study Hall was introduced by PEER in the second year (1988). This activity is provided five days per week, three hours per day for basic courses in mathematics, chemistry, physics, computing, and basic engineering.

MEW. The Math Excellence Workshop, introduced by the Minority Engineering Program in the summer of 1990, was designed for engineering students beginning their program at Clemson with the pre-calculus course. Students take the course as offered by the mathematical sciences department (five days per week for five weeks) and in addition attend a three-hour daily honors-level seminar, which follows the pre-calculus syllabus but emphasizes collaborative learning. This seminar is facilitated by a member of the mathematical sciences faculty. Tutors and mentors are provided in the late afternoon and evening.

The formal staff of the Minority Engineering Program consists of a director, a faculty liaison, a secretary, and the mentors. There are funds from a number of external grants and from corporate contributors. There is a program office with some study/lounge space and a few microcomputers for students to use during the day. The staff members encourage the students to use these facilities.

PEER and the MEW are effective at Clemson University. There are various measures of this success. For example, of 50 students who started in the pre-calculus course prior to the MEW only two graduated in engineering, but it is expected that at least 10 out of the 22 who participated in the 1990 MEW will graduate in engineering. This type of success has led to the development of a companion program in the College of Sciences (TEAMS) at Clemson.

The Mathematics Gatekeeper Courses. An important ingredient of the SC–AMP consortium project is organized activity for engineering and science students in their first year of mathematics courses. Some of this assistance occurs in summer programs for students entering college, and other assistance occurs during the regular academic year. It is felt that a solid foundation in these courses is a good indicator that the student is likely to graduate in science and engineering. The programs which are still developing at Clemson and the University of South Carolina will be sketched here.

The Math Excellence Workshop described above is being expanded to include students who major in the College of Sciences. A similar program is scheduled for this summer at the University of South Carolina. The USC activity consists of supplementary sessions in which the students are encouraged to work in groups of, say, four, and to develop solutions on paper and make blackboard presentations of problems-solutions which are an extension of regular classwork.

---

* Based on a talk by T.G. Proctor. From the Proceedings of an AMS Special Session at Knoxville, in March 1993, "Interventions to Assure Success: Calculus Through Junior Faculty" organized by Bettye Anne Case.

During the academic year there are programs to assist the participants in pre-calculus and in the first year of calculus. The students supplement their work in a regular section of their mathematics course by a laboratory which meets twice a week. Students are encouraged to work in small student groups on problems designed by faculty in the mathematics department. A teaching assistant is assigned to help supervise this activity. Part of this supervision is to see that each student participates in the solution of the problems.

At the University of South Carolina, the Calculus Excellence Workshop is conducted as MATH 151 for the first semester and MATH 152 for the second. The workshop meets four hours a week and carries two credits. Students are graded on presentations at the board, written solutions to selected worksheet problems, and occasional quizzes, as well as on participation and attendance. At the USC, 11 Calculus I students participated in the workshop during its first semester of implementation and achieved the following grades in their regular calculus courses: A: 4; B or B+: 3; C or C+: 3; D or D+: 1, for a group average grade of 3.1 in calculus. This compares very favorably with an overall average grade of 1.87 at USC in calculus I in Fall 1992. During the Fall 1992 semester, three calculus II students who had been awarded SC–AMP scholarships also participated in the workshop. They worked on the calculus I worksheets to deepen concepts gained in high school AP calculus and served as peer tutors for the other students. These students also worked as a group on calculus II problems prepared especially for them. Their grades in the regular calculus II were two A's and a B+.

At Clemson, the seventeen students in calculus laboratories also achieved good grades; however, the performance of students in the pre-calculus laboratory was disappointing. At that time in the project there was no mentoring or counseling for the nonengineering students. The spectrum of activity faced by students in the first semester of college distracts many from a good balance of study and other activity. Students' familiarity with the material in the pre-calculus course may give them a false sense of what will be required to make a good grade. This justification has been given previously to explain the widespread poor performance in pre-calculus. Plans are under way to create a better environment for students entering college at the pre-calculus level at Clemson.

Workshops in calculus II at both Clemson and the University of South Carolina are well under way and student interest remains high. Since there is considerable overlap of students who have participated in both the calculus I and calculus II workshops, a close network involving informal study groups in other subjects has developed. It is felt that the students will continue to support each other throughout their undergraduate years.

University Survival Skills. A key ingredient of the SC–AMP project is a freshman course to help college students develop academic survival skills. USC has offered a freshman seminar course for over 20 years, with demonstrable success in retention. Beginning in Fall 1991, specialized sections have been offered for students in science, engineering, and mathematics. The intent is to combine the proven elements of the traditional course with the analytical, quantitative, and computer skills needed to be a successful science student. Required course materials include a good calculator, a calendar/planner, and two computer disks. General requirements include class participation, membership in a student organization, a written report from a popular science or math book such as *Innumeracy*, an oral report on a science or mathematics topic, a class journal with weekly entries, attendance at the student organization fair, attendance at the undergraduate advisors panel discussion, completion of the Strong Campbell Interest Inventory, and completion of the Health Risk Appraisal. Syllabus topics are as follows: introduction to computers, university structure and requirements, qualitative skills development, time management, study and test-taking skills, library research, assertiveness training, ethics, cultural diversity, stress management, a forum with university officials and with outstanding scientists, advanced computer skills, career center, student presentations, critical thinking, and evaluations of the course.

The interest generated by students and faculty at the University of South Carolina has resulted in a similar course for students in the sciences which is scheduled at Clemson in the fall semester of 1993.

Concluding Remarks. A small number of scholarships have been awarded to entering students with good credentials and to first-year students who both participated in the gatekeeper course workshop and had good academic records in their first semester. These scholarships can be renewed throughout the first two years of college to reward good scholastic accomplishment. Plans call for students in the junior and senior years to earn funds for service as mentors. Juniors and seniors will also be encouraged to participate in undergraduate research projects at their home institution and at other locations with organized summer programs.

Students participating in SC–AMP have been encouraged to acquire a graphics programmable calculator for their work in mathematics and in other technical courses. Limited funds are available for such acquisitions.

Creation of the South Carolina Alliance for Minority Participation has brought faculty from the various scientific disciplines together to discuss the effectiveness of group learning experiences, the uses of technology, and other ways in which gateway courses can be restructured to enhance student success. Across the Alliance, plans are under way to implement summer pre-calculus bridge programs, to establish the freshman seminar on each campus, to expand the calculus workshops, and eventually to extend the workshop concept to chemistry and physics.

References

[1] Asera, Rose, The mathematics workshop: a description, professional development program at Berkeley, Department Report, University of California at Berkeley, March 2, 1988.

[2] Bronson, Richard and Judith Kauffman, Promoting student success in college science through structure and support, *Journal of College Science Teaching*, February 1992.

[3] Conciatore, Jacqueline, From flunking to mastering calculus, *Black Issues in Higher Education*, Special Report, March 2, 1989.

[4] Hrabowski, Frank, A. and Willie Pearson Jr., Recruiting and retaining talented African-American males in college science and engineering, *Journal of College Science Teaching*, February 1993, pages 234-238.

[5] O'Brien, Eileen, Berkeley model proves successful for blacks, hispanics' calculus performance, *Black Issues in Higher Education*, March 2, 1989.

[6] Treisman, Phillip Uri, A study of the mathematics performance of black students at the University of California, Berkeley, 1985.

[7] Watkins, Beverly, Many campuses now challenging minority students to excel in math and science, *Chronicle of Higher Education*, June 14, 1989.

Clemson University, Clemson, SC

What is Mathematics Education Research? *

Annie Selden and John Selden

Mathematics education research is a young field with its own character, history, methodology, journals, and professional organizations. Most work has been with elementary and secondary students; however, for about a decade, there's been an increasing amount of research on how university students learn, or do not learn, mathematics [1].

When mathematicians think of pedagogy, they tend to equate it with efforts at clear, logical presentations of subject matter. During the "new math" era it was widely thought, "If only we can present it clearly enough, they will understand." In the intervening years, cognitive science developed, and today there is a growing realization that, upon first meeting a topic, "a logical presentation may not be appropriate for the cognitive development

---

* From the Proceedings of an AMS Special Session at the 1992 Joint Mathematics Meetings in Baltimore organized for the AMS-MAA-SIAM Joint Committee on Preparation for College Teaching by Bettye Anne Case.

of the learner" [13, p.3]. Here no one's calling for illogical presentations, but for consideration of psychological factors in addition to logical development. Such comments hint at major teaching changes that are quietly, but surely, overtaking a largely unsuspecting mathematics community. While the exact nature of the coming evolution cannot be predicted, it will surely be substantial, with lectures playing a diminished role.

The Nature and State of the Field. Mathematics education researchers study how people learn and are taught mathematics. Research methods and background are borrowed from anthropology, sociology, and psychology, especially cognitive psychology, as well as from philosophy and artificial intelligence.

Research in mathematics education is not mathematics; however, on the university level, it is produced by, and used by, mathematicians. It is not curriculum development; however, research results can speak to curriculum development. Making major changes in curriculum or teaching methods with inadequate knowledge of how students learn is like designing flying machines with little knowledge of aerodynamics. It is possible, but requires a lot of time, patience, and test pilots and is prone to sympathetic magic. Conversely, expecting extensive how-to-do-it teaching information from a research project in education is like expecting the typical mathematics paper to affect engineering practice directly.

Results in mathematics education call for synthesis and interpretation in ways unlike mathematics. A mathematician, when using a result, typically does no more than check the argument. Unless an error is discovered, a result true one year is true the next. In mathematics education, the situation is less clear-cut. Even very careful observations only suggest, but do not prove, general principles. Corroboration of other studies can be very important. The discovery of additional information can make what previously appeared firmly established, less so later. Definitions do not have mathematical precision. Concepts may be only approachable rather than definable — try catching the meaning of "understanding" precisely.

There is no comprehensive theory underlying mathematics education, despite various attempts at theoretical frameworks; e.g., the work of Piaget, van Hiele levels in the learning of geometry, and the action-process-object view of concept development [3, pp. 426-437; 12]. The field is in Kuhn's pre-paradigm phase, and direct observational studies and classifications are useful, just as they were, early on, in anthropology and biology. Constructivism, in varying forms, is the philosophical starting point. The learner is viewed as an active participant, not a blank slate upon which we write or an empty vessel which we fill.

A Brief History. Mathematics education research as a field, separate from psychology or philosophy, is largely a 20th-century phenomenon. Early in the century important work was done by educational psychologists who studied the psychology of arithmetic and algebra, arguing that bonds between stimuli and responses are strengthened through repetition and reward. Subsequently, nothwithstanding John Dewey's progressive influence, behaviorist inquiry became the dominant research paradigm in the US, concentrating on drill and practice, diagnostic testing, and predictors of achievement, with teaching effectiveness measured by test scores. Classroom teaching practices were compared using the statistical methods favored in agriculture for studying the effects of fertilizer on crops — a technique unsuitable for many of today's questions.

By the middle of the century, studies by educational psychologists had largely excluded subject matter such as mathematics. However, in the late '50s, due to Sputnik, federal funds became available for curriculum development, and mathematicians like E. G. Begle and Patrick Suppes recognized the need for conducting research and evaluation studies. They consulted psychometricians and psychologists and became interested in the work of Piaget.

In the '70s and '80s, American mathematics education researchers were influenced by cognitive psychologists. Journals and professional organizations developed, along with a growing interest in theoretical underpinnings. There was a reaction against behaviorist scientific inquiry, in favor of more qualitative methods. Students — they had minds whose cognitive processes could be explored. (Of course, this was no surprise to the likes of Piaget.) Research was extended to include a wide range of collegiate topics beyond those concerning preservice teachers [3, pp.3-38].

Methodology and Topics Studied. A variety of quantitative and qualitative research methods are used [3, pp.56-63], and philosophical work is important — how mathematical concepts are learned involves what they are.

There are surveys — descriptive pictures of large populations — such as the NAEP studies, known as "the nation's report card." Statistical techniques are used to analyze experimental versus control group studies, where data is collected before and after some teaching intervention: There are also evaluation studies of teaching methods and curriculum reform[5]. In meta-analyses, data from a number of studies is pooled, compared, and re-analyzed [17].

Anna Sfard's work on the dual nature of mathematical concepts and Ed Dubinsky's similar action-process-object analysis are examples of philosophical/theoretical research [12; 4, pp.85-108]. Imre Lakatos's work on proof draws heavily on historical analysis [7]. Clinical research and case studies are used to answer difficult questions concerning mental processes that occur, for example, during problem-solving. Here researchers make detailed observations of small groups or individuals involved in realistic or complex mathematical situations. Techniques include interviews, video- or audio-taping, paired prob-

lem solving, and questionnaires. The results include detailed descriptions and models of individual cognitive processes. Schoenfeld's fine-grained characterization of one student's subject-matter understanding and changes in her knowledge structures is an example [10]. In his work at Berkeley, Uri Treisman used anthropological-style observations to discover why one minority population's (African Americans') success rate in mathematics was lower than another's (Asian Americans'). Starting with conjectures about motivation, he found the key factor was really informal cooperative study [15].

Topics of mathematics education research include aspects of mathematics, cognition, psychological and cultural factors, teaching methods, and effects of technology. Here are some examples: functions [4,8], calculus [5; 14; 13, pp.153-198], proofs, logic and reasoning [7; 13, pp.215-230], problem-solving [9], representations [6], visualization[18], misconceptions [3, pp.470-478], beliefs, attitudes, and emotions [3, pp.575-596], learning styles [16], gender and equity [3; pp.623-660], culture of the classroom [3, pp.101-114], teachers' knowledge and beliefs [3, pp.127-164], cooperative learning [17], writing, and technology [5; 4, pp.109-132; 3, pp.515-556].

Where is this Research Published? How Can a College Math Teacher Find It? Researchers publish in research journals in the style and detail required by the subject. A consumer of research, in this case a college math teacher, just wants the results. What did the study show and how might this affect teaching? What's needed for teachers is more good expository articles, a form of writing unfortunately not much rewarded in the academic community. One source, for now, is our Research Sampler column in *UME Trends*, although we do not give anything like a complete synopsis of results. We have found interesting research about collegiate mathematics education in NCTM's *Journal for Research in Mathematics Education*, which uses four or five blind referees per paper and has about a 15% acceptance rate. Other good sources include the *Journal of Mathematical Behavior, For the Learning of Mathematics,* and *Educational Studies in Mathematics* (a.k.a. Freudenthal's journal). Unfortunately, these journals cover all levels, and an article on calculus might be followed by one on counting. We have also occasionally found articles scattered through *Cognition and Instruction, Cognitive Science, Educational Psychologist, Instructional Science,* and AERA's *Review of Educational Research.* These journals are even broader and might have an article on reading next to one on problem-solving.

There is no one place that specializes in university-level mathematics education research; so it is very hard to find, especially in small libraries. This is changing; the Joint AMS/MAA Committee on Research in Undergraduate Mathematics Education (CRUME) is sponsoring a series of annual research volumes with Dubinsky, Kaput, and Schoenfeld as editorial troika.

Several professional organizations take an interest in

university-level mathematics education research. In addition to sessions at AMS/MAA meetings, every fourth year the International Congress on Mathematical Education (ICME) meets. Its program embraces mathematics education at all levels, but research, though represented, is not the main focus. However, an offshoot of ICME, the International Group for the Psychology of Mathematics Education (PME), is for researchers. It meets every summer, and its North American affiliate, PME-NA, meets every fall. PME contains a Working-Group on Advanced Mathematical Thinking (AMT), whose concern is students ages 16+. All have invited addresses, contributed papers, and proceedings. The atmosphere of these meetings is somewhat different from those of AMS or MAA. Working groups and discussion-sessions play a larger role. There is a strong feeling that knowledge can be created as well as transmitted in meetings. It's as if the informal conversations found at mathematics meetings were formalized. The NCTM's annual meeting has a research pre-session, which may sometimes focus on a college topic like calculus. In addition, their meetings have expository sessions which transmit education research results to teachers.

Who Does this Sort of Research? Mathematicians do. Who else could investigate how students come to understand abstract algebra? Here we are using the term "mathematician" inclusively — a PhD in mathematics is neither necessary nor sufficient. What is required is a thorough understanding of both mathematics and the psychological and sociological aspects of teaching and learning.

For some time, accomplished mathematicians (Lebesgue, Poincaré, Hadamard, Polya) have commented on education and the psychological mechanisms underlying mathematical discoveries. These days this interest is accompanied by a scientific approach. Researchers in this field may be mathematicians who have taken up education research techniques or those from education who have a special interest in mathematics. While the traditional PhD in mathematics can provide the required mathematical background and sophistication, any mathematician seriously wanting to take up mathematics education research, as opposed to curriculum development, faces a significant task, on a par with that of changing research areas within mathematics itself. On the other hand, college teachers have the advantage of insights born of experience, not to mention a built-in supply of subjects.

For a longer description of mathematics education research and additional references, see [11].

References

[1] Becker, Joanne Rossi, and Barbara Pence. "The Teaching and Learning of College Mathematics: Current Status and Future Directions." To appear in an MAA Notes volume of selected papers from the San Francisco Special Session on Research in Undergraduate Mathematics Education.

[2] Dreyfus, Tommy. "Advanced Mathematical Thinking." In Nesher, Pearla and Jeremy Kilpatrick (Eds.), *Mathematics and Cognition: A Research Synthesis by the International Group for the Psychology of Mathematics Education* (pp. 113-134). ICMI Study Series. Cambridge Univ. Press, 1990.

[3] Grouws, Douglas A. (Ed.). *NCTM Handbook of Research on Mathematics Teaching and Learning.* Macmillan, 1992.

[4] Harel, Guershon and Ed Dubinsky (Eds.). *The Concept of Function: Aspects of Epistemology and Pedagogy.* MAA Notes No. 25, 1992.

[5] Heid, M. Kathleen, "Resequencing Skills and Concepts in Applied Calculus using the Computer as a Tool." *Journal for Research in Mathematics Education* 19 (1988) 3-25.

[6] Janvier, Claude (Ed.). *Problems of Representation in the Teaching and Learning of Mathematics.* Lawrence Erlbaum, 1987.

[7] Lakatos, Imre. *Proofs and Refutations.* Cambridge University Press, 1976.

[8] Leinhardt, G. O., O. Zaslavsky, and M. Stein "Functions, Graphs, and Graphing: Tasks, Learning and Teaching." *Review of Educational Research* 60 (1990) 1-64.

[9] Schoenfeld, Alan H. *Mathematical Problem Solving.* Academic Press, 1985.

[10] Schoenfeld, Alan H., J. P. Smith, and A. Arcavi "Learning: The Microgenetic Analysis of one Student's Evolving Understanding of a Complex Subject Matter Domain." In R. Glaser (Ed.), *Advances in Instructional Psychology* (Vol.4). Lawrence Erlbaum, in press.

[11] Selden, Annie, and John Selden. "Collegiate Mathematics Education Research: What Would That Be Like?". To appear in *The College Mathematics Journal.*

[12] Sfard, Anna. "On the Dual Nature of Mathematical Conceptions: Reflections on Processes and Objects as Different Sides of the Same Coin." *Educational Studies in Mathematics* 22 (1991) 1-36.

[13] Tall, David (Ed.). *Advanced Mathematical Thinking.* Kluwer, 1991.

[14] Tall, David and Shlomo Vinner. "Concept Image and Concept Definition with Particular Reference to Limits and Continuity." *Educational Studies in Mathematics* 12 (1981) 151-169.

[15] Treisman, Uri. "Studying Students Studying Calculus: A Look at the Lives of Minority Mathematics Students in College." *The College Mathematics Journal* 23 (1992) 362-372.

[16] Turkle, Sherry and Seymour Papert, "Different Ways People Think: Epistemological Pluralism and the Revaluation of the Concrete." *Journal of Mathematical Behavior* 11 (1991) 3-33.

[17] Webb, Noreen M. "Task-Related Verbal Interaction and Mathematics Learning in Small Groups." *Journal for Research in Mathematics Education* 22 (1991) 366-389.

[18] Zimmerman, Walter and Steve Cunningham. *Visualization in Teaching and Learning Mathematics*, MAA Notes No. 19, 1991.

Tennessee Technical University, Cookeville, TN

Writing in Mathematics Classes:
Helping a Writing Expert Help You

Barbara E. Walvoord

"Writing about mathematics should receive increased attention" in mathematics instruction, says the National Council of Teachers of Mathematics (Curriculum and Evaluation Standards for School Mathematics, 1989). "Mathematicians write; mathematics students should too," is the title of Ann Stehney's lead article in *Using Writing to Teach Mathematics*, edited by Andrew Sterrett and published by the Mathematical Association of America (1992). The call for writing in mathematics is by now widespread.

But how do mathematics faculty find effective ways to integrate writing into their classes?

Many have turned to the so-called "writing-across-the-curriculum" (WAC) programs that exist on about 40% of U.S. campuses, according to a 1988 national survey (in McLeod, S. [Ed.], *Strengthening Programs for Writing Across the Curriculum*, San Francisco: Jossey-Bass). Typically spawned in English or composition programs and led by directors who are specialists in writing, these programs almost always offer workshops for faculty across disciplines, helping them integrate writing effectively into their courses in mathematics, philosophy, physical therapy, or what have you. For faculty whose campuses have no WAC program, such workshops are available regionally or nationally.

WAC workshops are potentially very valuable for mathematicians. As I look across the wide spectrum of disciplines represented on my own and other campuses, mathematics stands out because of its concern for pedagogy, voiced through its national associations and through the concerns of individual faculty members. Mathematicians themselves attest to the usefulness of writing for them and for their students. Yet few college mathematics faculty have any formal training in pedagogy, let alone in using writing in the classroom. So collaboration with writing specialists like me who have studied how students learn (or don't learn) to write and how writing can be used to enhance mathematical thinking, who know the literature in the field, and who lead workshops regularly are potentially very useful for mathematicians.

A workshop may be enormously useful or largely irrelevant to mathematics faculty, depending on how well the writing specialist who is leading it is able to address the concerns and characteristics of teaching mathematics. Many writing specialists do not know very much about mathematics or mathematics-teaching. I'm a good example—my last mathematics class was high school algebra in 19—well, never mind. Writing-across-the-curriculum programs, even if they do know how to speak relevantly to mathematics faculty, may not have good ways of reaching them. When was the last time you carefully read a brochure in your mailbox from a non-mathematician?

Having worked successfully with many mathematics faculty, and profited greatly from the good ideas of my mathematics faculty colleagues at the University of Cincinnati and elsewhere, I want to offer here a few suggestions about how the mathematical community might help itself get something useful out of a writing specialist like me and a typical WAC program like the one I direct.

1. *Make your needs known before the workshop.* Before I arrived to do a two-day workshop on WAC at one university a few years ago, the faculty committee who had invited me met with some selected colleagues, and polled others, to discover what questions and issues were particularly important to the people who would be attending the workshop. They sent me the list. On it were some questions specifically related to mathematics, reflecting the concerns of the mathematics faculty who would be attending. Whenever a university does that, it helps me, as workshop leader, arrive prepared to address their concerns.

2. *Send materials about writing in your discipline.* It was a mathematics-faculty member at the University of Cincinnati who introduced me to Andrew Sterrett's edited collection of articles by mathematicians, *Using Writing to Teach Mathematics* (Mathematical Association of America, 1992). Faculty help me enormously when they send me articles they've found and liked (or published themselves) about writing in mathematics. It's useful to include a reflection that helps me see how a mathematics-faculty member evaluates the particular article: "I think this article on writing in mathematics is especially good because....." After years of having mathematics faculty send me stuff, I now have a wonderful collection of such articles which I can use to educate myself, to draw upon in my workshops, and to distribute widely to other mathematics faculty.

3. *Help the WAC director understand mathematics and its teaching.* Mathematicians have talked to me at length about how they teach. They've shown me their syllabi and assignments, their students' papers, the articles they write and read in their field. On one occasion, I heard from

some of my freshman composition students that a professor in mathematics was a wonderful teacher and was making them write, even though he had huge, introductory-level classes. I called him just to ask him how he did it. He and a colleague who worked closely with him took the time to show me how they did it, and to share with me the fat booklet they were preparing to guide the TAs who worked with them. They helped me to do a better job in workshops with other mathematicians.

4. *Invite your colleagues.* One reason we've had a steady stream of mathematicians among the philosophers, psychologists, and physical therapists at our WAC workshops at the University of Cincinnati is that mathematics faculty who attended and benefitted then went back to talk about the workshop with their colleagues. Faculty attend workshops, not because some stranger from a WAC program or an English department sticks a brochure in their mailbox, but because a trusted colleague says, "This was really helpful; you should go."

5. *Sponsor a discipline-specific conference.* After some mathematics faculty had attended the interdisciplinary workshops at the University of Cincinnati, they worked together to sponsor shorter, on-campus workshops just for mathematics faculty. (The University of Cincinnati is a complex state university which has several branch campuses, some of them two-year programs. There are five mathematics departments in the various colleges and they are each independent in almost every way.) One small mathematics department met for an afternoon around a table with me, just to talk about their department's courses. In a general workshop I never could have gotten so many people from one department nor could I have been so specific about their concerns. Another time, the largest mathematics department in the university sent invitations to the other four mathematics departments and packed the room with about 70 mathematicians all at once. By myself, I never could have gotten that many mathematicians together in a million years. They served refreshments, making a collegial occasion out of it.

These workshops got very good reviews from the mathematics faculty who attended them. They had sent me relevant literature, so I could begin the workshop by quoting from their own Mathematical Association of America and American Mathematical Society publications about the importance of writing in mathematics. My mathematics colleagues had taken the time to explain their teaching goals and methods to me, so I could relate my suggestions about using writing directly to their everyday experiences. Since mathematicians had themselves extended the invitation to their colleagues, I met a receptive audience. In the workshop itself, faculty candidly raised issues and questions about what I had said, so I went away even more knowledgeable about mathematics, mathematics instruction, and the needs

of mathematics faculty. And subsequently, even more mathematicians attended the long, off-campus workshops.

Evaluations of the various kinds of workshops, both interdisciplinary and mathematics-oriented, indicate that my colleagues in mathematics believed I was able to help them integrate writing effectively into their classes. Truth is, they helped me help them. I hope these few suggestions will help other mathematics faculty also get some good use out of their faculty colleagues in writing-across-the-curriculum programs.

University of Cincinnati, Cincinnati, OH

Two-Year College Mathematics Teachers:
Does Their Education Match Their Job? *

Ann E. Watkins

Overview of Two-Year Colleges. The 1,200 community, technical, and junior colleges in the United States enroll almost six million students and account for about 38% of all post-secondary mathematics, statistics, and computer-science enrollment, up from 30% in 1985. (Unless another reference is given, all data and tables in this article are preserved here from [1] and the table numbers are from that document.)

The astonishing growth in enrollment has coincided with the evolution of the "junior" college of 25 years ago into the "community" college of today. The primary mission of the junior college was to prepare students for the university by providing a liberal arts education. Two-year colleges continue to provide the first two years of baccalaureate programs to students who want low cost, local schooling. In addition, they now offer vocational and technical programs in fields such as nursing and computer repair; courses for professional certification; courses for adults who want to broaden either their general education or to learn skills as specific as using a spreadsheet or growing fruit trees; and, most notably, instruction in basic subjects traditionally taught in secondary schools.

Today, a minority of two-year college students are enrolled in transfer programs, and transfer rates have declined. Consideration of transfer rates alone, however, underestimates the importance of two-year colleges as a source of mathematics students in four-year colleges and in universities. In many state colleges and universities, a large percentage of mathematics majors began their

---

* From the Proceedings of an AMS Special Session at the 1992 Joint Mathematics Meetings in Baltimore organized for the AMS-MAA-SIAM Joint Committee on Preparation for College Teaching by Bettye Anne Case.

studies in two-year colleges. In fact, "nearly 10 percent of U.S. students who receive a doctorate in the mathematical sciences began their undergraduate studies in a two-year college" [5, p.4].

The Job of the Two-Year College Mathematics Teacher. Increasingly, the job of the two-year college mathematics teacher is to teach remedial courses. In 1970, 33% of mathematics enrollment in two-year colleges was in courses at the level of intermediate algebra or below. In 1990-1991, the figure was 56%. Remedial courses are taught disproportionately by the huge part-time faculty, but constitute much of the teaching load of full-time faculty as well.

Two-year college mathematics teachers rarely teach college-level mathematics to mathematics majors. Fewer than 1% of two-year college students are math majors.

TABLE TYR.3 Enrollment (in thousands) in mathematical sciences and computer-science courses in mathematics programs at two-year colleges: Fall 1966, 1970, 1975, 1980, 1985, 1990.

	1966	1970	1975	1980	1985	1990
Remedial level						
1. Arithmetic	15	36	67	121	77	79
2. General math	17	21	33	25	65	68
3. Pre-algebra	na	na	na	na	na	45
4. Elementary algebra	35	65	132	161	181	262
5. Intermediate algebra	37	60	105	122	151	261
6. High school geometry	5	9	9	12	8	9

	1966	1970	1975	1980	1985	1990
Precalculus level						
7. College algebra	52	52	73	87	90	153
8. Trigonometry	18	25	30	33	33	39
9. College algebra & trig (comb)	15	36	30	41	46	18
10. Precal/elem functions	7	11	16	14	13	33
11. Analytic geometry	4	10	3	5	6	2

	1966	1970	1975	1980	1985	1990
Calculus level						
12. Mainstream calc I						53
13. Mainstream calc II	40	58	62	73	80	23
14. Mainstream calc III						14
15. Nonmainstream calc I	na	na	8	9	13	31
16. Nonmainstream calc II	na	na				3
17. Differential equations	2	1	3	4	4	4

	1966	1970	1975	1980	1985	1990
18. Linear algebra	1	1	2	1	3	3
19. Discrete math	na	na	na	na	L	1
20. Finite math	3	12	12	19	21	29
21. Liberal arts math	22	57	72	19	11	35
22. Business math	17	28	70	57	33	26
23. Math, elem tchrs	16	25	12	8	9	9
24. Elem statistics	4	11	23	20	29	47
25. Probability & statistics	1	5	4	8	7	7
26. Tech math	19	26	46	66	31	17
27. Tech math (calc lev)	1	3	7	14	4	1
28. Hand calculator use	na	na	4	3	6	L

	1966	1970	1975	1980	1985	1990
Computing						
29. Computers & society	na	na	na	na	na	10
30. Data processing (elem/adv)	na	na	na	na	36	21
31. Elem programming	3	10	6	58	37	32
32. Adv programming	na	na	na	na	5	8
33. Database management	na	na	na	na	na	4
34. Assmb lang programming	na	na	na	na	4	2
35. Data structures	na	na	na	na	2	1
36. Other comp science	2	3	4	37	14	20
37. Other math	8	14	32	27	14	23
TOTAL	**348**	**584**	**874**	**1048**	**1034**	**1393**

na: data not available; L means some but less than 500 enrolled

Mainstream calc is for math, physics, sci & engr;
non-mainstream for bio, soc & mgmt sci.

Prior to 1990 aggregate sums for Main Calc I, II & III
& for Non-Main Calc I, II & III were reported.

In Table TYR.3, notice the small enrollment in courses usually required in the first two years of the mathematics major: mainstream calculus, linear algebra, discrete math, and differential equations. In fact, only a third of two-year colleges even offered linear algebra during the 1990-1991 school year, and only half offered differential equations.

Full-time two-year college mathematics instructors teach about 15 hours a week, and about 44% teach an additional class or two, usually for extra pay. Many work outside the college. Although there is regional variation, most two-year college instructors are under no pressure to publish; promotion and tenure typically require adequate teaching and time in rank.

Class sizes are relatively small, averaging about 28 students per section. Ninety-four percent of two-year colleges report that most of their faculty use the standard lecture-recitation format.

An Inadvertent Education. Twenty-five years ago, the majority of two-year college faculty were recruited from the high schools. The typical route into two-year college mathematics teaching was bachelor's degree in mathematics, teach in high school while completing a master's degree in mathematics, then teach at night in a two-year college while continuing to teach high school, and, finally, be hired full-time by that two-year college. It was an opportunity for secondary-mathematics teachers to teach higher-level mathematics and an opportunity for the two-year colleges to raid the high schools of experienced, talented teachers. As recently as ten years ago, between 60 and 70% of the two-year college mathematics faculty had experience in high schools.

TABLE TYR.37 Source of new full-time faculty for mathematics programs at two-year colleges: 1989-1990.

	Doctorate		MA/BA		
Source	Math	Math Ed	Other		TOTAL
Graduate school	0	0	4	208	212
Employed by same 2-yr college in part-time capacity	0	0	0	195	195
Teaching in another 2-yr college	0	4	0	73	77
Teaching in a secondary school	0	0	0	64	64
Non-academic employment	0	0	0	56	56
Teaching in a 4-yr college or university	4	0	0	117	121
Otherwise occupied or unknown	0	0	0	6	6
TOTAL	4	4	4	719	731

This picture has changed. In 1990-1991, only 9% of new full-time hires in mathematics programs in two-year colleges had been teaching in a secondary school the previous year. Table TYR.37 shows where the new hirees were working the year before. The largest number, about 29% of the total, came directly from graduate school.

Only 2% of these newly hired two-year college mathematics teachers hold a PhD. A master's degree in mathematics continues to be the standard requirement for full-time employment (see Table TYR.26).

In the past, two-year college mathematics teachers got much of their training in secondary-teacher credential programs and on the job in high schools. Today the education of two-year college mathematics teachers has become a largely inadvertent and hidden function of graduate programs in universities.

Universities are increasingly aware that they have an obligation to prepare PhD graduates to teach in four-year colleges and are carefully supervising and training teaching assistants. See [2]. Although these programs are of some value to prospective two-year college teach-

TABLE TYR.26 Highest degree of full-time faculty in mathematics programs at two-year colleges by field of highest degree: Fall 1990.

	Highest degree				
Field	PhD	MA+1	MA	BA	TOTAL
Mathematics	8%	26%	31%	3%	68%
Mathematics Ed	6%	5%	6%	L	17%
Statistics	L	1%	1%	0%	2%
Computer Science	L	1%	2%	1%	4%
Other	2%	1%	5%	L	9%
TOTAL	17%	34%	45%	4%	100%

L: Fewer than half of 1%.

ers, teaching university students does not require the same skills as teaching remedial mathematics to two-year college students. In addition, the future two-year college teacher is unlikely to participate in such programs, which are largely designed for PhD candidates. In fact, in some universities, a student may not teach a course until the master's degree is completed.

Over the years, various plans have been proposed for master's degrees for two-year college teachers [3], [4]. Although some universities offer the DA or other advanced work for two-year college teachers who already have a master's degree, master's-level programs are rare. Too often the only thought given by university professors to two-year college teaching occurs when they advise PhD program drop-outs (and, in the 1992 job market, even PhD program graduates) to look for a job in a two-year college.

An Alternative Education. For two decades the Mathematics Department at California State University, Northridge, has been involved with the education of two-year college teachers. In 1968, a second option for the master's degree in mathematics was established to develop the mathematical background of secondary-school teachers who already have a bachelor's degree in mathematics. The degree requires an expository thesis about mathematics, not the mathematics curriculum, and graduate courses in the history of math, mathematical models, geometry, and other topics indispensable to teachers. The department was careful to keep the courses at the graduate level and use graduate textbooks; each course carries a senior-level mathematics course as a prerequisite. Students may elect to take one graduate course in research in mathematics-teaching from the School of Education. Graduates of the program have teaching experience either in secondary schools or in the remedial program at the university.

It soon became apparent that prospective and current two-year college teachers were gravitating towards this program and that secondary teachers with this degree were being hired by two-year colleges. In fact, not one of last June's seven graduates is teaching in a secondary school this year; they are either in PhD programs or teaching in colleges.

References

[1] Albers, Donald J., Don O. Loftsgaarden, Donald C. Rung and Ann E. Watkins, *Statistical Abstract of Undergraduate Programs in the Mathematical Sciences and Computer Science in the United States: 1990-1991 CBMS Survey*, Mathematical Association of America, Washington, DC, 1992.

[2] Case, Bettye Anne (Ed.), *Keys to Improved Instruction by Teaching Assistants and Part-time Instructors*, *MAA Notes* No. **11**, Mathematical Association of America, Washington, DC, 1989.

[3] Colvin, Charles R., The Fredonia plan for preparing two-year college teachers, *The Two-Year College Mathematics Journal*, Vol. **2**, #2, Fall 1971, 69-73.

[4] Long, Calvin T., The academic training of two-year college mathematics faculty. In Albers, Donald J., Stephen B. Rodi, and Ann E. Watkins, (Eds.), *New Directions in Two-Year College Mathematics*, Springer-Verlag, New York, 1985.

[5] National Research Council, *Moving Beyond Myths: Revitalizing Undergraduate Mathematics*, National Research Council, Washington, DC, 1991.

California State University, Northridge, CA

Excellence with Caring:
Florida A&M University Responds to National Concerns of Mathematics Education *

Roselyn E. Williams

In the report, *Moving Beyond Myths: Revitalizing Undergraduate Mathematics*, the National Research Council addresses the major crisis of educating students in the quantitative skills needed for the nation's scientific, technical, and managerial work force. Major universities and industries are making large scale efforts to attract and recruit gifted students for careers in mathematics through gradfests, special institutes, and internship programs. Yet this population of students is not large enough to supply the future labor force for which mathematical skills will be required. Through the motto, "Excellence with Caring," Florida A&M University (FAMU) is meeting this need with students of diverse backgrounds and is helping to develop in students positive attitudes toward mathematics.

FAMU has a population of approximately 10,000 stu-

dents, 200 of whom are undergraduate mathematics majors, and 15 full time mathematics professors. The student body consists of many students of low- to middle-socioeconomic backgrounds; yet FAMU leads the nation in enrolling National Achievement Scholars. Besides the use of computer technology and sponsoring programs for prospective scientists and mathematicians, FAMU takes a caring approach in helping students to develop their mathematical skills. In this article we examine this approach in assisting students to develop their greatest mathematical potential.

There are professors who support the idea of using challenging problems and theory in their lectures so that students can appreciate the complete picture and objectives of the theory. Such problems serve the needs of the gifted students and give all students a realistic perspective of the course level. Teachers provide solutions in the extensive detail required to capture the backgrounds of all students in the class. One teacher has developed an interesting way of providing remediation as needed through a collection of one-page modules. A module covers one concept, such as equations, properties of algebra, operations on functions, or exponents. Each module consists of a page of definitions, properties, examples, exercises, and references to the texts. For example, when a student is having problems with related rates, the concepts in which the student is deficient are combined to create a related rates module. A student can then develop a work book tailored to fit his needs.

Two problems that contribute to students' low performances are (1) not knowing how to study mathematics and (2) not allowing enough time to study. Many students are unable to interpret or grasp meaning from the textbook. Terminology, complexity of ideas, and examples with missing steps often become obstacles. Second, *comprehending* knowledge takes less effort and skill than does *communicating* or *transferring* knowledge. Fewer concepts are studied in a single class meeting than are evaluated on a test. Because of their ability to understand the concepts in class, many students falter by concluding that they can retain and communicate the concepts as needed, and fail tests. The students often underestimate the amount of drill and practice required to master mathematical concepts and, therefore, do not allow enough time for study. Including a list of suggestions for studying mathematics with the course syllabi benefits many students. As a result, students discuss study habits and techniques of self-evaluation with their teachers.

Grading daily assignments is believed to be an effective tool for developing mathematical writing skills. This is easily done in schools with small student-teacher ratios, but may be impractical for professors with large numbers of students. In order to give daily assignments, one professor hires mathematics majors through the university's work-study program as graders/tutors. As graders/tutors, the students benefit by strengthening their backgrounds through grading and by develop-

* From the Proceedings of an AMS Special Session at Knoxville, in March 1993, "Interventions to Assure Success: Calculus Through Junior Faculty" organized by Bettye Anne Case.

ing communication skills through tutoring. The students who have not developed good study habits are motivated by the opportunity to impact their grades with small evaluations at high frequencies. This helps them to develop discipline. The professor claims that these classes are livelier than those without daily assignments. Besides improving their writing skills, the students study more, participate more in class discussions, and perform better on tests.

We have found that mentoring relationships among professor and students are very beneficial. Students assist teachers in their research, and teachers supervise research projects for their students. Students have opportunities to attend and present talks at mathematical meetings. If students feel a part of the mathematical community, have professors who are sensitive to their needs, are learning techniques of self-evaluation, are stimulated to explore increasingly more difficult and challenging materials, they are more inclined to enjoy mathematics and become more successful learners.

Florida A & M University, Tallahassee, FL

You're the Professor, What Next?

Part V

Resources

Chapter 16 – Professional Societies and Governmental Agencies

Chapter 17 – Materials and References

This is a rich supply of information, references and resources which can be used for personal growth or to organize programs that prepare faculty for the full range of their academic responsibilities.

Professional Societies and Governmental Agencies

Developing a PROFESSIONAL LEXICON

AAUP
AMATYC
AMS
BMS
CBMS
COMET
CUPM
DAVID I & II
FIPSE
IPBM
MAA
MER
MSEB
MS 2000
NCTM
NRC
NSF
QED
SIAM
UMAP

Alphabet Soup. Nothing will trigger that secret, hidden feeling of uncertainty and inferiority in a young faculty member more quickly than overhearing successful oldheads throwing around academic alphabet soup in the coffee room. "NSF did not like the project, but a buddy I met at SIAM knows the chair of BMS at NRC and thought I could get the name of a project officer at ONR who might consider it. If that fails, I'll turn it into a COMAP module after I get back from the CSSP and CBMS meetings." Choke! Maybe I should have gone to law school after all.

The fact is that any competent, involved professional needs to know the professional territory. And that means some familiarity with the acronyms most commonly run across in mathematical circles. Obviously, it is not the acronyms themselves that are important, but the organizations, structures, and publications to which they point.

The groups identified below are an extraordinarily rich source of information, access, and support. Every collegiate faculty member should belong to and actively participate in one or more of the mathematics organizations we have listed below. These groups can help you find funding for research, assist you in getting published, put you in touch with others solving the same pedagogical problems, or help you establish a sharing network. If actively pursued, the friends and associations made in these organizations will form the core of a faculty member's mathematical family.

Many acronyms, of course, do not refer to professional organizations but to government agencies or to publications. These, too, are important sources of contacts and information for faculty. They frequently also are looking for your involvement as a contributor, as a writer or reviewer, or in some other fashion.

A list follows of many of the most important names, organizations, and publications that pertain to mathematicians in the United States. After the full names of organizations, we include for some their "alphabet soup" abbreviation. We have also included profiles supplied by some mathematical organizations that a beginning faculty member would be likely to encounter. The last paragraph in the description of CBMS lists its fifteen member societies, all organizations in the mathematical sciences. (See Chapter 17 for more about publications.) We apologize if we have left out someone's favorite group, but there were space limitations.

A more complete list of professional organizations and governmental agencies related to the mathematical sciences broadly defined can be found at the front of the *Mathematical Sciences Professional Directory* published annually by the American Mathematical Society. That listing goes into detail about organizational committees, committee chairs, and committee membership. More general reference works about organizations include the *Encyclopedia of Associations* (Gale Research, Inc.) and Congressional Quarterly's *Washington Information Directory*.

Now, study the list below carefully and for homework go back and translate the opening paragraph!

P.S. And don't forget the Internet. If you have not yet linked to your colleagues via e-mail on the Internet and/or joined an Internet users' group associated with your interest, do it today! Most colleges and universities are nodes on the Internet system, making access through your department easy. If your school is not a node, access may be available by a courtesy account through a neighboring institution. For an excellent short bibliography on getting started on the Internet, see the MAA Newsletter *FOCUS*, Volume 14, No. 1, Feb. 1994, p. 24.

American Mathematical Association of Two-Year Colleges (AMATYC)

The American Mathematical Association of Two-Year Colleges was formed in 1974. Membership is open to any teacher of mathematics or other person interested in the first two years of college mathematics education. AMATYC's current membership is approximately 3,000. AMATYC has 36 state or regional affiliates servicing 39 states. AMATYC is the only national professional association in the sciences that addresses two-year college issues exclusively.

The objectives of the association are the following:

a) To encourage the development of effective mathematics programs;

b) To afford a national forum for the interchange of ideas;

c) To develop and/or improve the mathematics education and mathematics related experience of students;

d) To coordinate activities of affiliated organizations on the national level;

e) To promote the professional welfare and development of its members.

AMATYC publishes the AMATYC NEWS three times yearly and its professional journal, THE AMATYC REVIEW, in the spring and fall.

It has six standing committees. They are:

Technology in Mathematics Education promoting the use of technology within the mathematics curriculum and the interaction of mathematics and computer science curricula.

Developmental Mathematics improving the quality of developmental mathematics programs in the two-year college.

Equal Opportunity in Mathematics enhancing the position of women and minorities in mathematics as students, teachers, researchers, or applied mathematicians.

Student Mathematics League encouraging student excellence at the two-year college via an annual mathematics competition; The top student receives the $3,000 Chuck Miller Memorial Scholarship Award to further that student's education. AMATYC is also a sponsoring organization of the American Mathematics Competition.

Technical Mathematics supporting mathematics courses in allied health and human services, business, computer/data processing, trade and industry, engineering, and emerging technologies.

AMATYC has week-long summer institutes serving as professional development for college faculty in Idaho, North Carolina, and Vermont.

AMATYC has published statements on the following matters: Guidelines for the Academic Preparation of Two-Year College Mathematics Faculty; Standards for Two-Year College Mathematics Departments; National Standards for Two-Year College and Lower Division Mathematics in the Courses Before Calculus.

AMATYC is a member of the Conference Board on the Mathematical Sciences (CBMS) and the Council of Scientific Society Presidents (CSSP).

American Mathematical Association of Two-Year Colleges, State Technical Institute at Memphis, TN 38134, 901-383-4643 (amatyc@stim.tec.tn.us; Fax 901-383-4503).

American Mathematical Society (AMS)

The American Mathematical Society is a non-profit professional society with nearly 30,000 members worldwide. In addition to its historic mission of promoting mathematical research and communication, the Society is concerned with mathematics education, public awareness of mathematics, the connections of mathematics to other disciplines and to technology, the status of the mathematics profession, and issues facing the international mathematics community.

The AMS is a major publisher of mathematial research through its books and journals, including the authoritative reference source *Mathematical Reviews*. The AMS also communicates mathematics by sponsoring national, international and regional mathematics meetings and conferences and through an expanding line of electronic products and services.

AMS electronic products include MathSci Online, an electronic interdisciplinary connection to literature in mathematics, statistics, and related fields, and MathSci Disc, a mathematics reference collection on compact disc.

As a service to mathematicians and scientists and professionals in other fields, the AMS has developed e-MATH, an electronic clearinghouse for timely professional and research information in the mathematical sciences. e-Math provides, in addition to a broad list of services, on-line access to the Science and Technology Information System of National Science Foundation (NSF) and access to the combined membership lists of the AMS, the Mathematical Association of America, and the Society for Industrial and Applied Mathematics. Access to e-MATH is free and available on Internet.

The Society offers its members a range of other services to complement its publications, electronic communications, and meetings. At its annual meeting in January, it operates an employment register, in which job-seekers and employers are matched for interviews. It publishes *Employment Information in the Mathematical Sciences*, listing open positions in mathematics, and *Assistantships and Graduate Fellowships*, providing information about support for graduate studies in mathematics. Information on job announcements, and curriculum vitae, can also be posted on e-Math.

For more information about membership in the Society, or the products or services it offers, call:

1-800-321-4AMS. AMS, P.O. Box 6248, 201 Charles St., Providence, RI 02904-2213, 401-455-4000, Fax 401-331-3842 (ams@math.ams.org).

American Statistical Association (ASA)

1429 Duke St., Alexandria, VA 22314-3402; 703-684-1221.

Association for Women in Mathematics (AWM)

The Association for Women in Mathematics (AWM) was established in 1971 to encourage women to study and have active careers in the mathematical sciences. Equal opportunity and equal treatment of women in the mathematical sciences are promoted.

AWM has more than 3,500 members, both women and men, from the United States and around the world, representing a broad spectrum of the mathematical community. As an organization promoting women in the mathematics arena, its efforts have led to greater participation of women in the mathematical community, especially as speakers at mathematical meetings and as members of committees of the mathematical organizations.

AWM members receive a bimonthly newsletter which contains informative articles, book reviews, and announcements of upcoming events of interest, as well as job announcements from academic and non-academic organizations.

AWM organizes panels and workshops at national meetings of the AMS, the Mathematical Association of America (MAA), and the Society for Industrial and Applied Mathematics (SIAM) to discuss issues of current interest in the profession and to highlight the achievements of women postdocs and graduate students. The workshops have been co-sponsored by the National Science Foundation (NSF) and the Office of Naval Research (ONR).

AWM offers travel grants three times a year for women to attend research conferences in their fields, thereby providing a valuable opportunity to advance their research activities and their visibility in the research community. These grants are co-sponsored by the NSF.

AWM awards annually two prizes — the Alice T. Schafer Award for outstanding undergraduate work by a female student, and the Louise Hay Award for excellence in mathematics education. In cooperation with outside funding agencies, AWM offers travel grants for women to attend research conferences and supports Sonia Kovalevsky High School Mathematics Days.

AWM annually presents the Emmy Noether Lectures to honor women who have made fundamental and sustained contributions to the mathematical sciences. A booklet profiling the women who have presented Noether Lectures is updated yearly.

AWM also produces and disseminates materials relating to careers in mathematics and encourages a variety of activities aimed at promoting the rights of women in mathematics. AWM, 4114 Computer and Space Sciences Bldg., University of Maryland, College Park, MD 20742-2461 (awm@math.umd.edu).

Board on Mathematical Sciences (BMS)

During its last year of operation (1984), the former Office of Mathematics of the National Research Council (NRC) completed work on a report that set the tone for the years to follow. This report, Renewing US Mathematics: Critical Resource for the Future — known as the David Report after former Presidential Science Advisor Edward E. David, who chaired the report committee — assessed the state of support for the mathematical sciences and recommended a five-year plan for renewal. This report painted a picture of the mathematical sciences as a dynamic resource for the nation. In the same report, action to maintain the health and quality of research in the mathematical sciences was recommended.

In response to the recommendations of the David Report, the National Research Council (NRC) established the Board on Mathematical Sciences (BMS) in 1984 to oversee activities formerly conducted by the Office of Mathematics and the Committee on Applied and Theoretical Statistics (CATS).

The Board consists of 15 members representing the areas of core mathematics, applied mathematics, statistics, operations research, and scientific computing. The Board has two standing committees, CATS and the US National Commission on Mathematics Instruction (US-NCMI). CATS, established by the National Research Council in 1977, concerns itself with issues affecting research and education in the statistical sciences. The chair of CATS is an ex officio member of the Board. USNCMI represents the US mathematical sciences education community internationally. Additional ad hoc committees, panels, and working groups are formed as needed to carry out individual projects. The Board, CATS, and the ad hoc groups established by the Board, typically comprise at any one time over 100 mathematical scientists, scientists, engineers, and medical personnel, including many members of the National Academy of Sciences, the National Academy of Engineering, and the Institute of Medicine.

BMS, c/o National Academy of Sciences, 2101 Constitution Ave. NW, Washington, DC 20418, 202-334-1597 (bms@nas.edu).

Consortium for Mathematics and Its Applications (COMAP)

An active producer and source of modules and other mathematical instructional materials. 60 Lowell St., Arlington, MA 02174; 617-641-2600.

Conference Board of the Mathematical Sciences (CBMS)

The Conference Board seeks to promote understanding and cooperation among the national organizations in the mathematical sciences so that they may work together in their various ways for the advancement of, the application of, and the dissemination of, mathematical knowledge.

It is the Conference Board's policy to engage primarily in the following kinds of activities:

1) to provide a forum for the discussion of issues of broad concern to the mathematical sciences community and a focus for mutual support among the member societies;

2) to organize and nucleate new functions for the mathematical sciences community;

3) to serve as an organization to which government agencies, professional societies of other disciplines, industry, and private foundations can turn for leadership and participation by the mathematical sciences, in the spirit described here, and for advice and counsel;

4) to serve as a point of representation for the mathematical sciences to these agencies, societies, and foundations.

It is the Conference Board's policy to minimize its engagement in long-term contract management. Specifically, projects begun by CBMS and deemed worthy of continuation are administered through member societies, possibly under the auspices of CBMS.

The 15 member societies in 1994 are: American Mathematical Association of Two-Year Colleges; American Mathematical Society; American Statistical Association; Association for Symbolic Logic; Association for Women in Mathematics; Association of State Supervisors of Mathematics; Institute of Mathematical Statistics; Mathematical Association of America; National Association of Mathematicians; National Council of Supervisors of Mathematics; National Council of Teachers of Mathematics; Operations Research Society of America; Society for Industrial and Applied Mathematics; Society of Actuaries; and The Institute of Management Sciences.

CBMS, 1529 18th St. NW, Washington, DC 20036, (202) 293-1170.

[US] Department of Education (DOE)

400 Maryland Ave. SW, Washington, DC 20202, 202-401-1576.

Fund for Improvement of Postsecondary Education (FIPSE)

This is an agency of the DOE. 7th and D Sts. SW, Washington, DC 20202, 202-708-5750 (Fax 202-708-6118).

Joint Policy Board for Mathematics (JPBM)

The Joint Policy Board for Mathematics is a joint venture of the American Mathematical Society, the Mathematical Association of America, and the Society for Industrial and Applied Mathematics. It was established to articulate and advocate sound public policy concerning the mathematical sciences and the field's ability to contribute to the public welfare. JPBM represents a wide spectrum of the professional mathematical community and promotes an integrated view of the discipline, encompassing mathematical education, research, practice, and application.

JPBM's Washington office serves as a liaison to the general public, the federal government, and other national scientific and educational organizations. Through this office, JPBM develops and promotes policy statements and works to improve public understanding of mathematics and its critical importance to a strong economy and a productive society.

Every spring, the Washington Office coordinates the annual National Mathematics Awareness Week, which was initiated by Presidential proclamation in 1986. Events are sponsored by mathematics organizations across the country. JPBM also presents a Communications Award to individuals who have been particularly effective at conveying mathematics to the public.

JPBM, 1529 18th St. NW, Washington, DC 20036, 202-234-9570.

Mathematical Association of America (MAA)

The Mathematical Association of America is the largest professional society of college and university mathematics teachers in the world. Today MAA's 33,000 members include college and university faculty, two-year college faculty, high school teachers, government and corporate workers, graduate school faculty, research mathematicians, and graduate and undergraduate students. The 29 sections of the MAA provide a regional base and inspire strong loyalty on the part of their members, a loyalty that extends to the national organization.

The original 1915 charter of the MAA declared that the purpose of the Association is to "assist in promoting the interest of mathematics in America, especially in the collegiate field." Three-quarters of a century later, that charter still shapes the mission of this organization.

Mission: To advance the mathematical sciences, especially at the collegiate level.

The traditions and mission of the MAA lead to four major program goals towards which most activities of the Association are aimed: education, professional development, students, and public policy. These goals express the mission of the Association in particular terms:

A. *Education.* Stimulate effective teaching, learning, and assessment in the mathematical sciences.

B. *Professional Development.* Foster scholarship, professional development, and a spirit of association among mathematical scientists.

C. *Students.* Enhance the interests, talents, and achievements of all individuals in the mathematical sciences, especially of members of underrepresented groups.

D. *Public Policy.* Influence institutional and public policy through effective advocacy for the importance, uses, and needs of the mathematical sciences.

These goals are carried out through an extensive program of publications and meetings. The Association makes available to a wide audience information on new developments in mathematics and mathematics education. It is actively concerned with teacher education and encouragement of students — especially those from under-represented groups — to continue their study in mathematics. Through its many committees, the Association provides models for collegiate curricula, guidance to teachers and institutions, and standards for collegiate programs in the mathematical sciences.

The national headquarters of the MAA is located at 1529 Eighteenth St. NW, Washington, DC 20036.

MAA is a nonprofit 501(c)(3) non-profit, independent, membership organization, each of its 29 Sections has 501 (c)(3) status under an MAA group exemption.

The telephone is 1-800-331-1622, and the e-mail address is maahq@maa.org.

Mathematical Sciences Education Board (MSEB)

This is another sub-organization of the National Academy of Sciences (NAS) and National Research Council (NRC). Its mandate is to concentrate on mathematics education issues from elementary school through the undergraduate level. See also BMS. 2101 Constitution Ave. NW, Washington, DC 20418; 202-334-3294.

National Academy of Sciences (NAS)

2101 Constitution Ave. NW, Washington, DC 20418, 202-334-2138.

National Association of Mathematicians (NAM)

The National Association of Mathematicians (NAM) is an organization whose purpose is to promote the participation of minorities, particularly African-Americans, in mathematics. The membership consists of professional mathematicians including teachers and other scientists. The organization welcomes as members all who share the goals of the Association. c/o Dr. Johnny L. Houston, Executive Secretary, Box 959, Elizabeth City State University, Elizabeth City, NC 27909 (nam@ecsvax.uncecs.edu).

National Council of Teachers of Mathematics (NCTM)

The National Council of Teachers of Mathematics' (NCTM) nationally-acclaimed *Curriculum and Evaluation Standards for School Mathematics* and *Professional Standards for Teaching Mathematics* have set demanding standards for what students must learn about mathematics and for what teachers must accomplish as professionals in the classroom. The Council's third Standards document, *Assessment Standards for School Mathematics*, is slated for release in April 1995. These documents support a vision of national expectations that employ mathematics as a means to solve problems, communicate, reason, and make mathematical connections.

In support of the *Standards*, NCTM has developed numerous projects, programs, and information packets that depict the value of mathematics education and the benefit of parental involvement in mathematics education. Working with schools, businesses, local communities, and other educational associations, NCTM is educating the public about the important role mathematics plays in our schools, on the job, and in the home.

With more than 100,000 members and 250 affiliated groups located in the United States and Canada, NCTM is the nation's largest organization dedicated to the improvement of mathematics education.

The national headquarters of NCTM is located at 1906 Association Dr., Reston, VA 22091-1593. The telephone is 703-620-9840, extension 113.

National Research Council (NRC)

The operating arm of the National Academy of Sciences (see above), the National Academy of Engineering, and the Institute of Medicine. See NAS, BMS, and MSEB.

National Science Foundation (NSF)

4201 Wilson Blvd., Arlington, VA 2230; 703-306-1234.

National Security Agency (NSA)

One of the leading non-academic employers of mathematicians. 301-688-0400.

Office of Naval Research (ONR)

Mathematics Division, Code 1111, 800 N. Quincy St., Arlington, VA 22217-5660.

[MAA] Placement Program (PTP)

This program assists two- and four-year colleges and universities with the development of on-campus placement programs to assess the mathematical skills of new students objectively and fairly when they enter college.

Colleges and universities pay an annual fee to subscribe to PTP. This program provides up-to-date tests and placement testing information to its subscribers. PTP materials and services include:

◇ A battery of placement tests that can be adapted readily to local needs. (You can customize the tests by "cutting and pasting" the PTP items.)

◇ A subscription to the *PT Newsletter* that contains interesting and useful articles on various aspects of placement testing.

◇ Sample tests to inform students about the kinds of questions they should expect on PTP tests.

◇ A User's Guide containing a wealth of information on how to initiate and administer a placement testing program.

The PTP packet of tests and materials changes annually, as old forms of tests are "retired" and new tests and other materials are added. New PTP tests and materials are normally added to the packets each year. Subscribers are authorized to reproduce unlimited quantities of any PTP tests and their items for use on their campuses during the 12-month period following receipt of the PTP packet.

The MAA's Committee on Testing also offers free consulting service to any college mathematics department that is planning to initiate a local mathematics placement program or that is seeking ways to improve its on-going program.

For information or to subscribe, write or call: MAA PT Program, 1529 Eighteenth St. NW, Washington, DC 20036, 1-800-331-1622.

Society for Industrial and Applied Mathematics (SIAM)

Inspired by the vision that applied mathematics should play an important role in advancing science and technology in industry, a small group of professionals from academe and industry met in Philadelphia in 1951 to start an organization whose members would meet periodically to exchange their ideas about the uses of mathematics in industry. This meeting led quickly to the organization of the Society for Industrial and Applied Mathematics (SIAM).

The goals of SIAM were to:

◇ Advance the application of mathematics to science and industry.

◇ Promote mathematical research that could lead to effective new methods and techniques for science and industry.

◇ Provide media for the exchange of information and ideas among mathematicians, engineers, and scientists.

These goals haven't changed; they are more valid today than ever before.

Applied mathematics, in partnership with computing, has become essential in solving many real-world problems. Its methodologies are needed, for example, in modeling physical, chemical, and biomedical phenomena; in designing engineered parts, structures, and systems to optimize performance; in planning and managing financial and marketing strategies; and in understanding and optimizing manufacturing processes.

Problems in these areas arise in companies that manufacture aircraft, automobiles, engines, textiles, computers, communications systems, chemicals, drugs, and a host of other industrial and consumer products, and also in various service and consulting organizations. They also arise in many research initiatives of the federal government such as those in global change, biotechnology, and advanced materials.

SIAM fosters the development of the methodologies needed in these application areas. It is fitting that the acronym SIAM also represents the society's slogan – Science and Industry Advance with Mathematics.

Just as applied mathematics has grown, so has SIAM membership – from a few hundred in the early 1950s to more than 8500 today. SIAM members are applied and computational mathematicians, computer scientists, numerical analysts, engineers, statisticians, and mathematics educators. They work in industrial and service organizations, universities, colleges, and government agencies and laboratories all over the world. In addition, SIAM has over 300 institutional members – colleges, universities, and corporations.

To serve this diverse group of professionals:

◇ SIAM publishes ten peer-reviewed research journals; SIAM News, a news journal reporting on issues and developments affecting the applied and computational mathematics community; SIAM Review, a quarterly journal of expository and survey papers; and 20 to 25 books per year.

◇ SIAM conducts annual meetings and many specialized conferences, short courses, and workshops in areas of interest to its members.

◇ SIAM sponsors nine (special interest) activity groups, which provide additional opportunities for professional interaction and informal networking.

◇ SIAM sponsors regional sections for local technical activities and university (student) chapters that bring faculty and students together in activities consistent with SIAM objectives. Recently, SIAM reinstated its visiting lectureship program.

3600 University City Science Center, Philadelphia, PA 19104-2688, 215-382-9800 (siam@siam.org).

Strengthening Underrepresented Minorities Mathematics Achievement (SUMMA)

c/o MAA, 1529 Eighteenth St. NW, Washington, DC 20036.

Young Mathematicians Network (YMN)

In the summer of 1993, the Young Mathematicians Network (YMN) was modeled on the Young Scientists Network. (Here "young" should be understood in terms of years past the degree.) Its purposes include helping to keep the mathematical community informed about the job market, publishing, and sources of grant support, and about other concerns of junior mathematicians. Though YMN sponsors programs at meetings of mathematicians, it operates otherwise almost exclusively through the Internet; its principal vehicle is a weekly

electronic newsletter called *Concerns of Young Mathematicians*. Subscription by e-mail is accessible to anyone and is free. To subscribe to the newsletter, send e-mail to cyeomans@ms.uky.edu. Back issues of the newsletter can be obtained via anonymous ftp to ftp.ms.uky.edu and looking in the directory /pub3/mailing.lists/ymn-list. Information is available from Charles Yeomans (809 Cooper Dr., Lexington, KY 40502, 606-268-3870, Fax 606-268-3958) or any other member of the board:

Mark Winstead, winstead@euclid.ucsd.edu;

Vic Perera, vperera@silver.ucs.indiana.edu;

Franklin Mendivil, mendivil@math.gatech.edu;

Stephen Kennedy, kennedy@stolaf.edu;

Kalin Godev, kalin@math.psu.edu;

Neil Calkin, calkin@math.gatech.edu;

Curtis Bennett, cbennet@andy.bgsu.edu;

Jeff Adams, adams@bright.uoregon.edu;

Edward Aboufadel, aboufade@scus1.ctstateu.edu;

Frank Arlinghaus, frank@math.ysu.edu;

Matt Hudelson, hudelson@math.washington.edu.

Chapter 17

Materials and References

In addition to the essays and reports written for this book, and the reprints included in the Appendices, there is much other material about college teaching, broadly interpreted, available for individuals or for proseminars with doctoral students. The resources are so vast that the Committee on Preparation for College Teaching thought it would be useful to provide a guide to some of them in one place. This is what we do here, although we omit in this chapter references to the papers written for this book or reprinted in its appendices. Some of the entries in this selected bibliography are discussed in these included and reprinted papers. (An Author Index is included at the end of the book which lists the authors of all essays, reports, and reprints.) This chapter begins with a description of some periodicals which regularly have information about teaching. Then there is description of some types of books which are of interest to mathematical sciences faculty members, with examples. Most of this chapter consists of an extensive bibliography of books and articles, many with short reviews written by Stephen B. Rodi. Obviously, this chapter can include only a sampling of available resources and materials.

Journals and Newsletters

The following is a partial listing of journals and newsletters which may be of interest to mathematics faculty. Most entries deal with professional issues of concern to faculty or with undergraduate instruction in mathematics. These are publications with which professional mathematicians and mathematics educators should be familiar. It is hoped that all active faculty members subscribe to or read regularly several of these publications. In a proseminar setting, information about these resources is itself a direct benefit; the substance of books and articles becomes, in turn, a rich source of current discussion about issues important to the profession.

The *Mathematical Sciences Professional Directory* listed below is an excellent source of even more journals and newsletters which are published for special content areas within mathematics or by professional organizations like the American Statistical Association or the Canadian Mathematical Society or various societies devoted to the applications of mathematics.

AMATYC News. Source: AMATYC.*

AMATYC Review. Source: AMATYC.*

American Mathematical Monthly. Source: MAA.*

American Statistician. Source: ASA.*

AMS Notices. Source: AMS.*

AWM Newsletter. Source: AWM.*

Chronicle of Higher Education. Source: Subscription Services Dept., P.O. Box 1955, Marion, Ohio 43306-2055; Corporate Offices, Suite 785, 1255 Twenty-Third St., Washington, DC 20037.

College Mathematics Journal. Source: MAA.*

Combined Membership List of AMS, MAA, and SIAM. Source: AMS.*

---

* See Chapter 16 for address.

Crux Mathematicorum. Source: The Canadian Mathematical Society, 477 King Edward, Suite 109, P.O. Box 450, Station A, Ottawa, Ontario, Canada KIN 6N5.

Math Horizons. Source: MAA.*

Humanistic Mathematics Network Journal. Source: Alvin White, 909-621-8867 (awhite@sif.claremont.edu).*

MAA *FOCUS: The Newsletter of the Mathematical Association of America.* Source: MAA.*

Maple Technical Newsletter. Source: Birkhäuser, 675 Massachusetts Ave., Cambridge, MA 02139.

Mathematica Journal. Source: Miller Freeman, Inc., 370 Lexington Ave., New York, NY, 10017.

Mathematical Intelligencer. Source: Springer-Verlag, 175 Fifth Ave., New York, NY, 10010.

Mathematical Sciences Professional Directory. Source: AMS.*

Mathematics in College: A Journal of the Instructional Resource Center and the CUNY Mathematics Discussion Group. Source: Instructional Resource Center, Office of Academic Affairs, The City University of New York, 535 East 80th St., New York, NY 10021.

Mathematics Magazine. Source: MAA.*

Mathematics Teacher. Source: NCTM.*

Minnesota Mathematics Mobilization: A Newsletter about Mathematics Education in Minnesota. Source: Editorial Coordinator, Department of Mathematics, St. Olaf College, Northfield, MN 55057.

Pi Mu Epsilon Journal. (A publication intended for college students and their teachers from the national mathematics honorary society.) Source: Secretary/Treasurer, Pi Mu Epsilon Society, Department of Mathematics, East Carolina University, Greenville, NC, 27858

PRIMUS: Problems, Resources, and Issues in Mathematics Undergraduate Studies. Source: Department of Mathematics, Rose-Hulman Institute of Technology, Terre Haute, IN 47803.

SIAM News: The Newsjournal of the Society for Industrial and Applied Mathematics. Source: SIAM.*

SIAM Review. Source: SIAM.

UMAP: The Journal of Undergraduate Mathematics and Its Applications. Source: COMAP*.

UME Trends: News and Reports on Undergraduate Mathematics Education. Source: MAA.*

"Types" of Books

Since groupings by category inevitably depend on the viewpoint of the one making up the list, they may be misleading to another user and may cause a reader to overlook something of interest. Categories are helpful if they alert the user to types of helpful material from

other sources. In using our list of resources, a reader might want to keep in mind that, generally speaking, four kinds of entries occur:

General State of Mathematics Education. First, there are sources in the bibliography which discuss from a general point of view the state of mathematics education in the United States. These include some well-known reports such as "David I and II" (*Renewing US Mathematics*) and, from the "MS 2000" Project, *A Challenge of Numbers, Moving Beyond Myths*, and *Reshaping School Mathematics*. The most extensive source of statistical information is found in the CBMS Survey Reports (see Albers *et. al*).

Pedagogy. The largest part of the extended bibliography deals with matters of pedagogy and instruction. A number of entries are about teaching in general. These would include Eble, *The Craft of Teaching*; Erickson, *Teaching College Freshmen*; McKeachie, *Teaching Tips: A Guidebook for the Beginning College Teacher*; and Weimer, *Improving College Teaching*. But many entries deal directly with teaching particular courses and methods (e.g. Krantz, *How to Teach Mathematics*). Often, these are specific to particular courses and methods. Hence, one will find sources like Ralston, *Discrete Mathematics for the First Two Years* or Sterrett, *Using Writing to Teach Mathematics*.

The *Notes* Series published by the Mathematical Association of America (of which we are pleased to have this volume be a part) is an especially rich source of instructional materials for undergraduate mathematics courses. As an example, see the five-volume set *Resources for Calculus* listed under Roberts, the project director. Another excellent source of mathematical instructional material, especially teaching modules, is the Consortium for Mathematics and Its Applications (COMAP) in Arlington, MA.

The Professoriate. A third kind of bibliographic entry deals with the general state of the professoriate at the end of the twentieth century and how it might be changing. Principal among these would be Ernest Boyer's *Scholarship Reconsidered*, its companion document from the Carnegie Foundation *The Condition of the Professoriate: Attitudes and Trends*, and Lynne Cheney's controversial and provocative *Tyrannical Machines: A Report on Educational Practices Gone Wrong and Our Best Hopes for Setting Them Right*.

Popular Commentary on Higher Education. Finally, the last kind of entry is more popular commentary on the state of American universities and the professoriate. These books by authors like Bloom, D'Souza, Getman, Hirsch, and Sykes are intended deliberately to be current and provocative. The Committee does not endorse any particular viewpoint presented in these books and we know that some of them will be dated in a decade. Nonetheless, the books listed contain many ideas which can form the basis of lively discussion, ideas about which reflective professionals young and old should

---

* See Chapter 16 for address.

be informed. This is consistent with the Committee's view that a mathematics faculty member's responsibilities and interests should go beyond the department to include the whole university and the community in which the university exists.

Selected Bibliography

With Reviews* by

Stephen B. Rodi

Essays written for this volume or material reprinted in the Appendices generally are not included in the listing of this chapter. For information about the papers included in this book, consult the Table of Contents at the front of the volume and the Author Index at the end.

The works below are listed alphabetically by author (or by publisher where no author is identified). There are short reviews for some entries; these are alphabetized along with those which have only bibliographic entry. We think the reviews will be especially useful in helping seminar planners, especially those on a limited budget, evaluate materials they may want to include in their programs. Addresses are truncated for publications from the organizations listed in Chapter 16.

Aboufadel, Edward. A sequence of articles — a "diary" — appeared in some issues of MAA *FOCUS* during academic 1992-1993 and 1993-1994. They first chronicle Aboufadel's experiences as a job-seeker and soon-to-be-PhD, and then as a new faculty member.

Albers, Donald J., Anderson, Richard D. and Loftsgaarden, Don O. *Undergraduate Programs in the Mathematical and Computer Sciences: The 1985-1986 Survey.* MAA *Notes* No. 7, 1987.

Albers, Donald J., Anderson, Richard D., Loftsgaarden, Don O. and Watkins, Ann E. *Statistical Abstract of Undergraduate Programs in the Mathematical Sciences and Computer Science in the United States: 1990-1991 CBMS Survey.* MAA *Notes* No. 23, 1992.

American Mathematical Association of Two-Year Colleges Education Committee, John Impagliazzo, Chair. *The Two-Year College Teacher of Mathematics.* Memphis, TN: American Mathematical Association of Two-Year Colleges (State Technical Institute at Memphis), 1985.

This hard-to-find but uniquely helpful pamphlet gives a comprehensive view of the role and responsibilities of mathematics faculty at two-year colleges. Its purpose is to help develop and support competent, respected professionals in what is the fastest growing and most

---

* An earlier version of some of these reviews was prepared for and presented at an AMS Special Session at the 1992 Joint Mathematics meetings in Baltimore organized for the AMS-MAA-SIAM Joint Committee on Preparation for College Teaching.

instruction-oriented segment of higher education in the United States. Chapters deal with faculty qualifications, classroom issues, student profile, professional development opportunities, and interaction with colleges and universities. [26 pp.]

Association of American Colleges. *Integrity in the College Curriculum.* Washington, DC: Association of American Colleges, 1985.

Atkinson, Richard and Tuzin, Donald. "Equilibrium in the Research University." *Change*, May/June, 1988, pp. 21-31.

Association for Women in Mathematics. *Careers that Count.* College Park, MD: 1991.

Barzun, Jacques. *Begin Here: The Forgotten Conditions of Teaching and Learning.* University of Chicago Press, 1991.

Fifteen essays and speeches about education in the United States, written by Mr. Barzun over a period of 40 years, are collected in this volume. Barzun himself has written a new introduction for each of them. As the book jacket highlights, Barzun laments the deadening effect of multiple choice tests, the misguided attacks on Western culture, and the transformation of history and geography into a make-believe world of "social studies." An essay titled "Math and Science Are Liberal Arts" is especially appropriate for reading by collegiate mathematics faculty. [xii + 222 pp., ISBN 0-226-03846-7]

Bloom, Alan. *The Closing of the American Mind: How Higher Education Has Failed Democracy and Impoverished the Souls of Today's Students.* New York: Simon and Schuster, 1987.

Bloom was a professor of social thought at the University of Chicago and a translator of Plato and Rousseau. These two careers come together in this dense (some say tedious) book in which Bloom argues that the moral and social problems of the modern age are an intellectual problem, in large part provoked by universities which turn out students who have no understanding of the past and no vision of the future. The jargon of liberation and the deification of "creativity" have replaced the discipline of reason. Students no longer have knowledge of the great philosophy and literature which gave them a sense of place in nature. They are reduced to living in a world of despair and relativism (even nihilism) which has its roots in Nietzsche. The book is heavy reading (verging on the incomprehensible in places) but, as one reviewer wrote, is recommended to anyone who "takes seriously the old question of the role of reason in the formation of the virtuous character." [392 pp., ISBN 0-671-47990-3]

Blum, Debra E. "Colleges Urged to Make Radical Changes to Deal With National Crisis in Mathematics Education." *Chronicle of Higher Education*, 37:31

(1991).

Board on Mathematical Sciences. *Actions for Renewing U.S. Mathematical Sciences Departments.* Washington, DC: National Academy Press, 1990.

Board on the Mathematical Sciences. *Educating Mathematical Scientists: Doctoral Study and the Postdoctoral Experience in the United States.* National Academy Press, Washington, DC, 1992. [64 pp., ISBN 0-309-04690-4]

Boyer, Ernest L. *Scholarship Reconsidered: Priorities of the Professoriate.* Princeton University Press, 1990.

In the second chapter of this timely report, Boyer argues that the intellectual work of the professoriate should be thought of as having four separate, yet overlapping, functions: discovery, integration, application, and teaching. He insists that all faculty at all colleges and universities should think and act as scholars and that these four kinds of "scholarship" form the basis for allowing them to do so by expanding the notion of scholarship beyond that which has dominated the universities since the late nineteenth century. He develops at length this new view of scholarship. Many believe it can become the basis for resolving the controversy in the nation's universities between research and teaching, between faculty priorities and society's needs. The report is presented in soft cover and has its roots in a 1989 survey by The Carnegie Foundation for the Advancement of Teaching about faculty attitudes toward scholarship. The survey was mailed to almost 10,000 individuals with about 5500 responses. The results of the survey, and technical details about it, are given in the appendices. [xiii + 147 pp., ISBN 0-931050-43-X]

Carlson, David, Johnson, Charles R., Lay, David C., Porter, A. Duane, Watkins, Ann, and Watkins, William (Eds.). *Teaching Linear Algebra.* (in preparation)

Carlson, David, Johnson, Charles R., Lay, David C., and Porter, A. Duane (Eds.). *Gems of Linear Algebra.* (in preparation)

Carnegie Foundation for the Advancement of Teaching. *The Condition of the Professoriate: Attitudes and Trends, 1989.* Princeton, NJ: The Carnegie Foundation for the Advancement of Teaching, 1989.

This is a technical report consisting of 103 tables and 40 charts which derive almost exclusively from The National Survey of Faculty conducted in 1989 by The Carnegie Foundation for the Advancement of Teaching. In a foreword, Ernest Boyer outlines three issues on which the survey focused: academic quality, teaching and research, and faculty attitudes toward the institutions at which they work. [xxii + 148 pp., ISBN 0-931050-37-5]

Case, Bettye Anne (Ed.). *Responses to the Challenge:*

Keys to Improved Instruction by Teaching Assistants and Part-Time Instructors. MAA *Notes* No. 11, 1989.

Committee on Teaching Assistants and Part-Time Instructors, with funding provided by the Fund for the Improvement of Postsecondary Education (FIPSE), conducted two surveys in the period 1985-1987 to determine how institutions incorporate their graduate students and temporary instructors into the mathematics department and how these individuals are helped in developing teaching competency. The results of the first survey appeared in an earlier unnumbered volume of the MAA *Notes.* Most of those results are repeated in this volume along with much new information. The first part of the report gives statistical data about class size, workloads, supervision, evaluation, and other matters. The remainder describes model programs in detail, including extensive excerpts from the written guides these programs provide to graduate students and others. Separate chapters are devoted to the special problems of international teaching assistants and to parttime faculty who are not graduate students. [vi + 266 pp., ISBN 0-88385-061-3]

Case, Bettye Anne and Huneke, John Philip. "Programs of Note in Mathematics." *Preparing Graduate Students to Teach.* Washington, DC: American Association of Higher Education, 1993, pp. 94-104.

Cheney, Lynne V. *Tyrannical Machines: A Report on Educational Practices Gone Wrong and Our Best Hopes for Setting Them Right.* Washington, DC: National Endowment for the Humanities, 1990.

The National Endowment for the Humanities has issued this report which "not only describes tyrannical machines — educational practices gone wrong — but also considers some of the most important work under way to set them right." Chapters are devoted to both schools (teachers, textbooks, and standardized tests) and to colleges and universities (research and teaching, examples of good practice). The title of the report stems from a 1903 essay on the PhD, the degree associated with the then new research-oriented universities, in which William James coined the phrase "tyrannical machine." Other observations include discussion-provokers from college faculty members like "I left [graduate school] with the idea that my main job was to do research, write books, and neglect undergraduates, because otherwise they would take all my time... My career has been in part an unlearning of what I learned in graduate school." [64 pp.]

Cobb, G.W. "Introductory Textbooks: A Framework for Evaluation." *Journal of the American Statistical Association,* 82(1987) pp. 321-339.

Committee on Preparation for College Teaching of AMS-MAA-SIAM, Bettye Anne Case, Chair. "How Should Mathematicians Prepare for College Teaching?" *Notices of the American Mathematical Society,* 36:10 (December 1989) 1344-1346. (See **First Report**, Chapter 1.)

Committee on the Undergraduate Program in Mathematics. (Bettye Anne Case, Chair, Subcommittee on the Undergraduate Major). *The Undergraduate Major in the Mathematical Sciences.* Mathematical Association of America, 1991. (Reprinted in *Heeding the Call for Change*, MAA *Notes* No. 22, 1992, pp. 225-247.)

The MAA Committee on the Undergraduate Program in Mathematics has issued this new report to update its 1981 recommendations about undergraduate programs in mathematics. Five of the principles proposed as program guides are very similar to those given in 1981 while four new tenets deal with the choice of tracks within the major, the resulting increased responsibilities for advising, the effects and applications of technology, and issues related to moving students through the so-called mathematical pipeline. The report does not call for revolution in undergraduate mathematics education but affirms the many good changes that have occurred in the past ten years and exhorts faculty and departments to focus on certain issues crucial to continued improvement. [24 pp., ISBN 0-88385-454-6]

Conference Board of the Mathematical Sciences, *Graduate Education in Transition*, Washington, DC, 1992. Reprinted in *Notices of the AMS* 39:5 (May/June 1992) 398-402 and *Newsletter*, Assoc. for Women in Mathematics, (July/August) 1992, 22-28.

David, Edward E. *"Renewing U.S. Mathematics: An Agenda to Begin the Second Century."* Reprinted in *Notices of the American Mathematical Society*, 35 (October 1988) 1119-1123. (Often called "the David Report.")

David, Edward E. *"Renewing U.S. Mathematics: A Plan for the 1990's."* Board on Mathematical Sciences, National Academy Press, 1990. Reprinted in *Notices of the American Mathematical Society*, 37:5 (May/June 1990) 542–546 and 37:8 (October 1990) 984–1004. (Often called "David II.")

Donald, Janet G. and Sullivan, Arthur M. (Eds.). *Using Research to Improve Teaching.* San Francisco: Jossey-Bass, 1985.

This volume is a collection of seven essays/chapters by prominent researchers whose studies focus on teaching in the university setting. The book has a professional educational research emphasis and is balanced between methodology and theory relevant to university teaching. The research reported in this volume is notable for resulting from long-range comprehensive research programs with replicated results and for repeated measures on large numbers of students or faculty members or direct observation of the faculty members themselves. [109 pp., ISBN 87589-773-8]

Douglas, Ronald G. (Ed.) *Toward a Lean and Lively Calculus.* MAA *Notes* No. 6, 1986.

D'Souza, Dinesh. *Illiberal Education: The Politics of Race and Sex on Campus.* New York: The Free Press, Macmillan, 1991.

D'Souza casts a critical eye on the philosophy and tactics of some who would use the university to establish a model "multicultural community." While sympathetic to the aspirations of racial and ethnic minorities, he concludes that the politics of race and gender on campuses can eat away at the tradition of scholarship and leave the university not a truly diverse but a balkanized community. Even a liberal spokesman for the American Civil Liberties Union like Morton Halperin, who no doubt would oppose D'Souza on many particulars, writes "This book should be read by anyone concerned with the American university and its commitment to the free exchange of ideas." [319 pp., ISBN 0-02-908100-9]

Dudley, Underwood (Ed.). *Readings for Calculus.* MAA *Notes* No. 31, 1993. (This is Vol. 5 of Roberts, A. Wayne, *Resources in Calculus.* See the comments under that listing.)

Eble, Kenneth E. *The Craft of Teaching: A Guide to Mastering the Professor's Art*, Second Edition. San Francisco: Jossey-Bass, 1988.

The AAUP Bulletin says "Eble's book is unmatched to date in its grasp of the essentials of effective teaching." In his preface, the author laments how little has changed with respect to better preparation of college teachers in the twelve years since the first edition in spite of "repeated recommendations to graduate schools to improve the preparation of future college teachers." Nonetheless, Eble again presents, as he did in the first edition, a practical and useful book which gives careful attention to details like choosing textbooks, class preparation, lecturing, testing, and grading as well as to those parts "of the soul and mind that commit worthy teachers to their calling." [xx + 247 pp., ISBN 1-55542-088-5]

Edgerton, Russell. "The Reexamination of Faculty Priorities." *Change*, July/August, 1993, pp. 10-25.

Erickson, Bette LaSere and Strommer, Diane Weltner. *Teaching College Freshmen.* San Francisco: Jossey-Bass, 1991.

This book is addressed primarily to college teachers. Its purposes, reflecting its three major subdivisions, are to understand the personality of the freshman student, to discuss effective teaching practices, and to provide practical suggestions for dealing with the special challenges of the freshman class. Many of the reflections and ideas come from the experience of the staff and faculty at University College at the University of Rhode Island. Students from schools as diverse as Duke University, Hampshire College, the Community College of Rhode Island, and the University of Delaware share their hopes, plans, and experiences as college freshman. The book is practically oriented with chapters about preparing syllabi, grading, and teaching large classes. It ends with a chapter about cultivating commitment to freshman teaching among faculty at a time when the job is more complex, difficult, and demanding than in the past. [xvii + 249 pp., ISBN 1-55542-310-8]

Fairweather, James S. "Faculty Rewards Reconsidered: The Nature of Tradeoffs." *Change*, July/August, 1993, pp. 44-47.

Fraga, Robert (Ed.). *Calculus Problems for A New Century.* MAA *Notes* No. 28. Washington, DC: The Mathematical Association of America, 1993. (This is Volume 2 of Roberts, A. Wayne, *Resources in Calculus.* See the review under that listing.)

Getman, Julius. *In the Company of Scholars: The Struggle for the Soul of Higher Education.* Austin, TX: University of Texas Press, 1993.

Getman, a professor of labor law at Yale, Stanford, Chicago, Indiana, and Texas and both general counsel and past-president of the American Association of University Professors, reflects on 40 years in higher education. He wonders why so many professors are so unhappy, in spite of what to the outside world seems like idyllic work and environment. One answer: professors are supposed to be researchers as well as teachers and can react extremely to research demands. On the one hand, some professors cannot face submitting manuscripts to the light of day where the possibility of rejection is high; on the other, some spend a lifetime never letting a single idea go unpublished, no matter how much posturing and intrigue that takes. He regrets that teaching often takes a second place to research and exposes forces in academia that pull in the direction of status-seeking rather than the simpler rewards of mentoring, collegiality, and community. Another source of psychic toll on the professor: teaching is both an elite and a populist enterprise. What is the balance between the aristocratic and the democratic? Getman, himself a liberal and director of Texas's Minority Orientation Program for law students, confesses to accommodating the post-sixties era and by dressing down to look like students and having rap sessions rather than giving lectures. In retrospect, he wonders about the value of too much egalitarian accommodation. [ISBN 0-292-72755-0]

Glimm, James G. (Ed.). *Mathematical Sciences, Technology, and Economic Competitiveness.* Board on Mathematical Sciences, National Academy Press, Washington, DC, 1992.

Gulette, Margaret Morganroth (Ed.). *The Art and Craft of Teaching.* Harvard University Press, 1984.

Halmos, Paul R. *I Want to Be a Mathematician.* Washington, DC: Mathematical Association of America, 1985. [xviii+421 pp., ISBN 0-88385-445-7]

Halmos, Paul R. *Problems for Mathematicians Young and Old.* Washington, DC: Mathematical Association of America, 1991.

This is a collection of 165 problems with hints, solutions, etc. that might serve the prospective mathematics professor as subjects for thought, material for classes to be taught in the future, and perhaps especially as models of enticing mathematical exposition. The first sentence is the tip-off: "I wrote this book for fun, and I hope you will read it the same way." *(This review provided by Donald W. Bushaw).*

Hirsch, E.D. *Cultural Literacy: What Every American Needs to Know.* Boston: Houghton Mifflin Company, 1987. [251 pp., ISBN 0-395-43095-X]

Hirsch, a professor of English at the University of Virginia, argues (as did Plato) that specific content transmitted to children is by far the most important element of education. In short, one must teach *something* and the student must learn *something.* Hirsch argues persuasively that good teaching at higher levels cannot occur if the professor cannot count on the students sharing a common body of fact, tradition, and symbol. The book ends with an appendix which gives the author's perception of information every literate American should know.

Hoaglin, David C., and Moore, David S. (Eds.). Perspectives on Contemporary Statistics. MAA *Notes* No. 21, 1992.

Jackson, Michael B. and Ramsay, John R. (Eds.). *Problems for Student Investigation.* MAA *Notes* No. 30, 1993. (This is Vol. 4 of Roberts, A. Wayne, *Resources in Calculus.* See the review under that listing.)

Jarvis, Donald K. *Junior Faculty Development: A Handbook.* New York: The Modern Language Association (10 Astor Place, 10003-6981), 1991.

This 128-page pamphlet contains many ideas useful to all academic departments in a university and not just those directly associated with the Modern Language Association. The booklet outlines programs which institutions or departments can set up to improve junior faculty teaching and research and to support and reward them for successful efforts. The booklet also teaches junior faculty themselves how to document scholarship and get involved in university service.

Kaput, James J. and Dubinsky, Edward (Eds.). *Issues in Undergraduate Mathematics Learning.* MAA *Notes* No. 33, 1994.

This volume is a collection of essays which investigate how students actually learn some of the most important ideas of mathematics: sets, functions, calculus, proof, problem solving and abstract algebra — to name a few. The book has a strong epistemological flavor but also delves into related areas such as the role of the emotions in problem-solving. The thesis of the book is that college mathematics faculty need to understand how students really learn if they are serious about offering effective instruction.

Karian, Zaven A. (Ed.). Symbolic Computation in Undergraduate Mathematics. MAA *Notes* No. 24, 1992.

Keith, Sandra Z. "Getting Hired at a Teaching College." MAA *FOCUS*, 14:1 (February 1994) 4-5.

Kenschaft, Patricia Clark (Ed.). *Winning Women Into Mathematics*, MAA, 1991.

The MAA Committee on Participation of Women developed this booklet to describe the current status of the Association's women members and prospective women members. Here, the committee endeavors to help members of the Association and others interested in increasing future participation of women in mathematics understand the present role of women, develop a vision of a world with more equal opportunity, and consider ways to achieve such a world. Two especially interesting essays are Patricia Kenschaft's "Fifty-five Cultural Reasons Why Too Few Women Win At Mathematics," with an extensive bibliography attached, and Frances Rosamond's fascinating summary (including many historical and current photographs) of women who have been and currently are active in the MAA and other professional organizations, covering a span of 100 years. It is an informal "who's who" of women in American mathematics which should be more familiar to all mathematicians. [72 pp., ISBN 0-88385-453-8]

Knuth, Donald E., Larrabee, Tracy, and Roberts, Paul M. *Mathematical Writing.* MAA *Notes* No. 14, 1989.

This report contains transcripts of lectures and copies of various handouts for the course CS 209, Mathematical Writing, taught at Stanford University during the autumn quarter, 1987. Some of the material in the course duplicated some of what would be discussed in writing classes offered by the English department. But the vast majority of the lectures were devoted to issues specific to mathematics or computer science. The material deals with the nitty-gritty of writing in these technical areas. Points covered include: superscripts and subscripts, documentation, the use of parentheses, and stylistic features of computer programs. The lectures are edited transcripts of class presentations and not polished essays on the topics. [115 pp., ISBN 0-88385-063-X]

Krantz, Steven G. *How to Teach Mathematics: A personal perspective.* AMS, 1993.

This booklet was written while the author was the site mentor at Washington University of the FIPSE-assisted project; see a description of that site program in Chapter 13. The Preface of the booklet is reprinted as one of the essays of Chapter 15 (see Krantz).

Leinbach, L. Carl, et al. (Eds.). *The Laboratory Approach to Teaching Calculus.* MAA *Notes* No. 20, 1991.

This useful volume is divided into two parts. The first contains a discussion of general issues related to teaching calculus in a laboratory format. These range from philosophical issues and choices that face an instructor who decides to use this approach to mundane, but important, details of designing labs and choosing equipment and software. The second part recounts the experiences of a wide variety of schools that have established calculus laboratory programs. Thirty-six different authors from 26 institutions contributed to the volume.

Mathematical Association of America. *Mathematical Scientists at Work: Careers in the Mathematical Sciences*, MAA, 1991.

Mathematical Sciences Education Board, National Research Council. *Reshaping School Mathematics.* Washington, DC: National Academy Press, 1990.

McCaughey, Robert A. "Why Research and Teaching Can Coexist." *The Chronicle of Higher Education*, August 5, 1992, p. A36.

McKeachie, Wilbert J. *Teaching Tips: A Guidebook for the Beginning College Teacher*, Eighth Edition. Lexington, MA: DC Heath.

This compact, handy, soft-cover volume is divided into 31 chapters, each of which treats a practical aspect of teaching. These range from tips on lecturing and managing student hostility toward examinations to the use of audiovisuals and the validity of student ratings of faculty. There is little about the day-in, day-out routine of college teaching that is not discussed somewhere in the book. It is an excellent idea-generator of issues experienced teachers learn about in the school of hard knocks and new teachers need to think about before stepping into the classroom. The author is a professional psychologist and Director of The University of Michigan Center for Research on Teaching and Learning. [xii + 353 pp., ISBN 0-669-06752-0]

Menges, Robert J. and Mathis, B. Claude. *Key Resources on Teaching, Learning, Curriculum, and Faculty Development.* San Francisco: Jossey-Bass, 1988.

This volume is a comprehensive, annotated guide to over 600 books and articles on teaching, learning, curriculum, and faculty development in four-year colleges and universities. Annotations vary in length. Most run about one-half page. Each attempts to "provide the reader with an understanding of the work's content in as few words as possible." [xviii + 406 pp.]

National Research Council. *A Challenge of Numbers: People in the Mathematical Sciences.* Washington, DC: National Research Council, 1989.

This second report issued by the Committee on the Mathematical Sciences in the Year 2000 describes the circumstances and issues related to human resources in the mathematical sciences, especially teachers and students. It provides in one place a wide range of demographic data from a large number of sources. Compiled in both essay and tabular form, the report deals with the US labor force, the pool of students interested in mathematics and science, the status of mathematics at the collegiate level, and the eventual placement of mathematical scientists in the workplace (many as collegiate faculty). The appendix contains 30 pages of numerical data appropriate to understanding these issues. [xv + 119 pp., ISBN 0-309-04190-2]

National Research Council. *Everybody Counts: A Report to the Nation on the Future of Mathematics Education.* Washington, DC: National Research Council, 1989.

Everybody Counts is the first of three reports issued by the Committee on the Mathematical Sciences in the Year 2000 of the National Research Council during l989-l990 dealing with aspects of mathematical education in the United States. It is described as a "public preface" to the work of three NRC units: the Mathematical Sciences Education Board, the Board on Mathematical Sciences, and their jointly sponsored Committee on the Mathematical Sciences in the Year 2000. The report describes how various forces — computers, research, demography, competitiveness — impinge on mathematics education and are a major force for change in an educational system peculiarly resistant to change. In a critically important section, the report describes undergraduate mathematics as "the linchpin for revitalization of mathematics education" in the United States. [xi + 114 pp., ISBN 0-309-03977-0]

National Research Council. *Moving Beyond Myths: Revitalizing Undergraduate Mathematics.* Washington, DC: National Research Council, 1991.

In a final report, the Committee on the Mathematical Sciences in the Year 2000 challenges colleges and universities to raise the level of their efforts in mathematics education to the level of the nation's mathematical research enterprise, already preeminent in the world. Common "myths" include belief that success in mathematics depends more on innate ability than on hard work, that women and members of some ethnic groups are less capable in mathematics, that most jobs require little mathematics, that all useful mathematics was discovered long ago, that to do mathematics is to calculate, and that only scientists and engineers need to study mathematics. These myths have led to critical "realities" that negatively impact the intellectual and economic life of the nation. The report suggests how such myths can be confronted by colleges, their faculty, and others as they revitalize undergraduate mathematics. [ix + 65 pp., ISBN 0-309-04489-8]

National Science Foundation. *Women and Minorities in Science and Engineering.* Washington, DC: National Science Foundation, 1988.

Neidinger, Richard. "Survey on Preparation for Gradu-

ate School." MAA *FOCUS*, 8:4 (September 1988) 4.

Oaxaca, Jaime and Reynolds, Ann W. (Eds.). *Changing America: The New Face of Science and Engineering, Final Report.* Task Force on Women, Minorities, and the Handicapped in Science and Technology, January 1990.

Quay, Richard H. *Research in Higher Education, Second Edition.* Oryx Press, 1985.

This is a bibliography of bibliographies about issues directly affecting scholarship on higher education. It is arranged by subject index as given in the *Thesaurus of ERIC Descriptors*, Tenth Edition. Major subdivisions include history and philosophy of higher education, students, faculty, curriculum and instruction, financial management, administrative behavior, and social and political issues. [x + 133 pp., ISBN 0-89774-194-3]

Ralston, Anthony (Ed.). *Discrete Mathematics in the First Two Years.* MAA *Notes* No. 15, 1989.

Roberts, A. Wayne (Project Director). *Resources for Calculus.* In five volumes: Vol. 1 — Learning by Discovery, Solow, Anita (Ed.); Vol. 2 — Calculus Problems for a New Century, Fraga, Robert (Ed.); Vol. 3 — Applications of Calculus, Straffin, Philip (Ed.); Vol. 4 — Problems for Student Investigation, Jackson, Michael B., and Ramsay, John R. (Eds.); Vol. 5 — Readings for Calculus, Dudley, Underwood (Ed.). MAA *Notes* No. 27-31, 1993.

This set is the product of an NSF-funded project involving two different consortia of liberal arts colleges, totaling 26 schools, The Associated Colleges of the Midwest and The Great Lakes Colleges Association. As the project director comments, faculty in these schools "lavish on calculus the same attention a graduate faculty might give to its introductory analysis course." The intent of the project was to give students a chance "to discover instead of always being told" by using calculators where appropriate, doing library research, and writing clearly reasoned conclusions.

The labs in Volume 1 cover 26 topics. They are written with a computer algebra system in mind but require no particular hardware or software. They can be done successfully using only a graphing calculator. Suggestions are provided for using the labs on Derive, Maple, and Mathematica.

The problems in Volume 2 emphasize conceptual understanding over rote drill. They are organized to parallel the development of a traditional calculus course. All problems have commentary which includes some history about the problem.

Volume 3 has 18 extended applications of calculus which can be used for either one or two class sessions or given to individual students for outside-of-class work. They include applications of calculus in the social sciences, business and industry, biology, and the physical sciences. The modules, each of which is multi-paged, are written for students, complete with exercises, solutions, and references.

Volume 4 proposes 30 different projects for students which require outside reading, consultation, and imagination. All are open-ended and are amenable to more than one method of solution. Each contains information intended only for the instructor to assist in guiding a student through the project.

Thirty-six different writings about the history of mathematics and calculus, its richness and diversity, make up Volume 5. Included are an excerpt from Newton's *Principia*; essays by well-recognized authors like Howard Eves and Eric Bell; and more recent pieces by Barry Cipra, P.R. Halmos, and Richard Feynman. There is even a message from *Ecclesiastes* for students who ask the proverbial question "Why Study Calculus?".

Schoenfeld, Alan (Ed.). *A Source Book For College Mathematics Teaching*, MAA, 1990.

Sponsored by the Committee on the Teaching of Undergraduate Mathematics (CUTM), this booklet offers recommendations on instructional goals, curriculum, the importance of resources, advising of students of all kinds (including advising of pre-collegiate students), instructional techniques, using computers in instruction, and evaluation. The discussion of each issue is compact, focused, and to the point. A long bibliography is provided. [ii + 68 pp., ISBN 0-88385-068-0]

Schoenfeld, Alan H. *Problem Solving in the Mathematics Curriculum: A Report, Recommendations, and An Annotated Bibliography.* MAA *Notes* No. 1, 1983.

This report is based on a Survey of Problem Solving Courses conducted in 1981 by the Problem Solving Subcommittee of the MAA Committee on the Teaching of Undergraduate Mathematics. The discursive body of the report develops suggestions for teaching courses oriented specifically to problem solving and for teaching problem solving within other courses . The bibliography is extensive and well annotated. [ii + 137 pp]

Schwartzman, Steven. *The Words of Mathematics: An Etymological Dictionary of Mathematical Terms Used in English.* MAA Spectrum Series, 1994. [262 pp. ISBN 0-88385-511-9]

Seldin, Peter. *The Teaching Portfolio: A Practical Guide to Improved Performance and Promotion/Tenure Decisions.* Boston: Anker Publishing Company, 1991.

This 95-page paperback provides a systematic way of recording real teaching-related activities that can be compared to what might be expected of an ideal teacher. Hence, as its subtitle suggests, it gives a structure for evaluating faculty at the time of promotion or tenure. Chapters discuss what a teaching portfolio is, items to include in a portfolio, how to initially set up a portfolio, and how a portfolio might be used as an evaluation instrument.

Senechal, Lester (Ed.). *Models for Undergraduate Research in Mathematics.* MAA *Notes* No. 18, 1990.

The material in this book grew out of the Second National Conference of the Council on Undergraduate Research held at Carleton College in July, 1988. Ten essays describe summer research programs for undergraduates, and five deal with programs which run during the academic year. Together, these fifteen pieces make up about half of the text. The remaining half consists of student descriptions of their research experiences and sample student and student/faculty research papers. The volume concludes with excerpts from the proceedings of the first such conference held at Carleton in 1961. [200 pp., ISBN 0-88385-070-2]

Sigma Xi, The Scientific Research Society. *An Exploration of the Nature and Quality of Undergraduate Education in Science, Mathematics and Engineering.* Triangle Park, NC: Sigma Xi, The Scientific Research Society, 1989.

Sigma Xi, with support from the Johnson Foundation and The National Science Foundation, convened a 40-person National Advisory Group at Racine, Wisconsin, January 23-26, 1989, (1) to explore the nature and quality of undergraduate education in science, mathematics, and engineering and (2) to identify the significant topics and issues that should be addressed in charting policy for undergraduate education in these areas. This report summarizes the various discussions of that group. The appendix contains the keynote address "A Profile of Undergraduates in the Sciences" by Kenneth A. Green of the Higher Education Research Institute at UCLA. [viii + 44 pp., P.O. Box 13975, Research Triangle Park, NC 27709]

Sigma Xi Committee on Science, Mathematics, and Engineering Education. *Entry-Level Undergraduate Courses in Science, Mathematics and Engineering: An Investment in Human Resources.* Triangle Park, NC: Sigma Xi, The Scientific Research Society, 1990.

These proceedings make up the "general consensus" report of a workshop sponsored by Sigma Xi and held at Racine, Wisconsin, June 21-24, 1990. The purpose of the workshop was to explore better ways to serve students in entry-level courses in science, engineering, and mathematics. The workshop was sponsored by the Johnson Foundation and by the Division of Undergraduate Science, Engineering, and Mathematics Education at the National Science Foundation. The report addresses the courses themselves, students, faculty, and collegiate institutions. Special attention is paid to the problems in undergraduate education faced by women, minorities, and individuals with physical disabilities. The appendix contains abstracts of four papers presented at the conference and short descriptions of eleven different creative initiatives in science education (broadly defined). [x + 22 pp. + 29 pp. appendix, P.O. Box 13975, Research Triangle Park, NC 27709]

Simon, Barry (Ed.). *Report of the Committee on American Graduate Mathematics Enrollments.* Washington, DC: Conference Board of the Mathematical Sciences, 1987. Summary in *Notices of the American Mathematical Society*, 34 (August 1987) 748-750.

Smith, David A. *et al* (Eds.). *Computers and Mathematics: The Use of Computers in Undergraduate Instruction.* MAA *Notes* No. 9, 1988.

This report is a project of the Committee on Computers in Mathematics (CCIME) of the Mathematical Association of America. A lead article gives a philosophical basis for using computers in mathematics classes. A useful essay discusses how to evaluate mathematical software. Another is devoted to computer algebra systems. Finally, a series of 15 contributions cover computer use in specific courses including calculus, linear algebra, differential equations, statistics, discrete mathematics, probability, geometry, abstract algebra and number theory, remedial mathematics, and problem solving. [xi + 147 pp., ISBN 0-88385-059-1]

Sobel, Max A. and Maletsky, Evan M. *Teaching Mathematics: A Source Book of Aids, Activities and Strategies, Second Edition.* Englewood Cliffs, NJ: Prentice Hall, 1988.

This book is addressed principally to mathematics teachers in grades 6 through 12. It includes extensive discussion of the use of manipulatives and multi-sensory aides in the classroom and of other laboratory and discovery techniques appropriate to mathematics. The second edition includes chapters directly related to teaching algebra, geometry, and probability and statistics as well as suggestions on the use of computers. Many of the ideas presented are as useful in college classrooms at the precalculus level as in secondary schools. [xi + 289 pp., ISBN 0-13-894148-3]

Solow, Anita (Ed.). *Learning By Discovery.* MAA *Notes* No. 27, 1993. (This is Vol. 1 of Robert, A. Wayne, *Resources in Calculus.* See the review under that listing.)

Steen, Lynn Arthur (Ed.). *Calculus for New Century: A Pump, Not a Filter.* MAA *Notes* No. 8, 1988.

Steen, Lynn Arthur (Ed.). *Challenges for College Mathematics: An Agenda for the Next Decade*, MAA *FOCUS*, 10:6 (November-December 1990).

Steen, Lynn Arthur (Ed.). *Heeding The Call For Change: Suggestions for Curricular Action.* MAA *Notes* No. 22, 1992.

The preface to this volume points out that the chapters cover a wide variety of issues related by the theme of changing the structure of mathematics instruction. Specific topics covered are disciplinary (statistics and

geometry), curricular (the undergraduate major), administrative (assessment), and philosophical (multiculturalism). The volume also contains the 1991 report from the Committee on the Undergraduate Program in Mathematics (CUPM) which makes specific recommendations about the mathematics major. Finally, recommendations are included from an MAA conference about collegiate mathematics education held jointly with the Association of American Colleges (AAC) in November, 1991. This conference was one of 12 held by AAC with discipline organizations as part of a comprehensive review of majors in the arts and sciences.

Steen, Lynn Arthur (Ed.). *Reshaping College Mathematics.* MAA *Notes* No 13, 1989. (Reprints of earlier CUPM reports.)

Steen, Lynn Arthur. "20 Questions That Deans Should Ask Their Mathematics Department." *Bulletin of the American Association for Higher Education*, Vol. 44 (9), 1992.

Sterrett, Andrew (Ed.). *Using Writing to Teach Mathematics.* MAA *Notes* No. 16, 1992.

This volume is a collection of 31 essays dealing with various aspects of writing in the mathematics curriculum. It is the outgrowth of contributed paper sessions at the 1988 and 1989 Annual Meetings of the Mathematical Association of America in which over 600 people attended presentations by 30 speakers. The essays are deliberately chosen for ideas that are easily transported to other institutions and to a variety of courses. An introductory section lays out reasons why mathematics students should be expected to write more. Seven essays deal with the hard part: getting started. Three reflect on the level to which grammar should be corrected. One section of papers talks about journals. The remainder of the pieces outline experiences in writing as attempted at a wide variety of schools and in a wide variety of specific courses. [xv + 139 pp., ISBN 0-88385-066-4]

Straffin , Philip (Ed.). *Applications of Calculus.* MAA *Notes* No. 29. Washington, DC: The Mathematical Association of America, 1993. (This is Vol. 3 of Roberts, A. Wayne, *Resources in Calculus.* See the review under that listing.)

Sykes, Charles J. *Profscam: Professors and the Demise of Higher Education.* Washington, DC: Regnery Gateway, 1988.

To the question "Who is responsible for the decline of American higher education?" Sykes has a simple answer: professors. He credits an article by his professor-father, "The Sorcerers and the 7.5 Hour Week," as the inspiration for this book. In it he lays out, no punches barred, his view that faculty have turned universities into their private clubs where the ruling culture is self-interest, good travel, lucrative side business deals, "incestuous obscurity" which very few ordinary citizens understand or care about, and as little teaching of undergraduates as possible. In a final chapter "Storming the Ivory Tower," Sykes lays out his views on what can be done to save the university from itself. [304 pp., ISBN 0-31203-916-6]

Thurston, William P. "Mathematical Education." *Notices of the American Mathematical Society*, 37:7 (September 1990) 844-850.

Tobias, Sheila. *Revitalizing Undergraduate Science: Why Some Things Work and Most Don't.* Tucson, AZ: Research Corporation, 1992.

Tobias, Sheila. *They're Not Dumb, They're Different: Stalking the Second Tier.* Tucson, AZ: Research Corporation, 1990.

Treisman, Philip Uri. "A Study of the Mathematics Performance of Black Students at the University of California Berkeley." In *Mathematicians and Education Reform.* AMS: CBMS Issues in Mathematics Education, Volume 1, 1990, pp. 33-46.

Tucker, Thomas W (Ed.). *Priming the Calculus Pump: Innovations and Resources.* MAA *Notes* No. 17, 1990.

This impressive softback book has been prepared by the subcommittee on Calculus Reform and the First Two Years (CRAFTY) of the Committee on Undergraduate Programs in Mathematics (CUPM). It provides detailed analysis of ten major calculus reform programs currently active (Clemson, Dartmouth, Duke, the Five Colleges, Illinois, Miami of Ohio, Michigan at Dearborn, New Mexico State, Purdue, and St. Olaf) as well as abstracts of more than 60 other projects. An extensive bibliography about calculus-related topics is included. The volume also contains an impressive collection of references and resources relating to funding sources, conferences, software, and graphing calculators. Many of the projects discussed have been supported by the National Science Foundation as part of a national calculus reform movement. The purpose of the volume is to disseminate rapidly and widely detailed examples of calculus reform in action. This volume is a must for anyone who wants to be in touch with the key people and places where calculus reform is under way and who wants easy access to a rich and varied source of calculus related information. [321 pp., ISBN 0-88385-067-2]

Weimer, Maryellen. *Improving College Teaching.* San Francisco: Jossey-Bass, 1990.

The author, who is director of instructional development at The Pennsylvania State University, writes in the preface that "faculty preparation in graduate school continues to offer very little or no instruction on how to teach... More often than not the choice of instructional method is a habitual one rather than a reasoned decision... College teachers teach pretty much as they were taught." The book discusses how to create an environment that fosters teaching excellence, and it presents case studies of improvement programs implemented at a variety of colleges. The book is written for individuals "who may not have much in the way of resources or training but believe instruction at their institution needs to be improved." [xviii + 232 pp., ISBN 1-55542-200-4]

Wilf, Herbert S. "Self-esteem in Mathematicians." *The College Mathematics Journal*, 21:4 (September 1990) 274-277.

Zimmermann, Walter, and Cunningham, Steve (Eds.). *Visualization in Teaching and Learning Mathematics.* MAA *Notes* No. 19, 1991.

This well-produced volume is the work of the MAA Committee on Computers in Mathematics Education (CCIME) and stems from that group's conviction "that visual thinking and the development of visual tools through computer graphics could make major contributions to mathematics education." Half a dozen essays discuss general issues like visual information and reasoning, our reluctance to visualize, and the difference between graphing software and educational graphing software. The remaining 14 contributions deal with visualization in specific courses (geometry, calculus, differential equations, analysis, differential geometry, linear algebra, complex analysis, numerical analysis, and stochastic processes). Practicing what it preaches, the book has numerous computer graphics, including four pages of color plates. [224 pp., ISBN 0-88385-071-0]

PROFESSOR

- Teaching
- Research
- Curriculum
- Advising
- Writing
- Speaking
- Recruiting
- Book Orders
- Faculty Senate
- Committee
- Committee

OUR GOAL

A well rounded
doctoral graduate

Appendices

Appendix A

Recommendations of the American Mathematical Association of Two-Year Colleges

Those interested in teaching in two-year colleges should consider carefully the recommendations in Appendix A. They will find helpful information elsewhere in this book and especially in Chapters 1, 3, and 14 through 17.

Appendix B

Reprints of Selected Short Papers

In addition to the essays written specifically for this book, short articles from a number of journals and newsletters have been selected and are copied here with permission. The intent is that the reprinted articles will be readily available to all participants in a doctoral or postdoctoral professional seminar. The wide range of topics and opinions represented help an individual reader sample ideas from the mathematical community. On issues under active discussion, we sometimes have been able to follow a particular article by one or several answering articles or letters to the editor. Many times such articles refer to other publications of professional societies or governmental agencies, and a number of those are reviewed in Chapter 17.

Since all of the reprinted articles deal generally with the role of the professor, there is no obvious criterion for classification. However, to help the reader find information on a given topic more quickly, the articles selected for reprinting are grouped loosely under three headings:

B-I. Careers and Graduate Education
B-II. Social Issues
B-III. Undergraduate Education

Appendix A

Recommendations of the
American Mathematical Association of Two-Year Colleges

Editor's Note: The latest recommendations for preparation of two-year college faculty members, as officially adopted by the American Mathematical Association of Two-Year Colleges, follow. It was a pleasure to work with Greg Foley on the general subject of college teacher preparation and to exchange ideas as the subcommittee which he chaired prepared drafts of the document for organizational consideration.

In addition to helpful general information, there are specific comments about two-year college teaching in Chapters 1, 3, and 14 through 17. Some articles in Appendix B in the sections on "Careers" and "Undergraduate Education" give helpful information. Sets of specific recommendations made in earlier years about preparation for two-year college faculty are found in:

"Qualifications for Teaching University-Parallel Mathematics Courses in Two-Year Colleges" (1969) reprinted in *A Compendium of CUPM Recommendations*, MAA.

"The Academic Training of Two-Year College Mathematics Faculty," by Calvin T. Long, and other papers in *New Directions in Two-Year College Mathematics*, Springer-Verlag, 1985.

"The Two-Year College Teacher of Mathematics" by John Impagliazzo, Chair of a subcommittee of the Education Committee of the American Mathematical Association of Two-Year Colleges, 1985.

BAC

Guidelines for the Academic Preparation of Mathematics Faculty at Two Year Colleges: Report of the Qualifications Subcommittee
A Subcommittee of the Education Committee
of the
American Mathematical Association of
Two-Year Colleges
*Approved by the AMATYC Delegate Assembly
on 7 November 1992*

Qualifications Subcommittee:

Gregory D. Foley, Chair 1986-1992

Pansy Brunson, 1987-1992

Robert L. Carson, 1986-1987, 1990-1992

Sharon Douglas, 1990-1991

David Ellenbogen, 1986-1988

Michael E. Greenwood, 1986-1992

Lou Hoezle, 1987-1988

Sue Parsons, 1990-1992

Preface

The following document has evolved over a period of years through the work of the Qualifications Subcommittee of the Education Committee, an academic committee of the American Mathematical Association of Two-Year Colleges. The Subcommittee was formed at the AMATYC annual meeting in San Francisco in 1986 largely in response to concern about reports that unqualified persons were being hired by some two-year colleges to fill mathematics teaching positions. Often these positions were of a part-time, emergency, or temporary nature. At some colleges, however, faculty from other departments were transferred into permanent mathematics teaching positions without appropriate credentials or provisions for retraining. Because of these reports, the Education Committee decided to form the Qualifications Subcommittee to investigate this problem and to make recommendations or establish guidelines to address it.

Subsequently, the Subcommittee met, discussed the issues, formulated a framework for the report, and responded to several draft versions of the report. These deliberations led to an original form of the report that was organized according to five categories of instruction: (1) adult basic education, (2) continuing education, (3) developmental and precollege vocational and technical courses, (4) college-level vocational and technical courses, and (5) university-transfer courses. Guidelines were established for each of these five categories.

The Subcommittee sought input from the chairs of related AMATYC standing committees: Developmental Mathematics, Equal Opportunities in Mathematics, and Technical Mathematics. Between meetings, drafts were composed and circulated to Subcommittee members and interested others. Lines of communication were established and kept open with the Committee on Preparation for College Teaching of the Mathematical Association of America chaired first by Guido L. Weiss and later by Bettye Anne Case. In late 1989, after the Baltimore AMATYC meeting, the finishing touches were put on the original version of the report. After approvals by the Subcommittee and the Education Committee Executive Committee, the report was sent to the AMATYC Executive Board for its consideration in the spring of 1990.

The AMATYC Executive Board returned the report for revision stating that in the best interest of the profession there should be one set of guidelines for all two-year college mathematics faculty rather than five sets of guidelines based on different categories of instruction. At the 1990 AMATYC meeting in Dallas the relative advantages of one set of guidelines versus five sets of guidelines were debated in an open forum at the Education Committee meeting and at a meeting of the Subcommittee. A consensus was never reached.

A version of the report that took into account the views of both camps was reviewed at the Seattle meeting in 1991. It laid out one primary set of guidelines, but since several members of the Subcommittee strongly believed that it was not realistic in all circumstances to have one set of guidelines for all mathematics faculty, the report included some disclaimers.

The present report is a slight revision of that previous version with fewer disclaimers. The most important recommendation of the report is that "hiring committees for mathematics positions at two-year colleges should consist primarily of full-time two-year college mathematics faculty." We are confident in the professional judgment of current two-year college mathematics faculty and know that they are capable of making appropriate decisions based on any local constraints that may exist. As the title of this report suggests, it is intended to guide decisions, not control them.

This report has undergone a rigorous review process and has benefitted greatly from it. The report now presents, as much as is possible, a shared vision of what the academic preparation of two-year college mathematics faculty should be. I thank the Subcommittee members for their efforts, and we thank the many reviewers for their comments and suggestions. We hope this report will serve as a catalyst for further discussion of the important issues it addresses.

Statement of Purpose

This document is addressed to two-year college professionals involved in the staffing and evaluation of mathematics programs for their colleges and to universities that have programs or will develop programs to prepare individuals to teach mathematics in two-year colleges. It is not intended to replace any regional, state, or local requirements or recommendations that may apply to hiring instructors, assigning them to classes, or evaluating their performance or qualifications. Rather, our goal is to provide guidelines that reflect the collective wisdom and expertise of mathematics educators throughout the United States and Canada regarding appropriate preparation for two-year college faculty involved in the teaching of mathematics, whether on a full- or part-time basis.

We strongly recommend that only properly qualified personnel be permitted to teach mathematics. Ill-prepared instructors can do much harm to students' knowledge of and beliefs about mathematics. Many two-year college students suffer from mathematics anxiety, and this should not be reinforced or exacerbated through inappropriate mathematics instruction. **Individuals trained in other disciplines should not be permitted to teach mathematics unless they have received sufficient mathematical training as well.** Moreover, individuals hired to teach mathematics at one level should not be permitted to teach at another level unless they possess appropriate credentials.

We are guarding the gates of our profession. This is our responsibility as the leading professional mathematics organization that solely represents two-year colleges. Staffing practices and procedures vary greatly from college to college and from region to region. We wish to ensure the integrity of our profession and the quality of mathematics instruction at all two-year colleges.

Motivating Factors

Disturbing Trends: Reports such as *Everybody Counts: A Report to the Nation on the Future of Mathematics Education* (National Research Council, 1989) document deep-rooted problems concerning mathematics education in the United States. Among these problems is the need to teach meaningful mathematics to individuals from all social, economic, ethnic, and racial backgrounds. This is imperative if our nation is to maintain a leadership role in the world of the future. The mathematics community should strive especially to increase participation of groups that are underrepresented in mathematics.

The two-year college can play a major role in turning our country around in this regard. A study conducted during the 1985-1986 academic year found that, among two-year college students, "one-fourth are minority students, and more than one-half are women" (Albers, Anderson, & Loftsgaarden, 1987, p. 112). Steen et al. (1990) reported that, "One-third of the first and second year college students in the United States are enrolled in two-year colleges, including over two-thirds of Afro-American, Hispanic, and Native American students" (p. 13). Two-year colleges are critical to the national effort to recruit and retain minority students and women as majors in mathematics and mathematics-dependent fields. Two-year college mathematics teachers must be prepared to help and encourage students from these underrepresented groups.

Many two-year college mathematics instructors are nearing retirement age (Albers, Anderson, & Loftsgaarden, 1987). We must work hard at recruiting and preparing the next generation of two-year college faculty. We must enable them to thrive as college mathematics teachers in our rapidly changing world.

Curriculum Reform Movements: The forces of curricular change have reached a relative maximum. The *Curriculum and Evaluation Standards for School Mathematics* (Commission on Standards for School Mathematics, 1989) and *Calculus for a New Century* (Steen, 1988) call for major changes in the content and methods of school and college mathematics. These and other related calls for reform (e.g., National Research Council, 1989, 1991) are due in part to the implications of the pervasiveness of computer technology in our society and in part to the sagging mathematics achievement among our students. It is appropriate that we now reexamine the preparation of two-year college mathematics faculty.

Guiding Principles

Two questions have guided the preparation of this report: What are the characteristics of an effective mathematics teacher? How can these characteristics be fostered and extended through academic preparation and continuing education?

There is a growing body of research related to effective mathematics teaching (Grouws, Cooney, & Jones, 1988). Effective mathematics teachers use their time wisely and efficiently both in and out of class; they present well organized lessons; and they know their subject. Effective instructors are reflective: They think about their teaching before they teach, while they teach, and after they teach. They are creative, resourceful, and dedicated. They use a variety of methods and respond to the needs of the particular class and students they are teaching. Effective mathematics teachers are skilled questioners who encourage and challenge their students. They are clear and careful communicators who recognize the importance of language in mathematics and mathematics as language. They model the behaviors they wish their students to exhibit, especially problem solving, exploration, and investigation. Effective mathematics instructors know a great deal of mathematics and understand the interconnections among its various branches as well as applications to other disciplines. They are continually developing their knowledge and understanding of mathematics, of teaching, and of how students learn. They are independent learners who can adapt and contribute to changes in collegiate mathematics curriculum and instruction. Effective mathematics instructors are active professionals. They read journals, attend professional meetings, and engage in other professional activities. Impagliazzo et al. (1985) elaborated on the activities and characteristics of professionally active mathematics instructors in *The Two-Year College Teacher of Mathematics*. The present report attempts to outline the academic preparation and continuing education that are necessary for a person to be an effective mathematics teacher at the two-year college level.

Organization of the Report

The remainder of this report is organized into four sections. The first concerns guidelines for the formal preparation of two-year college mathematics faculty. The second outlines important areas of mathematical and pedagogical content that should be included in such preparation. The third section discusses avenues for continuing education other than formal education. The final section, Closing Comments, addresses briefly the issues of part-time instructors and the desirability of diversity within a mathematics department. These sections are followed by a bibliography and an appendix that contains an outline for a course on college mathematics teaching. Such a course should be offered by universities that prepare two-year college mathematics instructors, and it should be included in the academic preparation of these instructors.

Guidelines for Formal Preparation

Mathematics programs within and across two-year colleges vary greatly. Mathematics instruction at a comprehensive community college may comprise adult basic education to prepare students for a high school equivalency examination; developmental and precollege vocational and technical courses designed to prepare students for college credit courses; courses for students in college-level vocational and technical programs; university-transfer courses through vector calculus, differential equations, and linear algebra; and continuing education courses that do not carry college credit. Other colleges may focus only on a subset of these types of instruction. Many two-year technical colleges, for example, focus on precollege and college-level vocational and technical courses. Because of this diversity, the standard for the mathematical preparation of two-year faculty must be sufficiently robust to guarantee faculty flexibility. This standard is divided into three parts: minimal preparation, standard preparation, and continuing formal education.

Definitions: All full- and part-time faculty should possess at least the qualifications listed under *minimal preparation*. All full-time faculty should begin their careers with at least the qualifications listed under standard preparation. All faculty should continue their education beyond this entry level. The *continuing formal education* section provides some suggestions. Continuing education of a less formal nature is not only valuable but essential. Avenues for informal continuing education are discussed later in this report. Continuing formal education that requires full-time university enrollment is best undertaken after several years of teaching.

The terms *faculty* and *instructors* are used interchangeably to refer to persons who hold teaching positions. No particular level within a ranking system is implied by either of these terms.

Courses in physics, engineering, and other fields can contain significant mathematical sciences content. Although there is no simple, set formula for doing so, such courses should be taken into account by two-year college mathematics hiring committees when evaluating a candidate's transcripts. Similarly, such courses should be carefully considered by university personnel when making program admission decisions and advising students who hold or may seek two-year college mathematics teaching positions.

Minimal Preparation: All full- and part-time mathematics instructors at two-year colleges should possess at least a master's degree in mathematics or in a related field with at least 18 semester hours (27 quarter hours) in graduate-level mathematics. A master's degree in applied mathematics is an especially appropriate background for teaching technical mathematics. Course work in pedagogy is desirable.

Standard Preparation: All full-time mathematics instructors at two-year colleges should *begin* their careers with at least a master's degree in mathematics or in a related field with at least 30 semester hours (45 quarter hours) in graduate-level mathematics and have mathematics teaching experience at the secondary or collegiate level. The teaching experience may be fulfilled through a program of supervised teaching as a graduate student. Course work in pedagogy is desirable.

Continuing Formal Education: All mathematics instructors at two-year colleges should continue their education beyond the entry level. Appropriate continuing formal education would include graduate course work in mathematics and mathematics education beyond the level of the individual's previous study. Such advanced study may culminate in one of the following degrees: Doctor of Arts in mathematics, PhD or EdD in mathematics education, or PhD in mathematics. For mathematics instructors at two-year technical colleges, taking courses in technologies served by the two-year college mathematics curriculum is also appropriate. Advanced studies may result in a second master's degree.

Evaluating Credentials: A great deal of specialized knowledge and judgment is required to evaluate a candidate's credentials. **For this reason, hiring committees for mathematics positions at two-year colleges should consist primarily of full-time two-year college mathematics faculty.** All staffing decisions related to mathematics instruction—whether full- or part-time—should be made by content specialists.

The Course Content of a Preparatory Program

Mathematics Content: The core of the academic preparation of two-year college mathematics instructors is course work in the mathematical sciences. The mathematics course work for individuals preparing to be two-year college mathematics instructors should include courses chosen broadly from the following areas. Graduate course work should fill gaps, broaden, and extend the undergraduate mathematics background of such individuals.

Discrete Mathematics

Computer Science

Mathematical Modeling and Applications

Calculus through Vector Calculus

Differential Equations

Real Analysis

Numerical Analysis

Complex Variables

Linear Algebra

Abstract Algebra

Probability

Statistics

History of Mathematics

Number Theory

Geometry

Topology

Combinatorics

Pedagogical Content: Course work in pedagogy is an important component in the academic preparation of two-year college mathematics instructors. Such course work should be chosen from the areas listed below. Courses in these areas should be offered by universities that prepare two-year college mathematics instructors.

Psychology of Learning Mathematics

Methods of Teaching Mathematics

Organizing and Developing Mathematics Curricula and Programs

Instructional Technology

Teaching Developmental Mathematics

Using Calculators and Computers to Enhance Mathematics Instruction

Measurement, Evaluation, and Testing

Teaching Mathematics to Adult Learners

Teaching Mathematics to Special-Needs Students

College Mathematics Teaching Seminar (see the Appendix)

Continuing Education

As noted earlier, effective mathematics instructors are active professionals. They read journals, attend professional meetings, and engage in other activities to continue their education. The American Mathematical Association of Two-Year Colleges (AMATYC), the Mathematical Association of America (MAA), the National Council of Teachers of Mathematics (NCTM), and other organizations sponsor conferences, offer minicourses and summer institutes, publish books and journals, and advertise other opportunities for continued professional growth. AMATYC, MAA, and NCTM workshops, minicourses, and institutes address many of the mathematical and pedagogical topics listed in the previous section. Participation in these activities is critically important for two-year college mathematics faculty to keep up-to-date in their field.

Closing Comments

Part-Time Faculty: Ideally, part-time instructors should possess the same level of preparation and commitment to quality teaching as full-time instructors. An MAA committee report entitled *Responses to the Challenge: Keys to Improving Instruction by Teaching Assistants and Part-Time Instructors* (Case, 1988) addresses this issue at length. We support the views of this report as they pertain to two-year college part-time mathematics faculty.

Variety of Expertise: A mathematics department should be composed of individuals possessing complementary strengths and areas of expertise. This is especially true within a comprehensive community college with a wide variety of degree programs. A mathematics department with experts or specialists in pedagogy, statistics, computing, applied mathematics, analysis, and history of mathematics is generally much stronger than one in which all members possess similar backgrounds. This together with programmatic needs and candidate qualifications should be taken into account when seeking and hiring full- and part-time faculty.

Bibliography

⋄ Albers, D. J., Anderson, R. D., & Loftsgaarden, D. O. (Eds.). (1987). *Undergraduate programs in the mathematical and computer sciences: The 1985-1986 survey* [MAA Notes No. 7]. Washington, DC: Mathematical Association of America.

⋄ Case, B. A. (Ed.). (1988). *Responses to the challenge: Keys to improving instruction by teaching assistants and part-time instructors* [Report of the Committee on Teaching Assistants and Part-Time Instructors, MAA Notes No. 11]. Washington, DC: Mathematical Association of America.

⋄ Case, B. A. (Ed.). (1991). *Preparing for college teaching* [Preprint; appearing in 1994 as *You're the Professor, What Next?* in *MAA Notes.*]

⋄ Commission on Standards for School Mathematics. (1989). *Curriculum and evaluation standards for school mathematics.* Reston, VA: National of Teachers of Mathematics.

⋄ Committee on the Mathematical Education of Teachers (COMET). (1988). *Guidelines for the continuing mathematical education of teachers* [MAA Notes No. 10]. Washington, DC: Mathematical Association of America.

⋄ Davis, R. M. (Ed.). (1989). *A curriculum in flux: Mathematics at two-year colleges: A report of the Joint Subcommittee on Mathematics Curriculum at Two-Year Colleges.* Washington, DC: Mathematical Association of America.

⋄ Greenwood, M. E. (1976). *A report on the training of community college mathematics teachers.* Unpublished manuscript, University of Illinois, Urbana-Champaign, and Clark College, Vancouver, WA.

⋄ Grouws, D. A., Cooney, T. J., & Jones, D. (Eds.). (1988). *Perspectives on research on effective mathematics teaching.* Reston, VA: National Council of Teachers of Mathematics, and Hillsdale, NJ: Lawrence Erlbaum Associates.

⋄ Impagliazzo, J. (Chair), Ayers, S. W., Lindstrom, P., & Smith, J. B. (1985). *The two-year college teacher of mathematics.* American Mathematical Association of Two-Year Colleges.

⋄ Leitzel, J. R. C. (Ed.). (1991). *A call for change: Recommendations for the mathematical preparation of teachers of mathematics* [Report of the Committee on the Mathematical Education of Teachers (COMET)]. Washington, DC: Mathematical Association of America.

⋄ Long, C. T. (1985). The academic training of two-year college mathematics faculty. In D. J. Albers, S. B. Rodi, & A. E. Watkins (Eds.), *New directions in two-year college mathematics: Proceedings of the Sloan Foundation conference on two-year college mathematics* (pp. 393-403). New York: Springer-Verlag.

⋄ National Research Council. (1989). *Everybody counts: A report to the nation on the future of mathematics education.* Washington, DC: National Academy Press.

⋄ National Research Council. (1990). *Reshaping school mathematics: A philosophy and framework for curriculum.* Washington, DC: National Academy Press.

⋄ National Research Council. (1991). *Moving beyond myths: Revitalizing undergraduate mathematics.* Washington, DC: National Academy Press.

⋄ Steen, L. A. (Ed.). (1988). *Calculus for a new century: A pump, not a filter* [MAA Notes No. 8]. Washington, DC: Mathematical Association of America.

⋄ Steen, L. A. (Ed.). (1989). *Reshaping college mathematics: A project of the Committee on the Undergraduate Program in Mathematics* [MAA Notes No. 13]. Washington, DC: Mathematical Association of America.

⋄ Steen, L. A. (Chair), Goldstein, J. A., Jones, E. G., Lutzer, D., Treisman, P. U., & Tucker, A. C. (1990, November-December). Challenges for college mathematics: An agenda for the next decade. Report of a Joint Task Force of the Mathematical Association of America and the Association of American Colleges [Center section insert]. *Focus*: The Newsletter of the Mathematical Association of America, 10(6).

Appendix
Outline for a Course on College Mathematics Teaching

Nature of the Course: The course should be a seminar focusing on timely and timeless issues faced by teachers of collegiate mathematics.

Participants: Enrollment should be open to all graduate students in mathematics and mathematics education.

Topics: Topics should be chosen chiefly from among those listed below:

1. *Teaching Issues:* Motivating ideas, motivating students, conveying the nature of mathematics, effective uses of calculators and computers to convey mathematical ideas, learning theory, teaching for understanding, teaching problem solving, characteristics of effective mathematics teachers, individualized instruction, the use and grading of written assignments, teaching adult learners, testing and grading.

2. *Program Issues:* Curricular trends, textbook selection, course and program development, course and program evaluation, student advising, placement of

students.

3. *Other Issues:* Writing for publication, committee work, professional meetings, service. This discussion will include (a) organizations and publications, (b) types of institutions, and (c) finding and retaining jobs.

Activities: Practice presentations and lessons, discussions of issues, outside readings, sharing of obtained information, writing, computer demonstrations, hands-on computer and calculator activities, guest speakers, videotapes, and films.

Suggested Requirements:

1. Attendance at all meetings, participation in all activities including discussions of assigned readings (a bound collection of readings can be made available for purchase at a local outlet).

2. Term paper within the area of the impact of new technology on undergraduate mathematics education, or other appropriate topic: one draft plus a final manuscript.

3. A 10-15 minute conference-style presentation with handouts and prepared transparencies.

4. Presentation of a classroom-style lesson with a computer-demonstration, workshop, or other innovative format.

5. Preparation of the following documents: (a) a biographical sketch, (b) a chronological list of graduate courses with date, instructor, and institution, and (c) a full curriculum vitae.

Textbooks could be chosen from:

◇ Albers, D. J., Rodi, S. B., & Watkins, A. E. (Eds.). (1985). *New directions in two-year college mathematics: Proceedings of the Sloan Foundation conference on two-year college mathematics.* New York: Springer-Verlag.

◇ Davis, R. M. (Ed.). (1989). *A curriculum in flux: Mathematics at two-year colleges.* Washington, DC: Mathematical Association of America.

◇ National Research Council. (1991). *Moving beyond myths: Revitalizing undergraduate mathematics.* Washington, DC: National Academy Press.

◇ Ralston, A., & Young, G. S. (Eds.). (1983). *The future of college mathematics: Proceedings of a conference/workshop on the first two years of college mathematics.* New York: Springer-Verlag.

◇ Schoenfeld, A. (Ed.). (1990). *A source book for college mathematics teaching.* Washington, DC: Mathematical Association of America.

◇ Steen, L. A. (Ed.). (1988). *Calculus for a new century: A pump, not a filter* [MAA Notes No. 8]. Washington, DC: Mathematical Association of America.

◇ Steen, L. A. (Ed.). (1989). *Reshaping college mathematics: A project of the Committee on the Under-*

graduate Program in Mathematics [MAA Notes No. 13]. Washington, DC: Mathematical Association of America.

◇ Sterrett, A. (Ed.). (1990). *Using writing to teach mathematics* [MAA Notes No. 16]. Washington, DC: Mathematical Association of America.

◇ Tucker, T. W. (Ed.). (1990). *Priming the calculus pump: Innovations and resources* [MAA Notes No. 17]. Washington, DC: Mathematical Association of America.

◇ Zimmermann, W., & Cunningham, S. (Ed.). (1991). *Visualization in teaching and learning mathematics* [MAA Notes No. 19]. Washington, DC: Mathematical Association of America.

Appendix B-I

Careers and Graduate Education

Some of the reprints in this section are of immediate interest to those who are looking for employment in mathematical fields and, more specifically, academic jobs. Most discuss broader issues related to the status of the profession, the varied responsibilities of faculty, the relationship of the mathematical community to industry and granting agencies, and implications for doctoral preparation to meet the resulting challenges. Most articles related to undergraduate mathematics education, its curriculum and pedagogy, are included in section B-III, Undergraduate Education. Many of these articles in B-III have direct implications for graduate education, or provide indirect information about academic careers.

Space constraints have limited the inclusion of series of articles, especially if not directly related to preparation for college teaching. For example, there are periodic articles in *SIAM News* related to mathematics and industry which are of much interest to those who work with undergraduate mathematics majors. Job-seekers may want to read the MAA *FOCUS* series of autobiographical articles published in 1992-1993 related to finding a first job and meeting its responsibilities. See the listings in Chapter 17 of this book for additional references.

Applying for Jobs: Advice from the Front

A. Crannell, Notices 39:6 (July/August 1992)

The introduction to the 1985 edition of *Seeking Employment in the Mathematical Sciences* flatly pronounces that "At the present time, job opportunities are somewhat plentiful in mathematics" [1]. Those who are facing the job market nowadays—either from the hiring end, where roughly 400 applications demand attention, careful reading, and sometimes pity, or from the applying end, where the battery of current statistics mingled with inexperience produces a wealth of anxiety—find a mirthless irony in those words. The situation is not easy for anyone: not for the applicants, not for the universities trying to find jobs for their graduating students, and not for the hirers. This article attempts to address, in as humane a way as possible, the issues facing applicants (especially recent graduates), and also to offer advice from one who has already "served time at the front."

I approach this subject from a personal angle; I have just graduated from Brown University and accepted (very contentedly) a job at Franklin & Marshall College. My knowledge of the application process comes through a variety of avenues: various articles [2–5], my work in Brown's Center for the Advancement of College Teaching, discussions with employers at the Joint Meetings in Baltimore last January and at my several job interviews, and other experiences. The information you'll find here is occasionally statistical, occasionally philosophical, and frequently anecdotal.

I will break my discussion into various parts: (1) a brief discussion of the job market; (2) advice for those who will be applying far in the future; (3) general aspects of the job application; and (4) issues of concern to women entering the job market.

What You'll Be Facing. Few of us have come this far in our careers without hearing the dire stories of the current job market. The reasons for this crisis are varied: the temporary reduction in the number of college-aged Americans, the increasing number of foreign mathematicians making their homes in the U.S., the swell in graduating doctorates, and the recession, which has forced severe budget cuts in state education as well as in smaller private schools. Preliminary reports [6] indicate that the number of advertised jobs is still decreasing (the January 1992 issue of *Employment Information in the Mathematical Sciences* contained 23% fewer positions than the January 1991 issue), and that half the nontenure track jobs are for one year only. The AMS Task Force on Employment will soon be publishing a report which includes more recent and complete statistics.

For a number of reasons, today's applicants apply to an incredible number of schools. They do so out of convenience—"print merge" has made 140 applications only marginally more difficult than seventy. They do so out of inexperience—many believe that blanketing the market is the most effective strategy for procuring a job. They do so out of terror—who has not heard about the excellent mathematician who applied to 200 places and received no offers? They do so out of pressure from their faculty—I initially applied to "only" sixty places and was urged to double that number. (I eventually applied to eighty, but subsequently withdrew many of those applications.) Finally, they do so out of peer pressure—while nonmathematicians are amazed that I applied to as many places as I did, for us it has become standard.

The sorry state for job applicants has not, however, resulted in a fiesta for employers. There is an incredible amount of work involved in sorting through the multitude of applications. Moreover, having a surplus of applications doesn't mean that it's easier to hire "superstars": the chair of one department told me that the best people are still fought over as fiercely as ever—much to his chagrin.

Long before You Graduate. All job search manuals begin with the timely advice, *Start early*. Unfortunately, the amount of work that is necessary to maintain a graduate existence keeps us from thinking about extraneous affairs before they are directly upon us, and so most people who see the words *Start early* have long since lost the advantage those words could have afforded. I am hardly a conformist, but I also present this advice with the optimistic and perhaps vain hope that it may do somebody, somewhere, some good.

The best way to get an interesting job is to have evidence that you have done interesting things. The best way to have interesting things to do is to have so many options that you can choose the most interesting ones yourself. Even a young, inexperienced graduate student with few connections and no reputation to speak of has a number of ways to open up those options, most of which essentially come down to advertising.

Volunteer. Go to departmental seminars. Go to conferences. Going to a local conference doesn't have to cost you anything—write a polite letter to your deans asking for a grant. They won't mind shelling out fifty or sixty dollars for a good cause. Getting grants, no matter how small, looks very good to employers. Giving talks to undergraduates or high school students is an excellent way to prepare for the bigger talks that follow, and it lets people know that you're out there (it looks good on your CV, too).

Most of all, talk about your interests. For young graduate students, it's often intimidating to talk to the faculty. However, making use of professors' knowledge, experience, and connections is one of the foremost reasons for being in graduate school. An appreciable benefit of talking to faculty outside of class is that, if the faculty know what you're doing, they'll feel much more comfortable writing letters or verbally recommending you to others, sometimes even before you ask them to. The two most exciting opportunities that came my way while I was in graduate school were both passed along by professors who'd received phone calls asking "We need somebody for such-and-such. Do you know anyone who might be interested?" For foreign graduate students, talking to faculty becomes an effective way to increase your command of English—and this will make a *big* difference when it comes time to apply for jobs.

Next, collecting and maintaining evidence of what you've done is of supreme importance. It's a wise idea to have a folder (mine was unabashedly called "Bragging" as long as it stayed in my drawer) where you can dump everything that's going to make you look good some day. You might keep lists of awards and honors you've received, invitations to speak or to teach external classes, brochures from conferences you've attended, copies of transcripts, copies of old CVs or resumes, interesting computer experiences, student evaluations, unsolicited comments from students (letters, notes on exams, etc.), statistics on student retention, letters—especially thank you letters—from faculty or administrators, and so on.

This folder can be used in various ways. You will almost certainly use it to prepare your CV. You can give it to your letter writers, who will be more than happy to have tangible things to say: "I've seen copies of her course syllabi and they're very good" is nicer to write and read than "I've heard she's a well-organized teacher." And finally, you can clean it up and carry it around to show to prospective employers.

Putting Together Your Application. First, figure out what kind of job you want to apply for. You'll have to do it someday, and doing it now will make your applications much more effective. What do you want out of a job? To learn more math? To work with the hot shots in your field? To have access to large computers? To work in a college? Four-year or two-year [7], liberal arts, community, or technical? To get out of academia altogether [8]? To live in a particular geographic area? Your "Career Services Office" or its equivalent probably has copies of books which discuss academic institutions and their various departments. Careful use of these books is a big help in deciding where to apply and in putting together well thought out applications.

Once you have these things in mind, you can begin to assemble your application. I am most familiar with applying to institutions that place a high emphasis on teaching (and with state budgets being cut, a lot of the hiring is indeed being done at private colleges), but I hope that this advice is generalizable to other institutions.

An application will include many of the following items:

◇ a curriculum vita, or CV, which is best prepared by looking at other people's CVs and deciding which style best suits your needs;

◇ a thesis abstract and research proposal;

◇ reprints or preprints of any articles you've written;

◇ two copies of your graduate and undergraduate transcripts. (These cost money and take time, so order early. You can send photocopies in your mass mailings, but once places get serious about you, they'll want the originals.)

◇ four letters of recommendation, one of which addresses only teaching (more on this later);

◇ a statement of your teaching philosophy (if you're interested in a teaching job); and

◇ a cover letter that includes the position you're applying for; your name, address, email address, and phone number; the names and addresses of your recommenders; your professional interests and aspirations; the reasons you're applying to that particular place (name people you can work with, programs which interest you, location, reputation, etc.); and the fact that you'll be attending the Employment Register at the Joint AMS-MAA meetings in January (if, in fact, you will be).

Keep the cover letter short. If you want to brag more, do it in ...

◇ a follow up letter, in which you brag like crazy and/or respond to questions the school has asked you.

There are a variety of places that advertise job openings, and a fairly complete list of these can be found by looking in [5], an excellent reference.

Here are some general strategies for arranging your application. First, if you are one of the many who have not been able to "start early", now would be the perfect time to invest in a good coffeemaker. When you sit down to put your application together, you must realize that today many schools are getting upwards of 400 applicants, a large number of which are obviously inappropriate (only one school I talked to in January had received as few as 250 applications, but they didn't start advertising until December). Reading hundreds of applications carefully without becoming jaded is strenuous (think about grading your exams), so the first sort merely verifies whether the applicant fits the advertised criteria. If a school advertised for a differential geometer and you're a topologist, you're out. If you apply to a two-year college that wants someone with computer expertise but your letters of recommendation all talk about nothing but your research, you're out. A lot of applicants still believe in the "safety school" approach—they want to do research, but they'll apply to a small college "just in case." Small colleges that advertise for teaching excellence want, believe it or not, teaching excellence, not researchers. So the first rule of thumb is: don't bother applying to places that are advertising for what you're not. It's a waste of time and money. (Some argue that an application is a form of advertising, of spreading one's name around. I believe there are more straightforward ways of achieving the same result.)

Another consideration to keep in mind is that a person high on one institution's list is likely to be high on another's list, and institutions are fully aware of this. Potential employers have to worry about not only whether the applicant is suitable for that school, but also whether that applicant is likely to accept the job if it's offered. Therefore, it's a good idea to try to convince the places to which you're applying that you know what you're doing. If you're applying to a new geographic area, for example, explain why you're doing so. (Employers are likely to be reluctant to interview people who are too far away—they're expensive to interview and less likely to accept.) Especially if you're applying to small places, pay attention to your cover letter. Larger schools may not pay them much heed, but smaller schools tend to emphasize the individual and read the letter fairly carefully. If you are "print merging" your letters, check them over: employers do not "read merge".

It's not a bad idea—and may even be a good one—to have some part of your application, clearly marked, that goes into depth about something that reflects your own strengths and interests. Your thesis abstract and preprints do this for the research side of you, but there may be another aspect you'd like to emphasize. It doesn't have to be teaching; it could be computers, or integrating music and mathematics, or getting grants for mathematical trips to the Caribbean. But there should be something about your application that makes a school think, "Wow. Wouldn't it be nice to have *this* person here?"

Some schools have started asking for a statement of teaching philosophy (which is why other schools have started seeing them even without such requests). I firmly believe this is a step in the right direction. A school that asks for a statement of teaching philosophy weeds out those not really interested in that job and also gains extra insight into each applicant.

I incorporated my description of my teaching directly into my CV. I kept my CV fairly standard for the first two pages—I was born, I went to various schools, I won awards, I did research, I taught courses, I went to conferences, I joined organizations—but then I added a third page called "Goals and Techniques in the Classroom." It was on this page that I mentioned my work with dyslexics, my use of computers and of writing assignments, the career advice that I give, and what students do after they leave my class.

This idea is based on one of the hottest new items in pedagogical circles, the "Teaching Portfolio," which is in turn modeled after the artist's or architect's portfolio. Teachers across the country are being encouraged to maintain artifacts that document their teaching effectiveness—course syllabi, student projects, external and self-evaluations, and so on. It is more comprehensive than a "4 out of 5" on a student evaluation, and is being used by institutions such as Stanford, Harvard, and the New Hampshire secondary school system. Those who are interested in more information should see Peter Seldin's *The Teaching Portfolio* [9]. (If your Career Services Office doesn't have a copy, ask them to get one.)

One last tip: Several schools told me that the small number of applications which are completed early get substantially more attention than the hordes that pour in at the deadline. It can also mean the difference in getting an interview during the Joint Meetings in San Antonio next January. So, give your letter writers plenty of time and push them to get things done early, and aim for getting things done early yourself.

Letters of Recommendation. When I talked to employers about the process of reading job applications, their second-largest gripe (next to the sheer quantity of applications) was bad letters of recommendation. Some "writers" can't. Some letters were, for various reasons, offensive. One example of such a complaint prompted a letter to the editor of the *Notices*: "We have seen letters of recommendation for job candidates that suggest anti-female bias on the part of the writer. The technique is subtle: the (female) candidate is compared only with other women; or statements are made such as 'she is the best female graduate student I have seen in the last five years'..." [10]. But most often, the letters did not at all take into account the type of institution to which they were being sent. One interviewer complained to me that letters for an applicant to a small, liberal arts college that emphasized teaching above all else often began, "Let G be a semi-abelian variety..."

You have more control over your letters than you might think. It is imperative that you tell your writers the kinds of jobs you're applying to, your top choices, as well as which aspects of your career you'd like them to emphasize. It's not unreasonable to ask for two letters, emphasizing different aspects. Neither is it pushy to show them your "bragging" folder—if you think about how hard it is to write letters of recommendations for your own students, you'll realize that your writers will appreciate it. (This is especially true for those writers who don't know you well.) For goodness' sake, give your letter-writers as much information as possible! Tell them your deadlines, both official and personal, and try to give them plenty of time to meet those deadlines.

Sending out letters can often be done through the department or through your Career Services Office. In fact, the folks at Career Services will often maintain a file of letters and other material you want sent out, and will even send them out free-of-charge.

The Interview. When and if you get an interview at a school, make the most of it. Your talk will be better if you've asked beforehand what types of people are going to be in the audience, which upper-level courses are being taught that year, and what kind of knowledge you should assume. If possible, choose a talk that allows you to highlight your own teaching style (use of computer, lots of pictures, whatever).

Your interview is the time to ask all those questions you thought you wouldn't ask until you accepted the job—Is there child care on campus? How much does it cost to live around here? Does the city have a square dancing club? Can I talk to some of the undergrads today?—as well as those that more carefully define the job—On what decisions is tenure based? What is the salary? What are the benefits? Are there tenure quotas? Do faculty "own" courses? What will I be teaching?. You'll probably be meeting about ten different people during the day, most of whom will ask you, "So, er, do you have any questions?" Feel free to ask the same questions over and over; you'll get a lot of different answers anyway.

The kinds of questions that you'll be asked are: What is your research? (This is invariably asked by a dean who hasn't had math since freshman year of college—practice now). What courses would you like to teach? Where's your research going? Um, er, do you have any questions?

You're going to make a much better impression if you are enthusiastic, energetic, and smiling. When you do get to the interview, enjoy it, and drink a lot of coffee.

Women in the Job Market. Graduating students, even in the best of economic times, are prone to bouts of uncertainty and anxiety regarding their futures. A market such as the one we're facing can only further erode their confidence. Many studies have shown that this crisis of confidence disproportionately affects women, who, although they tend to do better than their male counterparts at every stage of mathematical education, consistently undervalue their own skills (see, for example, [11]).

Often, women not only belittle their own accomplishments, but also believe others who belittle their accomplishments for them. I haven't yet met a woman in graduate school who wasn't told at some point (usually by friends) that her gender must have been a big help in getting in. Nor does it stop at graduate school; versions of the "gender boost" are prevalent in the job market.

Although these comments are intended to be encouraging, they further chip at a woman's belief in her own strengths, for they imply that a woman's success is not based on her qualifications alone. Instead, anecdotes and research abound that shows just the opposite: not only must women "make it" on their own merits, but many qualified women are turned away—or turn away—in spite of merits. The "turning away" takes many forms, from applying only for jobs for which she is overqualified, to leaving mathematics altogether.

At such a crucial stage as applying for jobs, it is vital for an applicant to have a realistic and even slightly idealistic view of her or his level of ability. This level may be higher than the applicant thinks, especially if she is a woman. I strongly urge all those who are in a position to advise students that they assure them of their abilities and encourage them to aim high.

To the women who are on the job market this year, I offer the following encouragement: For me, it was too easy to say, "Well, I've made it this far, but I don't know that I'm really any good." I learned pretty quickly that employers and colleagues alike believe that making it "this far" is a concrete indication that I *am* good. If the rest of the world is going to think you're amazing for doing all you've done, you might as well think so, too. Aim high.

For those who are interested in a more thorough treatment of this subject than I have space for here, I highly recommend the *Special Issue on Women in Mathematics* that appeared in the September 1991 issue of the *Notices* [12].

Conclusion. When you know how tough the job market is, it's hard not to send applications to every department that's advertising. Yet the sheer quantity of one-size-fits-all applications indicates that tailoring your applications for the jobs you really want is not only more considerate to prospective employers, but also a smart move on your own part. The application that stands out from the crowd is one that is intelligent and well-considered and that reflects the interests and aspirations of the applicant.

If I had to sum up my own experiences into one sentence of advice, I would say: "Start early, apply to the kinds of institutions where you'd really like to work, and do your best to convince them you're the perfect person for the job." If I were allowed two sentences, I'd add, "And drink a lot of coffee."

References

[1] *Seeking Employment in the Mathematical Sciences*, Mathematical Sciences Employment Register, Providence, RI, 1985, page 1.

[2] Allyn Jackson, "Hiring and Jobseeking in Academia: Anecdotes about the job market," *Notices*, December 1989, **36**:10, pages 1347–1351.

[3] Allyn Jackson, "The Mathematics Job Market: Are New Ph.D.s Having Trouble?", *Notices*, December 1990, **37**:4, pages 1349–1352.

[4] D.J. Lewis, "Employment of New Ph.D.s: Some Proposals," *Notices*, April 1991, **38**:4, pages 296–297.

[5] Bernard Madison, "Employment in the Mathematical Sciences: Advice To Applicants and Employers," *EIMS*, special issue, December 1991, vol. 114.

[6] Donald E. McClure, "AMS Task Force on Employment Report to the Mathematical Community Part I: Academic Hiring Survey," *Notices*, April 1992, **39**:4, pages 311–316.

[7] Ronald M. Davis, "Teaching at a Two-year College—Is It for You?", *EIMS*, May 1992, **117**:10.

[8] Fan R.K. Chung, "Should You Prepare Differently for a Non-academic Career?", *Notices*, July/August 1991, **38**:6, pages 560–561.

[9] Peter Seldin, *The Teaching Portfolio: A Practical Guide to Improved Performance and Promotion/Tenure Decisions*, Anker Publishing Company, Bolton, MA, 1991.

[10] Colin C. Graham, L. Thomas Ramsey, Letter to

the Editor, *Notices*, February 1992, **39**:2, page 100.

[11] The Wellesley College Center for Research on Women, "How Schools Shortchange Girls: Executive Summary," *AAUW Outlook*, **86**:1, 1992, pages 15–25.

[12] "Special Issue on Women in Mathematics," *Notices*, September 1991, **38**:7, pages 701–775.

Brown University and Franklin & Marshall College

The Endless Frontier
Meets Today's Realities*

R.H. Herman
Notices Amer. Math. Soc. 40:1 (January 1993)

What I tell you three times is true.
— "Hunting the Snark,"
— Lewis Carroll

Today, I shall speak to you both as a concerned member of the mathematical sciences community and as a dean. It is my view that the community needs to bring about substantive changes in itself before we are brought to an untenable position. We have missed opportunities which have been presented in the past. You, the department chairs, have always been in a position to play a pivotal role not just in the future of your own department but, as I will argue, in the future of the community. I served as one of you for five years—as a department chair at Penn State—and, contrary to popular wisdom, I am willing to admit that I enjoyed it.

I want to delineate the choices facing the mathematical sciences community, especially at universities. Broadly speaking, I will break my comments into three categories:

1) problems at universities—pressures from constituencies,

2) the mathematics department as part of the university, and

3) opportunities for the future.

The following quotation captures the essence of some substantial recent criticism of universities: "The discipline of colleges and universities is, in general, contrived, not for the benefit of the students, but for the interest, or more properly speaking, for the ease of the masters."

Of course the last word gives away the time of this remark. Indeed, the quotation is from the early 1700s and is due to Adam Smith. For a more up-to-date version of this comment just pick up your local newspaper. Here is one from mine, *The Washington Post*. Robert Samuel-

son, in an article entitled "The Low State of Higher Ed", writes the following: "Higher Education is a bloated enterprise. Too many professors do too little teaching to too many ill-prepared students. Costs can be cut and quality improved without reducing the number of graduates. Many colleges and universities should shrink. *Some should go out of business.*"

Samuelson's voice is bellicose, but when he lists his complaints—low admission standards, high dropout rates, decreasing teaching loads for faculty, and an explosion of graduate degrees—we all know that this is part of the reason we are in the dock accused of a lack of social responsibility. His comments are skewed to the extreme, yet the less polemical comments of former New Jersey Governor Thomas Kean do not provide a source of comfort: "Here is the reality, plain and simple. Our ivory tower is under siege. People are questioning our mission and questioning who we are. They claim we cost too much, spend carelessly, teach poorly, plan myopically, and when we are questioned, act defensively." To the last point we might examine the present "overhead wars" and Don Kennedy's response, and the allegations of price-fixing in the Ivy League.

In summary, we have lost the public trust. Now couple this with the recession and remember that in most states the funding of universities appears as a healthy portion of the discretionary funds. The result is that we are suffering severe cuts while being subject to calls from the state for greater accountability and visible signs that we are delivering on promises implicit to our existence.

Continuing with pressures on the university, let me turn to the government-university partnership. Most of us hearken back to the beginning of it all, Vannevar Bush's report *Science-The Endless Frontier*. This report, delivered to President Truman in 1945, and its progenitors are ultimately responsible for today's science policy. We need to turn to some recent policy statements to understand how things have changed.

Congressman George Brown, in an article for the *Los Angeles Times*, points out that "Science, *Inexplicably*, has come to occupy a place in American culture alongside Plymouth Rock, Johnny Appleseed, and the Bill of Rights." Brown says that society needs to negotiate a new contract with the scientific community. He suggests that we need to require the application of science to the critical problems in the civilian sector. In any event he points out in a carefully reasoned way that while science has arrived at a position of world leadership by many standards, other countries have radically different science policies which have afforded them greater success in linking research with national goals. The call for change comes from one of science's biggest supporters.

Additional evidence for change can be found in Walter Massey's remarks to the National Science Board. Hopefully, as George Brown proposes, there will first be "carefully defined and modular experiments to avoid throwing out the good with the bad." So, as a new contract is be-

---

* The following text is based on a speech presented by Richard H. Herman, Dean of the College of Computer, Mathematical, and Physical Sciences, and Chair of the Joint Policy Board for Mathematics, on October 16, 1992, at the Department Chairs Colloquium sponsored by the Board on Mathematical Sciences of the National Research Council.

ing drawn, and surely it will be, it is necessary for us to help guide the process. This will entail making some changes at universities, accepting further responsibilities in a way consistent with preserving "the taproot", the research effort that has brought us a justifiably high world reputation. Peter Likins, speaking on these issues, closed with a challenge to his audience "to draft their own rules and try to establish a new social contract."

For a reaction to some budget reversions, I will recount some of my own experiences as a dean at the University of Maryland. Over the past several years Maryland has suffered considerable cuts in its state budget. In an effort to improve communication, I went around to departments in the college, made a presentation on the finances, and then allowed for a discussion period. In one department I was greeted with a question by one of the best researchers—more of a complaint actually.

"A few years ago things were very different," he said, and asked, "When do you see a return to those times?" Actually he was pointing out that in the budget reversions a good deal of the traditional ability of the department to follow its own nose had been lost. Several responses came to mind. I tried to imagine what he was thinking; for example: a) in a couple of years things will get better (so let's ignore the situation), b) as soon as we get rid of this dean, c) never. In a very real sense "never" is the right answer, and perhaps that is optimistic.

Let us recall some major changes that have taken place over the years and how they have affected mathematics. In the late 1960s the Department of Defense, in response to the Mansfield amendment, increased emphasis on goal-oriented programs and decreased support of the mathematical sciences. At the same time, as pointed out in David 1, there were reductions in federal fellowships for graduate students and postdoctorals. Shortly thereafter, financial problems hit universities causing a tightening in the job market. At the same time mathematics enrollments doubled. We reacted by stemming the flow of Ph.D.s, and many university mathematics departments succumbed to the view of themselves as service departments by offering calculus in large, larger, and largest sections. Simultaneously, folks in English made a convincing case that composition should be taught in small sections. Perhaps we missed an opportunity. We certainly did not turn "the close identification of mathematics departments with teaching" to our own advantage. Likewise, we paid little attention to creating new opportunities for Ph.D.s.

What can you as a department chair do to ensure resources for the department when your entire university is under extreme pressure? Well, you can go to your dean and argue that the mathematical sciences are important in and of themselves, that applications of pure mathematics, applied mathematics, statistics, operations research, etc., are no surprise considering the founding of the discipline, and that you play an important role in the university. These arguments need to be made con-

tinually, but ask yourself if this differs from the physics chair who goes to the same dean and says virtually the same thing, points to the importance of superconductivity, and perhaps speaks to the exciting nature of the Superconducting Super Collider and the hopes of the field to make discoveries of a fundamental nature (and then argues for another condensed matter theorist). On the assumption that both departments are equally regarded, it is more than likely that the long-term distribution of resources will be split and equal attention (or lack thereof) will be given to both arguments.

Suppose we try a different approach. Try to imagine what the pressures are on this administrator. As Phil Griffiths has pointed out, "Universities will either emerge from this decade leaner, more focused on teaching, more socially relevant, and able to do fewer things better, or they will further erode their sources of support and end up weaker."

Instead of taking the view that your job as chair is to "protect the department", why not ask yourself how the department can help the university meet the pressures on it. Think then of the department as the "minimal corporate unit". So, if the mission of the university is teaching, research, and service, the department as a whole *must* address all of these issues.

Let us turn to one of these pressure points. Many students do poorly in early mathematics courses in college—with the overall withdrawal and failure rate being quite large at some institutions. And this says nothing of repeats. The causes are many and certainly include admission standards. Nevertheless, we can certainly do better with a very likely long-term effect. First, I would suggest that success in these courses translates into success elsewhere and ultimately could change the graduation and retention rates at your university. This has a long-term effect on funding for your (state) university and certainly on recruiting—think of the headlines when your graduation and retention rates go from 55% to 65%. Yes, it is labor intensive in the freshman and sophomore years, and perhaps it means shifting departmental resources a bit to accomplish this. However, your administration would dearly love to have this done (as should you), since efforts lavished in the first two years will reward the department with strong enrollments and strong students in the upper division and graduate courses. But what is the approach to this dean? (The view of deans is sometimes intriguing. Frank Warner tells the story of a phone call from an irate mother who called to complain about a grade for her son in a math course. On picking up the phone he was asked who he was. On informing the parent that he was chair of the Math Department, he was told that the mother expected that she would speak to no one lower than a dean. Frank, being quick of mind, offered that there was no one lower than a dean!) Let me tell you how I have been approached on occasion (not [on] this issue and not [about] the math department). "Well if it is so important, why doesn't the university provide the resources?" *Wrong!* The department is part of the university. Run a pilot program on your own, get it to succeed, and then appeal to the administration to share the long-term costs.

But you say resources are decreasing. You are right, but that is unimportant. There is always a certain amount of money for important things. (And the issue of graduation and retention is central to the university existence.) Recently, at a major state university, a decision was made by the administration to provide [Teaching Assistants] TAs based on overall enrollment—with a lag time built in. Math enrollments are decreasing, and the department is thus concerned. Why not offer to run a "Treisman-like program" with its success—using TAs. If that department only improved success rates in calculus, the argument for the extra TAs would be made.

Continuing on the education front, I would observe that leading mathematics departments have not often exercised effective leadership in the preparation of mathematics teachers. This is probably one of the hottest issues of the day, and history does not position us well. Moreover, the answer is not as simple as "just let them take the usual math major courses and everything will work out fine." If the teachers teach their students mathematics the way we teach them ("sit and git"), they will not succeed. There should be some changes. Again, here is an opportunity to make a real difference. Keep in mind that the half-life of school teachers is seven years, so that if colleges and universities produce strong teachers ready to involve students as recommended in the [National Council of Teachers of Mathematics] NCTM "Standards", we could turn around the entire school mathematics program in one generation.

Now, what type of graduates are we producing? By this I mean masters or Ph.D.s. We still produce for academe. There are some isolated instances of the contrary. Are we doing those who study mathematics a service by continuing to clone ourselves? Are we paying attention to areas of national need? Now do not, please, confuse my comments with the idea of discipline bashing. But over the years we have permitted the mathematical sciences community to become fragmented, with each of our parts attacking common problems separately. Why not make common cause again? There is one certainty. Many of the problems facing the country (environmental studies, for example) are interdisciplinary by nature. A good model for weather prediction requires physical modeling of atmospheric flows on various spatial scales, functional-analytic understanding of the equations of fluid dynamics on those scales, numerical analysis to create algorithms, and ability to match the algorithms with various sequential, vector, and parallel computer architectures. Adding the effects of the land into such models requires additional disciplines such as biology. Advanced materials, manufacturing processes, molecular manufacturing, biomanufacturing, statistical quality improvement, and design of flexible manufacturing systems are areas that need pure and applied mathematics, statistics, and operations research. The difficulty

is that we have, on the whole, only grudgingly accepted activities in this area. We have certainly not encouraged or rewarded them generally. But now, perhaps motivated by need, we will change. Perhaps tilt the axes and offer new types of degrees, such as professional masters degrees or something entirely new. Universities, themselves, are now viewing departmental structures and college structures as more than occasional impediments to "getting the job done". Mathematics (all of it), by virtue of its fundamental nature, is singularly positioned to address some of these issues.

Finally, and here I would risk calumny, suppose we decided that it was okay not to look like some of the "better known" universities. This might be called "diversity of mission". Some departments produce many masters degree students directly for local industry. If we judged them by their excellence in algebraic geometry or any one of a set of "hot" fields, the departments would not emerge high. Yet, they are doing something quite important. I would suggest that there is a community responsibility to address a certain number of issues, and maybe one does this by a division of labor with appropriate credit going to all *from all*.

Now history has it that we have been here before. In 1971 Gail Young wrote an article for the *Notices* bemoaning the employment situation and how all the predictions had been wrong. He spoke of the importance of "a steady supply of brilliant young scholars" and that 1971 had "the most talented, best trained young mathematicians in history," and went on to say that "if we are not careful the supply will be cut off." This is no less true today with all of mathematical research, pure and applied, flourishing as well as having the community significantly involved in education. But Young went on to say that only 15% of the Ph.D.s went into industry, and questioned whether we really know what industry needs. (We know a lot more now as the programs at the Institute for Mathematics and its Applications have shown us.) Here are Young's closing remarks: "What I do believe is that we must make fundamental changes in the nature of graduate work in mathematics which will prepare most of our students for something other than academic life."

Compare this to a comment from the 1992 AMS Task Force on Employment, chaired by Don Lewis. It says that the American Mathematical Society should "advocate for the broadening of doctoral programs in mathematics to recognize the value of nonacademic employment and the matching of talent with the teaching needs of the community, to produce doctorates with wider employment options."

Sound familiar? Arguably, the situation now is more serious than it was in 1971, and hopefully we will not have to wait as long as a generation to take some definitive action.

Jobs, Grants, and the New Ph.D.

A. Jackson
Notices Amer. Math. Soc. 40:6 (July/August 1993)

This article, the first of two parts, focuses on the experiences of new and recent doctorates facing today's tough academic job market. The second part, to appear in an upcoming issue, will explore the question of whether the community is paying enough attention to these problems and will present some suggestions for changes.

"Twenty years of schoolin'

And they put you on the day shift."

— Bob Dylan

"It's reaching crisis proportions," David Cruz-Uribe says of the job market in mathematics. It's mid-April, and he's just finished his Ph.D. at the University of California at Berkeley. "I don't think anyone has ever put together the full picture, except the people who are on the job market, but no one pays attention to us anyway. I'll talk to little, tiny schools that are nothing much, and they'll say, Oh yeah, we got 800 applications for one job... [Graduate students are] miserable. Most of the people I know who are on the job market are biting their nails back to their elbows. It's a very frustrating, frantic time... Every time I run into people I know who are on the job market, the question is the same: Have you heard anything? Have you heard anything? Have you heard anything? Know anybody who's gotten a job?"

Once again, for the third or fourth year in a row, the academic job market in mathematics is dismal. Views vary on whether this year is much worse than the previous few, but there seems to be little evidence that it's any better. The telltale signs are much the same—new Ph.D.s applying to 100 or more schools and getting no interviews, departments receiving up to 1200 applications, young mathematicians bouncing from one temporary position to the next, advisers calling around frantically to patch together some kind of work for their students. What seems to be new this year, though, is a current of discontent among recent doctorates. Frustrated by lengthy and anxiety-ridden job searches, disillusioned by earlier projections of a "scientist shortage", and irritated by federal funding battles over summer salaries when they don't even have jobs, many young mathematicians are wondering whether their problems are getting enough attention in the mathematics community.

Cruz-Uribe's situation was fairly typical. He finished his thesis this year before starting his job search. He applied for seventy positions, got five interviews at the Employment Register, and had one campus interview; as of mid-May, his prospects were bleak. Then, more or less out of the blue, a position that earlier in the year had been swallowed up by budget cuts was reopened: a two-year postdoc at Purdue, which is what he had hoped for all along. It took a while for the chair to "arm wrestle" the money out of the dean, but in the end Cruz-Uribe got the job, making for a happy ending to several months of anxiety and uncertainty. Like many interviewed for this article, Cruz-Uribe felt not triumphant but just lucky and relieved. "Getting this job was heavily conditioned on being in the right place at the right time, as well as knowing a number of people," he remarks.

Assessing the Market. The reasons for this year's poor job market are much the same as in previous years: the general economic downturn, cutbacks in state support of higher education, increased Ph.D. production, an influx of mathematicians from the former Soviet Union, and new visa regulations that allow Chinese doctoral students to stay in the U.S. as long as they like. Accurately assessing the job market is difficult, because data on unemployment across the entire mathematical community does not exist. The AMS-MAA Annual Survey provides information about the employment status of new doctorates, but does not track the employment of those who have had a doctorate for a number of years. According to the First Report of the 1992 Annual Survey (*Notices*, November 1992, pages 1026–1060), the unemployment rate for those receiving doctorates last year rose to 12.7%, up from 11.4% the previous year. The AMS Task Force on Employment is conducting a follow-up study on the employment status of a group of doctorates who been out a year or two (the results of the study will appear in the *Notices* when available).

Of the approximately twenty new or recent doctorates contacted for this article, about half had no jobs by early mid-May. This proportion is in line with estimates from students that about one-third to one-half of their friends were without jobs. Of course, after the last-minute wave of hiring over the summer, the unemployment rate will drop well below 50%. Still, these rough numbers imply not only an uncertain future for those seeking jobs but also a great deal of anxiety and declining morale.

And unemployment is only part of the story. "Underemployment" (some call it exploitation) is a persistent problem. Ben Lotto, a 1988 Berkeley Ph.D. who's finally landed a tenure-track position at Vassar, says, "People are out there teaching four courses a semester for $25,000 a year, and they're getting hired a year at a time to do this. That's no way to treat a colleague. That's what I consider the biggest problem, the high percentage of short-term jobs, one- and two-year jobs." Such positions are all too common as budget crises all over academia force mathematics departments to scramble at the last minute to cover their teaching commitments. One-year positions are particularly damaging to those without established careers. After taking one-year positions, says Lotto, people "have to apply again in October, and spend all their time in October, November, December, and January worrying about applications, and then spend all their time in February, March, and April worrying about whether they're going to have a job next year. There's no way to get a career started under those circumstances."

In addition, many who have been out for a year or

two continue to have difficulty finding permanent positions. Many young faculty bounce from one post-doc to the next, trying to land a tenure-track position. Stephen Kennedy got his Ph.D. from Northwestern in 1988, had four one-year positions in a row, and will finish a three-year temporary position at St. Olaf College next year. He believes that the career path of "migrant math worker" is becoming increasingly common. "I have several friends with similar histories, and the advertisements for one-year jobs. . . seem to become more numerous every year," he notes. In April, he says, the St. Olaf mathematics department advertised on e-Math a one-year leave replacement and received 150 applications.

One of the difficulties of the current job market is that the process is taking longer and longer to conclude. As a result, many well-qualified candidates appear to fall between the cracks. For example, Daniel Edidin got his Ph.D. in 1991 from the Massachusetts Institute of Technology. After a one-year position at Cornell University, he was back on the market, applied to about 100 institutions, and got no offers. Support from his adviser at MIT tided him over the 1992 fall semester, and then at the last minute he was able to stay on at Cornell in spring 1993. In this same year, when he could not even find a job for a full year, he was awarded a postdoctoral fellowship from the National Science Foundation. Says Edidin, "The NSF postdoc saved my career."

Some who can't find work are staying on at their doctoral institutions, hoping for a better market next year. Faculty and students say this practice is widespread. Andrew Mayer, who has finished his dissertation at Princeton this year, figures he'll be staying on another year; another student he knows at Princeton is in the same boat. Mayer applied to about thirty places and had no job by mid-May. He says he's not unhappy about the prospect of staying on another year at Princeton—it would give him the opportunity to expand his thesis and build a teaching record. Nevertheless, he says, when one sees others getting jobs, it's easy to blame one's self and not the job market. "It's a very discouraging process," he says. "I don't know what's denial, rationalization, or truth."

Teaching a Priority. This year, jobseekers report that demonstrated teaching experience is an important qualification. Mayer was on a full scholarship while at Princeton, so he never even served as a teaching assistant. He says this was a "tremendous disadvantage" in his job search. The few who did get tenure-track positions their first year out say that teaching experience was critical to their success. Randall Crist, a new Ph.D. from Texas A&M was one of the lucky few to land a tenure-track position—he'll be at Creighton University in Nebraska next year. "I wrote a detailed description of my teaching techniques and philosophy, asked for letters of recommendation that addressed teaching, and sent summaries of my evaluations to any school that indicated an interest in me," he says. "If I had tried for a research job, I know that I would have been out of luck—everyone

wants one."

David Atkinson landed a tenure-track job at Western Kentucky University. A doctorate from University at Illinois at Urbana–Champaign, Atkinson stayed on an extra year after finishing his dissertation because he couldn't find a job. He did get one offer from "a small, poorly funded state university," he recalls, and turned it down, assuming that he would have other offers. "My decision was actually applauded by most professors here who knew about it," he remarks. No other offers came through. This year, he had seven interviews and one offer. Atkinson says that one likely factor in his success this year was that he asked a faculty member at Illinois to sit in on one of his classes and write a letter of recommendation specifically focusing on his teaching.

"I think I got this job this year based on my teaching experience at Rutgers," says Edward Aboufadel, who finished at Rutgers in 1992 and now has a tenure-track position at Southern Connecticut State University. Aboufadel will likely be familiar to readers as the author of the "Job Search Diary" in the MAA newsletter *Focus*. "I think I sort of lucked into the teaching experience I had at Rutgers—I taught a course here, I taught a course there, and pretty soon I had six courses I'd taught, and that was enough to get my foot in the door."

Although he is happy to have gotten the job he did, Aboufadel wonders about the future. "Sometimes I worry that I've ended up going down this one road, and I'm never going to end up at a place like Rutgers or Michigan State or someplace, and be a research professor," he remarks. "It's too late now, that decision's been made. And partially it was just the way the job market was. . . Another concern with us young people is, there's this year, but what's going to happen in five?. . . By the time things get better, then the new hotshots will come out of graduate school, and where have I been? Well, I've been at Southern Connecticut for the last ten years, teaching four courses a semester and maybe I've published five papers in those ten years because I worked my butt off in the summer. Who are they going to hire?"

Some have noted that women Ph.D.s seem to have a significant edge in the job market, and some departments report that it was the women students who tended to get offers early on and that the offers were generally good. Departments seem to be serious about increasing the number of women faculty. (As it turned out, two of the four women contacted for this article were having little luck on the job market. But, on the whole, not many women were suggested as interviewees for this article, possibly because most of them were getting jobs.) Male jobseekers who noted this trend did not appear to be resentful at their women colleagues' success on the market, but some expressed frustration at university affirmative action policies.

For example, Atkinson says that departments he interviewed with told him "time and again that everything they do—including interview lists and hiring offers–have to go through this one individual," the affirmative action officer. He says that, at one school where he interviewed, the affirmative action officer would not approve the department's interview list, and the department had to change it. The message this affirmative action officer sent, says Atkinson, was that anyone who fit the minority criterion only had to meet the job qualifications. "It was not, 'Is this the best person you could find?'," he says, "but 'Is this person qualified at all? And if so, you should be interviewing and making the offer.' "

Asian Jobseekers Have it Tough. The job market is tough for all but a very few, but it appears that Asian jobseekers have special difficulties. Language problems and the perception that Asians lack an understanding of American culture conspire to narrow their opportunities. "Asian students have a fair chance among the research universities, but less chance the more teaching-oriented the university is," says Guoliang Yu, who was fortunate in landing a tenure-track position at the University of Colorado at Boulder. He finished his Ph.D. in 1991 at the State University of New York at Stony Brook.

"At least among the Asian students I know, few of them are hired by non-research universities," he says. "So, in a sense, they have less options available to them... Among people I know, for the first one or two years—because those jobs are just postdoctoral positions or temporary positions at research universities—it seems there is no major difference between Asian students and American students [in finding a job]. But after the postdoctoral period, there seems to be a difference, because lots of jobs come from teaching universities."

Examples of this problem abound, particularly with new visa regulations allowing Chinese students to stay in the U.S. for as long as they like. One postdoc, who asked not to be named, received a doctorate at a top department in the East, spent a year at the Institute for Advanced Study, and three years in a postdoc at a top research university. With ten publications to his name, he had no offer by mid-May. He was actually fortunate to have had four interviews—another Chinese postdoc in his department had had only one interview (and no offers), and another had no interviews at all.

Like other Asians interviewed for this article, Yu feels it is understandable that teaching-oriented institutions steer clear of those who are likely to have language problems or who are not familiar with the culture of liberal arts institutions. On the other hand, he notes, "the group of Asian students is quite diverse. Some like research, but there are quite a few who like teaching. So it's definitely not true that the whole group is just interested in research. Sometimes there are not many teaching jobs available to them, they have no choice, they have to concentrate on research." All the Asian students interviewed for this article said they had no evidence of outright discrimination agains Asians in interviewing and hiring. But they expressed bewilderment at the fact that even those Asians with outstanding research records were having so much trouble on the market.

Are People Leaving the Field?. Yu says that some of his friends are considering leaving mathematics because they cannot find work—some are looking for jobs in other sciences or in industry, while others are returning to school for degrees in what they believe are more marketable fields, such as computer science. Kennedy says that off the top of his head he can think of seven friends who were without jobs by mid-May, and at least two of these have decided to leave the profession. "The other five are still desperately looking," he says.

Conventional wisdom says that training in mathematics prepares one to solve problems in lots of different areas. However, it seems that most new Ph.D.s do not know how to look for a nonacademic position. Even if they did, it's not clear it would make a big difference in these days of research cutbacks at major companies, declining defense spending, falling computer industry profits, and general economic woes. According to a recent report from the *Wall Street Journal*, college placement officers say that 35% of 1993 bachelor's degree recipients are taking jobs that do not require a college degree.

Andrew Lazarus started out with three offers out of fifteen applications when he first got on the market in 1989, and as of mid-May of this year, he had no job. Like most who finished their doctorates a few years ago, Lazarus has lost touch with many of his fellow Ph.D. students. However, he says, "I have little doubt that many more of my classmates and colleagues are not employed in a mathematical way, but I have lost touch with them," he says. "My impression is that unemployed Ph.D.s vanish from view out of shame and frustration."

"It just seems like there are going to be a lot of people who are going to be genuinely crushed, they're going to leave the field," says Patrick McDonald, a 1990 MIT doctorate who's finishing a postdoc at Ohio State this year. As of April, he had had no offers. "They've got to feed themselves next year, they can't wait two or three years for the market to get better." And it's not even clear that things will improve in a couple of years. "We're just hoping that if we take a bad job now, we'll be able to find a good job later — maybe."

New Wave at NSF: Industrial Postdocs
IMA Workshop Explores Potential and Pitfalls of
Industry-University Collaborations

A. Jackson
Notices Amer. Math. Soc. 40:6 (July/August 1993)

Linking industry with academic research has become a high-priority item on the federal science policy agenda. A number of factors are at work: the perennial quest for "economic competitiveness" that will give the U.S. an edge in world markets, attempts to guide and improve the haphazard process of "technology transfer", the need to broaden academic training in science, industry's declining investment in research, and the simple fact that good science can come out of such collaborations. So it is no surprise that the Division of Mathematical Sciences (DMS) of the National Science Foundation (NSF) is launching a new program for industrial postdoctoral fellowships. On May 23, 1993, at the Institute for Mathematics and its Applications (IMA) at the University of Minnesota the NSF sponsored a workshop to discuss the ups and downs of academic-industrial partnerships.

The DMS program will begin in fiscal year 1994, which starts October 1, 1993. The basic idea of the program is that faculty would act as scientific mentors to postdocs and select appropriate industrial problems for the postdocs to work on. The cost would be shared by the DMS and the industrial partner. The industrial postdocs comprise one component of a proposal for a "national postdoctoral program" submitted to the NSF by the Joint Policy Board for Mathematics (JPBM). The industrial postdocs are in addition to the long-standing Mathematical Sciences Postdoctoral Research Fellowships. (For more information on the new Industrial Postdoc Program see the Funding News section of this issue of the *Notices* for further details.)

How to Get Started. How does an academic mathematician establish contact with industry? Avner Friedman, director of IMA and guru of mathematical-industrial linkages, kicked off the workshop with a step-by-step explanation. Others who spoke throughout the day reinforced his message, describing much the same steps:

◇ *Develop a list of science or engineering managers from local industry.* This can be done through the university vice-president for research, the dean of engineering, mathematics or engineering colleagues, former and present students, or any acquaintances with ties to industry. In addition direct contacts can be made by attending meetings of the Society for Industrial and Applied Mathematics, the Institute for Electrical and Electronics Engineers, or other professional societies with industrial connections.

◇ *Call and state your purpose.* Friedman says to "be frank and mean what you say". He suggests telling them you wish to develop interactions between your mathematics department and industry and that you would like to visit and learn about the problems they are working on. Emphasize that it's in your own interest to establish such contacts, "Otherwise, it will sound like you are trying to sell the *Britannica*."

◇ *Visit and ask questions.* Ask the person what he or she is working on, and guide your questions to the mathematical aspects of the problem, says Friedman. But, he cautions, you must find out what the person is interested in; don't present a lecture on your own interests. If some of the mathematical issues appeal to you invite the person to give a seminar talk.

◇ *Follow up the seminar.* Don't wait several months

after the seminar talk to contact the person again, says Friedman. In that time the person might leave the company or shift to another task. Give the person your suggestions and see what the response is, but "do not get carried away by tangential mathematical generalizations", Friedman warns. "Do not tell him or her how the problem can be generalized to n dimensions. That would be a disaster . . . Stick to the problem."

◇ *Mention the NSF program.* It is helpful to point out that a federal agency is supporting this kind of work and that other industries are taking advantage of it, says Friedman. "Mathematicians have a lot to bring to industry, but they have a lousy track record," notes Friedman. "You have to fight against that."

◇ *Be persistent.* If the response is negative, says Friedman, don't break off contact, because the person may soon switch to a different problem where you can help out. At the same time continue to develop other industrial contacts.

Some say that Friedman has a special affinity for distilling the mathematical essence out of problems from a wide variety of areas, but Friedman believes that any mathematics faculty member who is serious about interacting with industry and who is energetic and curious can be successful at it. There can be a letdown when industry just wants mathematicians to solve easy problems or crunch numbers. But, Friedman says, "if a problem is easy, then you solve it in half an hour." He says one can usually tell if it's boring number crunching, and in that case it pays to be frank and explain that there is no point in starting a project that will prove unsatisfying on one side or the other. In addition, "you have to sort through lots of problems," he remarks. "And you have to have a feeling for a problem *area*, not just one problem." He points out that the four IMA industrial postdocs have worked on problems that really took two years to make some headway.

An Example of Success. As an example of the kind of problems that such postdocs could work on, one of the IMA postdocs, David Dobson, described the research he did as a postdoc with Honeywell. Dobson worked in diffractive optics, developing ways to model diffraction gratings which reflect and transmit radiation. These devices have gratings with heights on the order of microns and are manufactured by the same techniques used for semiconductor devices. The grating pattern is biperiodic, so one must use a three-dimensional model. Essentially, the problem is to try to solve Maxwell's equations on this three-dimensional structure. Dobson posed the problem in a variational form, which led to a numerical finite element scheme. The code, written by Dobson and a programmer that Honeywell hired to help him out, is now in use at the company to make predictions for diffraction devices they design. Dobson used his model to do some extensive work on developing optimal designs for different kinds of devices, which raised a number of interesting mathematical questions.

Dobson reported that he learned a lot through the experience. His graduate work on inverse problems in partial differential equations proved very useful in this research, but he said he also ended up learning a whole new area. He published several mathematics papers as a result of the research and also broadened his research horizons considerably. In addition he notes, "It's exciting when you design something, they build a device, and it works." For the past year he has been on an NSF postdoc at the University of Minnesota, and in the fall he will start a tenure-track position at Texas A&M University.

J. Allen Cox, a senior research fellow at Honeywell Systems and Research Center in Bloomington, Minnesota, was Dobson's industrial contact. The $25,000 Honeywell invested in Dobson is a "drop in the bucket in our research budget", said Cox. Postdocs are "really cheap, from our perspective". Investing that kind of money in a consultant would not have resulted in the computer code they now have, he said. Combining the relatively low expense with a technical problem that's important to the company makes for an easy sell to management.

Cox also stressed that the problem must be sufficiently challenging for the postdoc; if it's trivial and the postdoc solves it in a month, not much is gained on either side. "It takes exploring on both sides to identify a mathematically challenging problem for the postdoc," he said. "It should be a postdoc-quality problem" that should take two years to work on. One measure of whether a problem is sufficiently challenging, he noted, is to ask whether or not the work will result in publications in mathematics journals.

There was a short discussion of proprietary issues, and Cox said that generally this has not been a problem because the mathematics can usually be abstracted in such a way that the details of the applications are not revealed. But companies are serious about such things. One participant, Ellis Cumberbatch, who directs the long-running industrial problems clinic at Claremont Graduate School, said that he had heard a story about a company in England which appealed to a problem-solving group at Oxford University with a problem about curry baked beans. It seems the curry in the beans was eroding the lining of the cans. The group looked at this as a diffusion problem on the inside of the can, came up with an algorithm to solve it, and gave it to the company, which used it with success. Two years later the company revealed that the application actually involved a home pregnancy kit they were developing. To keep their ideas secret they cloaked the chemistry in the guise of a problem about curry baked beans.

The workshop also featured presentations by a number of people who have been successful in establishing university-industrial liaisons: in addition to Cumberbatch were Robert Fennell of Clemson University, James G. Glimm of the State University of New York at Stony Brook, and Mary F. Wheeler of Rice University. Glimm warned that there is no "pot of gold" at the end of the rainbow, and that it takes a lot of time and effort before any monetary benefits are realized. He points out that faculty should pursue industrial contacts for intellectual reasons, out of curiousity about the problem areas involved. "If you can't find a motivation like that," he said, "then I don't think it's going to make much sense."

What are the Pitfalls? During the workshop a number of issues were raised about industrial postdoc programs in general and the DMS program in particular. Right now the details of the DMS program are being developed, and part of the motivation for the workshop was for the DMS to get reactions to and advice on their plans so that they can make further refinements.

One potential difficulty is how to review the proposals for the DMS program. DMS deputy director Bernard McDonald, who made a presentation about the DMS program during the workshop, said that the review mechanism was not settled yet, but that it is likely a panel will be used. Some of the workshop participants pointed out that reviewing would be difficult because of the rapidity with which industries change their focus—it takes months for a proposal to be conceived, written, and reviewed, in which time the industrial partner may have moved to another problem. In addition some wondered which criteria would be important in the review; the quality of the science involved, the importance to industry of the proposed work, the track record of the principal investigator (either in mathematics or in working on industrial problems), and demographic distribution are all possibilities.

Another thorny issue was how individuals who finish such postdocs will fare on the job market. Conventional wisdom holds that this kind of "broader training" for mathematics doctorates will widen their job options, but, as Greg Forest of Ohio State University pointed out, most industries are downsizing their research efforts. "We have to be sure universities will hire them as faculty members because industry can't absorb them, at least not in the short-term," he noted. In addition he said that culturally the mathematical community is not prepared to hire and give tenure to people who have worked on industrial problems because the work involved is outside of traditional applied mathematics and is seen as less valuable mathematically. "If you bring the vitae of these people to a hiring committee in a mathematics department, they wouldn't be interested," he declared. Although Forest supports the DMS program and actively promotes industry-university collaborations, he also believes that these issues, especially with regard to young people, must be considered carefully. In particular, he notes, the role of the faculty mentor in the DMS program is crucial to insuring that the postdoc does not end up wasting time on a mathematically pointless area. As David Levermore of the University of Arizona put it, "They could be doing the world's best engineering for the world's best scientist, but they wouldn't be marketable to a mathematics department."

One IMA industrial postdoc, William Morokoff, said that he had some trouble in the job market this year until he landed an NSF postdoc. He got his Ph.D. from the Courant Institute and then did a two-year IMA postdoc at Siemens in Munich. In remarks after the workshop he said that he's concerned that industrial postdocs might end up working on problems that are of little interest to the academic mathematical community. "They will then end up in the situation I did this year, where no university showed any interest in me," he said. He cautions that the faculty mentor must be deeply involved in the scientific aspects of the industrial problems the postdoc works on. In addition, "The industry must have a mathematically interesting problem and be interested in doing mathematical research, as opposed to applying standard techniques to solve their specific problems."

A number of comments were made about the proposed size of the industrial contribution to the postdoctoral fellowship, which McDonald said is tentatively set at $15,000 per year; by contrast the industrial sponsors for the IMA postdocs contribute $25,000. A $15,000 employee is very cheap, warned Morokoff, so there is a danger that the industry "may just want a programmer to run their software...Unless the problems are carefully screened, this could end up being a dead-end" in terms of mathematical research. Levermore noted that if the postdocs are too cheap, industry will view them as cheap labor, so it is very important that the faculty mentors and the reviewers of the proposals insure a strong intellectual, mathematical component. As one participant summed it up, "We might be putting some young people's careers on the line."

Most of the participants seemed to feel that programs such as the DMS industrial postdocs hold a lot of promise, but that great care is needed in the planning and the implementation so that young people don't get the short end of the stick. With Dobson, for example, Honeywell "gets a brilliant kid for $25,000 who brings new analysis and computation to their bottom-line efforts," Forest noted. "They will now follow up his work to its practical end and bring in another young kid to work on another troublesome area. That's okay—in that case everybody's happy—but we need to make sure there's a balance of everyone's best interests. I want some commitment from those who are promoting this new postdoc activity that they will, in parallel, lobby and promote academic departments to seriously consider the finishing industrial postdocs for positions."

profit; and some there are who seek no further advantage than to look at the show and see how and why everything is done. They are spectators of other men's lives in order better to judge and manage their own.

Michel de Montaigne (1533-1592)

There is more than one way to win the prize. In mathematics, of course, we think of prizes as big theorems and the recognition that goes with them. Being a great mathematician means doing great mathematics. Mathematics, however, is rather stingy in awarding such acclaim. Few win prizes—that way.

Such a simple view of prizes does a good deal of harm to our subject. It makes young mathematicians set one-dimensional goals: Prove great theorems, write great papers, win prestigious grants. Because mathematics is stingy in awarding talent as well as acclaim, all too often measuring success becomes slightly distorted: Prove theorems, write papers, and win grants. Young mathematicians (and old ones too) confuse form for substance, measuring the value of research by the number of pages it consumes or the dollars it delivers. They believe they are competing for the prize, while most are selling trinkets for only slight profit.

The tragedy in this is not that so much minor mathematics is published—trinkets have some value after all—but rather that so many mathematics have a narrow vision of mathematics. They view mathematics as research alone (*their* research), and they equate their own ability to contribute with their ability to publish papers. Instead of seeing mathematics as a broad cultural enterprise, that includes research and teaching and scholarship and history, they see mathematics as a single important but limited activity. They fail to provide service to mathematics because that's not what *real* mathematicians do.

Service is not just sitting on committees. (surely *some* committees should be classified as "disservice.") Service can be as simple as explaining an intriguing bit of mathematics to a student (or colleague), or it can be as complicated as setting up a new program on a national level. Service comes from an attitude about mathematics, a sense of history and culture, a passion for the subject rather than its rewards.

Occasionally everyone out to become a spectator, to step away from the busy crowd, and to look at other ways to compete. Looking at the lives of other shows there are many ways to be a mathematician. There are many ways to win a prize.

Comments

J. Ewing
MAA Monthly 99:2 (February 1992)

Pythagoras used to say life resembles the Olympic Games; a few men strain their muscles to carry off a prize; others bring trinkets to sell to the crowd for a

Educating Graduate Students to Teach

F. Evangelista, MER Newsletter 5:2 (Spring 1993)

In the Spring '92 issue of the MER Newsletter, Catherine Roberts and I wrote an article proposing that the training of graduate students be broadened to better pre-

pare us for jobs in academic institutions. In particular, we felt tha graduate school should help us to develop both as researchers and teachers, instead of concentrating only on the research aspect of a mathematician's career. We asked other graduate students for their comments. This article is a result of discussions with Julian Fleron and John Holcomb of SUNY, Albany, who have both attended MER national workshops, and Masanori Itai, who was a former graduate student at UIC, and taught three years at St. Lawrence University. Though no solutions are proposed, we hope that this will be a part of a continuing dialogue on the education of graduate students.

One can debate endlessly about whether or not graduate programs should provide training in teaching. There is a consensus from the group, however, that the current employment situation makes such discussions academic. There are now fewer positions in large research institutions. A large number of schools which advertise in EIMS (Employment Information in the Mathematical Sciences) or interview at the AMS-MAA Winter Meetings are smaller colleges and universities which are looking for faculty who have a "commitment to undergraduate teaching." As Catherine Roberts puts it: "...the *reality* is that academic jobs include both (research and teaching) components and that therefore graduate programs should address both components." A review of the applicant resumes at the San Antonio meetings reflect this reality - career objectives and employment experience tend to emphasize teaching, especially teaching in innovative settings.

The problem then is not whether a graduate student should have significant teaching experience - this is already mandated by the job market - but how one can gain this experience, given the already heavy burden of coursework and research activities. The dilemma a graduate student faces is aptly stated by Holcomb: "I am very much in favor of establishing programs for graduate students to prepare us to be faculty members of the future. I am baffled as to when graduate students would find time and energy for such endeavors even though it is absolutely necessary in finding a job." And from Julian Fleron, "Unfortunately, most graduate students enter graduate school with little serious mathematical training and seldom any teaching experience. Gaining anything above proficiency in either of these areas requires tremendous dedication." Our combined experiences have shown that oftentimes the only solution is to stay another year in graduate school, hoping that the extra effor will make us more competitive candidates in the job market. But as Holcomb comments, "This is a serious issue. Because of poor health insurance and stipend levels, spending an additional year as a student is not only intellectually toiling, but economically impossible as well."

Furthermore, any program for graduate students requires the supervision of faculty members, lessening the time they spend for their own educational and research activities. It is important then for departments to find

an efficient but effective program for their students. Already, there are mathematics departments who have designed graduate programs that incorporate training in teaching as well as discussions of educational issues. Dissemination of information about such programs can be a start.

In speaking about the tension between research and teaching, Fleron states: "Research and teaching will always be in competition with one another for one's limited time and energy resources, and the relative merits of one over the other will be judged differently by different individuals and different institutions." This seems evident but is seldom recognized by both graduate students and faculty. In graduate school, we make a decision on our field of specialization, and with which faculty member we are going to work with. Our decision may be based on our interests, abilities, personal style, and perhaps, the employment outlook for experts of that particular field. Seldom are we asked to think of what our career goals are, and to think of configuring our activities to fit our goals. Certainly, a person who wishes to work in an industrial setting or in a research university needs different qualifications from one who wishes to teach in a liberal arts college. And certainly, in deciding how to allocate our time between research and teaching, we have to take these goals into consideration. I believe that a good graduate program is one that maintains that all these goals are valid and legitimate choices for Ph.D.'s, and will then help a student to choosing and fulfilling their goals.

Industrial Mathematics: The Working Environment

P. Davis, SIAM News 26:3 (May 1993)

A faculty colleague of mine used to eat dinner regularly at our house. As Lou stepped out the door on his way home, his parting line was always, "See you tomorrow at the plant." We invariably smiled at the image of ourselves in chalk-covered overalls, carrying lunch buckets filled with books and pencils, on our way to a paper factory with ivy-covered walls.

Of course, for many industrial mathematicians "the plant" is just that, and the product is quite tangible, be it steel or software or strategies for pollution control.

What is it like to work in industry, away from the familiar patterns and the ephemeral products of academia? How do the habits and practices of those working environments compare with those of academia? How are those mathematicians supported? Interviews and visits with nearly 40 industrial mathematicians suggest some answers.

Working with Colleagues. The usual industrial working environment ranges from large groups similar in size to university mathematics departments to isolated individuals. In any case, much of the work is done jointly, sometimes in rather large teams. The challenge of teamwork continues throughout a career, in contrast with the personally directed research path a tenured faculty member can choose to follow.

One industrial mathematician explains that the dictates of teamwork "may mean you have to do what you don't want to do for a while." Another says, "You must have tolerance for a range of abilities and the wisdom to navigate the demands of teamwork and a diversity of personalities."

As one female mathematician makes clear, gender can be a factor in collegial relations: "Its tough for women if they are not aggressive. They must make sure people know what they did. They can't be afraid to say, 'That's *my* idea'."

One mathematician's prescription for success emphasizes the importance of teamwork and skill with people: "Learn how to work together in teams, have an openness of mind and people skills. Bring in customers, and understand what they want. But understand that neither you nor they can know everything."

Sources of Support. Broadly speaking, industrial practitioners of mathematics are supported in three ways. (The rare exception is the laboratory with a pure research charter only loosely related to corporate productivity.) They may be part of a staff whose mission is directly linked to the company's product, production cycle, or service. Examples would be a mathematician developing signal processing algorithms for a defense contractor or a statistician responsible for quality control in a manufacturing plant.

The other two modes of support hinge on consulting. Those who function as consultants may be funded either directly from the corporate operating budget, often called funding from overhead, or they may be supported by billing their time to sponsors inside or outside the organization. Regardless of the source of funding and of problems, most industrial mathematicians agree that "being in the middle of the action pays."

Much like their academic colleagues who worry about the way the dean is treating their department, industrial mathematicians find that their support is often closely tied to the apparent value placed by upper management on the contributions of mathematics. In any case, mathematics is seldom the dominant technical discipline. At the corporate laboratories of a major, diversified chemical manufacturer, "Mathematics is always in the background. It is never in front with the physical problem. It is never in the limelight."

In the extreme case in which management practices intellectual apartheid, favoring one or two disciplines above all others, mathematicians are best hidden under the cloak of some other discipline. In happier situations, continuing efforts to point out the concrete contributions of mathematics can coincide with competitive forces to strengthen the support for mathematics. One

major chemical company, for example, is expanding its group of mathematical consultants because of a competitive analysis showing that it can no longer afford unguided, Edisonian build-and-bust experiments. More intelligent modeling must inform its experimentation.

Breadth versus Depth. Although industrial employers do rely on narrow expertise, they often want breadth as well. In the pharmaceutical industry (and certainly elsewhere), "You do need years of experience to develop your craft," but practitioners also need breadth. "Industry wants breadth but relies heavily on narrow expertise as well," says one industrial mathematician. The interdisciplinary work that is so common in industry also demands balance between breadth of knowledge and depth of knowledge. The latter alone is often the measure of mastery in an academic setting.

At one major corporate research laboratory, "The range of disciplines is so broad it doesn't matter what you know. Can you talk to others?" A mathematician who is an internal consultant to a petroleum company says, "I serve as a consultant. I can't specialize." The size of the group with which the individual associates may determine the relative needs for depth and breadth. Larger groups of mathematicians can usually support a greater number of narrow specialists than smaller groups.

Cultural Barriers. Given the tendency of the corporate culture to favor certain disciplines (typically an engineering discipline) over mathematics, the introduction of mathematical approaches can be quite difficult. Moreover, a kind of glass ceiling in the management structure may allow those trained in one or two anointed disciplines to move into leadership roles while holding back mathematicians. Questions about the favoring of other disciplines over mathematics elicit explanations like "Nothing replaces the physical background."

Beyond those labeled as mathematicians, there is a larger community of users of mathematics and developers of computational tools. They are potential employers of mathematicians, and their work could be advanced by collaboration with well-educated interdisciplinary applied mathematicians.

Among such users, there may also be significant cultural barriers to the introduction of individuals who are trained primarily in mathematics. For example, engineers at a prominent defense contractor tell stories of lost competitive bids and design disasters that cry out for simple analyses and simulation. The corporate culture, however, is not ready for mathematics. Facing the strains of the end of the cold war, management has little interest in gambling on an unproven (and perhaps threatening) discipline.

From a different perspective, the vice president for research and development at a medium-sized manufacturer of precision optical instruments has a different reason for failing to make full use of mathematics: "We barely use arithmetic in our own quality studies. The central issue is culture — corporate and on the manufacturing floor — not mathematics.

Comparing Academia and Industry. A cultural gap also separates academic and nonacademic mathematicians. A day at the plant for an industrial mathematician is clearly much different from a day in the halls of the academy, and many industrial mathematicians think that academics don't appreciate the difference.

One industrial mathematician says, "My adviser and faculty treated me like I was lost when I decided to go into industry." An experienced independent consultant argues, "We may need to nurture attitude changes among ourselves that produce a comprehensive acceptance of a wide range of professional needs, not just those of the academic research mathematician."

For the industrial mathematician, interactions with both concepts and colleagues from a variety of disciplines are the norm. Such demands challenge industrial mathematicians to walk a tightrope between narrow expertise, the basis of academic research, and broader but necessarily more superficial knowledge. Much of the work of these mathematicians is determined by corporate policy, not individual research interests.

Such experiences, far removed from those of academia, are clearly the building blocks of satisfying, productive careers for many talented mathematicians. Perhaps those payoffs of productivity and personnel satisfaction are just as deserving of the label "good mathematics" as anything that is published.

Paul Davis is a professor of mathematics at Worcester Polytechnic Institute. He was assisted in gathering the information for this article by Peter E. Castro of Eastman Kodak Co. and I.E. Block of SIAM.

Training a New Generation of Applied Math Faculty

C.D. Levermore, SIAM News 25:6 (November 1992)

At the first SIAM Forum on Industrial and Applied Mathematics, held in Indianapolis, May 15-17, 1992, David Levermore participated in a panel discussion on "Training in Applied Mathematics: Preparation for Academic Careers." The following article grew out of the remarks he had prepared for the panel and the ensuing discussion by members of the panel.

The organizers of the Forum had posed the following questions to the panel: How should the next generation of applied mathematics faculty be trained? Should some industrial exposure be part of the training for an academic career? Can we delineate some elements of a canonical curriculum? What are the desirable characteristics (knowledge, skills, and attitudes) that future applied mathematics faculty should have? Are there any existing graduate programs that provide students with

the opportunities to acquire these characteristics? If not, how should we as a community effect changes? What roles should SIAM play to push for these changes?

The answers mathematicians give to these questions naturally reflect their definition of applied mathematics. My own view is that applied mathematics is a craft. Hence, we should train applied mathematicians as craftspersons, as masters of both the tools of mathematics and the process by which these tools impact the larger scientific community. This view is manifest in the following five central themes which I believe should pervade any training program for academic applied mathematicians.

1. Tool Technique — learning to use a suite of mathematical tools well. We must teach a functional knowledge of the tools of the trade. This means being able to identify appropriate tools for investigating a given problem, knowing the potential and the limitations of these tools, and using them effectively. We should include not only the classical tools of applied mathematics (complex variables, linear algebra, advanced calculus, ordinary differential equations, partial differential equations, perturbation, and asymptotic methods), but also a significant fraction of the modern tools (probability, stochastic equations, numerical methods, dynamical systems, integrable systems, integration theory, topology, and functional analytic methods).

Of course, the topics on this list are far too wide-ranging to be absorbed during a normal graduate career. However, the emphasis when teaching some of these topics to applied mathematics students for the first time should be completely different from that for less-applied mathematics students. For example, the statement of the Lebesgue-dominated convergence theorem might be taught through examples and applications rather than though the details of its proof.

2. Tool Applicability — learning where and when to use the tools. We must foster an awareness of the universal applicability of mathematical tools to a broad spectrum of scientific problems. Students must see the scientific importance of applying existing mathematical tools to new areas and must learn the difficulties of doing so. Tool applicability is best achieved in a strong interdisciplinary environment in which the students are regularly exposed to classes and seminars presented by nonmathematicians. More specifically, a menu of special-topics courses, some of which could be listed jointly with other departments, should be offered each term. By joint-listing, two departments could offer a course regularly, perhaps in alternate years, that neither department could sustain by itself.

Industrial exposure, either through summer employment or through off-campus dissertation research, could provide a significant component of this training for some students but should not displace its presence in the academic setting. SIAM could catalyze connections between industries and graduate programs by acting as a clearing house for information regarding such opportunities. Such connections would also help bridge the gap between the academic and industrial researchers.

3. Tool Development — *learning to improve the tools.* We must instill an appreciation that applied mathematics is an integral part of the discipline of mathematics. Indeed, major developments in the modern tools of applied mathematics have been centered at places with strong intellectual (if not always institutional) ties to the larger mathematical community. It is important that our students be comfortable with the vocabulary and ideas of less-applied mathematicians, so that they can better identify the new mathematical developments that will impact areas of interest to them and be better able to incorporate those developments into their own work.

Toward this end, a broad spectrum of mathematics courses should be offered as a component of an applied mathematics curriculum. A few colloquia each year should focus on new developments of the past decade or so in various mathematical fields like geometry, topology, dynamical systems, or algebra. Perhaps Jones polynomials, or something similar, will have an impact on biology. Of course, we cannot know what new ideas will prove useful in the future, but we can certainly give students the framework within which to adopt them.

4. Taste — *learning to choose good problems.* We must provide direction as to how both the mathematical and the scientific value of a problem can be judged. We should not do this dogmatically, but rather by exposing students to our own thought processes and judgments and then allowing them the freedom to question. There are many mechanisms for this: seminars that encourage lively interaction, smaller working groups that explore new research directions, or less- directed informal settings, such as lunch or a coffee break. The key element is that faculty be accessible to more students than their own dissertation advisees and in a broader environment than their offices or classes. In this way students can draw on a wealth of viewpoints when formulating their own mathematical values.

5. Teaching Skills — *learning how to be an educator at all university levels.* We must present the education of others as an integral component of our profession, one that requires teaching skills and creativity and is appropriately rewarded. This means doing much more than merely having students teach some sections of lower-division courses. Students should be given guidance through lectures that expose them to the insights of experienced teachers on topics such as lesson planning and presentation, the role of homework, quiz and examination design, the use of departmental and university resources, and the handling of difficult situations.

Continued consultation could be provided through a mentor program that matches students (and perhaps young faculty) who are first-time instructors with non-supervisory faculty who have experience teaching the course. Given the increased emphasis being placed on instructional skills in the hiring and promotion process at many universities, direct classroom experience should be considered as an important part of graduate training, even for those with fellowships. This experience would help ease the time demands of first-time lecture preparation during the critical period at the start of a career.

Most importantly, though, we must set a good example by ensuring that the promotion and salary reward system for our faculty recognizes excellence and creativity in instruction as well as in research and service. We must acknowledge those who receive grants to develop innovative instructional ideas just as we do those who receive grants to develop innovative mathematical ideas. In other words, our students must perceive that all aspects of our profession are valued.

It should be clear that these themes overlap those of any training program for industrial mathematicians, provided that "Communication Skills" (with nonmathematicians) is substituted for "Teaching Skills."

I know of no graduate applied mathematics program that satisfactorily addresses all these themes, although many good programs manage several of them particularly well. SIAM might provide the framework for a database through which members could learn how these themes are being addressed at other institutions.

C. David Levermore teaches in the Department of Mathematics at the University of Arizona.

JPBM Committee Initiates Dialogue on "Rewards" in the Math Sciences

SIAM News 25:4 (July 1992)

The Joint Policy Board for Mathematics (JPBM) Committee on Professional Recognition and Rewards was appointed to consider changes in the existing "rewards" structure in the mathematical sciences. Such changes are a key to some of the community's most important goals — among them the revitalization of mathematics education at all levels, the promotion of interdisciplinary work, and the reaching out by the mathematics community to other disciplines and to industry.

Creating a dialogue within the mathematical sciences community is viewed as one of the most important missions of the committee. The committee began this dialogue by making five site visits in April and May. In addition to 15-20 additional site visits during the 1992-1993 academic year, the committee will survey mathematical sciences departments and conduct panel discussions at meetings of the three JPBM member societies — the American Mathematical Society, the Mathematical Association of America, and SIAM.

The committee will also identify contributions that should be recognized and rewarded; determine how those involved (faculty members, department chairs, deans, and mathematicians and managers in industry) value the various contributions and determine how the rewards system works in practice; study methods for evaluating contributions of the types identified as important; articulate the ways in which contributions are, and can be, rewarded; and produce a plan for implementing the recommendations.

The problems faced by the community and the role of the rewards system have been documented in a number of reports, particularly the National Research Council's *Renewing U.S. Mathematics: A Plan for the 1990s*; the Carnegie Foundation for the Advancement of Teaching report *Scholarship Reconsidered, Priorities for the Professoriate*, by Ernest Boyer; and the National Science Foundation report *America's Academic Future, A Report of the Presidential Young Investigator's Colloquium on U.S. Engineering, Mathematics, and Science Education for the Year 2010 and Beyond*.

Calvin C. Moore, associate vice president for academic affairs, University of California, is the chair of the committee; the other members are:

Barbara T. Faires, Westminster College; Gene H. Golub, Stanford University; Phillip A. Griffiths, Institute for Advanced Study; Shirley A. Hill, University of Missouri at Kansas City; William E. Kirwan, III, University of Maryland; Peter D. Lax, Courant Institute, New York University; Carolyn R. Mahoney, California State University, San Marcos; Barry Mazur, Harvard University; Gary C. McDonald, General Motors Research Laboratory; Stephen B. Rodi, Austin Community College; Richard A. Tapia, Rice University; John A. Thorpe, State University of New York, Buffalo; Andrew B. White, Jr., Los Alamos National Laboratory; and Carol S. Wood, Wesleyan University.

Comments and suggestions from individuals should be brought to the attention of the project director, William Adams (Department of Mathematics, University of Maryland, College Park, MD 20742; wwa@math.umd.edu).

Ten Rules for the Survival of a Mathematics Department

G.-C. Rota, MAA FOCUS 12:6 (December 1992)

Times are changing, and mathematics, once the queen of the sciences and the undisputed source of research funds, is now being squalidly shoved aside in favor or fields which are (wrongly) presumed to have applications, either because they endow themselves with a catchy terminology, or because they know (better than mathematicians do or ever did) how to make use of the latest techniques in P.R. The present note was written as a message of warning to a colleague who insisted that all is well, and that nothing can happen to us mathematicians as long as we keep proving deep theorems.

1. Never wash your dirty linen in public. I know that you frequently (and loudly, if I may add) disagree with your colleagues about the relative value of fields of mathematics and about the talents of practicing mathematicians. All of us hold some of our colleagues in low esteem, and sometimes we cannot help ourselves from sharing these opinions with our fellow mathematicians.

When talking to your colleagues in other departments, however, these opinions should never be brought up. It is a mistake for you to think that you might thereby gather support against mathematicians you do not like. What your colleagues in other departments will do instead, after listening to you, is use your statements as proof of the weakness of the whole mathematics department, to increase their own department's standing at the expense of mathematics.

Departments at a university are like sovereign states: there is no such thing as charity towards one another.

2. Never go above the head of your department. When a dean or a provost receives a letter from a distinguished faculty member like you which ignores your chairman's opinion, his or her reaction is likely to be one of irritation. It matters little what the content of the letter might be. You see, the letter you have sent forces him or her to think on matters that he or she thought should be dealt with by the chairman of your department. Your letter will be viewed as evidence of disunity in the rank and file of mathematicians.

Human nature being what it is, such a dean or provost is likely to remember an unsolicited letter at budget time, and not verly kindly at that.

3. Never compare fields. You are not alone in believing that your own field is better and more promising than those of your colleagues. We all believe the same about our own fields. But our beliefs cancel each other out. Better keep your mouth shut, rather than making yourself obnoxious. And remember, when talking to outsiders, you shall have nothing but praise for your colleagues in all fields, even for those in combinatorics. All public shows of disunity are ultimately fatal to the well-being of mathematics.

4. Remember that the grocery bill is a piece of mathematics, too. Once, when spending a year at a liberal arts college, I was assigned to teach a course on what you like to call "Mickey Mouse Math". I was stung by a colleague's remark that the course "did not deal with real mathematics." It certainly wasn't a course in physics or chemistry, so what was it ?

We tend to use the word "mathematics" in a valuative sense, to denote the kind of mathematics we and our friends do. But this is a mistake. The word "mathematics" is more correctly used in a strictly objective sense. The grocery bill, a computer program and class field theory are three instances of mathematics. Your opinion that some instances may be better than others is most effectively verbalized when you are asked to vote on a tenure decision. At other times, a careless statement of relative values is likely to turn potential friends of mathematics into enemies of our field. Believe me, we are going to need all the friends we can get.

5. *Do not look down on good teachers*. Mathematics is the greatest undertaking of mankind. All mathematicians know this for a fact. Shocking as it may sound to us, many people out there do not share this view. As a consequence, mathematics is not as self-supporting a profession in our society as the exercise of poetry was in medieval Ireland. Most of our income will have to come from teaching, and the more students we teach, the more of our friends we can appoint to our department. Those few colleagues of ours who are successful at teaching undergraduate courses should earn our thanks, as well as our respect. It is counterproductive to turn our noses at those who bring home the dough.

When Mr. Smith dies and decides to leave his fortune to our mathematics department, it will be because he remembers his good teacher Dr. Jones who never made it beyond associate professor, not just because of the wonderful research papers you have written.

6. *Write expository papers*. When I was in graduate school, one of my teachers told me: "When you write a research paper, you are afraid that your result might already be known; but when you write an expository paper, you discover that nothing is known."

Not only is it good for you to write an expository paper once in a while, but such writing is essential for the survival of mathematics. Look at the most influential writings in mathematics of the last hundred years. At least half of them, from Hilbert's Zahlbericht on down, would have to be classified as expository.

Let me tell it to you in the P.R. language that you detest. It is not enough to for you (or anyone) to have a good product to sell, you must package it right and advertise it properly. Otherwise, you will go out of business.

Now don't tell me that you are a pure mathematician and therefore that you stand above and beyond such lowly details. It is the results of pure mathematics, rather than those of applied mathematics, that are most sought after by physicists and engineers (and soon, we hope, by biologists as well.) Let us do our best to make our results available to them in a language they can understand. If we don't, they will some day no longer believe we have any new results, and they will cut off our research funds. Remember, they are the ones who control the purse strings, since we mathematicians have always proven to be inept at all political and financial matters.

7. *Do not show your questioners to the door*. When an engineer knocks at your door with a mathematical question, you should not try to get rid of him or her as quickly as possible. You are likely to make a mistake I myself made for many years: to believe that

the engineer wants you to solve his problem. This is the kind of oversimplification for which we mathematicians are notorious. Believe me, the engineer does not want you to solve his or her problem. Once, I did so by mistake (actually, I had read the solution in the library two hours ago, quite by accident), and he got quite furious, as if I were taking away his livelihood. What the engineer wants is to be treated with respect and consideration, like a human being he or she is, and most of all to be listened to in rapt attention. If you do this, he or she will be likely to hit upon a clever new idea as he or she explains the problem to you, and you will get some of the credit.

Listening to engineers and other scientists is part of our duty. You may even occasionally learn some interesting new mathematics while doing so.

8. *View the mathematical community as a United Front*. Grade school teachers, high school teachers of mathematics, administrators, and lobbyists are as much mathematicians as you or Hilbert. It is not up to us to make invidious distinctions. They contribute to the well-being of mathematics as much or more than you or many other research mathematicians. They are right in feeling left out by snobbish research mathematicians who do not know which way their bread is buttered. It is in our best interest, as well as in the interest of justice, to treat all who deal with mathematics, in whichever way, as equals. By being united we will increase the probability of our survival.

9. *Attack flakyness*. Now that communism is a dead duck, we need a new Threat. Remember, Congress only reacts to potential or actual threats (through no fault of their own, it is the way the system works). Flakyness is nowadays creeping into the sciences like a virus through a computer system, and it may be the greatest present threat to our civilization. Mathematics can save the world from the invasion of the flakes by unmasking them, and by contributing some hard thinking. You and I know that mathematics is not and will never be flaky, by definition.

This is perhaps the biggest chance we have had in a long while to make a lasting contribution to the well-being of Science. Let us not botch it like we did with the other few chances we have had in the past.

10. *Learn when to withdraw*. Let me confess to you something I have told very few others (after all, this message will not get around much): I have written some of the papers I like the most while hiding in a closet. When the going gets rough, we have recourse to a way of salvation that is not available to ordinary mortals: we have that Mighty Fortress that is our Mathematics. This is what makes us mathematicians into very special people. The danger is envy from the rest of the world.

When you meet someone who does not know how to differentiate and integrate, be kind, gentle, understanding. Remember, there are lots of people like that out there, and if we are not careful, they will do away with us, as has happened many times before in history to other very special people.

And believe yours as ever,

Gian-Carlo Rota

Gian-Carlo Rota is Professor of Mathematics at MIT, Cambridge, MA 02139. The views expressed above are Professor Rota's and do not necessarily reflect the official view of the MAA.

Beefing-Up Graduate Education

MAA FOCUS 12:3 (July 1992)

In order to promote serious discussion of the issues of graduate education and to move the mathematical sciences community to action, the Conference Board of the Mathematical Sciences (CBMS) organized a three-day conference on Graduate Education in Transition in Washington, DC, 4–6 May 1991. Support for the conference was obtained from the Exxon Education Foundation. Twenty-four individuals participated in the conference, among them seven current presidents and six former presidents of CBMS member societies. They were asked two key questions: what are the nation's needs and how can the community best respond?

The report they produced, Graduate Education in Transition, just published, is based on the written summaries of the working groups and on the discussions which took place at the plenary sessions. It is organized into two main sections: Discussion of the Issues, and Recommendations.

Highlights from the report are reproduced below. For copies of the complete document, free of charge, contact: the Conference Board of the Mathematical Sciences (CBMS), 1529 Eighteenth Street Northwest, Washington DC 20036-1385; (202) 293-1170.

Highlights from the Discussion. The report of the Federal Coordinating Council for Science, Education, and Technology (FCCSET) accompanying the President's 1992 budget both deplores the present state of science and mathematics education and proposes objectives and priorities to guide future federal activities. Recommended budget increases run 28 percent at the precollege level and 14 percent at the undergraduate level, but only 2 percent at the graduate level.

Unlike the postsputnik reform, which was driven by military needs, the present effort is driven by economic concerns. The government sees leadership in science and mathematics as a critical element to regain American competitiveness in the international arena.

Short-term intervention programs in the schools will yield some temporary benefits but the attitudes and skills of school teachers are, in the long run, molded in colleges and universities where these teachers are instructed by the products of our graduate schools. One

does not have to subscribe to a domino theory to see that all parts of our education system are interdependent.

Although our graduate programs have brought US mathematics to world leadership in research, they have been less successful in preparing students for college teaching and for positions in industry. Indeed, the report emphasizes that *"without reform in graduate education no lasting change in school or undergraduate education is likely."* (their italics)

The importance of all the roles of faculty — research and graduate education, undergraduate education, service to a broad set of client disciplines, community outreach, and service to the department, the university, the local community, and the profession — must be recognized and some preparation for these roles should begin in graduate school.

One unfortunate consequence of the parochialism in graduate mathematics education is that much of industry and business still regards mathematicians with some suspicion. Few industries have career paths for mathematicians; contributions of a mathematical nature are often not recognized as such because they are made by physicists, engineers, and computer scientists. To remedy this situation, the report "endorse[s] the key principles of the Mathematical Association of America's Committee on Preparation for College Teaching, namely that doctoral programs should prepare students to meet a wide range of professional responsibilities and should not be limited to specialization in narrow areas, and should give systematic attention to promoting excellence in the teaching of mathematics." (See "How Should Mathematicians Prepare for College Teaching," Notices of the American Mathematical Society, 36:10 (December 1989): 1344–1346.)

The continued technological and economic health of the US depends on maintaining at least the present supply of graduate-level mathematical scientists.

American women and minorities, who will be a large fraction of new entrants in the work force by the year 2000, have traditionally not been attracted in sufficient numbers to careers in the mathematical sciences.

We must convey to students (and to the public at large) that mathematics is a lively, dynamic, and varied profession that has attracted inquisitive minds since the birth of civilization. It is important that both principal aspects of mathematics be appreciated; its useful, indispensable role in science and technology and its continuing intellectual fascination over the ages.

Some of the Recommendations. Among the nineteen specific recommendations proposed by the committee are the following:

⋄ Mathematical sciences departments should: Expose graduate students to some collaborative projects with oral presentations; and prepare all students to become effective teachers or communicators of mathematics.

◊ Universities should: Establish a climate in which a broad spectrum of contributions by the faculty is recognized and valued. Consider adopting guidelines such as those described in Boyer's *Scholarship Reconsidered*.

◊ Professional societies should: Develop and publicize a list of mathematical contributions to industry and government. Promote the industrial use of mathematics, mathematical models, and computational mathematics. Develop an industrial liaison group to help industry find appropriate specialists for particular problems.

◊ Government should: Provide stable funding for PhD students by balancing teaching, fellowship, and research support.

◊ Industry should: Cooperate with professional societies and universities to communicate the mathematical needs of the workplace and to develop suitable programs in industrial mathematics.

How Can a Research Mathematician Help Foster the Development of Future Secondary Math Teachers?

M.K. Smith, UME Trends 4:3 (August 1992)

Not long after I started teaching, I realized that most of my students were lacking in mathematical background I had acquired well before starting college. I saw that this was a vicious circle; many future high school math teachers left college without this background, and therefore could not pass it on to their students. I eventually decided that I could get the most impact from my teaching efforts by concentrating as much as possible on future math teachers. Here are some of the things I have done over the years to further this goal. Some of them are very easy, others take more time and energy. If you are interested in following this path, I encourage you to start slowly. Remember that an important part of solving any problem is understanding it. We understand the mathematics well enough. The challenge is in understanding our audience and how best to help them learn. I continually try to improve my understanding by observing and listening to my students and to experts in the field of mathematics education. One thing I do understand is that it is not enough to tell the students what I think they ought to know, even if I could give the most brilliant and mathematically coherent lectures possible. I have to explain why these things are important, and to do that I need to know where the students are coming from and link my explanations to their concerns. Several of the suggestions below are aimed at doing this.

1. Join the National Council for Teachers of Mathematics[1] (NCTM, 1906 Association Dr., Reston, VA 22091-1593, 800-235-7566). With your membership, you will receive a subscription to the *Mathematics*

Teacher. This is a good source to begin understanding the problems high school mathematics teachers face. I find it good reading while waiting for the bus or when I lunch alone. Keep old issues in a prominent place in your office where students can see them. Mention relevant articles to future teachers. Most prospective teachers take education courses in which they must write papers about various aspects of teaching; often the *Mathematics Teacher* is a good source of references. Urge your students who plan to teach to join also.

2. Read the NCTM's *Curriculum and Evaluation Standards for School Mathematics*[1]. This document outlines standards to guide the revision of school mathematics. You may not agree with everything in it, but remember that it was put together by people who know more about the situation than you do. The way I got myself to read the Standards was to include it as part of a reading course for a prospective teacher. That had the additional advantage of seeing a student's reaction to the ideas in the document, and how she learned from the process. I recommend the experience. As with the *Mathematics Teacher*, keep the *Standards* prominently displayed in your office, refer to them when talking with prospective teachers, and urge prospective teachers to read them.

3. Do the same for the *Standards'* companion volume, *Professional Standards for Teaching Mathematics* (NCTM, 1991) and *Thinking Through Mathematics: Fostering Inquiry and Communication in Mathematics Classrooms* (Edward A. Silver, Jeremy Kilpatrick and Beth Schlesinger, College Entrance Examination Board, 1990).

4. Organize an informal reception for students planning to teach mathematics. Include those majoring in elementary education with a mathematics specialty as well as those planning to teach high school. If possible, include current teachers from the local school district. The object here is twofold: to show the students that the math department cares about teachers and to help promote a sense of community among future teachers. I did this last semester for the first time and was encouraged by the results. Many of the students did not know anyone else who was planning to teach mathematics, and found at the reception that there were other students in their classes with the same intention.

5. Get to know the people in the mathematics education group at your university. Ask them how you can help.

6. Offer a one-hour a week seminar for prospective teachers. Choose a topic that complements what is normally in the curriculum. I have done this twice. The first time we went through the first few chapters of Robert Devaney's *Chaos, Fractals, and Dynamics*. Part of the object was for the students to have the experience of adapting the programs in the book to whatever com-

---

[1] See [NCTM], Chapter 16.

puter resources they had available. The second time the topic was applications. Students took turns presenting applications they had chosen and researched from various sources. For next semester, I am taking a cue from this year's Mathematics Awareness Week and making the topic Mathematics and the Environment. Other good possibilities include statistics (if my university is typical, most teachers never take a statistics course) and the history of mathematics.

This is quite feasible to do off load. You do not have to lecture; indeed, you should not. Part of the benefit of a seminar is that students get the experience of more active participation than in a normal class.

7. Familiarize yourself with the requirements for secondary certification in mathematics at your university. Offer to be an academic advisor for students planning to teach.

8. Volunteer to teach courses required for secondary mathematics teachers. Mention whenever appropriate why the course is required and where it fits into the students' mathematical backgrounds. Referring to the *Standards* is a good way to do this. In many states, there is a required geometry course which is heavily populated by prospective teachers. I have taught our geometry course at least once a year for the past three years. One reason I initially volunteered to teach it is that I heard several prospective teachers say that they hated geometry. I aim to turn that attitude around. I plan to write another article for *UME Trends* describing how I approach this course.

9. Learn about individual differences, especially as they affect teaching and learning. One framework I have found particularly useful is the Myers-Briggs personality classification. This is a classification divided into sixteen types, formed from four bipolar variables. I know of only one study of the personality types of research mathematicians. It confirms what I would suspect: that we are concentrated in a few personality types, most of them among the least common in the population at large, with certain other types, much more common in the population at large, not appearing at all among mathematicians in the study. Our students, however, come from a much wider variety of types. The Myers-Briggs classification has been helpful to me in understanding how I am different from many of my students and how what interests me or comes naturally to me may not be so interesting or natural to many students. It helps me maintain the patience I need to deal with the students constructively instead of giving up in frustration when they do not immediately see what is obvious to me. For a brief overview of the Myers-Briggs types, with emphasis on mathematics and science, see Mary H. McCaulley's *Personality Variables: Modal Profiles that Characterize Various Fields of Science.* (This paper was presented to the symposium Birth of New Ways to Raise a Scientifically Literate Society, 1976 Annual Meeting, American Association of the Advancement of Science. It is

available from the Center for Applications of Psychological Type, Inc., 414 Southwest 7th Terrace, Gainesville, Florida 32601, (800) 777-CAPT.) For a more extensive and general introduction, see Isabel Brigg Myers' *Gifts Differing*.

10. Model a variety of teaching techniques in your classes containing prospective teachers. We mathematicians have been de facto selected for being able to learn well by traditional lecture methods. However, we need to teach students who may not learn best by those methods. High school teachers encounter even more of these typical students. It will be much easier for them to adopt non-lecture teaching methods if they have seen these methods used. Almost every issue of *UME Trends* has ideas on how to do this. Suggestions 1, 2, and 3 can also help.

11. Work to eliminate gender bias from your teaching. I do not mean blatant things like announcing emphatically that women are not good at mathematics; of course you would never do that. I mean the subtle things that almost all of us, male and female alike, do: giving males longer wait time when we ask a question; asking women routine questions and saving the challenging ones for the men; praising women for neatness, men for originality; responding to the man who blurts out a question when a woman has politely raised her hand. Most high school teachers are women; we need to be sure we teach them equally with the men. The experts in this area of subtle gender bias seem to be Myra and David Sadker of American University. For a brief survey of their findings, see their article *Sexism in the Classroom: From Grade School to Graduate School*, Phi Delta Kappan, March, 1986. I hope to write another *UME Trends* article on this topic, as well.

12. Compile a list of resources for teachers. Hand it out to future teachers in your classes; make copies available to other future teachers. I would be happy to send you copies of the ones I have compiled.

13. Send for the *Dale Seymour Secondary Mathematics Catalogue* (P. O. Box 10888, Palo Alto, California 94303-9843; (800) 872-1100). This will give you plenty of ideas to help implement Suggestions 6, 8 and 12. Be sure to mention this source to your students, too.

14. Design and teach a course especially for prospective teachers. I have based one loosely on the non subject-specific parts of the *Standards*: problem solving, reasoning, communicating mathematics, and mathematical connections. Students find it very helpful in tying together their mathematical education and relating it to their interest in teaching. I plan to write another article going into more detail on this also.

Martha Smith is at the University of Texas at Austin.

What are Mathematical Colloquia For?

P.J. Davis, SIAM News 23:2 (March 1990)

When I was a graduate student years ago, an announcement went out that Harald Bohr of Copenhagen would be the speaker at the next week's Mathematical Colloquium. I was thrilled. Bohr was a world famous mathematician, the creator of the theory of almost periodic functions, which I happened to be studying at the time.

When I entered the lecture hall, I saw that Bohr had densely filled three large blackboards with miniscule writing that was hardly readable, and the lecture he gave consisted of meticulously going over each line in a painful recitative. Naturally, I was lost after the fifth line, and it became quite clear to me that Bohr was trying to present in one hour the mathematical accomplishments of a lifetime.

From this and other similar experiences, I concluded that a colloquium talk was a social ritual, the least of whose purposes was to communicate mathematical knowledge. I was delighted to have met the Great Names of my profession at Colloquium Coffees, and I was delighted to have had the opportunities to talk to the faculty and fellow graduate students that congregated. (I first met Ralph Boas at such a coffee; he later became my thesis adviser.) But I was disappointed over and over again that I was able to retain nothing (or close to nothing) from the lectures themselves, other than a very vague flavor of who was doing what. Some years later, as a substitute colloquium chairman, I had to arrange for a talk by Michael Fekete, a most excellent mathematician who was responsible for the theory of the transfinite diameter. The paltry audience turnout was absolutely and crushingly embarrassing to me. I resolved then and there that if I ever had to arrange for another colloquium, I would call up an agency that supplied claques, or I would hire a bunch of professional mourners and fill the lecture room with these warm bodies. Several hundred dollars spent on ringers would be a justifiable departmental expense. Why do I bring up these old and painful memories? The chairman of a department of mathematics I am familiar with has just sent out a memo to his faculty to the effect that since attendance at colloquia has been so uniformly and embarrassingly rotten, he was seriously thinking of cancelling the whole colloquium series for this year. While institutions vary in their inner spirit and experience, I wonder nonetheless whether the experience of this department differs substantially from that of other departments. I am not sure when the mathematical colloquium lecture was born—in Weierstrass' day? in Gauss's day?—but I conjecture that as a communicator of ideas it was dead upon birth. The main charge that I would lay against the typical colloquium lecture is that it is addressed to the Happy Few and not to the vast majority of professionals who are ignorant of the drift, the nomenclature, and the raison d'etre of the speaker's sub-subfield. Often, at such lectures, one sees the ardent, energized Happy Few sitting up front; with each statement that emerges from the lectern—or from the overhead projector—they periodically bob their heads in approval and in unison, like a row of perpetual motion penguins perched on the edge of their water glasses. The rest of the audience, overcome by massive boredom, abhors the incomprehensible symbols, suffers, fidgets, works on its own stuff, thinks about supper, and wonders what is the earliest moment it can slip out without incurring the perpetual wrath of the colloquium chairman.

I am not really concerned about the colloquium lecture as a mechanism for the transfer of information. With xerography, fax, e-mail, desktop publishing, specialized newsletters both of the intra- and extramural variety, and the traditional technical journals (which are more and more serving only an archival purpose), the various mathematical constituencies are getting all the information they need. In fact, they are overdosing on information.

The number of conflicting talks is large; the audiences are fragmented. Everyone is talking, and no one wants to listen or has the time to listen to talks from morning to night. It seems also that everyone is writing and no one is reading—except what is perceived to be on the cutting edge of one's sub-subfield.

Intersubject rivalry and animosity flourish within departments, leading to a vast overvaluation of the particularities, the accomplishments, and the intellectual status of subspecialties. Mathematics seems now to consist of 2,500 separate hobbies, all lining up at the public trough for financial support and all offering the public assurances of their individual cosmic importance. Each of the 2,500 demands, at the very least, the establishment of a separate center with its own seminar, colloquium, newsletter, and journal. The long-touted unity of mathematics seems to have been placed at risk.

The various mathematical constituencies are becoming more and more self-isolated by the very intensity of their respective specializations. If a speaker is talking about strawberry ice cream cones, the people who specialize in maple walnut are not likely to show up. Strawberry is not on their cutting edge.

Nonetheless, the threatened cancellation in the department mentioned was met with groans of "Oh no, you can't cancel. We need, we love, we require, we would be lost without our colloquium. I promise to show up next week."

This breast beating will prevent the immediate dissolution of the long established colloquium tradition. With some consideration on the part of the speakers for their audiences, with some prodding of the faculty, with the carrot and stick applied to graduate students, perhaps this comatose colloquium can be resuscitated.

At the bottom line, mathematics is a human language and a human institution that requires eyeball-to-eyeball contact for the transference of its deepest meanings and values. That proposition, I think, is what is really at stake.

Philip J. Davis is a professor in the Applied Mathematics Division at Brown University.

Must Fine Presentation Imply Less than Deep Insight in a Mathematical Colloquium?

S. Pincus, SIAM News 23:5 (September 1990)

To the Editor:

I found Philip Davis' commentary "What are Mathematical Colloquia For" (SIAM News, March 1990) timely and, sadly, an accurate description of an endemic problem within mathematics. Most colloquium talks are inaccessible to all but the Happy Few (Davis' term), who already are quite familiar with the content beforehand. The crucial question is, Can such talks be organized differently, to allow most of the audience to come away with more than they came in with? I believe that the answer is "yes" and would like to present some possible guidelines for talks directed to a general audience.

Before getting into that, we must consider what the speaker hopes to achieve in giving such a talk. The Tao of mathematics, to which almost all mathematicians (myself included) subscribe, is that smartness counts, that in a world of form versus substance, we can well address the unpolluted substance with a minimum of "fluffy" form to contend with. When talking of unmet peers, one hears questions like "How good is she? Is she as smart as . . . ?" These questions are not a problem in and of themselves, but hint at a larger malaise. We have been socialized to distrust quality form, perceiving that since form and substance rarely coincide, fine presentation implies less than superb deep insights. After all, the "best" people are too busy with deep, important thoughts to spend oodles of time preparing lucid, easily accessible accounts of their work. While that is an unfair generalization, I think we must admit that many of our peers and role models, who invariably include the Happy Few, fall into that niche. I contend that in many instances, the speaker's primary vested interest is that a crucial minority know that he is smart and has done good work; effective communication with members of the audience is happenstance. The unfortunate corollary is that since "the truth is in the details," in order to impress the gods enough minutiae is presented to remove the vast majority of the audience from the room.

To make colloquia more accessible to a larger audience, we need to ask the speaker to modify the Tao. For the time-period of the talk, smartness = clarity + perspective. Remember, the journal paper, the one-on-one discussion, and the (colloquium) talk are very different forms of communication. Speaking does not equal audible text; it is the opportunity to present the intuition and perspective behind the formalism. Papers give

fine detail, and generality, while one-on-one discussions afford opportunities for free thinking, rambling ideas, pointed dialogue, and response to questions. Colloquia should strike a medium. They should give a culled, well-thought-out, yet general sense of what is out there: more than a visceral feeling and a ramble, less than an onslaught of fine detail. That way, we may later be triggered, "Oh yeah. I need to do something like XYZ. I once heard a colloquium by Smith-Jones on XYZ. . . . Time to learn more, to see if it really is relevant. Or maybe I should speak with Smith-Jones to see if this connection is substantive." Later may be a month, or several years, so the need for the audience to acquire a fundamental sense of the basic problem and methods at hand is apparent. Furthermore, clarity and approachability on the part of Smith-Jones will encourage future contact, with a host of potential symbiotic benefits. In particular, many significant advances are produced not by lightning bolts, but rather by techniques borrowed from elsewhere and newly applied; credit belongs to both sides. It is in everyone's best interest to encourage such cross-pollenization. This revised Tao does not seem too much to ask. Clear communicators, who can extract and present the pith of important problems in a widely accessible manner, include top scientists in many fields. James Watson and the late Richard Feynman are two notable examples, and since they are de facto role models in molecular biology and physics, (their) expository capabilities are seen as a desirable part of the package of a top researcher within these fields. Recently, in the process of interdisciplinary work, I've attended medical talks to gain perspective and knowledge of the technical state-of-the-art, and have generally found physicians to be excellent expositors, although frequently less than thorough scientists. I find it sad that I come away with more new perspective from an hours discourse on kidney function than from an hours talk on low-dimensional topology. Mathematics clearly needs some top-notch researchers who are widely respected for excellent presentation skills; few come to mind.

While some of the suggested guidelines that follow, like "motherhood," are good for all talks, some seem more acutely appropriate for math talks.

1. Come equipped with a few crucial handouts – papers, recent submissions, plus some form of extended, less formal description of the lecture (e.g., copies of viewgraphs). If the papers do not fulfill the need to impress/show the details, what will? In addition, possession of a written synopsis of the lecture frees the audience from the timeless bondage of accurate transcription of every ϵ and δ. It is, after all, well known that if you mistranscribe an ϵ, you will be visited by at least seven of the 10 plagues of Egypt. One-page summaries – key examples, results, definitions – are great. I like to use viewgraphs because otherwise I worry too much, about ensuring that my handwriting is legible and that every ϵ is correct, to properly concentrate on providing perspective. Of course, many mathemati-

cians view the use of viewgraphs as gloss and avoid the medium; note that in most scientific disciplines, some form of premade talk is the standard.

2. Frame the central problem so that most of the audience knows why you are doing what is to follow. Most implies knowledge of the makeup of the audience. The more heterogeneous the audience, the more time one needs to spend here. Similarly, for very new or highly specialized areas, one will often need to allot extra time for this part of the talk. You want a catch: "Why should we care about this lecture?"

3. Give a roadmap of the talk. Where are we headed? Are we nearly there yet? Check periodically to remind the audience how far you've gone, and recap in a sentence the last few minutes.

4. Start with key historical results, and a give sense of the new key results to follow. Be brief here. Little is lost by first describing results casually and in abbreviated fashion to snare the audience; you can always give precise formulations later. One reason that James Gleick's book *Chaos* was a best seller was that Gleick gave an integrated historical perspective of the development of this widely cast field. Knowing the "why" of the steps that were made is crucial in reevaluating what has transpired.

5. Define and motivate terminology properly. A specialized seminar in probability can begin "Let X_i be i.i.d. Banach-space valued random variables with $2+\epsilon^{th}$ moment...," but this phrase contains at least three or four terms, with associated history, that must be well understood for appreciation of what is to follow. What are i.i.d. random variables? Why do we consider them? What is a Banach space? What are some examples? Why do we consider probability on Banach spaces? What is a moment condition, and in particular, why do we assume a $2+\epsilon^{th}$ moment? Probabilists spend years acquiring the intuition and training to appreciate the answers to these questions, but do combinatorialists or specialists in partial differential equations share this particular training? Not usually, but it would seem reasonable that researchers in either of these fields might know existing technology that could be brought to bear on problems about to be discussed.

6. Present as many salient examples as time permits. We are trained under the curse of Gauss: Once a computation or proof is complete, develop the most general or elegant form, and present only that. This is appropriate for a written paper, where the reader has the luxury of the time both to unravel details and to see how known examples fit into the general schema. It is usually inadequate for a talk, in that much of the crucial intuition that guided subsequent development is hidden. We live on intuition and communicate that way behind closed doors; relevant examples and pictures tell a major story. They are not a substitute for proofs but are heavily used to suggest, confirm, and reject possible theorems and methodology. Very early on

in the talk, present an essential example that the audience can work with. Some of the audience may have considered similar examples, from a different vantage point, and may provide useful insights. Also, incorporate as many insightful pictures as possible.

7. Minimize proofs, maximize references. Most members of the audience are here for first-order connections and insights, not fine detail. Tell those interested in the detail where to find it, when the detail is needed. Certainly give a flavor of the essential detail, but don't overdo. The one exception is when the main point of the talk is an advance in existing technical machinery. In that case say so, and flag the crucial new widgets. Also, distinguish between those theorems in which the result is the crucial point (e.g., Whitney embedding theorem, implicit function theorem) and those in which the manner of getting there is as important as the result.

8. Break the talk into bite-sized pieces. This especially goes for nasty proofs and computations. Recap, bring the audience back. I can recall innumerable lectures in which, hopelessly lost, I was reduced to waiting till the end of the topic, which too often coincided with the end of the talk. You can spend only so much time wondering if the Kung Pao chicken at the postcolloquium dinner will be a good batch. We all drift in and out, even in talks we want to attend. In math, if we lose the thread of an argument, we are dead. In most other fields, in which sets of results are presented, this problem is not so acute. In short, make the talk robust to an occasional fadeout.

9. Include anecdotes, and occasionally adopt a casual tone. This will encourage the crucial "stupid" questions without the questioners feeling that you might bite their heads off. I'm fond of slipping in something about a personal silliness. Even if the speaker adopts the most solipsistic viewpoint, remember that questions might spur rethinking and new insights.

10. Conclude with a short summary and a laundry list of open questions. You (should) want to encourage others to work in this area, and a well-thought-out list of unanswered problems, ideally some of them first-order, will expedite this process.

11. Do whatever you can do to encourage questions. Tailor some of your laundry list to known members of the audience. Get the ball rolling. Be informal in responding; try to acknowledge good questions. If a question is specialized, or aimed to impress, explain the question and the "why" to the greater audience. Attempt to answer the question asked, not a related question to which you might know the answer (a standard ploy); or acknowledge that you've ducked the intended question, explain why, and tell the audience how much your dodge makes a difference.

12. Follow up questions, contacts. Little is more impressive to me than a follow-up contact on a previous thought: "You remember that question you asked?

I thought it was trivial and stupid, but the more I thought about it, the less obvious it became. I think I've puzzled it out now, here it is." Both sides gain. A new potential colleague, and instant respect. And that's what it is ideally about. Both the speaker and audience could ultimately be richer as a by-product of the good colloquium.

Steve Pincus, Guilford, Connecticut

Non-NSF Funding for Undergraduate Mathematics Education

S.C. Ross, UME Trends 3:3 (August 1991)

The National Science Foundation has been an important source of support for curriculum and faculty development activities in undergraduate mathematics education. There are, however, a variety of other funding opportunities for these activities. For instance, the Fund for the Improvement of Post-Secondary Education (FIPSE), United States Department of Education, Washington, D.C., 20202 (202-708-5750), is another federal agency that will fund projects related to undergraduate mathematics education.

In the private sector, nearly 2 percent of the funds awarded by the major foundations in 1986 was given to mathematics and computer science projects; over half of that support was for program and curriculum development. Some of the major foundations that will make grants for programs in undergraduate mathematics education are the Carnegie Corporation of New York, the Exxon Education Foundation, the Hewlett-Packard Company Philanthropic Programs, the Howard Hughes Medical Institute, and the Pew Charitable Trusts. In the past, the Sloan Foundation has supported mathematics projects such as the Williamstown conference on discrete mathematics and the Tulane conference on calculus, but Sloan is not currently making grants for mathematics projects.

The Carnegie Corporation of New York (437 Madison Avenue, New York, NY, 10022, 212-371-3200) was established "to promote the advancement and diffusion of knowledge and understanding." It makes grants primarily to academic institutions and national and regional organizations. There is no formal application procedure. Some representative grants from the education, science, technology, and the economy programs include support to the Mathematical Sciences Education Board for *On the Shoulders of Giants*, Carnegie-Mellon University for a program supporting women and under-represented minorities in science and engineering, and the MAA to plan a project to strengthen minority achievement in mathematics.

The Exxon Education Foundation works to "improve and promote science education at the high school and college levels." The Foundation has supported MAA's development of student chapters, MSEB's state coalition initiative, and the AMS in nonspecific ways. The Mathematics Education Program focuses on three areas of mathematics education: mathematics specialists in K-3, instruction at the college level, and policy analysis. The Foundation is currently not accepting any more proposals for K-3 specialists projects. Inquiries for the other two areas should be directed to Manager, Mathematics Education Program, Exxon Educational Foundation, P. O. Box 101, Florham Park, NJ 07932 (201-765-3003). The first phase of the college-level component will be to collect information regarding the teaching of mathematics at colleges today. Also in the first stages is the policy analysis program. It is anticipated that in the later stages some grants will be made for the implementation of recommendations arising from the policy analysis projects.

Grants from the Hewlett-Packard Philanthropy Programs are mostly in the form of state-of-the-art equipment. There are two programs relevant to undergraduate mathematics education, the University Grants Program and the Community College Equipment Grants Program. The university program requires the sponsorship of a Hewlett-Packard employee. This sponsor might be a former student, a campus recruiter, or someone else who knows the department. Community college grants are made to schools in geographic areas where Hewlett-Packard has major sales or manufacturing operations. Interested colleges should first approach the local Hewlett-Packard entity.

The Howard Hughes Medical Institute solicits applications for its grants from colleges based on the college's rating by the Carnegie Foundation and its track record in preparing science students, especially minorities, for graduate and medical schools. It will consider unsolicited proposals that fit into its programs. The grants officer at your institution should know if the college is on the Hughes list, but may not have thought of notifying the mathematics faculty. The institute seeks to enhance the overall quality of undergraduate science teaching and research; it is particularly concerned with increasing the number of capable students, especially women and minorities, in scientific disciplines. To obtain information regarding the Undergraduate Science Education Program, contact Stephen Barkanic, Howard Hughes Medical Institute, Office of Grants and Special Programs, 6701 Rockledge Drive, Bethesda, MD 20817 (301-571-0324).

Education is one of six fields of concern supported by the Pew Charitable Trusts. Within the education program are two components of interest to faculty working in undergraduate mathematics education: Teachers and Teaching, and Access and Success in Higher Education. A wide variety of projects to improve the quality of teaching in schools, colleges, and universities fit into the guidelines for the Teachers and Teaching component. The goal of the Access and Success in Higher Education

Program is to increase the retention and success rates for disadvantaged students in higher education degree programs. More information is available from the Pew Charitable Trusts, Three Parkway Suite 501, Philadelphia, PA 19102-1305 (215-568-3330).

There are also many state-based foundations and trusts that may be sources of funding for mathematics education projects. Check with your grants officer or librarian for a listing of these organizations.

Do You (Or Your Students) Want to Teach in a Two-Year College? Some Guidelines for Prospective Two-Year College Teachers

T. Shell, UME Trends 2:4 (October 1990)

An undergraduate who chooses to teach mathematics at the high school level has a well-defined course of study. This is basically true of a graduate student who wants to teach at a university or four year college. But students who want to teach mathematics at a two-year college often find that no one in the mathematics department or mathematics education department really has a clear idea of how to prepare. Furthermore, neither student nor advisor may even have a clear idea if teaching at a community college is an appropriate career choice. The purpose of this note is to offer some guidance.

An obvious question is, what does a two-year college mathematics teacher do? Without a doubt the primary role is to teach, rather than conduct research. A typical teaching load is fifteen hours per week along with five office hours. In addition, many make arrangements with their students to be accessible at other times. Approximately 42 percent of mathematics enrollments are remedial classes. The course offerings range from arithmetic through differential equations, including linear algebra, statistics, and computer programming. Therefore the two-year college teacher must have an educational background that emphasizes breadth in mathematics. This includes knowledge of applied mathematics, because one of the best ways to motivate a topic is with an application. Many instructors also spend time in their college mathematics laboratory. Computers are typically utilized in "math labs" for different purposes such as self-paced instruction, tutorials, or homework assignments such as numerical integration, statistical analysis, or a complicated graph. It is important to understand that two-year college students vary significantly in preparation, ability, and motivation. There are bright, well-prepared students who are right out of high school - typical university students. There are also many students who are twenty- or thirty- something but have not studied mathematics since high school. These students are called reentry students; they are usually quite motivated yet often need encouragement. There are also quite a few students who don't particularly like mathematics, but must "get through" a mathematics class to satisfy career or major requirements. These students are called many things by my colleagues and above all need to learn how to study mathematics. Teaching college mathematics to a group with such diverse backgrounds poses quite a challenge, and not everyone has the ability or desire to meet this challenge! Providing high-quality instruction requires not only spending considerable time and effort in preparation, but also an ability to be innovative. A complete professional must be aware of current issues and developments in mathematics education. For example, instructors should be familiar with software such as Mathematica or Maple as well as how they are affecting the undergraduate curriculum. Consequently a two-year college teacher is expected to be involved in professional organizations (such as the American Mathematical Association of Two-Year Colleges, the Mathematical Association of America, or the National Council of Teachers of Mathematics), and he or she should be reading appropriate journals *(The College Mathematics Journal, The AMATYC Review, The Mathematics Teacher, Mathematics and Computer Education).* A two-year college teacher is also expected to serve on department and campus committees.

A master's degree in mathematics is by and large the minimal requirement of mathematical competency to teach at a two-year college. Many two-year colleges do not accept a master's degree in mathematics education or a master of arts/science in teaching mathematics as equivalent. As mentioned earlier, one's preparation should emphasize breadth in mathematics. Coursework should include statistics, numerical analysis, computer programming, probability, history of mathematics, and even physics. An extra graduate course in differential equations might be a better choice than a third or fourth course in abstract algebra. Familiarity with a new branch of mathematics or a new programming language would be beneficial.

The National Center for Education Statistics recently reported results of a 1988 survey of higher educational institutions. Department chairs at two-year colleges rated the importance of eighteen factors in hiring entry-level faculty. The following lists these factors in descending order of importance, and each factor includes the percentage of chairs who rated it as very important:

Teaching quality (90%)

How well they fit with this department (65%)

Programmatic needs (59%)

Related job experience (58%)

How well they fit with the student body (57%)

Extent of teaching experience (56%)

Affirmative action considerations (39%)

Highest degree (39%)

Reputation in their professional field (32%)

Academic record (29%)

Salary requirements (23%)

Community or professional service (15%)

Reputation of graduate institution/program (8%)

Quality of publications (2%)

Extent of research experience (1%)

Candidate's ability to obtain outside funding (1%)

Quality of research (1%)

Number of publications (0%)

These results indicate that obtaining a Ph.D. or doing research isn't nearly as important as gaining experience in teaching. A place to start is tutoring precalculus, especially in a "drop-in" environment where the students' backgrounds are diverse. Experience as a teaching assistant or teaching part-time would be helpful. One way to get full-time experience is to teach at a high school; teaching at a public high school may require a secondary credential.

How well a candidate will fit with a department depends, among other things, on how well the candidate understands the role of a community college and what is expected of an instructor. Also, to fit in with the student body, a candidate must understand what kinds of students enroll in two-year colleges. Experience in teaching isn't the only way to achieve this understanding. Attending professional meetings and conferences is an excellent way to learn about what two-year colleges are doing. Common pedagogical problems and a variety of teaching techniques are often discussed, so these conferences also provide a newcomer ample opportunity to improve her or his teaching. Probably the best way to prepare for a particular junior college department's programmatic needs is by emphasizing breadth in mathematics. This will also enhance teaching quality. Most important, a prospective teacher must demonstrate a commitment to teaching. It is not enough to be a competent mathematician or to love mathematics. Coursework should include some teaching methodology that discusses how topics are motivated and presented, or how technology is affecting curriculum, or simply how to teach students to solve problems.

Terry Shell is at Santa Rosa Junior College.

Educating Mathematical Scientists: Doctoral Study and the Postdoctoral Experience in the United States

Notices Amer. Math. Soc. 39:5 (May/June 1992)

In July 1990, the Board on Mathematical Sciences (BMS) of the National Research Council (NRC) convened the Committee on Doctoral and Postdoctoral Study in the United States to conduct a study and make recommendations for improvement and change in doctoral and postdoctoral education in the mathematical sciences. The Committee's report, "Educating Mathematical Scientists: Doctoral Study and the Postdoctoral Experience

in the United States," was released in March 1992. Presented here are the report's Executive Summary and Introduction, in addition to critiques of the report written by five members of the mathematical sciences community.

Reprinted with permission from "Educating Mathematical Scientists: Doctoral Study and the Postdoctoral Experience in the United States," 1992. Published by National Academy Press, Washington, D.C. To order the complete book, sixty-four pages, £19, call 1-800-624-6242, or send a check plus £3 for shipping and handling to National Academy Press, 2101 Constitution Avenue, N.W., P.O. Box 285, Washington, D.C. 20055.

Committee on Doctoral and Postdoctoral Study in the United States: Ronald Douglas, State University of New York at Stony Brook, Chair; Hyman Bass, Columbia University; Avner Friedman, University of Minnesota; Peter Glynn, Stanford University; Ronald Graham, AT&T Bell Laboratories; Rhonda Hughes, Bryn Mawr College; Richard Jacob, Arizona State University; Patricia Langenberg, University of Maryland; D. J. Lewis, University of Michigan; J. Scott Long, Indiana University at Bloomington; John Rice, University of California at San Diego; Donald Richards, University of Virginia; Karen Uhlenbeck, University of Texas at Austin; Mary Wheeler, Rice University.

Executive Summary

Although the United States is considered a world leader in mathematical sciences research and in doctoral and postdoctoral education, concern is growing about whether the needs of the profession and of an increasingly technological society are being met. Many doctoral students are not prepared to meet undergraduate teaching needs, establish productive research careers, or apply what they have learned in business and industry. This inadequate preparation, continuing high attrition, and the declining interest of domestic students, the inadequate interest of women students, and the near-absent interest of students from underrepresented minorities in doctoral study are problems that transcend the current difficult job market.

The charge to the Committee on Doctoral and Postdoctoral Study in the United States was to determine what makes certain programs successful in producing large numbers of domestic Ph.D.s, including women and underrepresented minorities, with sufficient professional experience and versatility to meet the research, teaching, and industrial needs of our technology-based society. The committee based its findings on site visits to a diverse set of programs in 10 universities carried out in late 1990 and early 1991. These programs were in both small and large, and in both public and private, universities. They were also geographically diverse. They were all in the "top 100" and included four departments in the "top 20."

The audience to which the report speaks is all U.S. doctoral and postdoctoral programs in the mathematical sciences, and, in particular, those programs that have limited human and financial resources. The report suggests that even with limited resources success can be achieved if, among other things, a program focuses its energies rather than trying to implement a "standard" or traditional program that covers too many areas of the mathematical sciences. It also notes that departments with the best faculty do not necessarily have the most successful doctoral and postdoctoral programs. A quality faculty is necessary for a good program, but of equal importance are students and researchers that can benefit from the program.

In this report, a "successful" program is understood to be one that accomplishes the following two objectives.

◇ All students, including the majority who will spend their careers in teaching, government laboratories, business, and industry rather than in academic research, should be well prepared by their doctoral and postdoctoral experience for their careers.

◇ Larger percentages of domestic students, and, in particular, women and underrepresented minorities, should be attracted to the study of and careers in the mathematical sciences.

In its site visits, the Committee on Doctoral and Postdoctoral Study in the United States looked for features that were present in successful programs as well as for elements that were detrimental to quality education. The committee noted that successful programs possessed, in addition to the *sine qua non* of a quality faculty, the following three characteristics:

1. A focused, realistic mission,
2. A positive learning environment,
3. Relevant professional development.

A positive learning environment is an environment that provides the assistance, encouragement, nurturing, and feedback necessary to attract and retain students and to give them an education appropriate for their future careers.

The findings of the committee are as follows:

◇ There are several different models (missions) for programs, including

   ⋆ the standard model, which supports research in a broad range of areas, offers depth in each one, and has as its goal preparation for careers at research universities, and

   ⋆ specialized models, such as the subdisciplinary model, the interdisciplinary model, the problem-based model, and the college-teachers model, which were seen to alleviate two large human resource problems, recruitment and placement, and to be conducive to clustering of faculty, postdoctoral associates, and students, a practice that helps create a positive learning environment and promote relevant professional development.

◇ Both standard and specialized programs can be successful. However, programs that do not have the human or financial resources to run a successful standard program should consider whether a specialized model might better fit their needs.

◇ New Ph.D.s with a broad academic background and communication skills appropriate for their future careers are better able to find jobs.

◇ Active recruiting increases the pool of quality students. It does not just reapportion the pool. It also increases the number of women and underrepresented minorities. Students with strong mathematical backgrounds have a choice of studying mathematical sciences, physical sciences, engineering, law, medicine, and other areas. More of them can be attracted to the mathematical sciences.

◇ Clustering faculty, postdoctoral associates, and doctoral students together in research areas is a major factor in creating a positive learning environment.

◇ A positive learning environment is important to all doctoral students but is crucial for women and underrepresented minorities.

◇ All departments, including those characterized as elite and selective, need to provide a supportive learning environment.

◇ Doctoral students and postdoctoral fellows should receive broad academic preparation appropriate for their future careers in research universities, teaching universities, government laboratories, business, and industry.

◇ Doctoral students and postdoctoral fellows should learn teaching skills and other communication skills appropriate for their future careers.

◇ The number of postdoctoral fellowships in the mathematical sciences should be greatly increased so that such positions can be viewed as the logical next step after completion of the doctorate for the good student, not as a highly competitive prize for a select few. More postdoctoral fellowships should have applied, interdisciplinary, or pedagogical components.

Changing the American doctoral and postdoctoral system in the mathematical sciences so that it responds better to the needs of the profession, students, and the society is a task that requires the cooperative efforts of faculty, departments, professional societies, and federal agencies. The departments at research universities have a special responsibility to raise the level and increase the knowledge of talented but underprepared entering American doctoral students. Federal agencies should continue their programs and also increase their awareness of the impact of their programs on the doctoral and postdoctoral system. Professional societies should be involved in monitoring change in the universities, the agencies, and the community. But action, if it starts at all, will start from the faculty. The faculty should be aware that creating and maintaining a successful doctoral/postdoctoral program will require additional effort and time. The long-term benefits to the department, the students, and the society are clearly worth the effort.

Introduction

In the period since World War II, research in the mathematical sciences has flourished in the United States. Large numbers of graduate students and researchers from around the world have come to this country to study and to work in pure and applied mathematics, in well-established and new fields. The American system of doctoral and postdoctoral study and research in the mathematical sciences is considered by many to be an unqualified success, in contrast to the system of pre-college and undergraduate education.

The current scarcity of highly qualified domestic graduate students, often attributed to mediocre pre-college education and to related problems in undergraduate education, is seen by many as an unfortunate circumstance but not as a major problem. In the 1990–1991 academic year, only 43% (461 out of the adjusted total of 1061 reported in McClure, 1991, p. 1093) of the recipients of Ph.D.s in the mathematical sciences from institutions in the United States were U.S. citizens (McClure, 1991), whereas during the 1960s, 82% of the recipients of such Ph.D.s from U.S. institutions were U.S. citizens (NSF [National Science Foundation], 1988). Among the U.S. citizen recipients of Ph.D.s in the 1990–1991 academic year, less than a quarter were women and less than a twentieth were from underrepresented minorities. American mathematical sciences departments, research laboratories, and industry are relying increasingly on students, faculty, and professional researchers from abroad because fewer and fewer American students are being attracted to study in the mathematical sciences and because the education that many of those students receive leaves them ill equipped to compete with their foreign counterparts.

Noting the scarcity of highly qualified domestic students and the current tight employment market, some maintain that the chief problem in the doctoral and postdoctoral system is overproduction of Ph.D.s, a problem that should be solved by encouraging students to choose other disciplines and by reducing the number of doctoral students. Our overreliance on academia for jobs for new Ph.D.s is often not considered to be a problem, nor is the matching of doctoral education with the positions that graduates take considered to be a priority. Increasing production of Ph.D.s since 1987, international events that have increased immigration of students and professional mathematicians to the United States, and a recession in the economy have indeed combined to produce what is now the most difficult employment market for Ph.D. mathematicians since the 1970s. Further complicating the current picture are the indications that the demand for mathematical scientists will rise as the many mathematical scientists hired in the 1960s start to retire over the next decade (NRC [National Research Council], 1990b). The long-term growth in demand for mathematical scientists in academia, government, business, and industry and the expectation that the wave of immigration of mathematical talent to the United States will eventually taper off suggest that the country will be best served by a positive outlook that emphasizes attracting more domestic students into the mathematical sciences and giving those students proper foundations for their future careers.

A positive outlook that serves the interests of the profession and the country can be translated into actions intended to achieve the following two broad objectives:

◊ All students, including the majority who will spend their careers in teaching, government laboratories, business, and industry rather than in academic research, should be well prepared by their doctoral and postdoctoral experience for their careers.

◊ Larger percentages of domestic students, and, in particular, women and underrepresented minorities, should be attracted to the study of and careers in the mathematical sciences.

In this report, a "successful" program is understood to be one that accomplishes these two objectives. The needed renewal of the profession, as pointed out in the "David I" report (NRC, 1984), *A Challenge of Numbers* (NRC, 1990b), and the "David II" report (NRC, 1990c), requires larger percentages of domestic students. Although statistics invariably oversimplify the situation, the following two "completion rate" statistics concerning percentages of domestic students are useful in judging a program's success: (1) the percentage of students who entered the program five years earlier and who have received their doctorates, and (2) the percentage of students who completed their second year of graduate study four years earlier and who have received their doctorates. The first of these two types of completion rate is an appropriate measure of the success of highly selective programs, while the second is appropriate for less selective programs. The committee observed a number of programs for which both rates were well above 50 percent.

Purpose and Scope of This Report: The charge to the committee was to determine what makes certain doctoral and postdoctoral programs in the mathematical sciences successful in producing large numbers of domestic Ph.D.s, including women and underrepresented minorities, with sufficient professional experience and versatility to meet the research, teaching, business, and industrial needs of our technology-based society. The mathematical sciences are considered to be pure mathematics, applied mathematics, statistics and probability, operations research, and scientific computing. Computer science, a separate discipline, is not included among the mathematical sciences.

The doctoral period considered in this report extends from the first year of graduate study through completion of the thesis, regardless of whether or not the student obtains a master's degree. The postdoctoral period is the first five years after receipt of the Ph.D. A postdoctoral associate, postdoctoral fellow, or, in common parlance, "postdoc," is a recent Ph.D. who has a fully funded position to do research. Since such a small number of new Ph.D.s in the mathematical sciences enjoy postdoctoral fellowships, this report concerns not only postdoctoral fellows but also junior faculty working during the postdoctoral period.

The purpose of this report is to present and disseminate information about types of mathematical sciences doctoral and postdoctoral programs that succeed in attracting large numbers of domestic students, including women and underrepresented minorities, and succeed in giving their students academic and professional experience that is relevant to their future careers. There are U.S. programs that provide high-quality doctoral education to student bodies that are 80% American, have nearly 50% women, or have 30% underrepresented minorities. This report, based on the committee's insights gained in site visits to 10 universities, describes characteristics of these programs. What these programs do differently and what they and others consider to be their successes and their frustrations is information that this report seeks to make available to the community so as to encourage doctoral/postdoctoral program models that are relevant to the needs not only of academic research but also of teaching, government, business, and industry, and to increase the quality and number of domestic Ph.D.s, especially women and underrepresented minorities.

This report follows on and is complementary to a number of studies by the National Research Council that examine the health of U.S. mathematical sciences research and education, including *Renewing U.S. Mathematics: Critical Resource for the Future* (NRC, 1984), *Everybody Counts* (NRC, 1989), *A Challenge of Numbers* (NRC, 1990b), *Moving Beyond Myths* (NRC, 1991b), *Renewing U.S. Mathematics: A Plan for the 1990s* (NRC, 1990c), and *Actions for Renewing U.S. Mathematical Sciences Departments* (NRC, 1990a). One includes the following pertinent summary.

Graduate and postdoctoral training programs offered by mathematical sciences departments are key to the successful renewal of the profession and reform of mathematics education. Successful programs can attract individuals to a career in the mathematical sciences and can develop highly qualified teachers and researchers to stimulate, nurture, and train future generations. Is our present graduate and postdoctoral educational system in mathematics working well? The answer seems clearly to be that it could be much better. The community could attract more students to the study of the mathematical sciences, and more students entering graduate programs could succeed in obtaining doctorates. With nurturing and continued attention through good postdoctoral pro-

grams, more of these young people could develop into good mathematicians—some as teachers, some as researchers, and many as both. (NRC, 1990a, p. 13)

Contents of This Report: Chapter 2 gives a brief historical perspective of the mathematical sciences in America, with emphasis on doctoral and postdoctoral training.

Chapter 3 describes how some programs in the present system achieve success. Three characteristics of successful programs—a focused and realistic mission, a positive learning environment, and relevant professional development—are introduced in this chapter. The issue of having a high-quality faculty—a *sine qua non* of a successful program—is acknowledged but not discussed in detail in this report.

The heart of the report is Chapters 4–7, which treat the three characteristics of successful programs and human resource issues that must be taken into account. Chapter 4 discusses human resource issues; in particular, those related to domestic students, women, and underrepresented minorities are examined. A number of specialized missions for doctoral/postdoctoral programs are described in Chapter 5. Chapter 6 discusses a positive learning environment. Relevant professional development is described in Chapter 7.

Chapter 8 describes how faculty, departments, professional societies, and federal agencies can work together to create more successful programs.

A guide for self-evaluation by departments forms Appendix A. Appendix B includes advice to prospective doctoral students on how they can best choose a doctoral program. Appendix C is a brief discussion of master's degree programs in the mathematical sciences, a feature that may form a part of a well-rounded graduate program in the future as doctoral programs become more oriented toward wider job markets, including business and industry.

References

McClure, D.E., 1991, "1991 Annual AMS-MAA Survey (First Report)" in *Notices of the American Mathematical Society*, **38**, 1086–1102.

National Research Council (NRC), 1984, *Renewing U.S. Mathematics: Critical Resource for the Future* ("David I" report), National Academy Press, Washington, D.C.

National Research Council (NRC), 1989, *Everybody Counts: A Report to the Nation on the Future of Mathematics Education*, National Academy Press, Washington, D.C.

National Research Council (NRC), 1990a, *Actions for Renewing U.S. Mathematical Sciences Departments*, National Academy Press, Washington, D.C.

National Research Council (NRC), 1990b, *A Challenge of Numbers: People in the Mathematical Sciences*, National Academy Press, Washington, D.C.

National Research Council (NRC), 1990c, *Renewing U.S. Mathematics: A Plan for the 1990s*, ("David II" report), National Academy Press, Washington, D.C.

National Research Council (NRC), 1991b, *Moving Beyond Myths: Revitalizing Undergraduate Mathematics*, National Academy Press, Washington, D.C.

National Science Foundation (NSF), 1988, *Science and Engineering Doctorates: 1960–86*, National Science Foundation, Washington, D.C.

In order to stimulate discussion of the report, "Educating Mathematical Scientists: Doctoral Study and the Postdoctoral Experience in the United States," the *Notices* solicited the following critiques of the report from five members of the mathematical sciences community.

S. Garfunkel

First, let me begin by saying that I am pleased that this study was done and this report written. It is well past time that we take a serious look at doctoral and post-doctoral programs and how they could and should be improved. The fact that this study was undertaken under the aegis of the National Research Council (NRC) is admittedly a good news/bad news situation. The good news is that the prestige of the NRC may help increase the awareness of the community and encourage more people to read and take its contents seriously. The bad news is that, as with other NRC reports, the review process has a "smoothing" effect and tends to take out strong and controversial positions. This is not a strong report. With one notable exception, it says nothing wrong and may in fact do some good. However, it manages to miss the central point and is therefore largely irrelevant to serious debate.

That central point is the reward system (for both faculty and students). Nowhere in the report is there any mention of reward. If faculty are only promoted or granted tenure on the basis of research, the real message to graduate students is loud and clear. Moreover, what about the graduate students themselves? If they are "trained" to teach through courses, seminars, evaluations, etc., what are their rewards? What if they fail to demonstrate adequate teaching ability or communication skills, but can write strong theses? Do we go ahead and certify them as college faculty, or do we tell prospective employers of their deficiencies? In short, (how) do we put our money where our mouths are? While I realize that these are extremely hard questions requiring a great deal of community debate, to completely ignore them here is a dramatic failure.

If this were the only flaw in an otherwise well-intentioned report, I probably wouldn't have written this critique. Unfortunately, there is another more serious problem, and, to be honest, one I found quite disturbing.

This study's treatment of the subject of foreign students is very poorly done. The report correctly speaks to the issues involved in increasing enrollment and grad-uation rates of American students in general and women and underrepresented minorities in particular. However, in the half page devoted to foreign students, the entire tone is about the problems they present to their fellow graduate students and undergraduates they may teach. Thus, we are told that "foreign students often have a higher level of mathematical experience" and therefore "when placed in the same introductory courses...often perform better than the American students." Moreover, "the committee believes that such a disparity can contribute to an increase in the dropout rate among domestic students, especially among women and underrepresented minorities." The analysis of foreign students concludes with

> "Colleges and universities have of necessity had to rely increasingly on foreign nationals to teach undergraduate mathematics. The long-term ramifications and impact of this practice need careful study."

This discussion is at best disingenuous. Foreign graduate students now make up about 45% of all graduate enrollments. They *are* our students. Moreover, most of those who get Ph.D.s stay in this country, teach our undergraduates, and work in government labs and in industry. They may very well have different problems than domestic students, but they need "a positive learning environment" just as much as any other student group. Foreign students cannot be seen as a problem for American students or, worse, be pitted against women and underrepresented minorities. This is dangerous ground.

I am sure that the authors of this study were well intentioned. However, to advance the discussion of graduate programs we must seriously address the university reward structure and take a much deeper look at the participation of foreign-born students throughout the system. It is my sincere hope that the AMS or the Joint Policy Board for Mathematics (JPBM) will undertake such a project in the near future spurred by this report and our reactions to it.

Garfunkel is Executive Director of the Consortium for Mathematics and Its Applications in Lexington, MA.

F. Y. Hunt

In April, the *Washington Post* reported that an influential study of the National Science Foundation forecasting serious shortages of scientists and engineers beginning in the 1990s was based on faulty extrapolations and that the predicted need may in fact be overstated. This is disappointing news indeed, as it comes in the midst of one of the most unfavorable job markets for mathematicians in more than a decade. Yet, there are many in the mathematical community who believe that the gravest risk to the future of mathematics in the U.S. is not the oversupply of mathematicians but the undersupply of future American mathematicians. This is one of an array of what might be termed "boundary value problems" arising from our failure to communicate the value and meaning of our mathematical work to certain groups at the boundary of the mathematics world. In this case, the group is prospective mathematicians, but one could add other groups, such as students and scientists working in other fields and, beyond these, the general public.

In recognition of the problem of declining percentages of domestic students in mathematics Ph.D. programs, the Board on Mathematical Sciences charged the Committee on Doctoral and Postdoctoral Study in the United States to produce a report that surveyed a variety of programs throughout the country. The Committee has successfully carried out this task, defining and describing successful graduate programs that produce large numbers of domestic Ph.D.s and, more generally, Ph.D.s that are adequately prepared to fill our research and educational needs in universities, government, and industry. The Committee's most significant and admirable achievement was to make attracting and retaining larger percentages of domestic students—and, in particular, women and minorities—a crucial part of the definition of a successful program. Further, in a section entitled "The Key to Action," the Committee stated that "the departments at research universities have a special responsibility to raise the level and increase the knowledge of talented but underprepared entering American doctoral students."

Fortunately, the Committee found that there are programs around the country that are successful by this definition and achieve this success with many techniques that do not seem to require a lot of money. What is required is a commitment on the part of individual faculty and departments to seek out students and provide sustained support. In a national atmosphere of growing racial and ethnic animosity, the call for active nurturing of minority students will not be an easy one for us to hear. It will be tempting for the community to simply build on the recent modest increases in the number of female Ph.D.s mainly coming from ethnic groups already well represented in the profession. The techniques described in this report will work for many minority students, and I am convinced that continuing efforts to increase both women and minorities will enrich the community not only socially but scientifically as well.

In nonacademic settings, mathematicians are far likelier to encounter nonmathematicians ranging from scientists who use a lot of mathematics to administrative types with very little scientific background. The report urges departments to prepare students to operate successfully in such environments. Mathematicians need to be able to talk about mathematics to nonmathematicians in a way that doesn't trigger unpleasant memories from their student days. The training graduate students (should) receive in teaching can also be used to sharpen communication skills. In addition, I would recommend that graduate faculty who aren't already collaborating with nonmathematicians go out of their way to talk with researchers in other departments at their universities and to researchers in government laboratories or local companies so that she or he can gain first-hand knowledge

of some of the barriers to communication and collaboration, many of which are not well understood by the mathematics community.

To insure the health of the mathematical enterprise in the twenty-first century, how must doctoral programs change if they must at all? The Committee has succeeded admirably in addressing this question and making the case for change, and they have pointed to the characteristics of programs that are successfully meeting this challenge. The report is a worthy contribution, and I hope its recommendations are implemented.

Hunt is associate professor of mathematics at Howard University. On leave from Howard this year, she is currently a research mathematician at the Computing and Applied Mathematics Laboratory at the National Institute of Standards and Technology.

A.J. Lazarus

The Committee on Doctoral and Postdoctoral Study in the United States has produced a report whose suggestions fall into two categories. The first set discusses improvement of doctoral programs within the framework of their current mission, the production of research mathematicians. Most of these proposals are meritorious and all are reasonable. I commend two of them specifically: research atmosphere and recruitment.

I believe that the training of graduate students is a fundamental obligation of senior faculty. Many world-famous mathematicians find ways to integrate supervision of graduate students into their own research program, leaving examples, subcases, and analogues to the apprentices. Faculty and students work together in a way similar to programs in laboratory science. In a successful program, everyone roots for everyone else. Although I do not agree with every assertion in the report's chapter "Positive Learning Environment," I was impressed with its suggested models for a productive research environment and the implicit repudiation of the lamentably common competitive paradigm in which the adviser orchestrates a trial-by-ordeal for his (or her) inexperienced rival.

My own observations about recruitment of talented undergraduates agree with the report's: most departments, shockingly, just don't. We seem to be content to let dozens of students who enjoy mathematics go on to careers in physics, computer science, or even law, without mentioning mathematics graduate school. The number of students pursuing bachelor's theses or other independent work is ridiculously small.

Hit-or-miss recruitment methods overlook students capable of doing mathematics, but, unfortunately, they find more than enough to fill the small number of research positions available today. The sad truth is that with the collapse of the academic job market last year and the even worse market this year, there are already many talented research mathematicians at liberty, and there will soon be more. The committee attempted to address this issue by revising the mission of doctoral programs to include better preparation for "college teaching" and industry. Unfortunately, this second set of suggestions is vague and incoherent. To quote the report:

"The college-teachers model is designed to prepare teachers at two- and four-year colleges. A college teachers program is to be distinguished from a program that confers doctor of arts or doctor of education degrees. Breadth of course work and an emphasis on professional development in pedagogy are, in addition to a research apprenticeship, parts of such a program."

The attempt to combine course breadth and research depth is a recipe for mediocrity—as the committee realized when dealing with *faculty* specialization, where it suggested that departments, except the very best, concentrate their efforts. The same *should* hold true for graduate students. Since the research component of the college-teachers model will be diluted—it is intended for the students who will be *furthest* from the research frontier in any one field—the difference between the new Ph.D. and the D.A. and Ed.D. is that the latter two have poor reputations already while the first will need time to develop one. The committee seems to have assumed without proof that training in pedagogy is more important for college teachers than university professors, who still teach, albeit less. Except perhaps for faculty closely involved in precollegiate teacher training, I don't see why this is so.

The report also fails to articulate why graduates of this program will be more attractive to colleges. Colleges are now choosing from literally hundreds of applicants. Some colleges want demonstrated teaching excellence; others see the job crisis as an opportunity to hire first-rate scholars who just miss a position at a major university. None are complaining about a lack of qualified applicants, quite the contrary. The positions for graduates of the college-teachers model are chimerical. Even the report avers only that need for them "could increase" by the end of the '90s. One wonders where this idea came from, since the committee contained only one member from a four-year college, an elite institution with teaching load and publication expectations similar to research universities', and no members from two-year colleges.

The suggestions about involving mathematics with industry do not appear to be any better informed. (The committee contained one member from industry, but he is from an atypical quasi-academic think tank which has, moreover, had a hiring slowdown for years.) Only a handful of nonacademic jobs are being advertised in the *Notices* and *Employment Information in the Mathematical Sciences*. If there truly are plentiful industrial positions for mathematicians, active solicitation of ads for them would be a great service the Society could perform for recent graduates. But if not, and we are driving taxicabs, let us at least have earned a quality degree.

Lazarus received his PhD from the University of California at Berkeley in 1989. Since then, he has been a visiting assistant professor at UC Riverside and UC Davis.

J. Lewis

Almost any department chair in the mathematical sciences will find *Educating Mathematical Scientists: Doctoral Study and the Postdoctoral Experience in the United States* to be a useful document in connection with a careful look at their own department. The current job market for Ph.D.s has caused most people in doctoral granting institutions to reassess the job they are doing and to look for ways to improve the employment possibilities of their students.

The report offers what appears to be sound advice built around the major themes that a successful department must have a quality faculty, a focused mission, a positive learning environment, and relevant professional development. Of particular use will be the list of questions in the appendix on "Doctoral and Postdoctoral Program Self-Evaluation." Indeed, the report will have a positive impact on graduate programs in mathematics if enough faculty use this list to analyze their own departments.

The goal of the report was to determine what makes certain programs successful. The approach was to conduct site visits to ten programs that had been identified as successful and report on the characteristics that made those programs successful. The committee that produced this report made the fundamental decision that in order to receive the cooperation of the departments visited, no program identification would be made. As readers of the report, we are assured that the ten programs visited were both successful and diverse (small and large, public and private, top twenty and top one hundred, geographically diverse, etc.).

Unfortunately, this approach makes it impossible for readers to decide for themselves whether they agree with the conclusions of the report. Built into the definition of a "successful" program is the idea that "larger percentages of domestic students, and, in particular, women and underrepresented minorities should be attracted to the study of and careers in the mathematical sciences." Given that this is a part of the definition of successful, one longs for information as to how successful these programs are in order to compare their success with one's own department. Providing such information, or offering what the committee viewed as appropriate goals for a department, would have strengthened the report.

Whether our institutions are producing enough domestic students, especially women and underrepresented minorities, was one of the concerns that led to this report, and success with domestic students was built into the definition of a "successful" program. We are told that, "departments at research universities have a special responsibility to raise the level and increase the knowledge of talented but underprepared entering American doctoral students."

How shall we fulfill that responsibility? Each department makes decisions as to which graduate students receive financial support. Even though active recruiting can increase the pool of quality students, most departments will still be faced with choosing between a domestic student and a foreign student who appears to have better qualifications for graduate work. Should department chairs reserve a certain percentage of their resources for American students? If so, how much?

Another part of the definition of a "successful" program also concerns me: "All students, including the majority who will spend their careers in teaching, government laboratories, business, and industry rather than academic research, should be well prepared by their doctoral and postdoctoral experience for their careers." This theme is repeated so often that one who did not know better might conclude that the five areas described above are five separate but relatively equal career paths for Ph.D.s in the mathematical sciences.

Are there jobs for mathematics Ph.D.s in government laboratories, business, and industry that are going unfilled? A careful look at the 1991 AMS-MAA Survey in the November 1991 issue of the *Notices* reveals that, if one considers only the graduates of Group I, II, and III departments and takes out the statistics students, only thirty-three individuals went into government, business, and industry jobs. Let us all wish it were not so. Let us attempt as a profession to open markets for our graduates in government, business, and industry. At the same time, let us realize that currently only about 6% of new Ph.D.s outside of statistics and the Group V departments obtain jobs in the U.S. and outside of academia.

Finally, I take issue with the implied distinctions between careers in teaching and careers in academic research. Even while trying to stress that we should pay greater attention to developing the abilities of our students to teach, the report's repeated references to careers in teaching and careers in academic research as if they were separate perpetuates an unfortunate separation. A colleague of mine likes to say that he considers himself a teacher-scholar. I think this is the correct perspective. As institutions vary, so will the distribution of effort between research and teaching. Perhaps the real point is that about two-thirds of all academic positions at or above the assistant professor level are in institutions that do not offer a Ph.D. This fact, together with the need to open nonacademic markets for our graduates, should influence how we prepare Ph.D.s in mathematics.

Lewis is chair of the department of mathematics and statistics at the University of Nebraska at Lincoln. A member of the AMS Committee on Science Policy, he chairs the Committee's task force on academic support for mathematics and connections between mathematics research and education.

C.S. Wood

For mathematicians, a graduate program is highly desirable, particularly a doctoral program; it is one measure of seriousness of purpose, and it allows teaching at a very rewarding level. When administering a graduate program, most of us build on our own graduate experiences, incorporating a few modifications which we would have liked. Novel programs have come into existence mostly in response to external pressures. It would be surprising if a report from within the community called for extreme upheaval, although the current document does put in a strong plug for professional masters programs. Since this report has gone through NRC's editing and stringent approval measures, it may be argued that the result is radical indeed, calling for significant changes in how the U.S. mathematics doctoral community is to be renewed in future years, and stressing the importance of a "positive learning environment." The existence of such an environment cannot be left to chance, but happens by design and by intervention on the part of forceful, respected members of the program. Nonetheless, some believe that only a few schools matter, and that their students are well served by the sheer excellence of the faculty and the resultant opportunities to sit at the feet of these masters. It is unfortunate that the best schools are the least likely to heed this report; some will point proudly to the achievements of their successful students, thereby dismissing any notion that a more "positive learning environment" is needed. Part of me wants to agree that nothing but the pursuit of mathematics should matter...until I see the results. We're not talking about ogres here, or even administrators; we're talking about our fellow (sic) mathematicians, who rightly demand enormous respect. Why then do American students leave graduate school disenchanted? Who do some of the best never apply? Why so few women and minority students? I believe graduate schools have failed the rest of the mathematics community when their completion rates are low and when they promote a definition of future mathematician whose narrowness excludes most talented young people growing up in the U.S. Perhaps even those programs that are doing a good job along the recommended lines will deny the importance of such issues. My own colleagues may have my head on a platter when this appears. I shudder at the thought of a senior mathematician's saying sarcastically to a young woman student, "I'm told you need a supportive environment. What does *that* mean?" But the current generation of students takes careful note of the graduate environment, regarding atmosphere and attrition rates as crucial factors in considering graduate study. This is the impression I gain from listening to the students I meet across the country, undergraduate and graduate. There is a generational/cultural gap deepening over time. Many of us over-forty mathematicians are experienced know-it-alls, fiercely competitive and eager to stand out as the best. My mother loved games and hated to lose; by the time I entered a classroom I was happy to compete under "boys' rules." This played a greater role in my choice of career than I usually want to admit. In time I found an intellectual excitement in mathematics much deeper and more satisfying than simply beating the next guy to a theorem (although sometimes I wonder . . . after all, my middle name *is* Saunders). But our craft is still presented to students as much more a competition for stardom than is healthy or even accurate. We invite a reaction of disgust or discouragement: "I love mathematics but just don't belong with these people." Surely we have more imagination than to believe that every talented student who exits in this manner "just doesn't have what it takes to be a mathematician!" A recent NSF report from a conference of Presidential Young Investigators indicates that throughout the sciences the over-forty set is out of step with the ambitions and views of the next leaders. A highly competitive setting has very limited appeal to today's brightest students, and is in fact a downright turn-off to many women and minorities. Their interest in hard work is hardly increased by the threat of having their weaker peers weeded out. The NRC report suggests that we take a careful look at how we recruit and educate our successors and make some adjustments in our culture. We dismiss it at risk.

Wood is chair of the department of mathematics at Wesleyan University and president of the Association for Women in Mathemaitcs.

Educating American Mathematical Scientists

R.G. Douglas, UME Trends 4:2 (May 1992)

The role of the United States in the mathematical sciences changed dramatically during this century from that of colonial outpost to imperial center. During the last few decades research achievement grew to unparalleled levels and American universities attracted students and researchers in the mathematical sciences from around the world. The effects of this transformation on graduate education, however, were not all benign. Moreover, this was not the only change that affected doctoral and postdoctoral education in the United States.

By mid-century strong graduate programs at a small number of universities educated and trained students, mostly American, to do research in the mathematical sciences, and the best were well prepared to compete internationally. The remaining students, plus most of the doctoral students from other programs, took positions at colleges and universities that involved mostly teaching. This system worked reasonably well because the mathematics profession was small and teaching was the principal occupation of almost all faculty, in mathematics as well as in other disciplines. Moreover, mathematics was viewed less broadly and industrial needs, except for statistics, were nearly invisible.

The growth of the mathematical sciences, both in size

and in breadth, coupled with the increasing dominance of research in American universities, caused the system of doctoral education to function less well. Almost all doctoral programs in the mathematical sciences began to concentrate exclusively on producing academic researchers. Little attention was given to graduate students learning to teach, despite the fact that the vast majority would occupy teaching positions, nor was the possibility of doing applied industrial research taken into account.

Evidence of the inadequacies of this system abound: from the continued decline in the percentage of domestic graduate students to the growing problems in collegiate mathematics and to the increase in areas of the mathematical sciences abandoned to other disciplines. Further, there is the dearth of women and minority graduate students and a high attrition rate in many programs. Knowing that something is wrong, however, does not always lead to improvements or suggestions of what changes might be needed. Since the Board of Mathematical Sciences is charged with overseeing and maintaining the health of the research enterprise, these problems were its concern. But what to do? How to proceed?

Numerical data is often useful in identifying problems or corroborating hypotheses, already formed, but less useful in pointing to solutions. Rather than proceeding in a conventional manner, the Board proposed an unusual study in which a small committee would identify successful programs, visit them, and then attempt to reach conclusions, based on common features that enabled these programs to work well. The Board chose this approach because it knew that programs existed that were different, that functioned well, and that had overcome or avoided many of the problems that exist most places. The study was funded by the Sloan Foundation and the report was published by the National Academy of Sciences in April 1992.

Conducting the study was an extraordinary experience. The committee received complete cooperation from the doctoral programs at the ten universities it visited. Despite the extensive academic experience represented by the committee members, all agreed that the study gave them their first opportunity to discuss and consider doctoral and postdoctoral education in depth. Moreover, although there was no apparent reason for the plan of study to work, it did. The committee reached agreement on what set these programs apart. The committee also identified three basic principles on which success rested.

Programs need clear, focused goals with faculty and graduate students that can accomplish them. Not all programs need have the same goals. A focus on research in a broad range of areas with depth in each is usually most common, although more specialized goals, concentrating on approach and focused on applied areas, are also possible.

Programs need to provide a positive, supportive learning environment in which all students that put forth the necessary effort can be expected to succeed. While it is unlikely that all graduate students will actually complete their doctoral programs, those that do not should not be labeled failures.

Programs need to attend to the overall professional development of graduate students. Opportunities to learn to teach and communicate need to be provided and a broad view of the mathematical sciences instilled.

Although these principles may seem obvious, merely common sense, not all programs embody them; the successful programs visited by the committee do. The committee visited programs that attract high percentages of domestic students, including large numbers of women and minority students. The overwhelming majority of all students in these programs complete their doctorates and go on to successful careers in academia, both teaching and research, as well as in business and industry.

The report, *Educating Mathematical Scientists: Doctoral and Postdoctoral Study in the United States* describes features of these programs in more detail, suggesting changes that could profitably be made at all universities. The report concludes with recommendations for reviewing doctoral and postdoctoral programs with suggestions on how to proceed. Most past reviews, if any, have been based on faculty wisdom supplemented by an outside committee. It is hoped that this report will bring about a systematic review across the discipline.

There is much that is good with American doctoral and postdoctoral programs in the mathematical sciences and no one wants to give that up. However, some parts are not working so well and need change.

Ronald G. Douglas is at the State University of New York, Stony Brook.

Why I Didn't Quit My Job

J. Smart, UME Trends 3:1 (March 1991)

To the editor,

This article is a reaction to "Why I Quit My Job," by C. B. Kessel, in the December 1990 issue of *UME Trends*, but is not meant to be a criticism of either the article or the author.

Teaching mathematics at the university level is much more than a job; it's a profession. I've been in this profession since 1954, and I'm not looking for another because I love my work. It is true that some students arrive in my classes woefully unprepared, so they need me to work harder to help them learn mathematics. Watching them achieve is worth the effort. On the other hand, some of my students are brilliant and will far surpass their teacher in future accomplishments in the field.

In most of my classes, the majority of the students are "minority students" of one kind or another. Checking the class list is like calling the roll at the United Nations. This cultural diversity is stimulating and foretells the future of mathematics education. Taking advantage of it keeps us young in spirit and focused on the years ahead.

How we teach helps determine the outcomes. My students get involved in discussions and are expected to understand why they do something. Fortunately, our department has managed to keep classes relatively small because we believe it results in higher quality instruction. Teaching can be very exciting if you know the names and something about your students, respect them, and enjoy watching them change.

Seeing students make a discovery of their own, arrive at a generalization, learn to do real proofs for the first time, or get a better grade on a test than ever before is supremely rewarding. These young people (and some not young at all) should be more mature and better equipped for their future after taking a course in mathematics. My former and present students–more than 7500 so far–are more than 7500 reasons I didn't quit.

San Jose State University

Evaluation of Mathematicians and Scientists Who Make Substantial Contributions to Precollege Education

S. Willoughby, UME Trends 4:6 (January 1993)

From time to time university scholars have examined precollege education and been upset both by the content and by the attitudes being developed through that education. Occasionally, such scholars recognize the problem as being so significant that they choose to devote a substantial amount of their scholarly activity to helping solve it.

When the scholar is already distinguished in her or his field and a full professor, this choice may cause only a few derogatory comments from colleagues (often having to do with approaching senility) and, perhaps, less beneficent future salary increases. However, for less distinguished young scholars the choice is likely to destroy a career.

The Faculty of Science at the University of Arizona, recognizing the need for better scholarship in precollege education, has recently created new procedures and criteria for evaluating those members of the faculty who choose, with their department's approval, to devote a substantial part of their scholarly activity to such endeavors. We hope by so doing to encourage more young scholars to help with this important and difficult problem, and we hope to assure appropriate rewards for people who do make significant positive contributions to scholarly education.

Our procedures require, first, that the faculty member have a durable agreement with the department regarding duties, evaluation procedures, and criteria. If the agreement involves substantial activity in education, requests for promotion and tenure, with supporting evidence, are sent to a special Faculty of Science Education Promotion and Tenure Committee. That committee chooses several outside referees who are distinguished scholars in the candidate's field and who have demonstrated a serious interest in educational matters. The committee reviews all the evidence, including comments from the referees, and makes a recommendation to the appropriate departmental head and Promotions and Tenure Committee. The report becomes part of the permanent record, and subsequently the usual university review procedures are followed.

In a perfect world such a special Promotions and Tenure Committee ought not to be necessary. There are, however, so few members of most research university academic departments with a special interest in education, and criteria for success in this area are relatively vague and vacillating. We believe, therefore, that the extra step, along with the criteria, will tend to make these important decisions more consistent in the future, and thus encourage more outstanding young scholars to devote a substantial portion of their scholarly activity to education.

The introduction and criteria for scholarly activity are reproduced here from the Faculty of Science approved document.

Introduction: The purpose of mathematics and science education is to improve the teaching and learning of mathematics and science. Evaluation of faculty members who play a substantial role in mathematics and science education should take into account the impact they have had, are having, and are likely to have on the teaching and learning of mathematics and science. Both the magnitude of the impact and its direction should be considered.

Written evaluations by distinguished colleagues and others, both within and without the university, will necessarily play an important role in determining the magnitude and the quality of a professor's impact. Efforts that will be evaluated for science and mathematics education should be directed towards the systematic improvement of science and mathematics education beyond the faculty member's own classroom and advising activities. Examples of such efforts might include: scholarly works that make a contribution to improving teaching and learning, innovative textbooks that substantially affect teaching and learning, leadership in service activities, etc. In all cases, the magnitude and quality of the impact is the essential issue.

Further evidence of achievement may come from initiation and development of educational programs, from obtaining and managing grant support, from service on advisory and policy boards that have substantial influence, and other similar activities.

Traditional categories (research, teaching, service)

may be inappropriate for evaluating science and mathematics educators because the lines between the categories are often blurred. If these categories are to be used, however, caution must be exercised to avoid assigning creative scholarly work to the service or teaching category (where it ordinarily receives less weight in the overall process) simply because it is different from traditional research.

Criteria for Evaluating Research or its Creative Equivalent: *The University of Arizona Faculty of Science Guidelines for Judging Stature and Excellence in Research* includes the following statements:

◊ "Excellence in research means, among other things, performance that earns international stature."

◊ "In evaluating research, Standing Committees should look especially for publications and other efforts that reflect existing or developing international stature, e.g., refereed publications, invitations to substantial conferences, grants, and awards."

The criteria are appropriate for mathematics and science education, but some of the specifics may differ from more conventional evaluations within the faculty of science.

Worthy contributions could include: scholarly books that make a significant contribution; textbooks that are substantially different from, and better than, previous textbooks (if any) on a worthy subject; articles in refereed, respected journals that describe and advocate better practice or that present research results relating to learning science or mathematics; improved methods and instruments for evaluation; computer software, movie or television productions that enhance education; and so on.

No one person, of course, will make contributions in all of these ways, but any of these activities, and many similar ones, should be thought of as legitimate research or creative activities. The quality and impact of the work must be seen as the important issues.

Evaluation committees must, of necessity, consider with some care the actual origin of materials. If a textbook, for example, was designed and largely developed by employees of a publishing company, the author' should receive little credit for it. If coauthored articles or books were written largely by the other authors, that fact should be considered. In situations where possibilities of this sort exist, the evaluation committee has an obligation to establish the nature and magnitude of the faculty member's contribution.

Since activities such as professional talks, service to professional groups, etc., often presage or demonstrate national recognition, the guidelines suggest that such service be taken more seriously for this group than is typical. Teaching criteria are very similar to those for other faculty.

Steve Willoughby is at the University of Arizona.

Solving Environmental Problems Where are the Mathematicians?

B.A. Fusaro and M.P. Sward,
MAA FOCUS 10:2 (April 1990)

Sunday, 22 April 1990, is the 20th anniversary of Earth Day. Twenty years ago, we were just beginning to assimilate the perspective of Earth brought to us with beauty and drama via photographs from space. And it was just dawning on us that we are wreaking severe, sometimes permanent, damage on the atmosphere, the seas, and the creatures of Earth.

Since then, environmental degradation has continued, even accelerated. We have been shocked as TV, newspapers, and magazines have chronicled the extinction of entire species, and reported of new threats to the ozone layer, Amazonian rain forests, and human health.

In the past 18 months, environmental issues have moved to the fore nationally and internationally. In an historic and eloquent address before the US Congress, President Vaclav Havel of Czechoslovakia stated:

> Without a global revolution in the sphere of human consciousness, nothing will change for the better in the sphere of our being as humans, and the catastrophe toward which this world is headed – be it ecological, social, demographic or a general breakdown of civilization – will be unavoidable. If we are no longer threatened by world war or by the danger that the absurd mountains of accumulated nuclear weapons might blow up the world, this does not mean that we have definitely won. We are, in fact, far from the final victory.

Where in all this ferment are the mathematicians? Have we yet volunteered the problem-solving capabilities of our community? Are we figuring out how mathematics education and research can contribute both to public understanding and to the solution of technical environmental problems? Are we showing our young people, many of whom regard environmental issues as *the* issues of their generation, how mathematics can help solve the problems they regard as critically important to their futures?

Ironically, Sunday, 22 April 1990 is also the first day of 1990 Mathematics Awareness Week, our community's effort to raise public awareness of the contributions that mathematics has made to society. Mathematicians, awake! The calendar is reminding us that problems of the environment call for mathematics at every turn.

The pollution of our rivers, lakes, oceans, land, air, and rain, the mountain of waste that is rapidly accumulating, and the small but deadly chemical and radioactive particles that invade our environment challenge us to find solutions. These solutions have social and economic trade-offs, costs and benefits which can be analyzed, perhaps even optimized, using techniques often requiring nothing beyond school mathematics.

There are a number of individual mathematicians who have made significant contributions to the study of solid waste management, global climate change, and air and water pollution. Bust most of the rest of us, interested as we may be in our civilian lives, are not actively involved as professional mathematicians.

Where can we start? We can begin by talking to our colleagues in such fields as biology, geography, and environmental science and asking them what they see as the most pressing problems, and then looking for how mathematical tools might help in the search for solutions. We can build environmentally- based problems, into our existing mathematics courses. We can offer new courses focused specifically on the role of mathematics in solving environmental problems, showing future citizens and leaders that mathematics is vital to the things they care most about.

Through its publications and meetings, the MAA can increase awareness in the collegiate mathematics community and point out specific actions that individual members and their institutions can take. The authors are organizing a panel discussion at the January 1991 Annual Mathematics Meetings in San Francisco featuring prominent environmentalists and mathematicians, to be followed by an open planning session on what the MAA can do. A special opportunity for community action will occur during 1992, as the international scientific community celebrates both International Space Year (the theme of which is learning about the Earth's environment from space) and the 500th anniversary of Columbus' voyage to the new world.

Readers are invited to communicate their ideas to the authors and to participate in the activities in San Francisco next January.

Appendix B-II

Social Issues

The reprints in this section deal mainly with information about mathematicians from population groups that are underrepresented among mathematical scientists and among collegiate mathematics faculties in relation to their proportions in the general population. (See also Chapter 2 of this book for a description of the incidence of these individuals among graduate students, for related information, and for statistical data.) A number of articles are reprinted from a section featuring concerns of women in mathematics which appeared in the September 1991 issue of the *AMS Notices*. All of the publications from SUMMA, the *AWM Newsletter*, and writings by mathematicians active in NAM (see Chapter 16) are also good sources of information on matters related to underrepresented groups in mathematics.

In addition to the articles reprinted in this section, a number of essays in Chapter 15 which were invited for this book describe successful interventions to increase the proportions of underrepresented groups among mathematicians. Reprinted articles which describe other social issues facing the mathematical community in the context of undergraduate or graduate education are included in the other two sections of this appendix.

SUMMA

MAA FOCUS 12:3 (July 1992)

The MAA's Strengthening Underrepresented Minority Mathematics Achievement (SUMMA) program is a national effort focused on increasing minority participation in mathematics at every level, from elementary through graduate school and beyond. The program has developed a series of projects to implement SUMMA's goals — increased representation of minorities in mathematics, science, and engineering and the improvement of the mathematics education of minorities.

MAA Executive Director Marcia P. Sward and the MAA Committee on Minority Participation in Mathematics (CMPM), cochaired by Manuel P. Berriozábal of the University of Texas at San Antonio and Sylvia T. Bozeman of Spelman College, oversee the SUMMA program. As a result of consultation between MAA leadership and the SUMMA staff and members of the CMPM, an increased number of minority mathematicians now serve on MAA committees and councils.

Carnegie Intervention Grants. The Carnegie Corporation of New York awarded the MAA a two-year grant (now in its second year) of $327,000 through 1993 to encourage college and university mathematics faculty to initiate intervention projects serving minority middle- and secondary-school students. A round of Small Planning Grants in April 1991 and again in March 1992 distributed $100,000 in twenty-four grants. Recipients included: California (two), Georgia (two), Illinois, Maryland (two), Massachusetts, Mississippi, Missouri, New York (two), North Dakota, Ohio, Pennsylvania (two), South Carolina, Texas (two), Virginia (two), and the District of Columbia.

These twenty-four grant recipients include thirteen minority institutions and a tribally controlled college. Three community colleges received awards. We should note that twelve nonminority mathematicians also received awards.

The Carnegie grant has also supported intervention workshops at the 1991 Summer Meeting in Orono, Maine and the 1992 Annual Meeting in Baltimore, Maryland. Furthermore, since fall 1991 (to continue through spring 1993), the grant has subsidized similar workshops at meetings of twelve MAA Sections: Florida, Maryland-District of Columbia-Virginia, Michigan, Missouri, North Central, Northeastern, Ohio, Rocky Mountain, Seaway, Southeastern, Texas, and Wisconsin.

The SUMMA staff assists the planning grant awardees and other mathematicians in developing proposals for intervention projects involving underrepresented groups. Nine of the twelve first-round awardees will launch projects in the summer of 1992. All will have projects in place by the summer of 1993. The first-round awardees have received funding from several sources including the Young Scholars Program of the National Science Foundation (NSF) (two), the Eisenhower Program of their state Departments of Education (two), the GTE Foundation, the National Security Agency (two), and the Puget Sound Water Authority. These proposals raised over $1 million, excluding cost sharing. The other first-round awardees await funding decisions. Awardees submitted a total of twenty proposals.

In addition, the SUMMA office has assisted three mathematicians who did not receive planning grants; two of their intervention project proposals received $300,000 in funding. SUMMA also assisted two other mathematicians with teacher enhancement proposals linked to intervention projects. One of these proposals has received $120,000 funding.

Nationwide, mathematicians directed sixty-three intervention projects in 1992. Despite attrition for various reasons, the number of projects expected for the summer of 1992 will grow to seventy-one. In fact, the introduction of projects SUMMA directly advised prevented a net loss. (For additional information on SUMMA intervention activities, see the December 1991 issue of FOCUS, pages six and seven)

Other Grants to SUMMA. SUMMA has received a National Science Foundation (NSF) grant of $703,000 for three years, through 1995, to network existing mathematics-based intervention projects with those projects SUMMA assists. Activities under the grant include annual conferences of project directors to seek support for level funding as well as to share information about what works; the publication and dissemination of a descriptive directory of projects; a directors' handbook; a quarterly newsletter; dissemination of new curricular materials; and development of a database on projects and their participants.

SUMMA also received a grant of $97,000 from the NSF for its project on Attracting Minorities into Teaching Mathematics (AMIT). It will conduct a study to determine the characteristics of undergraduate programs successful in attracting minorities into teaching secondary-level mathematics. SUMMA will particularly emphasize two-year colleges and articulation concerns because a large number of minorities attend these institutions. Upon its completion, the MAA will publish the study's results.

In another direction, the MAA and SUMMA, as part of its subcontract in a pending NSF cooperative agreement with the Charles A. Dana Center for Mathematics and Science Education, will develop a minority student database and recruit minority faculty and students for a research summer school at the University of California at Berkeley. The National Security Agency (NSA) has agreed to provide $50,000 in core support for SUMMA.

Funding Sought. Several SUMMA projects still seek funding. The program has submitted proposals for the Mentoring Minorities in Mathematics project to several foundations. This project involves minority professionals in mathematics-based careers and provides new opportunities for attracting minorities into mathematics and mathematics-based fields through school visits, Mathematics Awareness Week (MAW) activities, general information for students on career paths, and one-on-one mentoring. The professionals would furnish information on scholarships, intervention projects, and minority mathematicians.

SUMMA has also designed a project to enhance the proven ability of departments of mathematics at minority institutions to nurture minority mathematical talent. It will involve 258 departments of mathematics at historically black colleges and universities, Hispanic-serving institutions, and tribally controlled colleges. The project will emphasize their needs, concerns, and strengths.

In addition, SUMMA's Archival Record has gathered more than three hundred names of minority PhDs in mathematics or mathematics education. A gallery of photographs of some of these mathematicians occupies a wall of the SUMMA offices; the program seeks additional photographs for its collection. SUMMA also seeks funding to research and record the educational accomplishments of all known minorities who were US citizens at the time they received doctorates in the mathematical sciences; SUMMA will subsequently publish this information.

Impact on the MAA. The interaction between SUMMA and the structure and members of the MAA has affected the Association markedly. The list of minority presentations at the 1991 Annual Meeting in San Francisco, California filled a page of the program; the list for the 1992 Annual Meeting in Baltimore, Maryland filled two. Planning for the October 1990 Symposium on Underrepresented Groups, sponsored by the Eastern Pennsylvania-Delaware Section, began before the establishment of SUMMA, but two participants in follow-up meetings at Swarthmore College later received SUMMA Small Planning Grants to initiate intervention projects. The Pacific Northwest Section has planned its Section meeting in cooperation with a tribally controlled college in Montana. A volunteer "SUMMA Coordinator" acts informally as a liaison between the Ohio Section and SUMMA. Every day, from inside and outside the mathematics community, the SUMMA staff receives requests for information about minority participation in mathematics.

Clearly, several projects of the SUMMA program are well underway. As important as external funding is and will remain to its mission, however, the participation of the MAA membership and other members of the mathematical community is even more important. The changes in attitudes and practices necessary to effect SUMMA's goals must begin within the mathematics community. It is being said more and more often that making mathematics work for minorities is the only way to make it work for other students as well. It should certainly be recognized now by all that the old ways created the current morass and need to be drastically modified. For additional information on SUMMA, contact: William

A. Hawkins, Jr., Executive Director of SUMMA, The Mathematical Association of America, 1529 Eighteenth Street Northwest, Washington, DC 20036-1385; telephone: (202) 387-5200; email: maa@athena.umd.edu; fax: (202) 265-2384.

The Escher Staircase

J. Harrison
Notices Amer. Math. Soc. 38:7 (September 1991)

Mathematicians, be they men or women, beginners or stars, love mathematics. They endure perennial anxiety for the joy of occasional moments of discovery. It is only natural for people enamored of the beauty and perfection of mathematics to expect the mathematical community to reflect, in its behavior and ideals, the perfection of mathematical thought. And so we mathematicians expect to be able to admire our colleagues' honesty as much as their precision and enthusiasm. We assume we will find ourselves in a tolerant, trusting community, held together by a passion for mathematics. It would seem that anyone who loved mathematics would be welcome. Given this commonly shared belief, it is surprising that women drop out of mathematics in greater proportion than men.

Some mathematicians are not comfortable with this topic because it involves issues of social inequities that run counter to the basic assumption of collegiality. The majority of male mathematicians are decent people who find it difficult to believe that some of their colleagues do not welcome women into the profession as equals. And yet, because of social factors, women have a particularly difficult time developing their mathematical talents and pursuing their mathematical ambitions. Like the people on Escher's famous staircase, they feel as if they're climbing and climbing, but never quite reaching the top.

The scenes I will refer to are largely not my own story; most were told to me by students and faculty at Berkeley, Oxford, Princeton, Yale, and Warwick. Most women will not have experienced all of them, but most will have experienced some. Each incident, on its own, may sound minor, but over the years they can build up to leave deep feelings of isolation and alienation.

To the young women who will read this: I find myself in a dilemma. If I minimize the problems, there is a risk you will be unprepared for what might be avoided. On the other hand, I would urge you not to be discouraged in view of the increasing number of successful women in mathematics. With foresight, support, and luck, one can overcome most obstacles and have a rewarding and challenging career as a mathematician.

Childhood. Picture an enthusiastic, confident girl, brilliant in many subjects and with exceptional mathematical talent. At age nine she is taught how to calculate square roots and works out her own algorithm for computing cube roots. She is obsessive and loves to solve problems. But life is hard for geniuses, especially when they reach puberty, and especially when they are girls. She learns quickly that her friends distance themselves unless she sacrifices the path entirely or adopts a lightweight style to mask her brilliance. She learns that mathematics is not considered to be feminine at a time when her femininity means so very much to her and her peers. Mathematics is not for sociable people [1] and she, as a girl, has been trained to be sociable. Boys stay away from her, and, if she persists, she fears losing the relationships that she is taught are central to a woman's life. Still, she takes the risk—more mathematics courses.

Many studies have shown that in high school, teachers favor boys, asking them harder questions and giving them more encouragement and attention [2, 3, 4, 5]. Parents, teachers, and friends all expect boys to be better at math than girls [6]. Counselors discourage girls from taking advanced courses [7] and do not give them crucial information about mathematics requirements [8]. At home, fathers, not mothers, are authority figures when it comes to mathematics homework [9]. Many women mathematicians have told me that their fathers were important early mentors and taught them that a cute little girl also could be a scholar. This helped them to weather peer pressure so that their self-confidence and enthusiasm, essential for success in mathematics, survived for the next round.

College. In college, it is more acceptable for women to be smart, and nowadays about half of all bachelor's degrees in mathematics go to women. The peer-pressure problems greatly diminish, but now the difficulties center on the teachers. Women students have almost no role models and fear acting too silly, motherly, aggressive, flirtatious, talkative, or shy. Harassment, from inappropriate flirtation to outright sexual assault, is a major problem that sometimes forces women to transfer or leave mathematics entirely. After her teaching assistant kept a regular vigil outside her house, a Berkeley woman transferred to another university. Another teaching assistant offered a woman a preview of the final exam in exchange for sexual favors. She became severely depressed and dropped out of school. Sometimes male graduate students find themselves, as teaching assistants, in a position of authority over some attractive women and they take advantage of it. A department chair at a major university claims that the biggest problem he has with new male graduate students is their making inappropriate overtures to female undergraduates.

Faculty and teaching assistants fresh from foreign cultures sometimes express unacceptable views more freely [10]. One such professor handed back tests saying, "Even the women did well."

Some of the professors neglect the women as students. Warwick students complained about one professor who completely ignored the women in class—he would not even answer their questions. An Oxford professor would regularly address a mixed audience as "gentlemen." At Berkeley, the students noticed that some professors, when asking questions of the class, would not make eye contact with the women students. Some women withdraw into shyness and are ignored even more. Only a rare individual will excel in such circumstances.

It is possible to change classroom dynamics dramatically with subtle body language and voice cues. A woman professor saw a typical pattern in her undergraduate class in real analysis—the most vocal students were men. The women students sat at the back of the room; they seemed intimidated and said nothing. She decided to try an experiment: to use verbal and body language to encourage the women. To succeed, she felt she needed to make space for the women and quiet the men. For example, she made regular eye contact with women to show that she expected them to know the answers, and she toned down her enthusiasm for the men. If a woman responded, the professor tried to refer to the student's ideas later in the lecture. Invariably, the student would beam and be more eager to participate the next time. The professor told no one what she was doing, and no one seemed conscious of it. By the middle of the semester, the class had turned around. The women had moved to the front row and were avid participants in the class. She knew something was happening when her grader commented, "It's amazing—your women students do so well." The class average (based on tests and homework) was a C+ but the women all made As and Bs.

This experiment was not fair to everyone. But one has to bear in mind that ordinarily the social climate is the reverse—it favors the men. This experiment shows that the atmosphere and social interactions in the classroom make a big difference in how well and how confidently the students grasp the material and produce good work. It's not simply a question of talent and desire to learn—the environment has to be right.

One of the most critical times of a woman's college days comes when she discusses her future plans with her adviser. The paucity of women faculty [11], especially in highly ranked mathematics departments [12], deprives female students of mentors who could help direct their career decisions. When a first-rate Berkeley undergraduate discussed the options with her male adviser, she let him know of her self doubts, and he questioned the sincerity of her desire to be a mathematician. He advised her not to go to graduate school unless she was absolutely sure. What she needed was validation of her ability and approval to move forward. She came to me for advice, and with my support and encouragement, she went on to get a doctorate and a job at a major university. Another honor student reported her adviser's response, "If you persist in this graduate school idea you will make some young man very unhappy." Another Berkeley student was advised to take up nursing despite her straight As in mathematics.

The student and adviser often compromise on a plan to try out graduate school in the master's rather than the Ph.D. program. The student does not realize that this tentative choice will label her for the next few years as not serious. I have rescued several women from this trap and got them into Ph.D. programs. One is now a postdoc in a top research department. An extreme case is a woman who was finishing her undergraduate work at Cambridge, having placed at the top each year she was there. No faculty member had suggested that she go to graduate school. I advised her to see an Oxford professor who became her thesis adviser, and she is now a tenured member of the Oxford faculty. A tenured faculty member at Warwick was never advised to go to graduate school. She only went to Harvard because her husband was encouraged to go there.

Many studies have documented the high drop-out rate of women graduate students. To quote just one source, the National Research Council report, "A Challenge of Numbers," says that less than one-tenth of women continue on for a doctorate from a master's degree, while nearly one-quarter of the men do. In addition, the report notes that "the attrition of women along the path from the bachelor's to the doctoral degree is significantly higher in the mathematical sciences than in other science fields" [23].

Graduate School. In graduate school, only about 25% of the students are female. Men and women alike arrive full of hope, enthusiasm, and energy. I have vivid images of the bright young faces during their first few days of graduate school at Berkeley. The women do not know that many of them will give up within a couple of painful years.

Graduate school is difficult and demanding for both men and women. However, I would argue that circumstances conspire to raise the hurdles even higher for women students. In graduate school, more than at any other time, role models and a support system are crucial. Unfortunately, the community of female mathematicians is still too small and dispersed to be of much help to young women [12].

Male mathematicians can be mentors for female students but they cannot be role models. A man can encourage, inspire, and teach women, but a woman cannot identify with him in the countless ways that distinguish women from men. Furthermore, an aggressive speaking style, minimal social skills, and blatant egotism—acceptable for men—are not suitable for most women to adopt. Finally, because most men have not experienced the subtle prejudice and often long-term discouragement women mathematicians do, men are less likely to be able to counsel women with the perspective needed to survive and flourish.

Three ingredients are needed for success in graduate school—talent, training, and confidence. If a student has had first-class undergraduate training, the course material will not pose a problem. Consider, though, a student who comes with straight As from a college without a strong mathematics program. The student will likely know much less than someone with four years of training from a strong department. Whether male or female, the student is likely to have some self doubts. But a female student is more likely to internalize these feelings of inferiority and to believe they reflect a lack of ability, while a male student is more likely to credit his failure to bad luck [14]. Peers and professors probably will make the same judgments due to the same social influences. If a woman student has good counseling—and that is a big if—the effect can be defused, but it is nonetheless very difficult for most. Her precious confidence quickly subsides. Of the three ingredients needed for success in graduate school, she now has only one.

It is common for a female graduate student to get far more attention for her femaleness than for her mathematics. This occurs in part because of the imbalance of numbers. In addition, a male postdoc conjectured to me that a female mathematician represents an ideal to many male mathematicians: not only can they make love to her, they can talk mathematics to her. She can understand them. The power difference between her and a male professor can worsen sexual tension, and the onus is on the professor, who may be older and one hopes wiser, to minimize this. If the faculty member is flirtatious, the student has to be on her guard for any changes of behavior. The inevitable silences in their conversations may make her feel so uncomfortable she stops thinking clearly and wants to leave the room. Women students are disturbed by interest based on their sex rather than mathematical ability. (I have heard many complaints about this.) Kindness, warmth, friendliness are fine, but the vast majority of women don't want their lives complicated with romantic overtures from their professors. One student's adviser was emphatic that he would not work with her unless they became sexually intimate. The power difference put her into a terrible bind. If she refused him, where would she go, would he retaliate? If she complied, could she live with herself? "One event can have a devastating effect on a woman. One uncurbed man can affect the careers of many women" [16]. But even when a faculty member is not overtly attracted to a female student, subtle differences in attention can profoundly curtail the educational opportunities offered to women.

Sometimes the mathematical side of a woman is belittled. A graduate student at Warwick gave a ride to a visiting star in her field. She started to talk to him about his lecture and saw him chuckling to himself. She asked him what was funny and he bent over double, laughing, "A woman, talking mathematics, and foliations, it's too much!"

In an unfortunate bifurcation of reality and perception, some men who observe women getting attention for their femaleness become jealous. They feel that any attention is good attention and believe that attractive women have some unfair advantage over them. A male graduate student at Berkeley complained that female students have an easier time getting advisers because of course the professors (mostly male) would prefer to be working with a woman. This can be taken to an extreme: a worldclass mathematician justified his vicious opposition to a female competitor because of all the attention she received as a woman.

These problems diminish when there are enough women around. The men are more accustomed to their presence and the women have each other for support. Last year, a third of the graduate students in one seminar group at Yale were women and the atmosphere was quite genial. The women told me what a pleasure it was to look around the seminar room and sometimes find themselves a majority.

I observed this critical-mass phenomenon at Berkeley in our dynamical systems group which a male professor and I led. Over a period of seven years, a third of the students were women. Again, the atmosphere was healthy. Even when potentially intimidating guest speakers would arrive, we would all take them out for a beer and talk about mathematics. We noticed no generic differences in talent between the male and female students in this group. They all got good degrees and good jobs. Some of the women were outstanding (so were some of the men!).

Shyness is the biggest difference between female and male students, and I have seen it everywhere. None of a woman's training has prepared her for the combative, schoolyard games she encounters in graduate school, and she may adopt shyness as an escape. One Berkeley student would only talk to her adviser from his office doorway for most of a year. Her adviser worked around these problems with great care, and she wrote an excellent thesis. A Yale woman literally trembled during her weekly appointment with her adviser. He had observed her reticence (luckily he did not confuse it with a weak intellect) and discussed with me constructive ways around it.

Female students can be especially quiet in seminars. A student at Berkeley couldn't answer a question directed to her by the speaker, although she knew the answer perfectly well—she had recently proved it in her thesis. Sociolinguist Deborah Tannen has an explanation for quiet women [13]. She believes that men speak and hear a language of status and hierarchy whereas women speak and hear a language of connection and support. "Many men are more comfortable than most women in using talk to claim attention." She notes that most women who want to ask a question or make a comment after a lecture need time to muster their courage, formulate their words carefully, then wait to be recognized by the speaker. Men are more comfortable interrupting and saying whatever is on their minds when there is an audience. "For most men, talk is primarily a means to preserve independence and negotiate and maintain status in a hierarchical social order. This is done by exhibiting knowledge and skill, and by holding center stage through verbal performance." Linguist Marjorie Swacker recorded discus-

sion sessions at academic conferences. The length of the women's questions averaged less than half that of the men's. The men (and not the women) often began with a statement, asked more than one question, and followed up the answer with another question or comment [17].

Women can learn to be more assertive. When I visited Warwick in 1988, I found the female students and faculty regularly gave lectures to each other. I tried this at Yale, and it was remarkable how much more comfortable the women felt, both as speakers and as members of the audience. The woman who had trembled before her adviser gave an eloquent lecture to this group. Later she gave a similar talk in her adviser's seminar and found that her practice session enabled her to speak with clarity and confidence.

Faculty. A true colleague should be part of an academic family—never left out, never feeling left out, not suffering from a sense of isolation. Having just left the advisers/parents, postdocs' professional self-images are vulnerable. Their ideas need to be recognized and their thoughts validated as worthwhile. Women are too often ignored at this point in their careers. One Berkeley faculty member recounted the many luncheons at which her remarks were ignored unless a male present repeated them word for word: "Did you hear what she said? It was really interesting." Then and only then would her thoughts be discussed. Another faculty member at a prestigious department avoids faculty meetings because she believes that her male colleagues don't listen to her. She sends her comments to the meeting with a male friend, believing the department will listen to him. Karen Uhlenbeck wrote in 1988 that overt discrimination was only a small part of the problem. "One of the most serious problems women ... have is conceptualizing and acting upon the subtle non-articulated lack of acceptance." [18].

"Inclusion brings confidence. Exclusion brings emotional damage, withdrawal from discourse. Still, some succeed. They do so in less competitive departments or within supportive subgroups within competitive departments. Some manage by working in complete isolation, producing nothing for a few years and then announcing a major, innovative result" [19]. Uhlenbeck, speaking at the 1988 AMS meeting in Atlanta, said, "I cannot think of a woman mathematician for whom life has been easy. Heroic efforts tend to be the norm" [19]. Judy Roitman, at the same meeting, said, "Women's achievement in mathematics has been too often accompanied by heroic feats of character. Think of Julia Robinson, unsalaried, sharing a corner of her husband's office for so many years, and consider the strength of mind and will that kept her focused on her work, and unconcerned about her career" [19]. Many men would find it difficult, if not impossible, to be productive under the conditions in which most female mathematicians routinely work [15]; yet the comparatively small number of women mathematicians is often attributed to innately inferior talent.

Joint work can present unusual problems. Intense intellectual intimacy is necessary for success in mathematical collaboration [20]. When the collaborators are of the opposite sex, they may run up against social taboos—too often they are suspected of sexual intimacy and the man is credited for the work [14]. A female postdoc reported that her male postdoc collaborator got all the invitations to speak about their work. Male collaborators may suffer from unconscious bias. Two women told me their collaborators appeared not to notice their ideas but later claimed credit for them.

As a woman gets older, many of these cultural problems lessen in their impact on her and it is easier for her to be a mathematician. She gives more lectures, and men talk to her because they are primarily interested in her work. Her male competitors are also more relaxed. Some of them have stopped doing research.

At what age do women do their best work? I made an informal survey and found that ten years is typically added to the answer a male mathematician would have given: 35-50 instead of 25-40. When asked, the women said it was a matter of confidence. The inequities not only decrease but the older woman is less dependent on the approval of others. She has tenure, she has publications, she has prestige. Her salary may not be as high [21] as that of a man; it is probable that her department is not as prestigious as that of a man [12]; and she has to confront the myth that mathematics is for young people [22]. But this is small potatoes compared to what she has been through.

Periodically, over the last 15 years, many have predicted significant increases in the percentages of women in the top math faculties. Despite these predictions and hopes of young women of yesteryear, this has not yet come to pass. Today, out of 303 tenured faculty in the ten most highly ranked U.S. mathematics departments, only four are women; among assistant professors, one out of eighty-six is a woman.

To the young women mathematicians who read this article, I hope it gives you an opportunity to consider ways in which you might respond to the kinds of predicaments I have described, so that, should they happen to you, you will not withdraw into your shell or blame yourself. Finding kindred spirits with whom to discuss the problem and share your emotions will help you to prepare a swift, dispassionate, sophisticated response.

I am indebted to Patricia Kenschaft for her excellent booklet, "Winning Women into Mathematics" [16] from which many of the references in this article were taken.

References

1. Keith, S., "Interest Inventories and Mathematics," unpublished article, Dept of Math., St. Cloud State University, MN 56301, 1991.

2. Good, T.L., Sikes, J.N., and Brophy, J.E., "Effects of Teacher Sex and Student Sex on Classroom Interaction," *Journal of Educational Psychology*, 65, pp. 74-87, 1973.

3. Becker, J.R., "Differential Treatment of Females and Males in Mathematics Classes," *Journal for Research in Mathematics Education*, 12, pp. 40-53, 1981.

4. Fennema, Elizabeth and Peterson, Penelope, "Autonomous Learning Behavior: A Possible Explanation of Gender-related Differences," *Mathematics in Gender-related Differences in Classroom Interaction*, eds. L.C. Wilkinson and C.B. Marret, Academic Press, New York, pp. 17-36, 1985.

5. Fennema, E. and Reyes, L.H., "Teacher/Peer Influences on Sex Differences in Mathematics Confidence," University of Wisconsin, Madison, Dept of Curriculum and Instruction, 1981.

6. Ernest, J., "Mathematics and Sex," MAA, Washington, D.C., April, 1976.

7. Haven, E.W., "Factors Associated with the Selection of Advanced Academic Mathematics Courses by Girls in High School," *Research Bulletin* 72, Princeton Educational Testing Service, 1972.

8. Kenschaft, P.C., "Confronting the Myths about Math," *Journal of Career Planning & Employment*, 48, pp. 41-44, 1988.

9. Eccles, J., et al., "Self-Perception, Task Perceptions, Socializing Influences, and the Decision to Enroll in Mathematics," *Women and Mathematics: Balancing the Equation*, eds. S. Chipman, L. Brush and D. WIlson, Lawrence Erlbaum Associates, Inc, Hillsdale, NJ, 1985.

10. Sandler, Bernice R., "The Campus Climate Revisited: Chilly for Women Faculty, Administrators, and Graduate Students," *Project on the Status and Education of Women*, Association of American Colleges, 1818 R St., NW, Washington, DC 20009, 1986.

11. The Annual Report on the Economic Status of the Profession, 1989-90, ACADEME, *Bulletin of the American Association of University Professors*, 76:2, p. 24, March-April, 1990.

12. Selvin, Paul, "Does the Harrison Case Reveal Sexism in Math?," *Science*, 252, pp. 1781-1783, June 28, 1991.

13. Tannen, Deborah, *You Just Don't Understand*, William Morrow and Co., Inc., New York, 1991.

14. Keith, S. and Keith, P., editors, "Proceedings of the National Conference on Women in Mathematics and the Sciences," St. Cloud, MN, 1990.

15. Helson, R., "Women Mathematicians and the Creative Personality," *Journal of Consulting and Clinical Psychology*, 36, pp. 210-220, 1971.

16. Kenschaft, P.C., editor, "Winning Women into Mathematics," Mathematical Association of America, 1991.

17. Swacker, Marjorie, "Women's Verbal Behavior at Learned and Professional Conferences," *The Sociology of the Languages of American Women*, eds. Betty Lou Dubois and Isabel Crouch, San Antonio: Trinity Univer-

sity, pp. 155-160, 1976.

18. Private communication, Oct., 1988.

19. *Association for Women in Mathematics Newsletter*, 18:3, p. 1988.

20. National Research Council, "Mathematical Sciences: a Unifying and Dynamic Resource," National Academy Press, Washington, DC, pp. 1-10, 1986.

21. Vetter, B.M., "Women's Progress," *Mosaic*, 18, pp. 2-9, 1987.

22. Stern, Nancy, "Age and Achievement in Mathematics: A Case-Study in the Sociology of Science," *Social Studies of Science*, 8, pp. 127-140, 1978.

23. "A Challenge of Numbers," Committee on the Mathematical Sciences in the Year 2000, National Research Council, p. 47, 1990.

24. Helson, R., "The Creative Woman Mathematician," *Women and the Mathematical Mystique*, eds. L. Fox, L. Brody, and D. Tobin, Johns Hopkins Press, pp. 23-54, 1980.

Jenny Harrison is a visiting professor at the Mathematical Sciences Research Institute in Berkeley.

Letter to the Editor
The "Escher Staircase"

M. Kary
Notices Amer. Math. Soc. 39:2 (February 1992)

Jenny Harrison's article on the "Escher Staircase" that women must climb in order to succeed in the mathematics profession revealed some of the rotten behaviour that some male mathematicians have displayed towards their female students and colleagues. The case of the professor at Warwick who would not even answer questions from women stood out as particularly blatant.

But what about the pedagogical "experiment" Harrison described, by one of her female colleagues? At the beginning of the term, the men of the class were answering most of the questions, while the women sat passively in the back. So the professor deliberately changed her mannerisms and style so as to be especially encouraging to the women, and somewhat discouraging to the men. "... no one seemed conscious of it. By the middle of the semester, the class had turned around. The women had moved to the front row and were avid participants in the class. [...] The class average was a C+, but the women all made As and Bs. [...] This experiment shows that the atmosphere in the classroom ... makes a big difference."

This "experiment" tells us first of all no more than the famous truth that there really is a reason for double-blind experiments. It also shows that there is something to be said for leaving psychological and sociological investigations to the professionals responsible for those fields, who are bound by a code of ethical conduct which

amongst other things prohibits such "experimentation". Harrison seemed to condone her unnamed colleague's behaviour, noting merely that "this experiment was not fair to everyone." Let us come right out with it: this "experiment" was unfair to the men. It wasn't completely fair to the women either, who may find mathematics less interesting when they are not the beneficiaries of favouritism. And Harrison's colleague is being unrealistic if she thinks no one noticed her bias: the women were enjoying it too much to worry, while the men likely had neither the guts nor the conviction to complain.

Visiting the sins of some of your colleagues on the next generation of students (even "experimentally") is no way to achieve progress. Who knows, perhaps the boorish professor at Warwick was also misguidedly retaliating for someone else's past sins. Likewise, holding all-female seminars, or relying more generally on strength in numbers, does not really address the basic problem. It is unreasonable to believe that every profession will eventually have both strong male and female representation at all times. Even in mathematics, where women may eventually comprise a good 50% or more of the field, this will not happen soon enough to solve the problems of those currently in it. So if we want to avoid misery for those women now and soon to be in math—not to mention for those men in nursing, and other men and women in similarly skewed professions—we had best make sure that men and women are capable of dealing with each other as fellow human beings, and not as alien life forms. Being unable to speak or be comfortable in the presence of men (as were some of the female graduate students described by Harrison) betrays a socialization that, while less dangerous and more endearing, is in the end not so much less defective than that of a boorish man. Graduate school is a frightfully late time of life for any human being to be learning basic social skills. It is also rather late to be acquiring any distinctly male or female characteristics for which a role model "to identify with ... in the countless ways that distinguish women from men" might be required. (For 'countless' translate 'I couldn't come up with any universals'.) And that is a good thing, for in graduate school one hopes to find many individuals with qualities worth emulating, and one should feel free to do so without worrying over whether they are male or female.

Boston University

Response from Professor Jenny Harrison
Notices Amer. Math. Soc. 39:2 (February 1992)

There appears to be a gap between Michael Kary's understanding and what I wrote in my article, the "Escher staircase". I did not condone the pedagogical experiment and wrote that it was not fair. It should not be repeated. But, according to more carefully conducted studies, it is repeated daily in the opposite form—professors call on male students more often, make more eye contact with them, ask them more interesting questions and give them more time to respond (see, for example, Sadker & Sadker, *Sex Equity Handbook for School*, Longman, Inc., 1982). However, the experiment was interesting and provided a lesson for us—the belief that any group is better than any other can be self-perpetuating. Women are not asking for preferential treatment, only that, *a priori*, they be treated as seriously as men. In an ideal world, we, as teachers, would treat all students with equal enthusiasm and expectation. Realistically, we can at least avoid judging mathematical ability by stylistic differences.

It is likely that Mr. Kary has never lacked role models for himself. This may explain why he belittles their importance. A good role model will lead a student to think, "Not only do you inspire me to succeed, you make me believe I can succeed because we are so much alike." Success is possible without a confidence strengthened by this guiding belief, but surely it is easier to attain success with it.

Merging and Emerging Lives
Women in Mathematics

C. Henrion
Notices Amer. Math. Soc. 38:7 (September 1991)

In the nineteenth century, there was a common belief that "as the brain develops the ovaries shrivel," implying that women's participation in the life of the mind would impair their ability as mothers [1]. This was part of a long tradition of identifying intellectual pursuits, particularly math and science, with men, and domestic responsibilities with women. Inevitably, these two spheres were hierarchically ordered: the life of the mind was considered far more important than life of the home.

As Plato said in the *Symposium*, "Those whose creative instinct is physical have recourse to women, and show their love in this way, believing that by begetting children they can secure for themselves an immortal and blessed memory hereafter for ever; but there are some whose creative desire is of the soul, and who long to beget spiritually, not physically, the progeny [of?] which it is the nature of the soul to create and bring to birth" [2]. The dichotomy is clear: one pursues a life of the mind, or one has a family, but one cannot do both. The hierarchy is equally clear: "everyone would prefer children such as these [from the soul] to children after the flesh" [3]. Plato does not consider the possibility of a woman leading a life of the mind. Kant continued in this tradition, defining math as the realm of men, saying that "women should not worry their pretty heads about geometry—that they might as well have beards" [4].

The image that "as the brain develops the ovaries

shrivel" was one that feminists had to combat in establishing formal education for women at the college level in nineteenth century America. They argued that women's access to higher education would make them better mothers for their sons, the future leaders of the country [5]. Although this strategy was successful in opening the doors to a life of the mind, it did not question the deep-seated dichotomy between the intellectual sphere and the domestic sphere. Indeed, those women who worked in American women's colleges in the 19th and early twentieth centuries were forced to choose between a professional life (as teachers) and a personal life (if they chose to marry), for the two were not compatible. As Rossiter reminds us: "It went without saying that according to the mores of the time, all candidates [for professorship] had to be of good Christian character and not only single but in no danger of marrying. Married women were not even considered for employment at the early women's colleges, even, it seems, when they were clearly the best candidate available ... Male faculty at the women's colleges, on the other hand, were expected to be married" [6].

Those seem like ancient times, and we breathe a sigh of relief that things are different now. Women have access to all kinds of formal education, they are able to secure good jobs even in such traditionally male fields as math and science, and they can choose to marry without sacrificing their jobs. Not only do women have access to formal institutions, but their numbers at these places are beginning to represent their proportion in the population. For example, nearly 50% of the math majors in this country are now women [7] (though this trend is sometimes hidden because many of the students who take lower level math courses are from engineering, physics, and computer science, fields that are still predominantly male).

But in the upper ranks, the percentage of women in mathematics declines dramatically. Women make up only 20% of those receiving doctorates [8], and less than 6% of tenured professors in mathematics [9]. Do these declining percentages simply reflect problems of the past? Is it just a matter of time before women come through the ranks and assume equal representation in the mathematics community? Or do these data indicate persistent problems that create unnecessary obstacles to women's full participation in mathematics—subtle barriers that make it less likely that women will pursue mathematics in graduate school and beyond?

These less visible barriers are what I am interested in examining, to see why many women who have "succeeded" in mathematics often do not feel like equal and central participants in the mathematics community. What contributes to this sense of being an "outsider," experienced by many contemporary women in mathematics? To what extent is there still a tension between their lives as mathematicians and their lives as women? [10]

The very concept of a woman mathematician begins to break down the sharp dichotomy between the professional/public/intellectual sphere and the private/personal/domestic sphere—a dichotomy that was solidified in the 19th century, and that still influences much of our society today. As women "cross-over" into the world of the mind, and science in particular, tensions arise, both internal tensions that women experience as they try to balance their personal and professional lives, and external tensions as the mathematics community continually shifts and adjusts to a new population of inhabitants.

One response to these tensions for women in mathematics is to say that women must learn to adjust to this new environment, that conflicts arise because they have not entirely broken ties with their traditional responsibilities. Once they learn to do so, their lives as mathematicians will be easier. But this response assumes that it is possible and desirable to create and maintain a split between personal and professional life. I will argue that such a separation is increasingly unrealistic for both men and women. An alternative response to these tensions is to try to break down the barriers between the two spheres, acknowledging the interconnection and inseparability of personal and professional life.

This article draws on my research on contemporary women mathematicians involving intensive interviews with ten prominent women in mathematics. Their lives help make visible what has previously been invisible: the traditional reliance on a support structure that allows us to maintain the myth that it is possible to separate our personal and professional worlds. At the same time, their lives suggest ways of striking a balance between the two.

I approach this subject with caution for two reasons. First, the only thing that can be said with certainty about all women in math is that they are all different. Any attempt to generalize leaves out specific women and specific details. Nonetheless, there are themes that emerge often enough in interviews with women mathematicians that they warrant attention.

The second reason for caution is that, in any discussion of the difficulties for women in mathematics, there is a temptation to conclude that women should not go into mathematics—either because math is not a hospitable place for women (so they would inevitably be miserable), or because women are not cut out for mathematics. I reject both of these overly simplistic conclusions. The problems discussed in this article are not inherent in mathematics or in women. They are problems that can be remedied, and to do so would benefit the entire mathematics community. The first step towards change is to articulate the problems and make them visible.

Stereotypes of Women, Stereotypes of Mathematicians.

The subtle tensions between being a woman and being a mathematician arise in part from the images that, from childhood onward, are all around us. We are all influenced to varying degrees by images, stereotypes, and messages of our society. The degree to which we internalize these messages depends on many factors: family, friends, educational experience, interests, community, age, and life experience. But, to a certain degree we are what we read and we are what we see.

Media images of women traditionally fall into three categories: wife/mother, sex object, and girl. In the last fifteen years, the "career woman" image has emerged, but even this image makes some concessions to the traditional roles of women as wife/mother or sex object. Carolyn Heilbrun, in *Writing A Woman's Life*, conveys the power of what she calls the "romantic/marriage plot" that most women are raised with, whether or not they choose to pursue it. These images continue to influence both women and men even when they are trying to define new paths. As women's roles expand to new arenas like business or science, they are still expected to also fulfill their domestic responsibilities, giving rise to the modern "super-mom" syndrome.

What do these images have to do with mathematics? Absolutely nothing—and that is the problem. None of the images of women are compatible with images of a mathematician. First and foremost, mathematicians are portrayed as completely unconcerned with anything on the material plane. We are often reminded of mathematicians who would become so absorbed in their work that they would forget to eat for days. They certainly think nothing about their clothes or physical appearance, and while they might have family, it is seen as peripheral to the focus of their lives. Certainly one's image of a mathematician does not include changing diapers or comforting a colicky baby, much less cleaning house or making dinner. Their life follows what Heilbrun describes as the "adventure or quest plot," as contrasted with the marriage plot.

But what do mass media images and stereotypes have to do with reality? Though we may be tempted to once again respond "absolutely nothing," these images affect—and reflect—our lives more than we care to admit. Most women have not extricated themselves from domestic responsibilities. And many mathematicians still praise those individuals who transcend the material world and lose themselves in their work, dividing personal and professional life in a way not feasible for most women. As Halmos says in his "automathography," to be a mathematician, you must love mathematics more than anything else, more than family, more than religion, more than any other interest [11].

> I do not mean that you must love it to the exclusion of family, religion, and the rest ... A spouse unsympathetic to mathematics demands equal time, a guilty parental conscience causes you to play catch with your boy Saturday afternoon instead of beating your head against the brick wall of that elusive problem—family, and religion, and money, comfort, pleasure, glory, and other calls of life, deep or trivial, exist for all of us to varying degrees, and I am

not saying that mathematicians always ignore all of them. I am not saying that the love of mathematics is more important than the love of other things. What I am saying is that to the extent that one's loves can be ordered, the greatest love of a mathematician (the way I would like to use the term) is mathematics. I have known many mathematicians, great and small, and I feel sure that what I am saying is true about them.

This passage illustrates that men mathematicians do indeed have personal lives and responsibilities. At the same time, the message is clear: although family and other interests may be tolerated, they are secondary to one's mathematics. However, this ordering is only possible if there is someone else who can take care of the children while the mathematician does mathematics. It assumes a traditional family structure of a professional and his supportive wife. Since it is extremely rare that a woman mathematician can rely on a supportive spouse to assume the domestic duties, the kind of ordering that Halmos suggests may not be possible, or even desirable. For women, such a vision can lead to a decision of exclusion: family or mathematics, rather than a decision about priorities.

Even, however, when women mathematicians do observe the priorities of their profession, they are still judged by society's standards and evaluated in terms of stereotypically female attributes. In Weyl's memorial to Emmy Noether, for example, he remarks on her appearance that "the graces did not stand by her cradle." A common issue that arises in discussions of Sofia Kovalevskaia's life is her performance as a mother, and whether she neglected those responsibilities. How often do we read a memorial of a male mathematician that discusses whether he was attractive, or whether he spent enough time with his children? [12]

Navigating Personal and Professional Life. As mentioned earlier, for the nineteenth-century women with academic careers in science, the professional and domestic spheres were completely disjoint. Almost all academic jobs open to women were in the women's colleges, and it was assumed that a female professor was single. If she married, she had to quit her job. This was not challenged until 1906, when a physics professor at Barnard College refused to resign when she announced her engagement to be married. "I think it is a duty I owe to my profession and to my sex to show that a woman has a right to the practice of her profession and cannot be condemned to abandon it merely because she marries. I cannot conceive how women's colleges, inviting and encouraging women to enter professions, can be justly founded or maintained denying such a principle." But the trustees countered that a married woman should "dignify her home-making into a profession, and not assume that she can carry on two full professions at a time" [13].

For contemporary women, the story is of course quite different. A wide variety of women have pursued mathematical careers, each with a very different story to tell. Each has navigated a distinct course through her personal and professional life. Some have had children, some have not. Some are single, some are married, many have had more than one spouse or partner. They came to mathematics in various ways and at different points of their lives, from as early as elementary school, to as late as graduate school. Most have experienced both supportive mathematical environments and less hospitable ones. But for almost none was there an obvious, natural path, one that easily fused their professional and personal lives. Virtually none had role models or examples of women who had "made it" in mathematics. In this way, most of these women were pioneers, forging a path that would accomodate the multiple aspects of their lives. For a few this was not problematic, but for most, being a pioneer meant dealing with periods of alienation, confusion, doubt, conflict, and compromise.

What is most striking in studying the lives of women in mathematics, now and in the past, is the lack of a traditional pattern. Few followed the standard path that is clearly outlined for male mathematicians: undergraduate major in math, graduate work, post-doc, tenure track job, tenure, full professor. There are certainly cases of men who do not follow this norm—notable examples include Persi Diaconis who began as a magician, and skipped undergraduate work; and Ramanujan, who had very little formal training—but these are exceptions. With women, the exceptions are the ones who follow the traditional, linear path. For a variety of reasons, women's lives are more accurately characterized by a kind of veering and tacking [14]. Although from the outside this is often seen as a lack of commitment, from the women's perspective, it is their way of accomodating the many pressures, needs, and desires of their lives. Often personal issues must be resolved before a woman is ready to immerse herself full-time in research. For some, this means entering a long-term relationship, for others it means having and raising children, or caring for dependent adults.

In addition to personal issues, professional factors have also prevented women's careers from following a traditional pattern, factors that women were not in a position to control. These include overt obstacles, such as nepotism rules, as well as subtle ones, such as not being seen as a serious mathematician because of one's sex. One prominent research mathematician was not able to work with the professor most suited to be her advisor because he thought she should be a high school teacher and would not take her seriously as a mathematician. In addition, a woman is often invisible in the math community and can have more difficulty forming connections with the main network of researchers in their field [15].

All of these factors, both personal and professional, affect the timing of women's lives. If we look, therefore, at the "time-line" of a woman's life—what she accomplishes when—it can look quite different from that of her male colleague's. Such differences in time-lines can give rise to difficulties in being accepted as a "real" mathematician.

In studying lives of women mathematicians, what emerges is a picture of a wide variety of time-lines, rather than a single standard. Joan Birman, a topologist at Columbia University/Barnard College, did not get her Ph.D. until she was forty years old. Lenore Blum returned to mathematical research in her forties, after years of teaching at Mills College and involvement in national programs to promote women in mathematics. Mary Ellen Rudin, who managed to stay professionally active even while raising four children, is finding that she is doing some of her best work in her fifties and sixties, now that most of the children are grown. She worked part-time as a lecturer until she was almost fifty, when the University of Wisconsin promoted her from a lecturer to a full professor. Judy Roitman did not decide to pursue mathematics until she was already enrolled in graduate school in a logic and methodology program. Though she had always enjoyed math, she had been given messages all her life, both subtle and not so subtle, that women didn't do math. Vivienne Malone-Mayes taught in a small Black college for years before having the courage to pursue a Ph.D. in math, a path that many in her community thought was absurd for a black woman, and certainly not practical for getting a job.

Clearly, each of these women had to "compose a life" of her own. These are examples—and there are many others—of women who succeeded. But there have also been many talented women who were not able to fit their unique lives into the world of mathematics, often because their life time-lines did not mesh with what is expected of a mathematician.

Integrating Children with Professional Life. Having children and integrating them into one's professional life provides a vivid illustration of how women's life time-lines differ from that of their male colleagues, and of the conflicts that can result. I choose this topic *not* because it is a given that all women will choose to have children, or can have children. Many women in all walks of life, including mathematics, have rich and rewarding lives without children. However, this topic brings into focus issues that arise for women with respect to many aspects of their lives—issues of timing, relationships, connection to math community, personal and professional conflicts—all of which apply to women with and without children [16].

Simply deciding whether or not to have children is difficult for many women, but timing is particularly problematic. Women hear three strong, conflicting messages. They are told that, biologically, the ideal time to have children is as early as possible. However, the present social climate dictates that fewer people are marrying or having children early in life; there is social pressure to wait until when one is established in a relationship and a career. Professionally, the ideal time is to wait until

after tenure. So these three pressures—biological, social, and professional—must be considered in turn.

Many women mathematicians did indeed have children early in their lives and felt that was a good decision. For example, two mathematicians, Lenore Blum and Fan Chung, each had a child in their later years of graduate school. And like many women at this time, they played down having a child for fear of not being taken seriously in their professional lives. In fact, when one of Lenore's professors saw her with her four-month old baby, he said "where did that child come from? Whose is it?" He had been oblivious to her pregnancy and birth. When Fan had her second child in her second year at Bell Labs, her manager wondered what she was going to do. Would she quit now that she was having a child? He was unaware that she already had a child who was two years old who was clearly not interfering with her work. In both cases, it was crucial that they had access to full-time child care and supportive husbands. Joan Birman had three children before and during graduate school. She returned to graduate school at New York University later in life, starting part-time in a Master's program. Realizing her ability and desire to work full-time towards a Ph.D., she was able to get graduate support, most of which went toward caregivers for the children.

These women found ways to have children early in their lives and still continue with their mathematical development. This was during the 1960s and 1970s, a period when most women had their children early in life. But today, those who marry tend to do so later in life. And those pursuing higher education rarely think in terms of having children early in their lives.

If, however, one waits until one is settled personally and professionally, other problems can arise. Women in their late 30s and early 40s have more trouble conceiving, more complications with pregnancy, and higher incidence of Down's syndrome or other genetic disorders, and are likely to be more physically exhausted once the child is born. It is also more difficult to adopt a child after 40. This is not meant to be alarmist; many women have children later in life without problems. Nonetheless, many women do experience the profound disappointment and frustration of having waited to have children and discovering at this later stage of life that they are unable to do so.

Given the biological issues of having children late and the changing social realities that make having children early very unlikely, only the middle period remains—after graduate school, but before tenure. But, as everyone knows, this is professionally the most pressured period of all. In a few short years, one has to establish oneself in one's field, make connections, go to meetings, publish, and teach many courses for the first time. Very often women also assume a disproportionately high administrative and service workload. Having children during this period is clearly risky business. If the pregnancy is easy, the birth smooth with no complications, and the child a happy, healthy one who likes to sleep a lot, and if the parents are willing and able to put their child in full-time day care, then one's professional career can stay on track. However, if any one of these factors goes awry, the consequences can be extreme because the cost of not staying professionally productive is very high.

In addition, it is still common to be perceived by colleagues as not fully serious about one's work if one has a child. At the early stages of one's career, judgment by one's peers and colleagues can have enormous impact. The implicit message—that either one is a mathematician, or one is a mother, but one cannot do both—is tied to the assumption that it is men who do the mathematics and women who do the mothering.

So, from the perspective of a young woman who wants to become a mathematician, there seems to be no period of her career that would be favorable for having children. This is why a career in mathematics and having children seem to be in conflict. These problems are not unique to mathematics or even to academia. Still, the mathematical community needs to fashion for itself ways of dealing with this conflict, for there are at least two aspects of the discipline of mathematics that exacerbate this problem.

First, there is the pervasive myth that mathematicians do their best work at a very young age. Philosophy professors may be entering their prime in their 50s or later, but this is rarely the image of a productive mathematician. As G. H. Hardy says in *A Mathematician's Apology*, "If then I find myself writing, not mathematics but 'about' mathematics, it is a confession of weakness, for which I may rightly be scorned or pitied by younger and more vigorous mathematicians. I write about mathematics because, like any other mathematician who has passed sixty, I have no longer the freshness of mind, the energy, or the patience to carry on effectively with my proper job" [17]. He goes on to say, "No mathematician should ever allow himself to forget that mathematics, more than any other art or science, is a young man's game" [18]. This powerful myth of the young, virile mathematician contributes to the pressure young women (and men) feel, despite the fact that there are many examples of prominent mathematicians who did excellent work in their later years [19]. In fact, most of the women I interviewed found that their work improved as they got older.

Second, academic careers in general, and mathematics in particular, exacerbate the problem because of the linear trajectory of career development: graduate school, postdoctoral study, assistant professor, associate professor, full professor. Any deviation from this norm is suspect. In particular, people strongly believe that to take a couple of years off in mathematics makes it very difficult, if not impossible, to return. As a result, there are very few reentry points to a career in mathematics. The consequences are more severe for women than for men since women are more likely to take a year or two off, for example, to have children.

As more and more men assume an equal share of domestic responsibilities, the more these problems will affect men as well as women. Increasingly, men face serious conflicts between personal and professional life. For this reason, the entire mathematical community should be concerned with these issues. In general, though, women still assume more of the domestic responsibilities and are still the ones that bear children. Traditionally, men who have pursued careers in mathematics have not had to choose between their professional life and personal life. Even now, as the traditional structure of "wife at home, husband at work" becomes rarer, we still do not expect a man to choose between his career and having a family. We should not ask a woman to make that choice either.

Looking to the Future. How can the mathematical community address these problems? As I see it, several options must be explored simultaneously.

◇ *Multiple entry and reentry points into mathematics.* For example, the Ada Comstock program at Smith College allows older women who left school in order to raise a family to finish their "bachelor's" degree. Certain graduate programs, like the one at New York University, are receptive to older students or those who have taken some time off. Joan Birman would not have been able to get a Ph.D. at a school like Columbia, where she is now a professor, because her personal circumstances necessitated starting out on a part-time basis, and Columbia does not allow part-time graduate students in mathematics. The National Science Foundation has a program for women in mathematics who are returning to research.

◇ *Part-time options.* There should be ways for mathematicians to have a part-time status during certain periods of their careers, perhaps in graduate school or as a professor. This is one way of allowing people to have children and yet remain professionally active, even if it is at a reduced pace for a few years.

◇ *Optional extension of tenure clock.* For extenuating personal circumstances, such as having children, the tenure-track period could be lengthened. Many colleges and universities are already beginning to institute such policies.

◇ *Support systems.* Day care at mathematics meetings, flexible teaching schedules, and regular day care at colleges and universities are important.

◇ *A change in attitude in the mathematics community.* Informal factors, such as attitudes, can be as important as formal policies in determining the feasibility of women returning to mathematics. As long as taking time off is frowned upon, women who attempt to return will have a very difficult time being accepted or succeeding.

When the mathematics community conveys a clear message that having children is not in conflict with a career in mathematics, we will have gone a long ways toward fully embracing women in mathematics.

Conclusion. Living in a world which sends strong messages about the roles of women and of men, we often internalize these messages unwittingly. We must recognize our hidden assumptions and bring them into open discussion. Only then can we make conscious choices about how to live our lives and define new images of what it means to be a woman and what it means to be a mathematician. There is no inherent reason for these images to conflict.

Balancing personal and professional life is a challenge for everyone, both men and women, and there is no one right way to strike that balance. Given that there is no longer a single prevailing model—in which the man is the professional and the woman stays home with the children—we need to be more flexible in our structures and recognize a multiplicity of models.

In focusing on access to the public roles that were once the almost exclusive domains of men, the women's movement of the early 1960s and 1970s failed to deal with the tensions of combining this public/professional life with the continued demands of personal life. The next stage, therefore, involves taking down the barriers that make these two spheres disjoint, seeing the interactive nature of personal and professional life and discovering how they can be effectively interwoven. We must recognize that personal life is a professional matter and professional life is a personal matter.

References

[1] Fausto-Sterling, Anne. *Myths of Gender.* New York: Basic Books. 1985.

[2] Plato. *The Symposium*, translated by Walter Hamilton. London: Penguin Books. 1951. 208e.

[3] Ibid. 209.

[4] Paraphrase from Immanuel Kant's *Observations on the Feeling of the Beautiful and the Sublime.* Berkeley: University of California Press. 1960, p. 78-79.

[5] Soloman, Barbara Miller. *In the Company of Educated Women.* New Haven: Yale University Press. 1985.

[6] Rossiter, Margaret. *Women Scientists in America.* Baltimore: Johns Hopkins University Press. 1982. p. 16.

[7] "Everybody Counts," Washington, D.C. : National Academy Press. 1989. (National Research Council document).

[8] "Everybody Counts."

[9] AMS *Notices*, November 1988, p. 1310-1312. According to this survey, only 5.4% of the full professors in the mathematical sciences are women. For doctorate granting departments, Group I-III, the percent drops to 2.9%.

[10] I examine these and other questions in depth in my forthcoming book on contemporary women in mathematics.

[11] Halmos, Paul. *I Want to Be a Mathematician.* New York: Springer-Verlag, p. 400. 1985.

[12] There are a few instances where such issues are raised in biographies of male mathematicians. See, for example, Constance Reid's biography of Hilbert.

[13] Rossiter.

[14] Aisenberg, Nadya and Harrington, Mona. *Women of Academe.* Amherst: The University of Massachusetts Press. 1988.

[15] These ideas are developed in more depth in my book.

[16] Other topics such as women caring for elderly parents, or sick or dependent adults are also very important and give rise to similar conflicts as having children.

[17] Hardy, G. H., *A Mathematician's Apology.* Cambridge: Cambridge University Press, 1940, 1985. p. 63.

[18] Hardy, p. 70.

[19] See for example the AWM Newsletter, Vol. 21 #2, p. 11.

Claudia Henrion is a professor of mathematics at Middlebury College in Middlebury, VT.

What Still Needs to Change for the Good of Women in Mathematics and for the Good of Mathematics

J. Roitman
Notices Amer. Math. Soc. 38:7 (September 1991)

The situation of women in the mathematical community has improved remarkably. Unfortunately, the situation of women in the mathematical community still needs remarkable improvement. Some of the problems lie within the mathematical community, and some lie outside it. Here is my short list of problems, with some discussion and suggestions for solutions. Some of the problems in this list are specific to women, but others are rooted in the general isolation of the mathematical community from the rest of society, with the resulting misleading myths and stereotypes—of mathematics, of mathematicians, and of mathematical employment. Many of these problems are being actively worked on by the national organizations from their Washington offices, but Washington is not the nation, and unless we all pitch in there will be little progress.

Women still are undervalued and less visible than men. This is the root of the problem. That it is not limited to the mathematical community only makes its impact worse. It has been studied (in broad settings) in many guises—the same paper or vita sent out under male and female names to be evaluated; painstaking documentation of verbal behavior in groups; painstaking documentation of the interplay between teacher and student. All of these studies come to the same conclusion: both women and men undervalue and over-ignore women. It would be hubristic and foolish to believe that the mathematical community is immune; anecdotal evidence abounds to the contrary. Even established women mathematicians have stories of insults endured, contributions ignored or misattributed, patronizing comments whose nature was not recognized by the speaker. Even the most prestigious researcher has had the experience of being in a crowd of mathematicians without her nametag on and finding her words undervalued, misunderstood, or even not heard by people who don't know her. Do men—do white men—have similar experiences? Of course. But do they have them with the consistency and predictability that women do, or bearing the particular weight of women's general social vulnerability? Of course not. If it is this way for women whose careers are established, how much worse is it for women students and for young Ph.D.s?

The recent calls for a vast infusion of non-whites and non-males into mathematics appear to have suspect origins. The mainstream mathematical community got excited about the need to recruit girls and "minority" (read non-Asian non-white) boys into mathematics exactly when it was noticed that white boys (especially U.S. citizens) won't do it any more. Is this entirely coincidental? If there is a perception that women are being invited in because the big boys are off somewhere else, the invitation will not be accepted. We must be completely clear about this: We want women and minorities doing mathematics because there is mathematical talent in those communities and we do not want to see that talent go to waste.

We must be clear about something else: A girl has no reason to go into a career that boys don't think is good enough for them. To make mathematics attractive to talented girls we must first make mathematics attractive to talented people.

The larger culture thinks that math is just about impossible to learn, especially for girls. We have valuable allies fighting this misconception, in the National Council of Teachers of Mathematics with its new *Standards*, which is trying to foment a much needed revolution from kindergarten through high school. We need to learn from and work with them. The resulting alliances, whether formal or informal, have tremendous potential, providing we have enough humility to not reinforce stereotypes of those elitist professors in (what else?) their ivory towers. The message that all children are capable of learning mathematics includes within it the message that girls are capable of learning mathematics.

The larger culture thinks that mathematicians are geeks anyway, and if they are girls they're ugly. Just about every professional mathematics organization is fighting this image with excellent promotional materials, and books like *Mathematical People* do a good job, but the news hasn't made it down to Mary Worth (who, as I write this, is giving advise to an uptight mathematician), or your average popular kid's book. Grassroots response is necessary here. How do you identify yourself at parties? How do you respond to the in-

evitably prideful "I'm lousy at mathematics"? How do you respond to passing comments in the media, or bigger blow-ups (such as the national columnist recently devoting a whole column to why nobody needs high school algebra)? Smart kids aren't always geeks; our lousy image is a major reason pubescent girls turn off to mathematics.

Unfriendly subcultures exist within the mathematical community. Here is a dirty little secret that, like most dirty little secrets, needs to be talked about. Certain subcultures encourage outright misogyny (e.g. the nerd subgroup in high school, which is where many teenage boys start to think of themselves as mathematicians) or, at the least, encourage social values with which many women, and not a few men, feel uncomfortable—the monkish ideal of the researcher who lives only for mathematics; the false dichotomy between research and teaching; a teaching ideal in which students are not listened to, but lectured at. None of these attitudes—not even misogyny—is exclusive to men; none of these is shared by all men; and, most importantly, none of these is intrinsic to good mathematics. But if these are the values which are perceived as dominating all others, we can expect most women and many men to be turned off to mathematics.

Students come to us with emotional baggage that works against them. The bright young woman sitting in your class did not spring full-blown from the head of Zeus. Unless she is quite unusual, she is experiencing pressures of gender and self-image which need active intervention on the part of faculty to keep her going. Even at a young age, girls tend to attribute their success to luck and boys their success to talent, while boys tend to attribute their failure to bad luck and girls their failure to stupidity. Many, if not most, successful women feel as if their achievements are illusory, that they can be taken away in an instant.

So tell her when her work is good. Show her how her ideas and questions can lead to further mathematics. Suggest she major in mathematics. Suggest further courses for her to take. Suggest graduate school. Find apprenticeship opportunities. Invite her to seminars, to departmental functions. Encourage her through personal hard times. Don't just sit back, give her an A, praise her to your colleagues, and call that enough.

Family leave policy barely exists; family friendly policies are rare. My cynical view is that nothing will be done until a lot of men start complaining about how childcare responsibilities are hurting their careers. Someone has to take care of the kids—usually women do substantially more than 50% (have you seen any nursing men lately?). Good policies encourage long-term productivity; bad policies are blinded by short-term vision. Unfortunately, most of the academic world is dominated by short-term vision (and not just in mathematics). Family leave may be the issue that forces academia away from its current near-sightedness, and allows academics to get off the short-term treadmills that many of us are currently on. The analogy with what's wrong with much of American business is obvious; as with American business, looking to the longer term is simply good policy.

The employment situation still needs improvement. Do you have a two-tiered situation? Is your pre-calculus taught largely by smart women in dead-end positions? There are two good reasons why these women should not be exploited: Simple justice, and the message that their current exploitation sends to young women.

As for regular faculty positions—see the May/June *Notices* for the current situation. If you say that a good woman is hard to hire, here is Marcia Sward's rejoinder: "I've found that offering more money never hurts."

Yes, the mathematics community has improved remarkably in its treatment of women. But we have a lot further to go. As a community, we should not become complacent. What needs to be done is good for both women and mathematics. We can start slipping backwards, or we can use the infusion of energy from our progress so far to move further towards the elusive goal of true equity, when gender becomes as irrelevant to mathematics as hair and eye color.

University of Kansas

Mathematics and Women: Perspectives and Progress

A.T. Schafer
Notices Amer. Math. Soc. 38:7 (September 1991)

The 1989 National Research Council report, "Everybody Counts," says (page 23) that, "gender differences in mathematics performance are predominantly due to the accumulated effect of sex-role stereotypes in family, school, and society." Of course, such a statement would not have been surprising had it appeared in the *AWM Newsletter*—women have been saying this for years. But it was refreshing to see it in such a report. The report also quotes Workforce 2000 (page 18) as saying, "White males, thought of only a generation ago as the mainstays of the economy, will comprise only 15% of the net additions to the labor force between 1985 and 2000." The report identifies women as one group that will be needed to fill a gap left by the absence of white males. What is being done to welcome women into mathematics and keep them there?

Some of the statistics are, unfortunately, depressingly familiar. According to *Science* magazine (28 June 1991, page 1781), there are 303 faculty in the "top ten" mathematics departments (identified as Berkeley, Caltech, Chicago, Columbia, Harvard, MIT, Michigan, Princeton, Stanford, and Yale), and the women can be counted on one hand. One is Joan Birman, who is actually tenured at Barnard College, a women's college of Columbia Uni-

versity. Another is Sun-Yung Alice Chang, who was offered a tenured professorship at Berkeley, but is currently at UCLA. The third is Berit Stensones, who has been appointed to an associate professorship with tenure at Michigan. And the fourth is Marina Ratner, who is tenured at Berkeley; for a history of her original appointment, see the *AWM Newsletter* from 1974 and 1975. The situation among non-tenured faculty is equally dismal: one woman out of eighty-six.

According to the October 1990 issue of *Notices*, there were 991 doctorates awarded in mathematics by institutions in the U.S. and Canada in 1989-1990, 18% of which were awarded to women. From that crop of doctorates, the thirty-nine "Group I" institutions employed 101 men, but just twelve women. Such statistics are often explained away by saying that there are no qualified women "out there." This is difficult to believe when one looks at the percentage of women receiving doctorates in mathematics, which has been plus or minus 20% for nearly 10 years now (with many of them coming from the "top ten"). And in recent years, many women have received postdoctoral fellowships in mathematics and have been invited speakers at national and international research conferences.

Once I had a conversation with a male mathematician who said he would never hire a woman mathematician because she would probably sue if she were not granted tenure. I know of no woman mathematician who has ever advocated that a woman be appointed to a position for which she was not qualified or that, once appointed, she be judged on any basis different from that of a male member of the department. Indeed, I was once asked by a man at one of the "top ten" institutions what I would do if faced with the following situation he encountered in his own department. A man and a woman were being considered for promotion to full professor. The woman's research was inferior to that of the man, but some members of the department felt that if one were promoted, the other should be also, for personal reasons. My answer was absolutely not! The woman's research should be judged on the same basis as any man's in the department. I suspected that my answer was a disappointment to the man who asked me; I think he had expected me to say that the woman should be promoted despite inferior research.

On a different occasion, when I was talking to a mathematician at another of the "top ten," I asked why there were no women on the faculty. His answer was that if the department could find anyone as good as "X," a woman at a less prestigious university, that his department would hire her. "What about hiring X?" I asked. No response—end of conversation. An answer I have heard many times from men at research universities is that women have children and cease to do research, but there are so many counterexamples that the argument is fallacious. And anyway, how many men, with or without children, have short "research lives"?

There has been a great deal of discussion in recent years about attracting more students—and, in particular, more women—into graduate school in mathematics. Programs and funding are not enough. There must be women on the faculties, and the women students must see their work evaluated on the same basis as that of their male colleagues. When it comes to mathematics, male and female students should be treated the same. But when it comes to certain kinds of social factors, it is, unfortunately from my viewpoint, sometimes necessary to treat women differently. During my years teaching undergraduates, I told my female students that I would not write a letter of recommendation for them for entry to graduate school unless they promised to complete the work for the doctorate. I do not tell my male students this, and some of them did not complete the work for the doctorate. One of my Wellesley students now jokingly tells me that the reason she has a Ph.D. is that I had refused to write a recommendation for a National Science Foundation (NSF) fellowship unless she promised she would complete the work. She is now married, has children, and does research.

It has been well documented that many capable girls and women have reacted to the myth that females cannot do mathematics by avoiding mathematics courses and steering clear of careers in mathematics and science. If schools, colleges, and universities have failed here, women and some men have worked to eradicate this injustice and have established organizations and programs for this purpose. There is space here to mention only a few.

I believe that by now all mathematicians know of the existence of the Association for Women in Mathematics (AWM) founded in 1971 as an independent organization and with offices at Wellesley College since 1973. A history of AWM, written by former AWM President Lenore Blum, appears in this issue of *Notices*.

Two years after AWM was founded, due to the efforts of Cathleen S. Morawetz, aided by Isadore M. Singer, the AMS created the Committee on Women in Mathematics, with Morawetz as its first chair. Under her direction, a Directory of Women in Mathematics was published in order to show the mathematical community that there were women who were qualified to be faculty members at research institutions, to be speakers at mathematics meetings and research conferences, and to be appointed members of important national committees. During my tenure as the third chair of the Committee, a second Directory of Women was published, and I. N. Herstein, a member of the Committee, wrote an article in *Notices*, "Graduate Schools of Origin of Female Ph.D.s" (April 1976, page 166). His idea was that, if women preparing for graduate school in mathematics were aware of the departments which had in the past been hospitable to women, they might want to consider those schools. At that time, the Committee also submitted several proposals to the NSF for funding for programs that would benefit women mathematicians, but, unfortunately, none

of them was funded. (I am happy to say that in recent years the NSF has begun to fund programs almost identical to the ones the Committee recommended.)

The Committee was later expanded to the AMS-ASA-AWM-IMS-MAA-NCTM-SIAM Committee on Women in the Mathematical Sciences and is currently chaired by Susan Geller of Texas A&M University. Geller reports that the Committee has developed a questionnaire to be distributed at Ph.D.-granting institutions in an attempt to determine why students in the mathematical sciences leave graduate school. The questionnaire has already been tested at six cooperating institutions, and as soon as funding is available, the study will include all the Ph.D.-granting institutions. The Committee has also been collecting statistics on the relative acceptance rates of male and female authors in various journals. (Journal editors interested in this study should contact Geller.)

In 1975, the MAA established the Women and Mathematics Program (WAM), the first program in the country designed to encourage female students to continue to study mathematics and to seek careers in fields requiring the use of mathematical tools. WAM participants are women from business and industry whose career choices involved a strong background in mathematics and science. They serve as mentors, role models, career counselors, and classroom visitors to elementary, middle, and high school students in sixteen regions throughout the United States. Many of these WAM participants also arrange plant tours for groups of students. The current director of WAM, Alice Kelly of Santa Clara University, says, "We work with both male and female students in our classroom visits, career counseling, and tours, thus exposing young males to the woman of today."

Many colleges and universities have instituted programs to encourage girls in elementary school and to show them how exciting mathematics can be as well as instituting programs for high school women students. An excellent reference which has descriptions of many of these programs is the proceedings of the National Conference on Women in Mathematics and the Sciences, held in 1989 and organized by Sandra Z. Keith of St. Cloud University.

In 1987, the MAA established a second committee on women in mathematics, known as the Committee on the Participation of Women, chaired by Patricia Clark Kenschaft of Montclair State University. Among the Committee's recent endeavors is the MAA publication "Winning Women into Mathematics," which includes a list of fifty-five cultural reasons why women are underrepresented in mathematics. During the national meetings in the past couple of years, the Committee has presented skits using mathematicians as actors to dramatize "micro-inequities" that have actually happened within the mathematical community. Kenschaft describes micro-inequities as "small slights that are often humorous in themselves but chip away at women like water dropping on a rock."

Charlene Morrow and her husband James are Directors of SummerMath at Mount Holyoke College. Describing the program, she writes: "SummerMath, now in its tenth year, was designed to address the underrepresentation of women in mathematics-based fields. It is an intensive, six-week program for high school age females that provides new perspectives and new experiences in mathematics, computing, and science. We emphasize greater conceptual understanding, affirmation of young women as capable members of a learning community, and the importance of constructing one's own understanding of complex ideas .... The atmosphere of the program is one of challenge with support: the challenge of rigorous study and hard problems with the support of a community of teachers, residential staff, and peers ... . Students learn to take charge of their mathematical education, gain a mathematical voice, and experience increased success in mathematics classes upon returning to school."

The Sonia Kovalevsky High School Mathematics Days began in 1985. They were initiated by Pamela Coxson and Mary Beth Ruskai, who at the time held Office of Naval Research science fellowships at the Mary Bunting Institute of Radcliffe College. They suggested that the twenty-fifth anniversary of the Institute and the fifteenth anniversary of AWM be celebrated together, and the Sonia Kovalevsky Symposium was the result. As part of the Symposium, Coxson organized a Sonia Kovalevsky High School Day for high school women and their teachers in the Boston area. The Days have continued on a national scale with some funding from AWM and the remainder from local businesses and industries.

According to Donna Beers of Simmons College, the Sonia Kovalevsky Days held at her institution "celebrate the beauty and uses of mathematics. The goal of these programs has been to show students women professionals working in attractive and challenging fields in which their mathematical preparation has proven indispensable. Above all, organizers aim to encourage young women to persevere in their study of mathematics throughout all four years of high school and beyond. The basic ingredients of these programs have included: hands-on workshops on cutting edge applications of mathematics, e.g. percolation theory, genetics, cryptology, chaos theory, and fractal geometry; career panel discussions led by women professionals, e.g. accountants, actuaries, aerospace engineers, statisticians, computer software engineers; and a lunch-time keynote speaker, often a woman scientist or mathematician who shares her mathematical biography, stressing the hard work required as well as the satisfaction and confidence that developing one's mathematical potential can bring. Student and teacher evaluations of the Sonia Kovalevsky High School Mathematics Days have been uniformly positive, urging that more be held more often, even at the middle school level in order to have a wider impact on young women students early on."

As mentioned above, the NSF now funds programs similar to those first suggested in the early 1970s by the Committee on Women in Mathematics; for example, the Visiting Professorships for Women and the travel grants for women to attend professional meetings and conferences. The NSF also funds many other programs for women, such as research planning grants, Career Advancement Awards for experienced women scientists, and Faculty Awards for Women for those who are tenured but not yet full professors. The latter two programs aim to recognize the nation's most outstanding women scientists and engineers in academic careers of research and teaching and to retain these women in academia.

Another NSF program, Research Experiences for Undergraduates (REU), has proved to be of great value to women undergraduates. For example, the nomination papers for the AWM Schafer Prize, which recognizes outstanding undergraduate women mathematics majors, show that many of the nominees spent summers in REU programs, and some became coauthors of research papers. This past summer, the NSF funded a six-week Summer Mathematics Institute at Mills College in Oakland, California for twenty-four women students, who worked intensively on advanced topics in a seminar setting. The aim of the Institute was to encourage these talented students to go to graduate school in mathematics. For a more complete description of opportunities at NSF, consult the brochure "Opportunities in the Mathematical Sciences," available from the NSF.

At present, fewer than one-fifth of the nation's mathematicians and scientists are women, and the prediction is that between 1991 and 2000 more than half of those entering the workforce will be women. We need to develop more ways of attracting and retaining women in mathematics. Women should be held to the same mathematical standards as men and should be judged on the same basis as men. *Everybody Counts* urges us to increase the pool of students who are successful in mathematics. Let's take that challenge seriously.

Alice T. Schafer is chair of the Mathematics Department at Marymount University in Arlington, Virginia.

Women and Mathematics

MAA FOCUS 12:3 (July 1992)

The Women and Mathematics (WAM) program encourages female students, primarily in grades six through twelve, to explore mathematical and scientific topics and to develop their talents in these areas. The program seeks chiefly to free female students from the "women-can't-or-don't do mathematics" stereotype. WAM provides contacts with role models, career and academic counseling, workshops, corporate tours, and mentors as well as student-parent-teacher association meetings and classroom presentations. WAM participants are all women pursuing careers that require an extensive foundational knowledge of mathematics. The program was founded in 1975 with initial funding from IBM.

Active regions in the Women and Mathematics program include: Baltimore-Washington, Boston, Chicago, Greater Philadelphia, Hawaii, Kansas City, Michigan, Montana, New York-New Jersey, North Carolina, Northern California, Puget Sound, Texas, and Utah. The Connecticut and Southern California regions are reorganizing under a new director and the program will consider new regions in Georgia, New Hampshire, and South Carolina. WAM needs new coordinators to reopen the Central Ohio and Oregon regions. A region in Florida also implemented the WAM program, but that state's success with an aggressive career and academic advising program minimized its demand for WAM activities. WAM is always interested in developing new regions where coordinators and funding can be established.

The WAM program and the Speakers Bureau of the Association for Women in Mathematics (AWM) continue to explore forming an alliance. Combining the school contact activities of these two organizations would benefit both groups and the students they serve. For example, volunteers available through AWM would enable WAM to reach a audience both geographically and academically more diverse. WAM's regional organization would assist AWM speakers to contact schools and arrange visits.

A WAM member has begun compiling a book of problems along with career descriptions and brief biographies of the WAM volunteers who submitted the problems. The book will answer typical questions from a secondary school audience concerning the uses of mathematics and various careers. The program will used any profits from the book to support its activities.

WAM has scheduled a Strategic Planning Workshop for 1992. After seventeen extraordinarily active years, the program must reevaluate its mission, goals, and activities to revise for the current decade. WAM will utilize advice from experts in the needs of female students and schools to direct future activities. The program seeks funding for this workshop.

In 1992, WAM contacted over 27,000 students, 1,500 teachers, and 800 other adults through more than 400 presentations. Currently, two national directors, 26 coordinators, and 520 role models administer and implement the program.

WAM receives funding from corporations and from contributions from both individuals and the MAA. In 1991, IBM, Hewlett-Packard, and Northern Telecom provided grants. Yearly, WAM receives more than one-third of its funding from inkind contributions from participants and coordinators, as well as contributions of time and other support from their employers. In addition, Hewlett-Packard contributed thirty 28S calculators for distribution as awards throughout the regions. Chevron reproduced over 1,000 career brochures for distribution

to schools and students.

WAM regions, independently and in cooperation with other women's groups, also organize and participate in such career conferences as Sonya Kovalevskaya Day, Expanding Your Horizons, and Math Options. These conferences feature workshop leaders who first received encouragement to study mathematics at a similar conference or WAM presentation.

If you would like to participate in, or contribute to the Women and Mathematics program, or, if you wish to receive additional information on its activities, contact: Alice J. Kelly, National Director, Women and Mathematics (WAM) Program, Department of Mathematics, Santa Clara University, Santa Clara, California 95053-0001; (408) 554-4525; Akelly@scu.bitnet. Fax: (408) 554-2700.

Strategies for Making Mathematics Work for Minorities *

B.J. Anderson
AMATYC Review 14:3 (Spring 1993)

By the year 2000, minorities will constitute one in every three American students. It has also been projected that from 1985 to the year 2000, over 21 million new jobs will be created (US Department of Labor, 1988). These new jobs, even those not requiring a college education, will require basic skills in mathematics and the ability to reason. More than half of these jobs will require some education beyond high school and almost a third will require a college education. Thus, over the next ten years, Americans must take significant steps to keep minorities in school and focused on the appropriate academic areas those jobs will demand. Hence, it is no longer just an educational issue but, indeed, it has become an economic one as we consider who ultimately will be supporting American systems, such as social security.

A Vision

In the year 2000, as we take that flight into the future, I see the two-year colleges having strong articulation programs with four-year institutions, and those four-year institutions will include Historically Black Colleges and Universities (HBCU), Hispanic Serving Institutions (HSI), and four-year tribal colleges. I see two-year institutions setting world-class standards, especially in the area of mathematics, that will guarantee unchallenged acceptance of two-year college students for the continuation of higher education in any university. I see four-year institutions accepting graduates from two-year colleges

---

* Editor's note: These remarks were the concluding part of Dr. Anderson's keynote address at the 1991 AMATYC Annual Conference in Seattle. Much of the earlier portion of the address was devoted to history and statistics concerning minority students and the two-year college. The interested reader should consult the various reports listed under "References."

with marked enthusiasm, knowing that an influx of these graduates will not devalue their institutions. I see remedial courses designed to ensure student success, i.e. small classes staffed with experienced, well-motivated and talented teachers, as well as with student mentors. I see two-year colleges in partnerships with HBCUs, HSIs, and majority institutions to produce more teachers of mathematics prepared to teach in urban school systems, and in heavily minority populated schools. I see two-year colleges identifying potential teachers of mathematics, chemistry, and physics, as well as potential engineers, and scientists, and working hard at strengthening these students, especially in mathematics. I see these students, perhaps during their second year, having joint enrollment in both the two and four-year institutions that are in joint partnership. I see two-year institutions having numerical targets for transferability and numerical targets for minorities to transfer into teacher education programs in the mathematical sciences, as well as in engineering, and other mathematics-based programs. I see mathematics faculty at two-year colleges asking themselves the question posed by Dr. Tilden Lemelle, the new president of the University of the District of Columbia, in his first address to the faculty: How does what we do prepare our students for living and for making a living? I see mathematics faculty in two-year institutions serving as mentors for students, especially minority students to help them see what is and can be for them — to show them a future in the mathematical sciences. I see mathematics faculty in two-year institutions working closely with school teachers and faculty in four-year institutions to strengthen programs and facilitate student transition. I see mathematics faculty providing good educational advising, serving to create and sustain mathematics clubs, and serving as the core change agents at the two-year institutions.

So then, *what more can two-year colleges do* to make mathematics work for minorities? I will recommend ten strategies on what we can do to make mathematics work for minorities:

Strategy I: Shift our paradigm, if necessary, to one which allows us to behave under the belief that all students can and must learn mathematics and that minorities can succeed in mathematics and mathematics-based fields. Set high expectations for all students and, most of all, make sure that students know these expectations.

Strategy II: Set up articulation and collaborative programs with HBCUs, HSIs, as well as minority institutions, to facilitate smooth transferability. Also, work closely with faculty in schools and four-year institutions to strengthen programs and facilitate student transition. You may want to examine the program, Exploring Transfer, directed by Dr. Janet Lieberman of Fiorello H. La-Guardia Community College, and Dr. Colton Johnson of Vassar College. A collaborative, voluntary program, this program emphasizes experiential education, collaborative structures, and the power of the site.

Strategy III: Make the teaching profession a glamorous and rewarding one — one worthy of the best students! Identify the best potential teachers among your ranks, provide incentives for them to go into the teaching profession, and provide them with the strongest possible program to prepare them for the four-year institutions. Have your pre-teacher education program designed like those in the connecting universities and with appropriate support for student success. Develop mechanism for joint enrollment in both the two-year and the four-year institutions. Bear in mind that today only 9% of the public senior high school teachers of mathematics are minorities, which is evidence of a wide disparity between the supply of minority mathematics teachers and the proportion of minority students in virtually all states. Also, less than one-half of all public senior high school teachers of mathematics (47% with primary assignment) actually have a college major in mathematics.

Strategy IV: Intensify minority recruitment, showing students the advantages of beginning their college work in a two-year institution and of the ties that the two-year institutions have with four-year institutions. Be sure to let them know about ties to HBCUs and HSIs and majority institutions without steering them into specific schools. HBCUs may be especially receptive to developing strong ties in teacher education in mathematics and science.

Strategy V: Promote and communicate with appropriate minority communities' research on accomplishments of minority mathematicians and scientists who began their college training in two-year institutions. Also, follow your own students, and have successful ones come back to the college and talk to the students.

Strategy VI: Restructure remedial courses for success, incorporating co-operative learning, peer tutoring and computer assisted instruction as supplements to traditional teaching methods. Have experienced, well-motivated, and talented teachers work with students in small classes, and promote group study inside and outside of the classroom setting.

Strategy VII: Set numerical targets for transferability of minority students. Also, prepare this group of students with the capability of meaningful choice for immediate employment in a technological society, bearing in mind that all workers should be prepared to adapt to emerging technology, or prepare them for the continuation of higher education. You might want to examine closely the success of Austin Community College in transferability. Approximately 9 out of every 10 students who transfer from Austin Community College to public colleges and universities are still enrolled a year later (survivability is nearly 40% higher than the statewide average). They claim that transfer success comes from paying attention to curriculum; courses at Austin community College are similar in content, emphasis, and difficulty to those in neighboring universities. Two-year college faculty have frequent contact with faculty at nearby universities. A program in place for nearly a decade brings Austin Community College onto the University of Texas campus two nights a week to university students. For those students, the university counts hours of enrollment at Austin Community College toward a student's minimum full-time enrollment obligation at the university.

Strategy VIII: Establish partnerships with industry, and give all students an opportunity to benefit from programs developed under these partnerships. In response to an industrial need, Seattle Central Community College developed a two-year biotech program. Together, industry leaders and the college faculty planned the course of study. Industry leaders also sat on the community college advisory boards and gave instruments and equipment to the college. College President, Dr. Charles Mitchell, said that it is hard for two-year colleges to keep up with advancing technology in isolation; hence, it is necessary to form partnerships with industry.

Strategy IX: Seek financial and human resources from government, especially agencies that are in need of mathematicians and scientists and those charged with improving mathematics and science education. Develop programs for increasing minority participation in science and mathematics such as mentoring and career awareness programs with agencies such as The National Science Foundation, Department of Education, and National Security Agency.

Strategy X: Promote the teaching and learning function in mathematics. According to the Mathematical Association of America's document, *A Call for Change*, "to adequately prepare students for the 21st Century, the nation's mathematics educators must create classrooms that recognize students and teachers as thinkers, doers, investigators, and problem solvers." Walter Massey tells us, "The most important factor listed by minority students at successful institutions was a supportive environment: the presence of mentors, study groups, science and mathematics clubs, good advising and remedial courses when needed."

Conclusion

The role of two-year institutions to increase minority participation in the mathematical sciences cannot be overstated. This group of institutions is serving, and will continue to serve, a majority of the minority student population in college. You are uniquely positioned to determine, to a large degree, the success that we as a nation will have in increasing minority representation in the mathematical sciences and for preparing minorities to make a productive life for themselves in the 21st century. Two-year colleges have a wonderful opportunity to provide our country with a second wave of students who will become prepared in the mathematical sciences and in education to take on the many jobs that will require mathematics and/or their ability to use current and emerging technology. It certainly is clear that any effort to recruit more majors in mathematics-based fields,

to strengthen the undergraduate major in mathematics-based fields, and to prepare students for life, and for making a living, must be carried out in a manner that includes two-year colleges as a full partner. It is no longer a matter of good-will to merely allow for the emergence of talents from the growing population of minorities, but rather, it is in our best interest for two-year institutions to actively encourage and sustain such an emergence for the survivability of our country and of our way of life. The future rests with you. Thank you.

Office of Minority Affairs The Mathematical Sciences Education Board of the National Research Council and Professor of Mathematics, University of the District of Columbia

References

Action Council on Minority Education (1990). *Education that Works: An Action Plan for the Education of Minorities*, 1990.

American Association of Colleges for Teacher Education (AACTE), (1990). *Teacher education pipeline: Schools, colleges, and department of education enrollments by race and ethnicity*, Washington, DC: AACTE Publications.

Anderson, Beverly (1990). "Minorities and Mathematics: The New Frontier and Challenge of the Nineties." *Journal of Negro Education*, Vol. 59, No. 3, Howard University Press.

Carl, Iris (1990). "Mathematics Education Today: What Are We Doing And Where Are We Going." *Making Mathematics Work for Minorities, A Compendium of Papers Prepared for the Regional Workshops*. Washington, DC: Mathematical Sciences Education Board.

Council of Chief State School Officers (1990). *State Indicators of Sciences and Mathematics Education: Course Enrollment and Teachers*. Washington, DC: CCSSO.

Hodgkinson, Harold L., Outtz, Janice Hamilton & Obarakpor, Anita A. (1990, November). *The Demographics of American Indians: One Percent of the People; Fifty Percent of the Diversity*. Institute for Educational Leadership, Inc., Center for Demographic Policy. Washington, DC

Johnson, William B. and Parker, Arnold E. (1987, June). *Workforce 2000: Work and workers for the twenty-first century*. Indianapolis, IN: The Hudson Institute.

Mathematical Association of America (1989). *A Curriculum In Flux: Mathematics at Two-Year Colleges*. Washington, DC.

Mathematical Association of America (1992 in print). *Statistical Abstract of Undergraduate Programs in the Mathematical Sciences and Computer Science: The 1990-1991 CBMS Survey*. Edited by Donald J. Albers, Donald O. Lottsgaarden, Donald Rung, and Ann Watkins: Washington DC.

Mathematical Sciences Education Board (1990). *Making Mathematics Work for Minorities: A Framework for a National Action Plan.* Washington, DC: National Academy Press.

National Council of Teachers of Mathematics (NCTM) (1989). *Curriculum and Evaluation Standards of School Mathematics.* Reston, VA: NCTM, 1989.

National Research Council, Board of Mathematical Sciences and Mathematical Sciences Education Board (1989). *Everybody Counts: A report to the nation of the future of mathematics education.* Washington, DC: National Academy Press.

National Research Council, Board of Mathematical Sciences and Mathematical Sciences Education Board (1991). *Moving Beyond Myths: Revitalizing Undergraduate Mathematics.* National Academy Press, Vol. 18, No 13, Cox, Mathews and Associates, Inc.

Quality Education for Minorities Project (1990). *Education That Works: An Action Plan for the Education of Minorities.* Cambridge, Massachusetts: Massachusetts Institute of Technology.

Rodriquez, Roberto (1991, August). "Ghetto/Barrio Label Persists Despite Community Colleges' Efforts to Transfer More Black/Latino Students to Four-Year Institutions." *Black Issues in Higher Education.*

Task Force on Women, Minorities, and the Handicapped in Science and Technology (1989). *Changing America: The new face of science and engineering.* Washington, DC: The Task Force.

US Department of Education (1991, April). *America 2000.* Washington, DC: US Government Printing Office.

US Department of Education, National Center for Education Statistics (1989). *Digest of Education Statistics,* US Government Printing Office.

United States Department of Education, National Center for Education Statistics (1990), *Digest of Education Statistics,* Washington DC: US Government Printing Office.

US Department of Labor, Bureau of Labor Statistics (1988, March). *Projections 2000 (Bulletin 2302).* Washington DC: US Government Printing Office.

University of California, *Report of the 1990 All-University Faculty Conference on Graduate Student and Faculty Affirmative Action.*

Vetter, Betty (Computer, 1991). *Professional Women and Minorities A Manpower Data Resource Service.* 9th edition, Washington, DC: Commission on Professionals in Science and Technology.

Honors Mathematics Courses Flourishing in "HBCUs"

R.J. Newman, UME Trends 4:5 (December 1992)

The past decade has seen a marked increase in honors programs in Historically Black Colleges and Universities (HBCUs). Many of these institutions now offer honors work at least at the freshman level. Since HBCU's tend to require some mathematics of all students, the honors programs usually include honors mathematics. This is significant because the honors mathematics courses stand a good chance of producing good mathematics majors.

Because of a lack of funds needed to run small, multiple section courses, few honors programs extend far beyond the freshman year. Southern University of Baton Rouge, Louisiana, and Prairie View University of Texas are exceptions, with Honors Colleges administered by deans. Southern offers advanced courses on what it calls an honors option basis. This means that a student enrolls in a regular course and contracts with the professor to do an honors paper for the course. Upon completion, the student receives credit for the course in the Honors College.

In the past, honors programs have had their detractors. There was a school of thought which held that honors programs were elitist, that they caused an unwholesome separation among African-Americans, and promoted competition when they should be working together rather than working against each other.

This sentiment seems to be largely muted today. Says Dr. Carla Robinson of Spelman College, "I view honors programs as comprising a very positive component of any undergraduate program. They give serious and talented students the opportunity to extend themselves academically and to develop more fully their abilities." Robinson says that she does not see any negative impact of her honors program on either the academic or social life at Spelman. "Our honors students go about their work as do all other students," Dr. Robinson says. "And there is absolutely no tension between our students and the regular student body, as there is none between, say, biology majors and English majors."

The same sentiments are expressed by Dr. Donzell Lee of Alcorn State University and chairman of the National Association of African American Honors programs. Lee says that the students in his honors program get challenges that they probably would not get during regular matriculation. Yet there is no feeling of elitism on the part of anyone.

Dean Beverly Wade of Southern University's Honors College is more provocative. "I try to give my students as much recognition as possible," says Wade. "I don't think that academics get into the campus spotlight enough," she says. "Just think of all the media attention that athletes get!"

Southern has two honors dormitories, one for men and one for women. Wade strongly believes that Honors College students need living quarters where quiet time is observed. This is not an attempt to separate the honors students from the general student population, according to Wade, but is simply an effort to give these students as much support as possible.

The attitude of nonhonors program students is perhaps expressed well by Dwight Bonnet, a senior mass communications major. "I have nothing against the honors program. In fact, it's good for those who want to bust their butts," says Bonnet.

The National Collegiate Honor Counselor (NCHC) lists 19 HBCUs as members. This is the association of all honors programs. There are probably a few more. For example, Florida A&M and Morris Brown are not listed but they have honors programs.

Complete enrollment figures are not available at this time. However, citing Southern University as an example, which may not be typical, the Honors College boasts an enrollment of 530, 250 of whom are freshmen. This figure has grown from less than 350 two years ago, in spite of the fact that the academic requirements have increased from an ACT of 21 to 25 and a GPA of 2.5 to 3.2 on a four point scale.

What of honors mathematics courses? As mentioned earlier, most if not all of the honors programs at HBCUs include a mathematics course at the freshman level. These courses are often of the precalculus variety peppered with nonroutine topics and problems. However, some of the mathematics courses are honors calculus. Both kinds of courses can develop creativity and originality in mathematics. At any rate, honors programs give mathematics departments a crack at some of the best minds of young people.

American Indian Student Successes in Mathematics and Computers — AISES/Caltech, Summer 1991

C. Bradley, UME Trends 4:6 (January 1993)

This four-week program enhanced the computer science and engineering literacy of tenth grade American Indian Students. The academic classes were: Mathematics Class with the Binary Number System and Boolean Algebra; Computer Lab with `LOGO`; physics lectures; Engineering Lab with Construction of Circuit Boards; and Writing Skills Class with Literature of Indian Authors. The field trips included visits to the Jet Propulsion Lab, TRW, McDonald Douglas, and Disney Engineering Group.

The cultural component included attending a local California pow wow, weekly talking circles, Indian dancers at the closing banquet, creating Navajo rug patterns with the computer, reading literature by American Indians, discussions with Indian staff members and Indian guest speakers, and learning about Indian communities. Indian students being, learning, and laughing together is very important. They learn to work, have fun, and accomplish ordinary things together, which in the long run become extraordinary.

The mathematics instructor lectured on ratio, proportion, the Pythagorean theorem, and trigonometric functions. She helped students use graphics calculators. Concurrently, the physics instructor covered proportional reasoning and the ratio approach to the trigonometric functions is, while in the computer lab students created 30-60-90 degree triangles for a star pattern. When studying binary numbers with their operations, octal, hexadecimal, and binary coded decimal numbers in mathematics class, students used the binary coded decimals in their group projects in engineering lab. The mathematics instructor lectured on logic gates and truth tables, which overlapped some of the logic gate lectures in the physics class. This not only reinforced understanding of logic gates, but also connected mathematics with physics and engineering. During the fourth week, the mathematics lectures covered Boolean algebra with truth tables to verify laws of commutativity, associativity, distributivity, and other properties, as well as simplifications of Boolean Algebra statements using a sequence of laws and deduction.

To highlight the connection between binary number systems, Boolean algebra, and design of circuits, a Caltech student demonstrated his handmade speech digitizer during mathematics class. He pulled out his breadboard and explained the circuit design drawing, the logic gate circuit, and waveforms on the chalkboard.

`LOGO` is a computing language developed by psychologists and computer scientists. `LOGO` appears to be very easy to learn at the onset, but is also very sophisticated and diverse. In the AISES/Caltech program, students used `LOGO` Writer, which is `LOGO` plus word processing. They created a Navajo rug pattern using exclusively 90 degree angles for their first project. Their second project, Navajo Star Pattern, utilized 30-60-90 degree right triangles, which required knowledge of the Pythagorean theorem and trigonometric functions. Color procedures made their designs attractive. They tried making games such as the Race game, which required different shapes for the turtles found in the Shape Page, and the Maze game, which required students to draw mazes and race a turtle through the maze until it escaped. The `IF` statement detected when the turtles had reached the finish line. The `RANDOM` statement picked distances for each turtle to move toward the goal. Upon learning new strategies, several students returned to their rug patterns and added new features or rearranged parts.

The class worked for hours on an adventure game of an old castle with a hallway, a cave, a dungeon, a tower, and a wooded forest. During the game the player tries to retrieve all the treasures and avoid all the traps. To add animation, three shapes and two turtles created a man walking across the screen. For a mathematics project students wrote procedures to display the Fibonnacci sequence and made golden rectangles. They inserted quarter circle arcs inside squares of the golden rectangle to create the golden spiral. These projects require recursion, which is `LOGO`s way of making procedural loops.

Other mathematics activities included two simple fractal projects: the snowflake and nested triangles. For a literary project, procedures generated poetry using the RANDOM statement to pick a word in a variable list for the end of the poem. The final project was Pascals triangle displayed on the screen. The procedure required two loops and a continuously increasing variable list.

The dorm room of the computer lab assistant became a drop-in computer center for students. His room had the computer lab: an Amiga computer, the video camera, and editing facilities. Two weeks into the program, the students borrowed an IBM computer to add to the collection of equipment in the room.

The primary goal of the engineering/physics instructor was to expose students to design, the essence of any engineering discipline. He wanted students to explore the creative process in engineering. He chose elementary digital circuits for projects because little study time is needed before one is able to design a new project. As a group project, students developed logic circuits for a seven segment digital decoder. For individual projects they were given a digital output and had to design the logic circuit using three inputs. The instructor helped to wire and insert logic gates into their breadboards.

The engineering lab included exploration of direct proportion and Hookes Law by connecting masses to wires and comparing the stretch and the mass. It also included a hand generator and light bulb experiment to observe circuit assembly basics. The afternoon physics lecture covered circuits with capacitors, batteries, series and parallel connections, electromagnetic force, and capacitors. The engineering lab included currents in resistors, resistance, and color codes, while the physics lecture covered properties of basic electronic components and their application in data collection and analysis. The lecture also covered field effect transistors and their function in building integrated circuits and the lab involved integrated circuits and explorations using electronic breadboards.

Students became familiar with the engineering processes of design, test, redesign, and troubleshoot. Their individual projects entailed the designing, building, and testing of logic circuits. The students third-week laboratory and class activities consisted of laboratory pin placement explorations for integrated circuits; assignment of individual logic design projects; lectures on rules for combinatorial logic, binary addition, subtraction, and multiplication instruction. The engineering lab and physics lecture exposed students to engineering applications of concepts learned in the previous three weeks and the usefulness of the team approach in solving difficult problems. In the last week, student teams explored logic circuits within calculators and using half- adders. Everyone prepared a display board, including a breadboard, for the banquet.

In the writing and study skills class, students read literature written by American Indian authors and wrote essays on ideas expressed by the author. They read poetry and short stories on environmental issues, cultural aspects, and community problems, such as alcoholism. They wrote reflective essays on ideas in literature which were read anonymously to the class and ellicited quite meaningful class discussions.

The field trips accomplished several purposes. They provided students with a range of products that engineers design and the scope of the environments that engineers work in. They enabled students to meet American Indian engineers in their work settings and to learn of the opportunities the students may expect to be open to American Indian engineers in the future. Engineers at TRW showed students scale models of satellites which are now flying in the atmosphere, their office environment, and the labs where satellites are assembled and tested thoroughly before being sent for launching. At McDonald Douglas, students explored the scale model of the proposed first space station, scheduled to be assembled by the year 2000. At Disney Corporation, an engineer gave a slide-sound presentation of Disneylands engineering projects. She explained the entire Engineering Departments responsibilities, organization, and processes at Disneyland. Two Navy personnel led the students on a bus tour explaining Navy preparations to handle the cleanup after war and natural disasters as well as Navy maneuvers designed by engineers and practiced by Navy personnel at Port Hueneme. Students observed engineers monitoring the space shuttle mission in the Central Control Lab.

Students had opportunities for social life in the dormitory with dorm socials, swimming at Malibu beach, a walk along Hollywood Boulevard, amusement park rides at Magic Mountain and Disneyland Park, volleyball games, and campus picnics. Talking circles provided a powerful spiritual and social experience for the students and staff. The circles enabled them to share feelings and develop a feeling of unity and comradeship. Some students attended a local sweat lodge in the evenings during their own time.

The entire camp culminated in a banquet. Students used LOGO Writer to type up autobiographies, experiences in the Caltech program, and reports on their circuit project to display on poster boards. Everyone included three graphics printouts from LOGO projects with the display. They placed their engineering projects in front of the poster boards. Staff and invited guests had an opportunity to view the posters and projects, while having punch and conversation with the students before the meal was served. The opening prayer was said by a medicine man from the community. After dinner American Indian dancers performed in honor of the students. Each student was given a graduation certificate, a medallion, and a few gifts to remember the program.

Students enjoyed most the field trip to the Navy base and the experience of designing, building, and testing individual circuits. Trips to McDonald Douglas Space Station and Jet Propulsion Lab were liked almost as much. The more academic, lecture-type, mathematical/science experiences were liked the least. This data indicates the students preferred the hands-on, experiential, and personable events over the lecture-type, academic, mathematical/science ones.

In general, AISES students left the Caltech program with improved attitudes. Students took the Fennema-Sherman Mathematics Attitudes Questionnaire at the start and end of the program. Statistically significant changes in attitudes were found in both male and female students. Mathematics and science were no longer perceived as a male domain. Confidence in learning mathematics and science had increased. Mathematics and science were viewed as useful and worthwhile to study. Motivation to pursue mathematical, scientific, and engineering careers had increased.

Claudette Bradley is at University of Alaska, Fairbanks.

Enhancing Female Participation in Programs for Mathematically Talented Students

H. B. Keynes, UME Trends 1:3 (August 1989)

The Talented Youth Mathematics Program

Recent studies have linked the critical role of early mathematical preparation to success in science, engineering, and other technologically dependent careers. Attitudes and interests at the middle school level influence important decisions about mathematics courses. The opportunity for an intense learning experience in a serious but supportive environment can be extremely beneficial fo the most capable mathematically talented students. The thirteen year old University of Minnesota Talented Youth Mathematics Program (UMTYMP) is a statewide program designed for these students. Supported by both private sources and state funds, UMTYMP provides an intensive academic experience for mathematically talented students via a sequence of specially designed accelerated mathematics courses. This article describes new activities to improve participation of females in the Program.

Students in grades five through eight are selected for the 5-year program by a specially designed pre-algebra mathematical aptitude test. While continuing other subjects in their local schools, students come to the University and several regional sites after school one day a week for a two-hour mathematics class. The two-year high school component covers Algebra in the first year, and Geometry/Analysis in the second. Significant homework is required, and self motivation is an important attribute. Full credit towards high school graduation is mandated by state law.

The University component of UMTYMP covers a college honors-level sophomore calculus program in the next three years. These idea and problem oriented courses are individually designed and taught by senior mathematics faculty in the School of Mathematics. Sophomore calculus is taught using a linear analysis approach. Courses carry university as well as high school credit. After calculus, individualized advanced mathematics courses are then offered to students until they graduate from high school. UMTYMP develops a long-term culture of mathematics which has a major influence on its students.

The postsecondary record of UMTYMP students is striking. College opportunities for both admissions and scholarships are impressive. Several major universities give preferential treatment based on the prior successes of UMTYMP graduates. Female graduates fare especially well. Most choose difficult engineering and science majors at leading schools, and easily persist in their programs. In alumni surveys, they testify to the importance of UMTYMP in their intellectual growth and in developing self confidence. The Program appears to patch the severe "pipeline leakage" too often seen for female students at the undergraduate level. Initial data indicates that significant numbers of UMTMP alumni are continuing with graduate studies in science, mathematics, and engineering, and considering academic careers.

Issues Facing Female Students

Based on comparisons with other programs which involve serious selection criteria and intense personal commitment, historical participation by female students in UMTMP has been quite good. However, considering both statistical data and anecdotal information, it was felt that greater involvement could be achieved.

Prior to 1988-89, females comprised approximately 40% of the 1300-1500 students testing, 30% of the 100 students qualifying, and 22% of the first-year enrollments. Moreover, participation was heavily weighted towards the first two years of the Program, with about 17% overall female enrollment.

The issues associated with these statistics are quite complex, but several main features stand out. Students were primarily chosen to participate in the qualifying examination by school identification. It appears that the schools do not do as well at identifying mathematically talented female students as they do with male student. Even when a female student qualified, she was more likely to turn down admission than her male counterpart. Informal analysis indicated lack of encouragement in both the schools and the home as important factors. Once in the program, given equal ability and equal grades, female persistence was lower. The lack of an appropriately supportive school, and especially family, environment and the lack of other girls in the class were important issues here. Having a family indifferently request withdrawal of a girl performing near the top of her class while another tenaciously pleads to continue a boy who is struggling, occurred all too frequently. Cultural issues such a socially coping with being a smart girl and lack of realization of the impact of mathematics were other important aspects. Also, inconsistent and sometimes negative messages from the schools for participation in UMTYMP affected girls more severely. Finally, issues affecting personal esteem - lack of self confidence in abilities, lack of involvement in even cooperative competitions - played a prominent role.

The Bush Foundation Intervention Project

With the support of a major 16-month grant from The Bush Foundation, the Program has designed several interventions to address these concerns. The interventions are basically addressing informations, counseling and support issues rather than the program structure itself. They involve working with the families and the schools as well as the students. Activities to encourage and support girls who nearly qualify are included to enhance retesting, and a program targeted at select schools encourages key teachers to focus on increased participation of more mathematically talented females. New retesting procedures and special orientation meeting for girls who qualify were established. Regular academic year workshops discuss study skills and sex equity issues, and frequent social activities are offered. Regrouping of classes and regular counseling contacts with the home are now in place. Together with support from the NSF Young Scholars Project, special summer institutes for both current UMTYMP students and girls interested in participating will be offered fo a two week period in June, 1989. They will provide mathematical enrichment, career information, and opportunities for socialization with other smart girls. The overall aim is to create a supportive environment which encourages capable females to seriously involve themselves in mathematics.

Initial Results

Although the project began in September, 1988, already some initial results are very promising. The new testing procedures resulted in a much stronger applicant population. Using a more difficult qualifying test and a higher cutoff from previous tests, a class 20% larger than usual was admitted. The testing population was 44% female, and 32% of the entering class was female. Despite the higher standards, female scores were better overall and more evenly distributed than in the past.

Based on the literature and prior experience, it is considered that a class with 50% enrollment of each sex is desirable. This year, three classrooms are 50%–50% and two are all male. Mixed classes have led to much improved and more supportive classroom dynamics for the female students In fact, one class is clearly dominated by its strong and vocal female group. This structure seems to have led to improved retention of girls. Of the few students (about 8%) who have withdrawn, only 30% are female. Socialization among females is now developing. Some of the issues of isolation and significant friendships with other UMTYMP girls are being resolved.

The Program will continue to try new approaches and modify its old ideas. We are convinced that with these types of efforts, the quantity and quality of female par-

ticipation in UMTYMP will improve.

Harvey Keynes is at the University of Minnesota.

Involving Minorities in Mathematics

J.A. Jones, UME Trends 3:1 (March 1991)

The following article is excerpted from the keynote address of the organizational meeting of the Alliance to Involve Minorities in Mathematics (AIMM) held in Washington, D.C., on December 6-8, 1990.

Mathematics is mankind's greatest endeavor. Everything ever built–from the most primitive tool to the most sophisticated computer system–has involved the use of knowledge and concepts from mathematics. Mathematics is the indispensable underpinning for all of our science and technology. Throughout the ages, we have counted, we have measured, we have examined and compared objects of various sizes and shapes; we have made logical inferences, we have developed mathematical models for real-world events, and we have developed very profound relationships among apparently unrelated concepts. Today, pure algebraic concepts, including group theory and field theory, are used to derive new knowledge and understanding of concepts in topology and in real and complex analysis. Also, new knowledge in analysis and topology are used to reveal greater understanding of seemingly unrelated topics in algebra.

There is no doubt that among some people, mathematics will continue to flourish in the U.S. The major questions that we must consider are, "Who will these people be; who will decide, and how and when will these decisions be made?"

By its very nature, mathematics should be and is, intrinsically interesting to the human mind, and no racial or ethnic group should lay claim to having special ownership of mathematics and its propagation. However, there is a strong myth in the U.S. that only a few people can do well in mathematics and that these few people generally will not include women, African Americans, Hispanics, and American Indians. This myth reigns supreme from kindergarten through graduate school and is perpetuated through the kinds of curricula our youth receive in mathematics at all grade levels.

By its very nature, mathematics (even at an elementary level) involves geometry (both plain and solid), arithmetic, probability, statistics, and problem solving applications to real world phenomena. Nevertheless, we usually present mathematics to our elementary school children as a boring series of basic arithmetic facts to be learned in a rote fashion through paper and pencil drill and practice exercises. We then "find out" what the students know by giving them quick recall paper and pencil tests. Geometry, probability, statistics, and problem-solving are rarely mentioned to our children, and we wonder why most of them hate mathematics after only a few years of over-exposure to arithmetic.

I would like to set the record straight and say that there is nothing inherently wrong with mathematics that should cause it to be hated and feared by young people and adults alike. The problems involving the achievement of our young people in mathematics must rest with us. There is nothing about mathematics which dictates that it should be limited to the few or that only a few people need ever aspire to engage fully in the charms and pleasure it has to offer.

Mathematics is the nectar of the mind, but many of us have never been encouraged to taste it. Many women and minorities have been told to stay away from real mathematics because it will not be nectar to their minds but rather will give them a terrible headache from which they might never recover.

But all is not lost. Some good things regarding mathematics and minorities have happened and, hopefully will continue to happen. Some programs have worked, and excellence has been achieved. The big problem we face today is not that nothing good has been done, but that the good things have been rather limited. Sometimes these good things have resulted from the work of one teacher at one school; sometimes they have resulted from National Science Foundation-supported programs, only to fade away once the funding was gone. They have sometimes resulted at magnet schools that serve only a small part of the school system's population; sometimes they have resulted at special schools of mathematics and science that serve only a few select students, and sometimes they have resulted from special ongoing national programs that serve only a limited number of students. In many cases, these are outstanding efforts but, unfortunately, they serve collectively only a small fraction of the students who could benefit from them. We can and we must do better!

If the Alliance to Involve Minorities in Mathematics (AIMM) is to make a difference, it must help to significantly broaden our local, state, and national efforts to bring in those youngsters who previously have been left out of all our best efforts. The Alliance should be viewed as an integral but critical part of the national education reform movement, in general, and mathematics education reform, in particular. Indeed, we cannot have real education reform in this country unless we deal forthrightly with the issues involving the achievement of minorities in mathematics.

We have now reached a point where our failure to provide educational equity to all our youth is no longer merely a question of equal opportunity and civil rights, but also a question of the national security and continued economic viability of the U.S.

By the year 2000 it is estimated that one in every three American students will be minority (mainly African Americans and Hispanic Americans), and by 2020 demographic projections indicate that today's minorities will become the majority of students in the U.S.

Already, the ten largest school systems in the U.S. are 70 percent minority.

During the past year I served as chairman of the steering committee for the "Making Mathematics Work for Minorities" project under the auspices of the Mathematical Sciences Education Board. In this capacity, I had an opportunity to attend six regional workshops and a national convocation, all designed to help develop a national action plan to make mathematics work for minorities. I am pleased to report to you this evening that the framework for a National Action Plan is now in place and will form the basis upon which our new alliance can move forward toward great improvements in the next decade and beyond.

More than 1,200 individuals participated in our deliberations. They represented groups that are essential for any successful education reform efforts. Each group was asked to formulate what it could do to make mathematics work for minorities. These formulations provide the core of the National Action Plan. What must be done? Among others, teachers should:

◇ Incorporate teaching strategies and new assessment tools that develop reasoning skills in all students.

◇ Nurture the expectation that all students can and must be given the opportunity to learn mathematics.

School administrators should:

◇ Develop positive and productive school climates that facilitate students' learning of mathematics.

◇ Eliminate low-end tracking while promoting a core curriculum in mathematics for all students.

Institutions of higher education must:

◇ Institute financial incentives to encourage minorities to pursue mathematics-based majors, including teaching.

◇ Launch a widespread and visible effort to eradicate erroneous beliefs that inhibit the performance of minorities in mathematics.

◇ Increase the allocation of resources in support of mathematics education for all students.

◇ Follow national standards and recommendations that require strong mathematics content for teachers at all grade levels.

Professional organizations should:

◇ Implement creative outreach programs that will encourage minority students to seek careers in mathematics.

◇ Work to increase public awareness of the need for more minority participation in mathematics.

Parent and community groups should:

◇ Work together to ensure that school budgets are adequate.

◇ Encourage children to study mathematics from preschool through high school and in college.

The media should:

◇ Publicize the many examples of success in teaching mathematics to minority students at all grade levels.

◇ Determine the mechanisms for giving exposure to the achievements of minorities–both students and mentors–who can serve as role models.

Business and industry should:

◇ Sponsor mathematics mentoring programs for students.

◇ Become a part of state coalitions and work with leaders, parents, and legislators to facilitate more participation by minority groups in the mathematical sciences.

Local and state governments should:

◇ Expand efforts to attract and retain first- rate teachers.

◇ Provide funds to those individuals and groups who have a track record of success in working with minorities in mathematics for support of massive teacher and staff development activities at all levels of education.

Federal agencies should:

◇ Identify and highlight programs that are successful in improving the performance of minorities in mathematics, and fund the replication or adaptation of those programs across the country.

◇ Establish teacher institutes for the purpose of enhancing teaching skills in mathematics, broadening teachers' knowledge of various related subject matter, and providing guidance in relevant issues regarding the mathematics achievement of minorities.

The U.S. president should:

◇ Provide strong national leadership and support for achieving his stated goal that, "By the year 2000, U. S. students must be first in the world in math and science achievement."

If this country is to maintain its position as a world leader, it must provide a system in which education in general and mathematics education in particular will work for everybody.

◇ This education must work for the poor as well as for the rich.

◇ This education must work for children in inner city schools as well as for children in suburban schools.

◇ This education must work for children who attend public schools as well as for those who attend private schools.

◇ This education must work for children in nonmagnet schools as well as for those in magnet schools.

◇ This education must work for those children with a single parent as well as for those children with both parents.

◇ This education must work for African Americans, Hispanics, American Indians, and women as well as for Euro-American males.

As an alliance, we have both an opportunity and responsibility to make mathematics education work for everybody. We must say "no" to an inferior education for any of our children. We must say "no" to those who say we are doing the best we can do and we should wait until next year to do something. We must say "no" to those who say that most of our minority youngsters can't learn mathematics and science at the highest levels. We must say "no" to ignorance because a mind is a terrible thing to waste.

But we must move from saying to doing! We must move from using our jawbones to using our backbones. For if we, as a nation, do not make education work for everybody, we will not only waste the minds of our youth, but we will eventually lose the vitality and soul of our country.

J. Arthur Jones is President and Chief Executive Officer of Future Technologies, Inc. of Reston, Virginia.

Colleges and Universities Give Attention to Minority Students

R.J. Newman, UME Trends 3:3 (August 1991)

An increasing number of majority institutions of higher learning are devoting resources to recruiting and retaining minority students. I receive calls and mail regularly from colleagues at majority institutions with graduate programs who are eager to enroll capable black students. Not a few of these have attractive financial inducements. This is indeed an excellent trend, one that must continue and even be expanded if the national goal of bringing significantly more minority students into mathematics and science study is to be reached.

Graduate schools are not the only ones beckoning. Undergraduate programs are also in the act–and often more innovatively–with special programs designed to attract and serve minority students. Often, students are brought in through special programs for high school students.

Five of these programs were highlighted at a recent one-day conference in Washington, sponsored by the National Science Foundation. Represented were Berkeley, Cal Poly, UT Austin, SUNY, and Xavier, New Orleans.

The Berkeley, Cal Poly, and UT models emphasized the development of a sense of community among selected minority and women students. Under this rubric, selected students form a closely knit "community" of students who may live, take courses, study, and socialize together. Thus these students are, in some sense, set apart from the general population. The idea is to help them overcome any feeling of individual isolation and at the same time receive whatever help they may require in making up deficiencies. The goal is to retain as many students as possible.

The program at the University of Texas is described

in more detail in an accompanying article by Jackie Mc-Caffrey on this page.

Xavier University, which is itself a minority institution, boasts that 47 percent of its students graduate with mathematics and science degrees, compared with the national average of 7 percent. It also takes pride in producing, according to a 1987 report, the second largest (after Howard University) number of black students entering medical school. The list of competitors includes Morehouse, Spelman, Michigan, Berkeley, Northwestern, Yale, and Harvard.

According to Xavier officials, its success is due to hard work by students and firm commitments by faculty. For Xavier, "community" is not crucial since the entire student body is the community. The emphasis here is on disciplined hard work. The no-nonsense approach is tempered by a caring spirit for student success by the faculty and administration. The students know that the university cares. This is not to say that there are not students who fail and even flunk out of school. But all have been given a chance and every bit of help that the institution can offer. Perhaps other institutions interested in getting involved with minority students could profitably emulate this model.

Persons interested in receiving more information about these programs may contact Uri Treisman at Berkeley, J. W. Carmichael of Xavier, Joseph Griswald of City College of New York, Jacqueline McCaffrey of UT Austin, and the SEES (Science Educational Enhancement Services) program at Cal Poly.

Emerging Scholars at UT Austin

J. McCaffrey, UME Trends 3:3 (August 1991)

Cristina Villalobos, Jennifer Ellison, Rey Rivera, and Mati Gaona have just completed their sophomore year in college. James Scott is a senior; Tiffany Maultsby, a rising sophomore. All are undergraduate mathematics majors at the University of Texas at Austin, and, along with a growing number of under-represented minority students at UT, intend to continue their mathematics education beyond the baccalaureate level with a view toward pursuing academic careers. As freshmen, these students, along with some 100 others since fall 1988, participated in the Emerging Scholars Program (ESP), a faculty-sponsored project aimed at increasing the numbers of under-represented minorities and women of all ethnicities excelling in freshman calculus and pursuing careers in mathematics and related disciplines.

Modeled on the Mathematics Workshops created by Uri Treisman at the University of California at Berkeley, the Emerging Scholars Program initially was developed by the UT math department in response to the high rate of failure (approximately 60 percent) among minority students in UT's freshman calculus program. Like the Berkeley program, remedial work is carried out, when necessary, in the context of a mathematically rich program that stresses high performance and provides students with the means to attain it.

At the heart of the Emerging Scholars Program is an intensive discussion section attached to a standard calculus lecture. Typically, calculus students at Austin attend three one-hour lecture classes and two one-hour discussion sections per week. Emerging Scholars attend the same lectures, but, in lieu of the regular discussions, attend three two-hour intensive laboratory sections per week. The typical lecture section contains 120 students, the standard discussion sections about 40. The ESP sections are limited to 24 students.

Each section is led by a hand-picked teaching assistant and by two undergraduates who are ESP alumni. During a typical session, the students work individually and in groups on carefully crafted problems designed to deepen their understanding of the calculus. Some of the problems have been designed to reveal deficiencies in student preparation, which can then be addressed on the spot. Others were created to challenge and stimulate independent and group investigation. In the sections students talk energetically about mathematics, and sometimes about parties and football as well. The atmosphere is welcoming and respects the students' affection for mathematics. Students are encouraged to attend departmental functions, especially colloquium talks by visiting mathematicians. Faculty members frequently visit the ESP sections. The goal is to draw students into the life of the department and to advocate careers in research and teaching.

An adviser from the dean's office is on permanent assignment to the math department to work with ESP students, bringing student services directly into the department. This is supplemented by informal advising. The students regularly consult the undergraduate assistants and each other about such issues as which teachers to take or not to take for a particular course, or just for general advice about the university.

The Emerging Scholars are a diverse group. They enter the University of Texas with math SATs ranging from as low as 420 to as high as 710: the majority are in the top 20 percent of their high school graduating class. Some students have taken AP calculus, others have gone only through pre-calculus. While approximately three-quarters of the students are African American or Chicano, the ESP classroom also includes white students and students of Asian descent. The majority of the students have declared a science or engineering major, but any student with an interest in mathematics is encouraged to participate.

The program was piloted in Fall 1988 with twenty-one participants. Eighteen of these earned grades of A or B in first semester calculus. Most students went on to earn A's in Calculus II. In Fall 1989 the program grew to serve 34 freshmen, 30 of whom received A's and B's, with the majority earning A's. In Fall 1990, 52 students participated. The combined mean calculus GPA for Emerging Scholars, three-quarters of whom were African American and Latino, was 3.39 (on a 4.0 system) as compared to the mean grade of 2.32 earned by non-ESP students attending the same calculus lectures and taking the same exams. Since the program's inception, more than 80 percent of participating students have earned grades of A or B in Calculus I; about 75 percent of the participants (through Spring 1990) earned A's in Calculus II.

Of the 34 students who began the program in Fall 1989, five have changed their major to mathematics and six others have declared mathematics as a second major. Of the five ESP students who entered the university in Fall 1989 planning to pursue mathematics degrees, all are continuing. With only two exceptions, all of the ESP students who have chosen mathematics as a major plan to continue their academic work beyond the baccalaureate level. Interestingly, one African-American woman who is presently completing a major in business intends nevertheless to pursue the Ph.D. in mathematics. For the last two years, the top winners of the prestigious Bennet Examination, a competitive examination for freshmen calculus students at UT, have been Emerging Scholars. This has helped to build substantial support for the program in the department.

Jackie McCaffrey is at The University of Texas, Austin.

Appendix B-III

Undergraduate Education

The reprints in this section provide a sampling of the lively discussion in the mathematics community concerning pedagogy and curriculum, the former often classified as mathematics education reform. Some of these articles were stimulated by the publication of several reports under the auspices of the National Academy of Sciences, in particular the so-called "David Reports" (see David, Chapter 17). A large project at the NAS called the "Mathematical Sciences in the Year 2000" produced three major reports which have occasioned much comment: *Everybody Counts*, *Challenge of Numbers*, and *Moving Beyond Myths*.

Other articles in this section deal with such matters as collegiate mathematics education research, curriculum issues, the "calculus reform" movement, and student advisement and placement testing. A brief sampling reprinted here may encourage the reader to start keeping up with the many varied ideas and teaching methods included in two regularly featured sections of *UME Trends*: Research Sampler and Exchange on Teaching Innovations.

Mathematical Education

W.P. Thurston
Notices Amer. Math. Soc. 37:7 (September 1990)

Mathematics education is in an unacceptable state. Despite much popular attention to this fact, real change is slow.

Policymakers often do not comprehend the nature of mathematics or of mathematics education. The 'reforms' being implemented in different school systems are often in opposite directions. This phenomenon is a sign that what we need is a better understanding of the problems, not just the recognition that they exist and that they are important.

I am optimistic that our nation will find solutions to these problems. Problems arising from failure of understanding are curable. We do not lack for dedication, resources, or intelligence: we lack direction.

Symptoms

There are many symptoms of the problems in mathematics education. The number of undergraduate mathematics majors is about half what it was 15 years ago. The number of U.S. graduate students is less than half what it was 15 years ago, although foreign students have taken up some of the slack. The performance of our students at all levels, as measured by standardized tests, is below that of other industrialized countries.

The typical response of American adults, on meeting a mathematician, is one of dismay. They apologetically recall the last mathematics course they took, which is usually the one where they lost their grip on the subject matter.

Many companies recognize lack of mathematical competence as a major problem in their workforce. Technology has removed the need for elementary mathematical facility in a few jobs, such as ringing up hamburger orders, but it has also eliminated many unskilled jobs, such as assembly-line work, while creating many others requiring considerable mathematical sophistication.

Students in college mathematics courses are unresponsive. They are afraid to speculate and afraid to reach into themselves for ideas. When one visits classrooms at different grade levels, one sees a dramatic decline in liveliness and spontaneity with age. One gets the impression that the natural interest and curiosity of young children in mathematics has been weeded out.

In most places, it is harder to get students to do homework or to study outside the classroom than it was twenty years ago.

Even those college students who have been successful in the high school mathematics curriculum, including calculus, have a narrow base of knowledge of mathematics.

Stratification and Compartmentalization

A major source of problems, and a major barrier to solutions of the problems, is the stratification and compartmentalization of the immense mathematical education enterprise, from kindergarten through graduate school. In particular, there is very little communication between high school and college teachers of mathematics. There is also less than optimal communication between the people who are involved with curricular reform at the college level and research mathematicians. This splintering partly results from the existence of several different professional associations, including the American Mathematical Society (AMS), which primarily represents mathematical research; the Mathematical Association of America (MAA), which primarily represents undergraduate mathematics; and the National Council of Teachers of Mathematics (NCTM), which primarily represents the most committed high school mathematics teachers. Additional organizations represent two-year colleges, applied mathematics, statistics, and operations research, as well as computer science.

The membership of the AMS and the MAA overlap by only about 1/3. Most members of the AMS are not even aware of the existence of the NCTM. Conversely, few members of the NCTM are aware of the AMS. Very few academic mathematicians have any contact with school mathematics teachers except through their children. Most teachers are trained at institutions with few if any research mathematicians. A number of universities are strong both in mathematical research and teacher training, but, even at those universities, most research mathematicians are not involved in teacher training.

Even more severe is the compartmentalization due to the division of primary and secondary education into many individual school districts and schools where the real decisions are made, and the division of college education into many different academic departments where the real decisions are made. Many people are thinking about the problems and making individual efforts to solve them. Many of these local efforts are quite successful. Unfortunately, there is far too little coordination and sharing of the insights gained.

Mathematics is a Tall Subject

One feature of mathematics which requires special care in education is its 'height,' that is, the extent to which concepts build on previous concepts. Reasoning in mathematics can be very clear and certain, and, once a principle is established, it can be relied on. This means it is possible to build conceptual structures which are at once very tall, very reliable, and extremely powerful.

The structure is not like a tree, but more like a scaffolding, with many interconnected supports. Once the scaffolding is solidly in place, it is not hard to build it higher, but it is impossible to build a layer before previous layers are in place.

Difficulties arise because students taking a particular course are in different stages of mastery of the earlier learning. They also tend to be secretive about exactly what they know and what they don't know. For instance, many calculus students don't correctly add fractions, at least not in symbolic form: the typical mistake is that

$$\frac{a}{b} + \frac{c}{d} = \frac{a+c}{b+d} \quad \left(\text{much simpler than } \frac{ad+cb}{bd} \right).$$

However, students feel guilty that they are shaky on addition of fractions and are slow to admit it and ask how and why it works, even to themselves.

Addition of fractions is a very boring topic to someone who already knows it, but it is an essential skill for algebra, which in turn is essential for calculus. It is not so hard, when talking with students individually, to find out what parts of the structure need shoring up and to deal with those parts individually. But it is quite difficult to find a level of teaching which is comprehensible and at the same time interesting to an entire class with heterogeneous background.

Mathematics is a Broad Subject

Mathematics is also very broad. There are many subjects that are never discussed in the mainline curriculum which culminates in calculus. The subjects that are discussed have many interesting side-branches that are never explored.

In my parents' generation (during the 1940s), the standard first college mathematics course was college algebra. Soon afterward, the standard first college course was calculus, until the early 1960s, when calculus became standard for the best high school mathematics students. By now first year calculus has largely migrated to high school in affluent school districts, so that most of the better mathematics and science students at our best universities have already taken calculus before they arrive. At Princeton, for instance, two-thirds of entering students placed out of at least one semester of calculus last year.

The acceleration of the curriculum has had its cost: there has been an accompanying trend to prune away side topics. When I was in high school, for instance, it was standard to study solid geometry and spherical geometry along with plane geometry. These topics have long been abandoned. The shape of the mathematics education of a typical student is tall and spindly. It reaches a certain height above which its base can support no more growth, and there it halts or fails.

These two trends (lengthening and narrowing of school mathematics) have been hastened by the growing reliance on standardized tests. Standardized tests are designed to cover topics on the most standard curriculum: if only half the students study some topic, it is unfair to ask about it on a standardized test. This is not so bad as long as tests are used in a disinterested way as one of several *devices for measurement*. Instead, higher test scores are often treated as the *goal*. Legislators, newspapers and parents put superintendents and school boards under pressure, superintendents and school boards put principals under pressure, principals put teachers under pressure; and teachers put students under pressure to

raise scores. The sad result is that many mathematics courses are specifically designed to raise scores on some standardized test.

We don't diagnose pneumonia with only a thermometer, and we don't attempt to cure it by putting ice in a patient's mouth. We should take a similarly enlightened attitude toward testing in mathematics education.

The long-range objectives of mathematics education would be better served if the tall shape of mathematics were de-emphasized, by moving away from a standard sequence to a more diversified curriculum with more topics that start closer to the ground. There have been some trends in this direction, such as courses in finite mathematics and in probability, but there is room for much more.

Mathematics is Intuitive and Real

Students commonly lose touch with the reality and the intuitive nature of mathematics. From kindergarten through high school, they often have teachers who are uncomfortable with anything off the beaten path. Young children come up with many ingenious devices to work out mathematical questions, but teachers usually discourage any nonconventional approach—partly because it is not easy to understand what a child is thinking or trying to say, and the teachers don't catch on, partly because the teachers think it's not okay to use an alternative method or explanation.

By the time students are in college, they are inhibited from thinking for themselves and from admitting out loud what they are thinking. Instead, they try to figure out what routines they are supposed to learn. When there is any departure in class from the syllabus or the text, someone invariably asks whether it's going to be on the test.

Unless mathematics makes a real connection to people, they are unlikely ever to think about it or use it once they have completed a course.

Precocity and Competition

Along with the emphasis on tests has come an emphasis on precocity and acceleration in mathematics. It is relatively easy for a bright student to work through the mathematics curriculum far more quickly than the usual pace.

There are several problems associated with precocity. People who skip ahead in the curriculum often have gaps in their background which only show up later. At that point, the person may be too embarrassed to admit the gap and tries to fake understanding. This regularly leads to disastrous results.

Another problem is that precocious students get the idea that the reward is in being 'ahead' of others in the same age group, rather than in the quality of learning and thinking. With a lifetime to learn, this is a short-sighted attitude. By the time they are 25 or 30, they are judged not by precociousness but on the quality of work. It is often a big letdown to precocious students when others who are talented but not so precocious catch up, and they become one among many. The problem is compounded by parents in affluent school districts who often push their children to advance as quickly as possible through the curriculum, before they are really ready.

A third problem associated with precociousness is the social problem. Younger students are often well able to handle mathematics classes intellectually without being able to fit in socially with the group of students taking them.

Related to precociousness is the popular tendency to think of mathematics as a race or as an athletic competition. There are widespread high school math leagues: teams from regional high schools meet periodically and are given several problems, with an hour or so to solve them.

There are also state, national and international competitions. These competitions are fun, interesting, and educationally effective for the people who are successful in them. But they also have a downside. The competitions reinforce the notion that either you 'have good math genes', or you do not. They put an emphasis on being quick, at the expense of being deep and thoughtful. They emphasize questions which are puzzles with some hidden trick, rather than more realistic problems where a systematic and persistent approach is important.

This discourages many people who are not as quick or as practiced, but might be good at working through problems when they have the time to think through them. Some of the best performers on the contests do become good mathematicians, but there are also many top mathematicians who were not so good on contest math. Quickness is helpful in mathematics, but it is only one of the qualities which is helpful. For people who do not become mathematicians, the skills of contest math are probably even less relevant.

These contests are a bit like spelling bees. There is some connection between good spelling and good writing, but the winner of the state spelling bee does not necessarily have the talent to become a good writer, and some fine writers are not good spellers. If there was a popular confusion between good spelling and good writing, many potential writers would be unnecessarily discouraged.

I think the answer to these problems is to build a system which exploits the breadth of mathematics, by allowing quicker students to work through the material in greater depth and to take excursions into related topics, before racing ahead of their age group.

Mystery and Mastery

Mathematics is amazingly compressible: you may struggle a long time, step by step, to work through some process or idea from several approaches. But once you really understand it and have the mental perspective to see it as a whole, there is often a tremendous mental compression. You can file it away, recall it quickly and completely when you need it, and use it as just one step in some other mental process. The insight that goes with this compression is one of the real joys of mathematics.

After mastering mathematical concepts, even after great effort, it becomes very hard to put oneself back in the frame of mind of someone to whom they are mysterious.

I remember as a child, in fifth grade, coming to the amazing (to me) realization that the answer to 134 divided by 29 is 134/29 (and so forth). What a tremendous labor-saving device! To me, '134 divided by 29' meant a certain tedious chore, while 134/29 was an object with no implicit work. I went excitedly to my father to explain my major discovery. He told me that of course this is so, a/b and a divided by b are just synonyms. To him it was just a small variation in notation.

One of my students wrote about visiting an elementary school and being asked to tutor a child in subtracting fractions. He was startled and sobered to see how much is involved in learning this skill for the first time, a skill which had condensed to a triviality in his mind.

Mathematics is full of this kind of thing, on all levels. It never stops.

The hard-earned and powerful tools which are available almost unconsciously to mathematicians, but not to students, make it hard for mathematicians to learn from their students. This puts a psychological barrier in the way of listening fully to students.

It is important in teaching mathematics to work hard to overcome this barrier and to get out of the way enough to give students the chance to work things out for themselves.

Competence and Intimidation

Similarly, students at more advanced levels know many things which less advanced students don't yet know. It is very intimidating to hear others casually toss around words and phrases as if any educated person should know them, when you haven't the foggiest idea what they're talking about. Less advanced students have trouble realizing that they will (or would) also learn these theories and their associated vocabulary readily when the time comes and afterwards use them casually and naturally. I remember many occasions when I felt intimidated by mathematical words and concepts before I understood them: negative, decimal, long division, infinity, algebra, variable, equation, calculus, integration, differentiation, manifold, vector, tensor, sheaf, spectrum, *etc*. It took me a long time before I caught on to the pattern and developed some immunity.

Teachers also are frequently intimidated about mathematics. High school teachers are often timid about approaching college and university teachers. They also question, with some justice, whether university professors are in touch with the problems they have to deal with. There is so little general contact between those who teach mathematics in high schools, colleges, and universities that few professors know much about the educational problems in high school or elementary school. Elementary school teachers are often quite insecure about their grasp of mathematics and timid about approaching anyone.

The Problems and Solutions are the Same

There is much in common about the problems—and the solutions—of mathematics education at different levels, from kindergarten through graduate school. The failure to communicate is a real loss, and the potential gain from opening of two-way communications is great.

Over the past two years, I have met many people involved in mathematical education at all levels, partly as a member of the MSEB (Mathematical Sciences Education Board), a national board for mathematics education, composed of people from diverse backgrounds in mathematics and education, including a number of teachers. I have learned a lot.

During the spring semester of 1990, John Conway, Peter Doyle, and I organized a new course ('Geometry and the Imagination') at Princeton which borrowed a good deal from the ideas I learned. We taught as a team, shunning lectures and emphasizing group discussions among students. We emphasized manipulables, cooperative learning, and problem solving. We asked students to keep journals and to write out their ideas in good and complete English. Each student did a major project for the course. These ideas are all borrowed from ideas current in K-12 mathematics education, focused in the NCTM curriculum standards.

To culminate, we held a 'geometry fair': like a science fair but with a popcorn machine and without prizes. It was great fun.

The course came alive, qualitatively more than any course we had taught before. Students learned a lot of mathematics and solved problems we wouldn't have dared ask in a conventional college class.

One topic we discussed was mathematical education. Students were given an earlier draft of this essay and wrote 70 thoughtful essays based on their own experiences. This essay has benefitted considerably from their comments.

Socially Excluded Groups

Why do so few women become mathematicians, and so few of the non-Asian minorities?

I am convinced that the poverty of the school mathematics teaching and curriculum has a lot to do with it. The way mathematics is taught in school does not address the real goals of a mathematical education. It is very hard to get a sense of the depth, liveliness, power, and breadth of mathematics from any ordinary experience with mathematics in school. I believe that most students who really master the subject matter, and eventually become scientists, mathematicians, computer programmers, *etc.*, are those who have some other channel for learning mathematics, outside the classroom. Some-

times it is the home, sometimes it is books, sometimes it is an unusual teacher, but often it is the 'nerd' social subgroup in school. When I was in high school, I certainly belonged to this subgroup (although the name 'nerd' was not yet current), and I appreciated it very much. However, it is a very different matter for white and Asian males to join this social subgroup than for women, Hispanics or blacks.

Since channels of learning outside the classroom are currently dominant in mathematics education for the top group of students, improvement of the general quality of mathematics within the classroom therefore should act as an equalizer, particularly helping blacks, Hispanics, and women. For people who are not already tuned in to the mathematical style of discourse, it is especially important to teach in a way that is not watered down, but that begins from a person's real experience.

Intimidation also has a lot to do with it. The emphasis on precocity, high test scores, and competition work to amplify the small differences which arise from other sources.

Goals and Standardized Tests

What is mathematics education good for?

Mathematics in life. First, mathematics is a basic tool of everyday life. When coffee comes in 13 oz. packages and 16 oz. cans, can you take that information in stride (walking slowly by the shelf), along with the prices and the prominent signs claiming 'contains 23% more' on the cans, to help decide which you'd prefer to buy? In buying a new car with various gimmicks in financing, rebates, and features, do you understand what is going on? If most people did, the gimmicks would be pointless.

Second, mathematical reasoning is an important part of informed citizenship. Can you understand the reasoning behind studies of health risks from various substances, and can you judge how important they are? When listening to politicians, can you and do you use your reckoning powers to help decide how important some statistic is, and what it means? Can you measure and calculate adequately for simple sewing and carpentry? Can you plan a budget? When you see graphs in newspapers and magazines, do you understand what they mean, and are you aware of the several devices frequently used either to dramatize or to play down a certain trend?

Third, mathematics is a tool needed for many jobs in the infrastructure of our increasingly complex and technological society. These uses are pervasive and varied. The dental technician, the fax repairperson, the fast food manager, the real estate agent, the computer consultant, the bookkeeper and the banker, the nurse and the lawyer, all need a certain proficiency with mathematics in their jobs.

Fourth, mathematics is intensively used (and sometimes abused) in most branches of science. Much of theoretical science really *is* mathematics. Statistics is one of the most common uses of mathematics. Many scientists use the widespread computerized statistical packages, which alleviate the need for computation. However, people who use statistical packages are often shaky in their understanding of the basic principles involved and often apply statistical tests or graphical displays inappropriately.

Mathematics is alive. To me, these utilitarian goals are important, but secondary. Mathematics has a remarkable beauty, power, and coherence, more than we could have ever expected. It is always changing, as we turn new corners and discover new delights and unexpected connections with old familiar grounds. The changes are rapid, because of the solidity of the kind of reasoning involved in mathematics.

Mathematics is like a flight of fancy, but one in which the fanciful turns out to be real and to have been present all along. Doing mathematics has the feel of fanciful invention, but it is really a process of sharpening our perception so that we discover patterns that are everywhere around. In his famous apology for mathematics, G.H. Hardy praised number theory for its purity, its abstraction, and the self-evident impossibility of ever putting it to practical use. Now this very subject is applied very widely, particularly for encoding and decoding communications.

My experience as a mathematician has convinced me that the aesthetic goals and the utilitarian goals for mathematics turn out, in the end, to be quite close. Our aesthetic instincts draw us to mathematics of a certain depth and connectivity. The very depth and beauty of the patterns makes them likely to be manifested, in unexpected ways, in other parts of mathematics, science, and the world.

To share in the delight and the intellectual experience of mathematics—to fly where before we walked—that is the goal of a mathematical education.

Testing and "accountability". Unfortunately the goals of school mathematics have become incredibly narrow, much narrower even than the first set of utilitarian goals listed above, let alone the others. It is popular lately for politicians and the public to demand "accountability" from the school systems. This would be great, except that educational accounting is usually based on narrowly-focused multiple-choice tests.

It is as if students were considered to have mastered Shakespeare if they could pass a timed vocabulary test in Elizabethan English, or that they had learned to write when they could correctly choose the grammatical form of a sentence from four possibilities.

The state and regional boards of education these days hand out a laundry list of skills which students are supposed to know at a certain age, rather than a curriculum: horizontal addition versus vertical addition, addition of 2 digit numbers to 2 digit numbers with a 2 digit answer versus addition of 2 digit numbers to 2 digit numbers

with a 3 digit answer, *etc.*

A front-page article in the New York Times Metropolitan section of July 24, 1990 contrasted the elementary schools in two similar difficult neighborhoods of Brooklyn. The first was a 'successful' school, with two reading lessons a day, the second lesson in 'test-taking skills' and practice for the standardized reading test. In this school, 80.5% scored at or above grade level. The other school was an 'unsuccessful' school where they prepare for the test for 'only' 3 months. In that school, 36.4% score at or above grade level.

The reporter cited an example of how the principal sets the tone in the 'successful' school:

> She is not satisfied with just the right answers; she wants the right steps along the way. In one fourth-grade class, she noticed that pigtailed Keanda Snagg had made a wild, though accurate, stab at a problem requiring her to average which of two stores had lower prices. "It looks like you were going to do it without doing the work," she told Keanda. As she watched Keanda go through the calculations, she stressed fundamentals beyond arithmetic, like putting numbers in clearly ordered columns.

I can't tell which school is actually more successful without seeing them for myself, but one thing I know: neither the test scores nor the cited incident are demonstrations of greater success.

Thinking and Rote

Narrow goals are stultifying.

People are much smarter when they can use their full intellect and when they can relate what they are learning to situations or phenomena which are real to them.

The natural reaction, when someone is having trouble understanding what you are explaining, is to break up the explanation into smaller pieces and explain the pieces one by one. This tends not to work, so you back up even further and fill in even more details.

But human minds do not work like computers: it is harder, not easier, to understand something broken down into all the precise little rules than to grasp it as a whole. It is very hard for a person to read a computer assembly language program and figure out what it is about. A computer reads and executes it in the blink of an eye. But the most powerful computer in the world is not clever enough to drive a car safely, or control a stroll along the sidewalk, or come up with an interesting mathematical discovery.

Studying mathematics one rule at a time is like studying a language by first memorizing the vocabulary and the detailed linguistic rules, then building phrases and sentences, and only afterwards learning to read, write, and converse. Native speakers of a language are not aware of the linguistic rules: they assimilate the language by focusing on a higher level, and absorbing the rules and patterns subconsciously. The rules and patterns are much harder for people to learn explicitly than is the language itself. In fact, the tremendous and so far unsuccessful attempts to teach languages to computers demonstrate that nobody can yet describe a language adequately by precise rules.

It is better not to teach a topic at all than to attempt teaching it in tiny rules and bits.

Answers and Questions

People appreciate and catch on to a mathematical theory much better after they have first grappled for themselves with the questions the theory is designed to answer.

There is a natural tendency, in teaching mathematics, to use the logical order and to explain all the techniques and answers before bringing up the examples and the questions, on the supposition that the students will be equipped with all the techniques necessary to answer them when they arise.

It is better to keep interesting unanswered questions and unexplained examples in the air, whether or not the students, the teachers, or anybody is yet ready to answer them. The best psychological order for a subject in mathematics is often quite different from the most efficient logical order.

As mathematicians, we know that there will never be an end to unanswered questions. In contrast, students generally perceive mathematics as something which is already cut and dried—they have just not gotten very far in digesting it.

We should present mathematics to our students in a way which is at once more interesting and more like the real situations where students will encounter it in their lives—with no guaranteed answer.

What Can We Do?

This depends on who we are.

In our compartmentalized system, it is hard for a single organization or individual to do very much to affect the overall system directly. But addressing the local situation will indirectly influence the global situation. I will address the question from the point of view of college and university mathematicians.

First, college and university mathematics departments should develop courses which can give students a fresh chance in mathematics. Remedial courses are widespread, but their success is limited: going over the same material one more time is tedious and boring, whether you understood it the first time or you didn't. There is a built in handicap to enthusiasm and spontaneity.

Instead, there should be more courses available to freshmen and nonmajors which exploit some of the breadth of mathematics, to permit starting near the ground level without a lot of repetition of topics that students have already heard. For instance, elementary courses in topology, number theory, symmetry and group theory, probability, finite mathematics, algebraic

geometry, dynamical systems (chaos), computer graphics and linear algebra, projective geometry and perspective drawing, hyperbolic geometry, and mathematical logic can meet this criterion.

Second, we should work to create better channels of communication between the compartments of the educational system. We need to find devices so that the educational accomplishments of professors are visible within the profession, not merely within the classroom or within the department. We need to find vehicles for exchange of interesting ideas between different departments: for instance, exchange visits between directors of undergraduate studies and chairs of departments. We should visit each other's classrooms. We need more talks and special sessions related to education at our professional meetings, and more prizes for educational accomplishments.

The newsletter, *Undergraduate Mathematics Education Trends*, and the MAA newsletter, *Focus*, are two such vehicles, and there are several other publications which carry articles on undergraduate education. The annual chairman's colloquium organized by the Board on Mathematical Sciences of the National Research Council is another vehicle of communication, although it has a broader agenda than education. Still, these existing channels are very small compared with what we could establish.

Even more important and more difficult is the creation of channels for communication between the strata: most important for colleges and universities is communication between high school, college and university mathematics departments. This communication must be two-way: college and university professors can learn a lot about how to teach from school teachers. The MSEB is one such channel, along with the state mathematics coalitions that they have helped to stimulate, but what we need is a much more massive exchange.

How can the senior professors, who are at the top of a system which is clearly not doing such a great job, presume to teach their juniors how to do better? The graduate students and the junior faculty often do a better job at teaching mathematics than the senior faculty, who have sometimes become resigned to the dismal situation, settled into a routine, and given up on trying any new initiatives. Even when they do a pretty good job in their own classrooms, against the odds, they do not usually get involved in improving the overall system.

Often other professors are suspicious of the professor who does take an interest in education. They tend to assume that research is the only activity which really matters and that turning to education is a sign of failure in research. Senior professors sometimes explicitly advise junior faculty not to waste too much energy on teaching, or they will never be promoted.

We must recognize that there are many different ways that we can make important contributions to society and to our institutions. It is dumb to measure mathematicians against the single scale of research. Education is an important and challenging endeavor, which many people engage in by choice, not necessity. We should judge them by what they accomplish, not by what they might have accomplished if they spent their time and energy elsewhere.

What urgently needs to change is the system of professional rewards. We need something better than the current situation within university mathematics departments where there is lip service to the importance of teaching, but, when it comes to the crunch of hiring and tenure decisions, teaching and service count only in the marginal cases where the candidates cannot be differentiated by the quality of research.

People are socially motivated. As we discuss education with each other, we put more energy into it, and it becomes more important to us. The academic culture *can* change, and it has changed. The process of change is mostly an informal one (what you talk about at lunch), not controlled by organizational decisions. But when the time is ripe, as I believe it is now for mathematical education, a little nudging by organizations can help stimulate a huge change.

The needed reforms will take place through collegial, cooperative efforts. Good mathematical ideas spread very rapidly through informal channels in the mathematical community. As we turn more of our attention to education, good educational ideas will also spread rapidly.

William P. Thurston is professor of mathematics at Princeton University. A Fields medalist and internationally-known researcher in topology and geometry, he has in recent years also become interested in issues in mathematics education. He currently serves on the Mathematical Sciences Education Board of the National Research Council and on the Executive Committee of the AMS Council. Thurston has been a judge in the Westinghouse Science Talent Search, and, in June of this year, he made a presentation at a workshop organized by the Mathematicians and Education Reform Network. He also serves as mathematics editor-in-chief for Quantum, a mathematics and science magazine for young people. Thurston says: "The [following] essay was stimulated by a study on the teaching scholarship connection, sponsored by the Pew Charitable trusts, on how graduate education and junior faculty 'apprenticeship' does or doesn't prepare us to be teacher-scholars. I was the mathematician on the panel, and the contrast between the different fields was brought home to me."

Can Mathematicians Be Involved in the Profession?

H.B. Keynes
Notices Amer. Math. Soc. 37:4 (April 1990)

I. The Issues. Mathematics education is undergoing intensive examination at all levels, and major reforms in

curriculum and direction are now in the process of being implemented. The need for mathematicians to be extensively involved in these endeavors has been widely recognized. A few have organized major innovative projects in teacher training, networking for curricular reform, and innovative programs for underrepresented groups and gifted students. Yet for the most part, mathematicians are ignoring these educational opportunities, and avoiding any real level of involvement. Despite excellent educational funding opportunities at a time when research support is declining, and constant pleas for increased involvement from university, government, and business leaders, academic mathematicians are clearly questioning whether their participation can be part of their professional activities. We need to examine and question some of the reasons for their hesitancy and change many of the perceptions if we hope to raise the educational involvement of the mathematics community.

At the heart of the issue is the belief that educational issues, concerns about teaching, and involvement in school curriculum are really not matters that are professionally attractive to most university and college faculty. One program officer at NSF described the typical reaction from chairs and senior faculty at *all* levels of state universities to be that while education and teaching are important for faculty at schools which should emphasize undergraduate education, their own school is primarily interested in research. This attitude is even seeping into some of the best small colleges, where emphasis on providing undergraduate research opportunities and competition for faculty have led to increased interest in faculty research, especially for younger faculty. So virtually all of the leading universities and colleges dismiss involvement in K–12 mathematics education in the name of various research activities. The deeply held conviction that research interests and educational involvement are incompatible continually permeates discussions on these matters. In fact, most mathematicians generally believe that involvement in educational activities at any level is a sign of disinterest or incapability in research activities. All too often, a faculty member who has lost interest in research and scholarship activities is described as automatically being interested in teaching and education. Even the few mathematicians with significant educational interests and research activities frequently feel the stigma of being viewed as less dedicated and valuable than colleagues with only research activities. We must examine ways to change this attitude of our colleagues and peers if we expect to see more mathematicians considering educational involvement.

Closely related to this view of education is the perception (and in most cases the reality) of the reward system in our profession. In discussions with younger full professors who may be more willing to consider educational involvement, their opinion of the reward system in mathematics is a major deterrent to their participation. They believe that even significant and highly successful efforts in education will not provide merit salary increases and other professional respect and opportunities. Despite lip service to the contrary, this is the reality in many departments. In a recent essay on mathematics education, Bill Thurston states "What needs to change urgently is not so much the system of advice, of teacher training or of mentoring, as the system of professional rewards." In a few departments where such activities are rewarded, chairmen need to keep these considerations virtually secret for fear of disapproval among the senior staff. Some highly respected mathematicians would find it unacceptable if anything other than a theorem or an outside offer was to play a role in a merit increase. Various presidents, provosts and deans publicly lament how faculty are not involved in educational activities, yet seem to be unwilling to take on the conventional wisdom and advocate changes in the reward system.

So the impediments to more involvement in education are clear: professional respect among colleagues and peers, and professional rewards. Since the forces to change these issues reside primarily in our mathematics community, we need to address ways to influence our departments and professional societies.

*2. **What Can We Do?*** The key to any lasting reform will be changes in professional attitudes of most mathematicians. This does not mean that very many will change their professional goals or their personal value systems. What it does mean is that professional tolerance for colleagues who want to become involved in education should be encouraged at all levels in our community, and professional rewards for high quality, successful and significant activities become embedded in the system. An important aspect is not to trivialize a good educational project as any non-routine activity outside of research. The efforts required to organize and implement a multi-year educational project can be enormous and the chances for no real success can be quite high. Education is unlike traditional mathematical research, and more like many experimental sciences. It is frequently not the conception of the project, but the enormous organizational and administrative efforts, together with sensitivity to classroom cultures in the context of mathematics content, that makes for successful educational programs. While these efforts may be intellectually less challenging than standard research, the emotional, physical and organizational strain are usually far more intense. We need to respect these levels of involvement, and reward those faculty that meet high standards similar to those of good research.

An important aspect of providing high-quality professional involvement in mathematics education is to encourage participation when personal circumstances are right. Many active researchers might find a particular educational program appealing, and wish to be involved for a short period or even a few years. Because of fresh ideas and higher levels of enthusiasm, the involvement of research mathematicians is critical to both new and established programs. Many times, a reasonable research program is maintained during this initial involvement,

and in some cases, a new professional balance between research and education is reached. We should encourage these elective shifts of emphasis. In the context of maintaining an appropriate research and scholarly program for their department, younger faculty could be encouraged to have some exposure to educational activities. There are sufficient numbers of senior mathematicians who are re-examining their professional directions so that without really compromising tenure and promotion standards, significant educational activities could still take place. For certain senior mathematics faculty, professional respect for educational issues might encourage them to seek new and productive involvements in education when their research activities have peaked. With the current demographics of the community, this could provide a growing source of mathematicians to meet the escalating needs in education.

Which organizations are best equipped to influence these changes? Because the heart of the problem resides in the attitudes of research mathematicians, our research societies must play a leading role. The AMS is seriously beginning to examine its role in education. Discussions at the 1990 meeting in Louisville could set the tone for a significant official level of recognition of educational activities. SIAM has also maintained a serious interest in education, and would likely support any new AMS position. The MAA has always maintained its own educational agenda, and their excellent ongoing efforts would clearly support and supplement any changes within AMS and SIAM. Finally, the Mathematical Sciences Education Board (MSEB) and the NSF-funded Mathematicians and Education Reform Network (MER) can play appropriate advocacy roles for shifts in educational interests within the mathematics community. But the real change must take place within the rank-and-file of the mathematics profession, most specifically among the academic research mathematicians. This will clearly be a long-term project, with many dissenting voices and calls for historical standards. Such debate is healthy, and perhaps should take place in our professional journals and at professional meetings. Some courageous department heads and deans should take the initial steps in changing the reward systems at their schools, and begin the process of professionally integrating educational activities into the life of their mathematics departments. Most importantly, we need the research leaders in our community to actively support these new roles in mathematics education, and be willing to challenge other leaders with less tolerance.

*3. **Some Final Observations.*** Especially in academic environments, mathematicians have historically juggled two aspects of their professional lives - their personal esthetics and research interests on one side, and their educational and service components on the other side. The balance is always shifting, but in the post-Sputnik years moved heavily towards the research end. We currently have a culture that is generally comfortable with this role, and has built a professional structure to support it. But recent changes in our society now question the wisdom of the mathematical community to continue to accept the current balance. Among these is the widespread recognition that the need to reform the K-12 mathematics curriculum necessitates the involvement of mathematicians at all levels to help develop alternate approaches and curriculum. Undergraduate and even graduate mathematics curriculum is also under careful review and revision. Because of the central role of mathematics in our changing society, mathematics and mathematicians are now linked to two central concerns of our society - economic competitiveness and human resources. Future resource allocations to all sciences, including mathematics, will probably be heavily dependent on the country's perception of how effectively they contribute to the improvements of these essential areas. To insure the vitality of our profession and to maintain attractive professional rewards, we as a community need to shift the professional balance at this time and re-affirm our interest in educational activities. We need to electively make this change within our own professional culture, and truly support our peers and colleagues who engage in high quality and exciting educational programs.

University of Minnesota

More on Researchers and Education

H. Rossi
Notices Amer. Math. Soc. 37:6 (July/August 1990)

I read with interest and enthusiasm the articles in the first FORUM (*Notices*, April 1990). The question, *What is the role of mathematical researchers in mathematical educational reform?* is now being asked at all major conferences and meetings, inspired, I suppose, by the insistence of the leadership of the National Science Foundation (NSF) some years ago that there just has to be some such role. Election to the Council last year drew me from the relative peace of science in Utah and dropped me in the AMS' Committee on Long Range Planning which is now asking me the same question. As I must now answer it, I choose to do so with a contribution to the FORUM.

There is a role for research in education reform. Drs. Herb Clemens and Harvey Keynes are right; but I disagree with them. Herb's plan for direct involvement of research mathematicians in grade school education is a credible and creditable one. Where this was done with imagination, ingenuity and patience (and Herb excels with all three) it has been exceedingly successful. And it is best to get to them in grade school before *their interest and excitement yields to other compelling influences*. However, I do not believe that the central issue is the preservation of the species. Harvey argues, and (as a Dean) I support him, that attitudes must change: *professional tolerance for colleagues who want to become*

involved in education should be encouraged at all levels. But I find fault with what he says later: *It is frequently not the conception of the project, but the enormous organizational and administrative efforts, together with sensitivity to classroom cultures in the context of mathematics content, that makes for successful educational programs.* While I agree with him, I do not see this as the business of the research mathematician who might best be left out of those efforts.

What then is the central issue, and how can the research mathematician be useful in addressing it? Before giving you my response, let me confess that I have been unfair to the writers of April's FORUM articles. For the purpose of my own axe-grinding, I have attributed to them interpretations of their writing to which I know they do not ascribe. Dr. Clemens is not saying that cultivation of future mathematicians is the central issue of educational reform; he is saying that it is an issue, and the one to which research active mathematicians can address themselves without compromising that activity. Dr. Keynes is not saying that research mathematicians should applaud, support, and even become, those among us who are prone to meddle around in education as if we knew better. Through his own work, Dr. Keynes has amply demonstrated his very clear conception of the role of the researcher in educational reform, and he makes us understand that the projects to which he refers require the combined effort of educators and professionals at all levels, each playing their appropriate role. What then is that role of the researcher in education, and how is it to be implemented, and in what ways can the AMS be involved?

The central feature of research is that the worker must live, intellectually, in a highly ambiguous environment. Problems are not really well formulated, the data is either insufficient, buried in irrelevancies, or both, approaches are obscured, and outcomes, although guessed at, are highly uncertain. To put it bluntly, most of the time we feel stupid. The issue in education for the researcher is that the purpose of our educational system seems to be to drive out, at the earliest age possible, any ability to tolerate the feeling of stupidity. This is an observed phenomenon: men and women entering college do not display the same eager inquisitiveness, curiosity and tolerance of their own intellectual innocence which is displayed by boys and girls entering first grade. It is not my purpose to go into the causes of this although I believe that the intellectual standards of our culture play a significant role.

Why don't girls and minorities do as well in mathematics as white boys? First of all, although this is a popular question, the assumption underlying it is mistaken. The correct observation is that girls and minorities do as well as boys at the same educational level, but drop out of the math/science curriculum at markedly higher rates. Why? Because being unsure of yourself doesn't look good. The difference between underrepresented and overrepresented groups is that of acceptance; if you are a member of the majority, it is automatic; if not, it has to be achieved. These are obvious but not trivial statements. The discomfort of intellectual uncertainty is very difficult to tolerate, all the more if you are on the outside looking in. If you are accepted, it's not so bad to take risks; recovery (often by bluff) comes easy. If you are aspiring for acceptance, risks and displays of uncertainty and weakness are to be avoided, and attempts at bluff are disastrous.

What is the role of the researcher in education? To turn things around; insist that it's not only OK to live with uncertainties, but that it is essential to progress. Insist that having to work in ambiguous situations is not restricted to research mathematics, or the academic environment, but is a reality throughout the professions and today's workplace wherever it is. Instruct that there is more value in the thrill of the hunt than the size of the trophy; help create endless illustrations of this at all educational levels, and above all, crusade to get this program adopted. (We have had several years now of pilot and model educational projects, some of which are exceptional. There is less need today for pilots than there is for stewards.)

How is the role played out? Clemens and Keynes have two excellent and important ideas: 1) stop being so smart, go out and be stupid for the kids, show them what it is about scientific research that attracts you; 2) struggle to change our academic culture so that not only do we notice that there are students around, but we know how to shepherd them through intellectual challenges. These ideas are important, not only because they address the problems directly, but because they address the culture; they make a statement.

As I have unfairly paraphrased the content of other articles, I have unfairly represented my own thesis. I can do a better job of expressing my ideas only by citing examples which illustrate them in operation. (First suggestion: let that last sentence be our guide in teaching).

Phil Wagreich, University of Illinois at Chicago, and coworkers have developed a sequence of "experiments" (basically elementary qualitative physics) for primary school children. They have worked directly with the teachers, using illustrative experiments at their level as well, so that the teachers themselves get to experience the thrill of uncertainty before introducing it into the grade school environment. This program has had remarkable success with the disadvantaged student in the Chicago area. Here, in the state of Utah, some of us are hoping to move part of the grade school curriculum away from workbooks by similar techniques. We have begun to train "teacher-leaders" for the school who can instruct their colleagues in experimentation and provide tools for it. I am teaching calculus to a small group as part of this program. No, I'm *not* teaching calculus; I am showing them how to analyze data graphically, that dynamical processes are understood by formulating the rules of change, how to test hypotheses by means of specific data and spreadsheet manipulations. Last week we worked on inhibited growth, and by studying the US census from 1790 deduced that the US population looks like it will stabilize toward the end of the next century at around 299 million.

The NSF programs, *Young Scholars* and *Research Experiences for Undergraduates*, are superb programs from this point of view. Although they are motivated by preservation of the species and directed toward the most promising young scholars, they have profound effects in other ways.

First of all, we the researchers are introduced to the idea that there is something to the way we do our work which must be, and *can be*, communicated at all educational levels. Secondly, these students become conduits of scientific method to their colleagues. In effect then, although it doesn't say so in the NSF announcements, these programs are direct attacks on the national culture.

What is the role of the AMS in mathematics education? If I am right, that the role of the researcher is to insure by illustration that the technique and ideas of the process infect the entire educational enterprise, then the role of the AMS is clear: it must explain to the schools, the public, the government and, above all, *its own members* what mathematics research is doing, how and why. It must do so in its own meetings, its publications, and through its interaction with other societies. We need expository journals, conferences and panel discussions. Above all, our research leaders need to feel in their joints and eyesockets the absolute necessity of explaining to that world out there, the one we live in, what they are doing. The age of the mathematician who declares, "I do what I do because I do it" is over.

Do I have a good idea? Will it work? Do they do it in Japan and Korea? I don't know the answers to the first two questions, but the answer to the last is: *No*. What they do in Japan and Korea is endless workbooks, drill and exercise. They do require everybody to take a lot more mathematics, but their technique is about the same as ours; their classes are as full, and their texts as dull. Well then, why should we do something different? Because what we (and they) are doing isn't working here. The Japanese culture (as well as the Jewish, Chinese and other cultures which produce disproportionately large percentages of research scientists) has a deep respect for inquisitiveness, contemplation and intellectual probing; it is built into their religion, their society and their education. We have no such asset, and we're not going to get it by paying for it or proclaiming an education Presidency. All we can do is give the kids a viable alternative to the press of instant gratification prevalent in our culture, and let them run with it.

University of Utah

Comments

J. Ewing
MAA Monthly 99:6 (June/July 1992)

An associate dean (a member of the English Department) recently began an interview with a young job candidate with a short speech. "Everyone knows the teaching of mathematics is a disgrace," he said. "What are *you* going to do about it?" At a party, a member of the Biology Department walked up to me, introduced himself, and began his conversation with a question. "Why is it that no one can teach in the Mathematics Department?" Not long ago, an acquaintance in the Education School called me on the phone: "Why are all our students failing your courses?" he demanded. "Can't you *do* something about teaching in your department?"

All these people are convinced that there is a crisis in Mathematics Education, that mathematicians are a sorry lot in the classroom, and that the scoundrels aren't doing much to fix things. And where did they learn all this? By listening to us. They read that "innovations in undergraduate teaching lag far behind advances in research" and that "both in instructional methodology and in curricular content, undergraduate mathematics is far below what it should be..." They read that "interest in teaching college mathematics has declined significantly at both undergraduate and graduate levels." And they read that the consequence "is a dysfunctional system of undergraduate mathematics beset on all sides by inadequacies and deficiencies..." They read all this in *Moving Beyond Myths*, a recent report from our community.

Enough! I know, I know, some of my colleagues are not always conscientious teachers; but many others are creative and able instructors at every level. These are people who care about students and think about the courses they teach. I know, I know, much can be improved in the curriculum; but there is also much to recommend a curriculum that has some historical roots. Those roots help us to set standards and to compare one generation to the next. And I know, I know, we ought to experiment with new and innovative learning techniques (I think that's what "instructional methodology" means); but many of my colleagues *already* experiment with courses, and indeed *like* to teach new courses in new ways rather than the same stale course year after year.

Should we be satisfied with teaching in mathematics? Of course not. But we ought to realize that the problems of education go far deeper than flawed instructional methodologies and curricula. That dean who complained about mathematics teaching resides in the English Department, where 84% of the grades are A's and B's; the figure in Mathematics is 47%. The Biology professor teaches only students who choose his courses as electives. And the Education School friend? Most of his colleagues in the Education School *want* to use mathematics as a screening device for their students. The Education School awards 87% A's and B's.

Mathematics, like most disciplines, has poor teachers; it also has some great ones, and lots of people in between. We can do better; we *are* doing better. Let's experiment and innovate and be creative teachers. But let's not exaggerate our problems. Mathematicians have a reputation for honesty. When we go about wringing our hands and moaning about the dismal state of mathematics teaching, people begin to believe us.

Some Thoughts on "Everybody Counts"

E.G. Effros
Notices Amer. Math. Soc. 37:5 (May/June 1990)

In a recent letter to *Notices*, Eleanor Palais eloquently expressed her doubts regarding the latest "Everybody Counts" movement for reforming mathematical education. Specifically she asked:

> Why hasn't the focus of 'Why are the students' scores so poor?' been more on the students themselves, rather than on the schools, the teachers, and the curriculum?

Ms. Palais' letter touches upon concerns that many of my colleagues and I have regarding the National Research Council (NRC) publications "Everybody Counts" and "Reshaping School Mathematics", and the National Council of Teachers of Mathematics (NCTM) "Curriculum and Evaluation Standards for School Mathematics". These reports have failed to relate the sorry state of mathematical education to the general context of educational failure in the United States. Instead, the authors have attempted to remedy the problem with a *livelier* mathematical curriculum and with a much greater emphasis on calculators and computers. In this article I will first address certain aspects of the proposed curricular changes, and I will then go on to consider possible approaches to the real problems that were avoided in these reports.

For many of us, the weakest component of the "Everybody Counts" program is the great stress placed on calculators and computers. Contrary to the claim in the "Standards" that

> There is no evidence to suggest that the availability of calculators makes students dependent on them for simple calculations,

many of our students have already become *calculator idiots*. At UCLA we are now witnessing the appearance of students who are so hooked on their calculators that they cannot multiply 9 by 7 in their heads. It is difficult to see how such students will be able to do the *estimations* emphasized in the new program if they cannot instantly do such simple calculations. If indeed the problem with our current elementary school curriculum is that the

current goal in most elementary school classrooms is

far in excess of what is needed for tomorrow's society, then why is it that many of our students are unable to do even the simplest *paper and pencil* calculations? Banks advertising positions in the UCLA *Daily Bruin* now feel it necessary to indicate that candidates "should know fractions".

The "Everybody Counts" de-emphasis of manual calculation in high school mathematics is even more problematical. In these reports it is suggested that owing to the development of calculators that can do algebraic calculations, algebraic facility is no longer as important as it once was. But this misses the point that we are not concerned with whether or not our students can find subtle factorizations of fifth degree polynomials. Many of our calculus students cannot factor quadratics! When you teach calculus, the students must have the simple algebraic manipulations in their heads and not in their calculators.

Few would question that calculators and computers can be used to great effect at all levels of education. But they will have little or no effect on the basic *innumeracy* that we are currently witnessing. Our students are simply not fluent in the mathematical language. As in any other language, drill and practice remain the most important tools at our disposal for learning the first principles of mathematics.

Of course advocates of "Everybody Counts" will reply that safeguards against overdependence on calculators and computers are included in these proposals. Nonetheless, the statement in "Reshaping School Mathematics" that

[one of] the two fundamentally important issues discussed in 'Everybody Counts' and in the NCTM Standards ... [is the] ... changing role of calculators and computers in the practice of mathematics

will almost surely mislead many educators and politicians into regarding calculators and computers as a panacea. To illustrate what is already happening, consider the following story. Boston University recently took over the Chelsea school system in an attempt to reverse the educational decline of that district. In a letter to the editor of Science, Chairman of the Management Team of that project, Peter R. Greer, recently reported that

the University is in the process of installing $600,000 worth of computer hardware and software in Chelsea's classrooms at minimal future cost to the Chelsea taxpayers.

This would be laughable if it were not so tragic. The director of the UCLA mathematics department computer lab, used by a relatively benign population, tells me that maintaining equipment generally costs 10% of the initial investment, and that one would have to have a full time person to advise the faculty on software, etc. In impacted school systems the administrators are not even able to keep the plumbing going. What are the

chances that installations like this will still be running after a year? A teacher with only a quarter of Jaime Escalante's ability and dedication would have been worth far more than all the gimmicks.

It is well known that American students are failing in all of their academic pursuits, and in particular, *innumeracy* is highly correlated with illiteracy. But the most disturbing development is that many of our students seem unable to concentrate, regardless of the subject. This was considered in an editorial written by Leonard Gillman, then President of the MAA, in 1987. After discussing the primary pursuit of students today (watching TV), he wrote:

Mathematics requires intense concentration; television encourages nonconcentration. I sometimes wonder how many of my students are capable of concentrating on one idea, uninterrupted, for ten full minutes.

The same alarm is also sounded in Eleanor Palais' letter:

Students are not motivated and want only easy ways to get good grades without lifting a pencil. One has simply to walk into a public school math class and observe students with bored looks, slouched in chairs with arms folded. Often homework has not been done and the class must be invited to take notes, or even to open their books or to get a pencil ... I believe our young people today are doing poorly in their school work because thay have been brought up as passive listeners in a TV generation. When I am teaching, my students are often glassy-eyed and watching what is happening as if I were a TV performer. The students do not receive what is being said!

The simple facts are that many of our students are not studying, and that we have allowed the entire educational apparatus to decay. No form of mathematical curriculum fiddling or new techniques of *presentation* will have an appreciable effect on these fundamental problems. The societal causes for this situation are well-known, and some of them were briefly summarized in "Educating Americans for the 21st Century":

... a fundamental improvement in K-12 mathematics can be hoped for only within the framework of a general impovement of the total school environment. ... difficulties facing the teaching community ... [include] ... low teacher salaries, low prestige, lack of support by society, lack of discipline in the classroom, irregular attendance, etc.

The basic flaw in the NRC and NCTM reports is that they have failed to address these basic issues. Perhaps the disclaimer stated in "Educating Americans for the 21st Century" that these problems are societal in nature and fall outside both the mandate and the competence of this group was appropriate for the NCTM, but was that really the case for the National Research Council committees? With such impossible constraints, the NRC committee members might have tried for some sort of holding action. Instead they have sought ways of making mathematics more attractive to students, and to look for "technological fixes". This defeatist attitude is reflected in a letter by Ron Douglas published recently in the AMS *Notices*. Replying to Eleanor Palais, he wrote

> ... although we are somewhat reluctant to admit it, kids are different. While the wistful comment that the real problem is getting students to work may have some validity, it is analogous to stating that the problem with night is that the sun doesn't shine.

Ironically, this pessimism regarding our students co-exists with an unbounded optimism in the efficacy of curricular change. I cannot resist recalling one of the blurbs in "Educating Americans for the 21st Century":

> A plan of action for improving mathematics, science and technology for all American elementary and secondary students so that their achievement is the best in the world by 1995.

There is certainly much of value in the "Everybody Counts" movement. If one is able to deemphasize the *calculator-computer* aspects of the reports, there are many positive suggestions for reinvigorating the curriculum. Some of the unimaginative approaches that are currently being used in the reaction to *the new math* should be eliminated. Nonetheless, these steps are not likely to have much effect in the current educational environment. The disappointment that will inevitably ensue will even further erode the credibility of mathematical educators. That is why it was so unfortunate that the members of the NRC committees did not address the need for much more drastic reforms.

Educational failure is not the only problem that our Democratic society faces, and in some of our other difficulties we have developed more promising strategies. As in the case of drugs, crime, disease, and environmental destruction, we must have grass roots movements for social change if we are to have even a chance of success. Specifically we need organizations and lobbies that will insist on a change in the attitudes of parents and children alike. Even simple ad campaigns would have an effect. We should have messages on TV stating that "If you don't read a book, how can you expect your child to read a book?", or "Keeping this TV set on can damage the educational potential of your children". We must also revise our educational philosophy by making it clear to everybody that the best way to "feel good about oneself" is to be good. We must make it our goal to stir our somnolent fellow citizens into stemming the further decay of our culture.

In conclusion, although "Everybody Counts" and the "Standards" have many useful recommendations, they are flawed by an overemphasis on calculators and computers, and they provide little hope for reversing the continuing decline.

References

1. "Everybody Counts, A Report to the Nation on the Future of Mathematics Education," National Research Council, National Academy Press, Washington, D.C., 1989.

2. "Reshaping School Mathematics," Mathematical Sciences Education Board, National Science Education Board, National Research Council, National Academy Press, Washington, D.C., 1990.

3. "Curriculum and Evaluation Standards for School Mathematics," National Council of Teachers of Mathematics, 1989.

4. "Educating Americans for the 21st Century," National Science Board Commission on Precollege Education in Mathematics, Science and Technology, National Science Foundation, Washington, D.C., 1983.

5. L. Gillman, "Two Proposals for Calculus," *FOCUS*, September 1987.

6. E. Palais, "A Differing View on Mathematics Education Reform", *Notices of the AMS* (36) 1989, 1189–1191.

7. R. Douglas, "A Differing View on Mathematics Education Reform", *Notices of the AMS* (37) 1990, 263–264.

8. P. R. Greer, "Boston University/Chelsea Project", *Science* (247) 1990, 1167.

Everybody Counts – Another Point of View

A.A. Cuoco
Notices Amer. Math. Soc. 37:7 (September 1990)

Edward G. Effros is unhappy with the emphasis that the National Council of Teachers of Mathematics (NCTM) "Standards" and the National Research Council (NRC) reports "Everybody Counts" and "Reshaping School Mathematics" place on computers and calculators. He argues that these "curricular changes" will make the problem of innumeracy (a new term that refers to the condition of not being able to calculate 9×7 in your head) worse than it already is and that they do not put the problems of mathematics education in the more general context of general educational decline.

The emphasis that the NCTM and NRC put on technology is *not* calls for curricular change. Instead, these recommendations are part of a demand that we change the way our students *learn* and *do* mathematics.

Indeed, the "Standards" outlines a curriculum that is topically quite conservative. If the "Standards" are implemented in schools, students will continue to study algebra, geometry, elementary functions, and perhaps calculus. There will be an infusion of mathematical induction, some combinatorics and probability, and a few topics from discrete mathematics, but these are not fundamental changes. There are no calls for courses in linear algebra, number theory, or logic, even though these topics are within reach for many high school students. I've talked with many high school teachers about the proposed curricular changes, and none of them feel that the suggested curriculum is a radical departure from the high school curriculum of 1990. What *is* different, and what almost every high school teacher feels will be an extremely difficult change, is the emphasis that the NCTM and NRC places on having students construct mathematics for themselves.

The new reports are constuctivist documents. They call for a decreased attention to the rote memorization of facts, the active involvement of students in the construction of mental models for mathematical abstractions, and the change in role for the teacher from one of transmitter of knowledge to one of intellectual coach. It is becoming clear that the traditional mathematics classroom, in which a teacher explains some mechanical algorithm and then has the students work a page of practice problems, just doesn't work. Effros wonders why his university students can't do even the simplest paper and pencil calculations. Does he honestly think that it's because they didn't practice the calculations *ad nauseam* in their school days?

Of course they practiced; some of them practiced the same kinds of drill for eight years. But all that drill didn't help them understand the underlying mathematics, and it didn't even help them remember how to do the calculations. Recently, I taught a semester course in "remedial" mathematics to high school students. I discovered early on that (with a simple example to jog their memory) the students could easily combine fractions, do long division, convert decimals to fractions, and the like. In fact, they fully expected that this course would be another five months of mindless drill, and they were perfectly willing to spend their time that way. Instead, I asked them to solve problems, very simple arithmetic problems about cost, pay, size, and distance. One of the problems was this:

In 1978, 664 families out of every thousand families had a color TV. How many families per thousand did not have color TVs in 1978?

One student's answer was 1314. I asked to see her work, and my worst fears were confirmed: She had subtracted 664 from 1978.[1]

The argument, then, isn't about whether students should have paper and pencil skills or calculator skills. It isn't about whether students should draw pictures with a ruler and protractor or with a mouse. And it isn't about whether students should manipulate polynomials

---

[1] When I described this amazing incident to some middle school teachers, several of them wanted to know if my student did the subtraction correctly.

with a pencil or with a computer. The argument is about whether or not the kind of mathematics instruction that we all experienced in our school days, the mathematics instruction of drill and technical expertise, can help students *do* real mathematics. Effros describes banks that advertise for people who "know fractions". These banks probably couldn't care less about how people *manipulate* fractions; they want people who can apply arithmetic with rational numbers to the situations that come up in a bank, and it's not at all clear that proficiency in, say, "borrowing" in subtraction will help the tellers with what they need to know.

Perhaps Effros actually believes that drill and practice help people become mathematicians (or, at least, "numerate"). He says, "As in any other language, drill and practice remain the most important tools at our disposal for learning the first principles ... ." That's not how most people become proficient in their native tongue, nor is it how people learn how to express themselves in a foreign language, in a programming language, or in mathematics. A new medium for expression is learned by *immersing* onself in the medium, and that's exactly what the NCTM and NRC are asking for in their recommendations about how mathematics instruction should change. They are asking us to create a *culture* of mathematics in schools, a culture in which students can have at their disposal high-powered tools (including pencils) for doing mathematical experiments, and a culture in which students can work together on problems that are developed in collaboration with their teachers.

The fact that Effros' criticism of the national reports centers around what he sees as the bad effects of calculators and computers is curious indeed. It is something like a criticism of miter boxes in cabinet making: the condemnation of a tool is as silly in mathematics as it is in carpentry. There are certain *uses* of tools that might be detrimental to students' mathematical development, but Effros never gets that specific. Is he opposed to the kind of activity that Muench describes in his review of *ISETL* in [**Mu**]? Is he opposed to using a computer to manage the very drill and practice exercises that he considers so important? How about using calculators as trigonometric tables? Using computers to graph equations? The point here is that computers (and to some extent calculators) are general purpose machines, and it's impossible to give a generic argument that shows how *all* uses are bad for *all* students.

To be sure, there are many issues that remain unresolved in the use of technology in mathematics education. Here are some that deserve the attention of the mathematical community:

1. Does the writing of computer programs in a programming language that is a close approximation to the language of mathematics help students build mental models for mathematical constructs? If so, why?

2. In spite of years of study in analytic geometry, many

students do not understand the connections among the representations of a function as a process, an object, a graph, and a table. Can certain kinds of software help students use multiple representations for functions?

3. Can the computer modeling of recursively defined functions help students understand mathematical induction?

4. Most mathematicians believe that new results spring from capturing subtle patterns in many concrete calculations. Will such patterns be as easy to find if the calculations are carried out with a symbol manipulator?

5. One of the big differences between research mathematics and school mathematics is that the latter doesn't include experimentation as a primary focus. Will it be easier for students to do mathematical experiments if they use mathematical software? Exactly what kinds of uses will make experimentation easier? What kinds of experimental hueristics will students develop if they investigate mathematical phenomena in a computer environment? What kinds of phenomena can we ask them to investigate if a computer with appropriate software is in their repertoire of tools? Can technology help students make the leap from data-driven discovery to theoretical understanding? Using _Mathematica_, for example, it is fairly easy to develop a conjecture about which n have the property that $x^n - 1$ splits into exactly two factors over Z. But what about a proof?

These are the kinds of questions we should be discussing. We should not be quibbling about non-issues like whether or not the advocates of the use of computers as mathematical tools think that young people should know their multiplication tables. _Everyone_ agrees with Effros that high school graduates should be able to multiply 12 by 5, that they should be able to estimate half of 752, and that they should know that a difference of two squares is not irreducible. Some of us believe that technology will actually help students, not only with these low-level tasks, but with real mathematics as well.

Effros maintains that the reports he cites fail to put the problems of mathematics education in a societal context. I agree, but I don't agree with his assessment of this context or with his suggested remedies.

Are young people different from us? Of course they are. I'm different from my father, and he's different from my grandfather. Today's students look at mathematics through a different lens than the one we used in school. Young people will use existing tools in spite of us; it makes absolutely no sense to them when we prohibit them from using the machines they see all around them. If you ask a high school trigonometry class to give you the exact value of $\sin \frac{\pi}{3}$, many students will say that $\frac{\pi}{3}$ "equals" 60°, enter 60 on a calculator, hit the **sin** key, see .866, and know that the teacher must want $\frac{\sqrt{3}}{2}$ for an answer.

Are young people less capable at mathematics than us? Of course not. Because they have grown up with technology that was new in 1970, we can't expect students to respond to classroom methods from 1960. Sure, students become glassy-eyed when exposed to an hour lecture on mathematical dogma. But they respond very well in lab-like settings, or when they are asked to work on what seem like open ended problems. My experience is that today's best students are more creative, more eloquent, and more facile with abstraction that the best students of a decade ago.

What the NCTM and NRC do not address is the societal resistance to the teaching methods they propose. Already, we are hearing that making mathematics a subject in which students construct notions for themselves is too expensive. The kinds of activities required cannot take place in classes of 30 (or even 25) and the equipment involved is costly. The question, then, is whether or not people want to pay for a mathematics curriculum in which young people do more than memorize outdated algorithms. In Massachusetts, the answer is no. We are struggling here with tax-cap legislation that makes it impossible for schools to adequately fund their programs. So, while teaching loads should be decreasing so that teachers can study new mathematics or take on student apprentices, teachers are being laid off and class sizes are going up. Our school hasn't bought any new equipment since 1983. This complete lack of adequate funding is supported wholeheartedly by local business and industry; it makes one wonder about how much the banks want students to _really_ know about fractions. From the beginning, public schools have been pulled in two directions. On one hand, working people see schools as institutions for social change, as places where their children can gain the insights they'll need to "move up the ladder". On the other hand, business and industry see schools as instruments for perpetrating inertia, as meta-factories for the production of workers. To the extent that workers today must know more about mathematics than their counterparts of a decade ago, schools will find support from the business establishment. But if a change in emphasis in mathematics education (or in any other field) threatens to put students in an environment where they are encouraged to question everything, we can expect little backing from those whose interest is served by the maintenance of the _status quo_. The NCTM "Standards" are asking schools to empower students in a way that will never find broad based support in the industrial complex. Effros' "simple ad campaigns" will have little effect on the way our schools are supported.

The "Standards" outlines a broad program for closing the gap between they way mathematicians and students of mathematics work. It remains to be seen whether or not the American public wants the gap closed, but clouding the waters with criticisms of a technology that will not go away contributes little to the debate.

References

[E] Effros, E., Some Thoughts on "Everybody Counts," *Notices of the American Mathematical Society* **37**, 1990, 559–561.

[Mu] Muench, D., ISETL - Interactive Set Language, *Notices of the American Mathematical Society* **37**, 1990, 276–279.

[NCTM] *Curriculum and Evaluation Standards for School Mathematics*, National Council of Teachers of Mathematics, 1989.

[NRC] *Everybody Counts*, National Research Council, National Academy Press, Washington, D.C., 1989.

Woburn High School, Woburn, MA

MS 2000 Final Report Focuses on Undergraduate Mathematics

M.H. Clapp, MAA FOCUS 11:3 (June/July 1991)

The final report of the Committee on Mathematical Sciences in the Year 2000 (MS 2000) was released on 9 April 1991 at a public policy briefing at the National Academy of Sciences in Washington DC. In it the Committee calls for fundamental changes in undergraduate mathematics as it presently exists on most college and university campuses in the United States. The report, *Moving Beyond Myths: Revitalizing Undergraduate Mathematics*, examines the health of undergraduate mathematics education, identifies certain myths and deficiencies, and presents recommendations and an *Action Plan* for reinvigorating the quality of collegiate mathematics, using existing successful programs and strategies as the starting point.

Moving Beyond Myths thus becomes the latest in a sequence of documents issued by the mathematics community over the last several years which have examined the overall health of the discipline in this country and cited the need for change. This report is the first to address specifically the broad issues of undergraduate mathematics education and to recommend change in how mathematics is taught; full use of the mathematical potential of women, minorities , and the disabled; greater involvement of mathematics faculty with the preparation of teachers and with school mathematics.

The Committee found many examples of effective programs and dedicated faculty on individual campuses across the nation. The report's format features sidebars throughout the narrative which document successful initiatives and which, in the words of the report "...provided both grounds for optimism and models for more widespread improvement." The Committee concludes that in view of this evidence, the mathematics profession knows what is working and what needs to be done; it is now time to take the next steps needed to revitalize undergraduate mathematics.

The report extablished four goals for mathematical sciences faculty:

◇ Effective undergraduate mathematics instruction for all students.

Acknowledging the lack of interest and of success experienced by so many students who take mathematics, the report calls for the identification and use of effective alternatives to lecture as an instructional technique greater uses of computers, and the linking of mathematics content and instruction to students' own experiences and to other disciplines. It urges faculty to become more knowledgeable about how mathematics is learned and to elevate the importance of effective mathematics instruction.

◇ Full utilization of the mathematical potential of women, minorities, and the disabled.

Noting that college mathematics attracts far too few African-American, Hispanic, and Native American students, the Committee points to examples of successful programs which reverse this pattern and which should be expanded and replicated. While women have about equal representation with men in undergraduate mathematics, their numbers decline precipitously between undergraduate and graduate school. Parity for women, minorities, and the disabled in mathematics education must be achieved, the report states, because failing to do so constitutes "...an appalling waste of human potential, denying to individuals opportunity for productive careers and to the nation the resources for economic strength."

◇ Active engagement of college and university mathematicians with school mathematics, especially with the preparation of teachers.

Because teachers of school mathematics receive their own instruction in colleges and universities, the report urges faculty to model effective instructional styles, since most teachers will eventually teach mathematics the way they themselves have been taught it. Mathematics faculty are also encouraged to become involved with issues of mathematics teacher preparation with K–12 mathematics education.

◇ A culture for mathematicians that respects and rewards teaching, research, and scholarship.

The report quotes Ernest L. Boyer's *Scholarship Reconsidered: Priorities of the Professoriate*, "What we urgently need today is a more inclusive view of what it means to be a scholar — a recognition that knowledge is acquired through practice, and through teaching." Teaching and research ought not be viewed as competitive activities, it notes, and scholarship associated with how mathematics is taught and learned is needed and valuable.

The report's *Action Plan*, according to MS 2000 committee chair Dr. William E. Kirwan, II "challenges our institutions of higher education to bring their mathematics education efforts up to the standard set by the nation's mathematical research enterprise, which is pre-eminent in the world." To meet that standard, the mathematical community is called upon to make fundamental changes in the content, context, and culture of undergraduate mathematics. Moreover, the report points out, that while leadership for mathematics education reform is the responsibility of the mathematics faculty at colleges and universities everywhere, others also have roles to play: colleges and universities, business and industry, professional societies (particularly the MAA in view of its role in undergraduate mathematics), governmental agencies, and educational policy makers. "For those who have worked hard on educational issues," the report states, "it is time for redoubled effort; for those who have not, it is time to begin."

The *Action Plan* identifies the roles for each of the players in helping to achieve the revitalization of undergraduate mathematics. The *Plan* contains three components, each containing multiple tasks for each of these groups to undertake. The components are:

◇ Develop and Promulgate Effective Instructional Models.

Mathematics faculty and their departments are the principal focus of this component; it also details actions for colleges, universities, professional societies, and the government. It emphasizes mechanisms to improve teaching and learning in undergraduate mathematics, to increase the recruitment and retention of students currently underrepresented in the mathematical sciences, and to encourage experimentation and innovation in instruction and program development. It calls for the establishment of a journal dealing with undergraduate mathematical research and practice and for government support for undergraduate mathematics, program dissemination, and increased budgets for student support programs and fellowships.

◇ Establish and Disseminate National Guidelines on Standards.

The professional mathematical societies and the faculty and departments they represent are key players in this component, followed closely by college, universities, and the government. The *Plan* calls for development of new advisory national guidelines dealing with curriculum, teaching, and evaluation in undergraduate and graduate programs, aligned as appropriate with the NCTM *Standards* for school mathematics. The need for planning, top priority for teaching, and government support for emerging guidelines are also listed as basic elements for action.

◇ Build and Sustain Supportive Attitudes and Structures.

Here the *Plan* calls for the President and the governors of the fifty states, federal and state agencies

and legislatures, and boards of regents and trustees to make primary contributions to reform efforts. These contributions are both policy oriented ("Retain the national education goal of being 'first in the world'") and resource related (creation of a network of regional centers for excellence in the teaching of mathematics). Universities and colleges are asked to provide the resources needed for effective mathematics instruction, including the use of technology, and to speak out about the importance of support for research funding and scholarship in creating healthy undergraduate education. Faculty and their departments are urged to establish networks with colleagues on campus, on other campuses, and with mathematics teachers in local schools to strengthen the infrastructure of mathematics education.

The report points out that mathematics represents a major component of undergraduate instruction. It constitutes about 10% of higher education enrollment in the United States, and is the second largest discipline taught (next to English). Each year over three million students are enrolled in mathematics classes in some 3,000 two-year colleges, liberal arts colleges, comprehensive universities, and research institutions. Recently released figures show that in 1988, remedial enrollments accounted for 21% of this total; 48% of the enrollment was in (non-remedial) introductory courses below the level of calculus, 22% was in calculus courses, 8% was upper division, and 2% was in graduate courses in mathematics. Mathematics accounts for approximately one-third of all science and engineering enrollment in higher education. There are about 40,000 full and part-time faculty, half of whom hold the doctorate, teaching in these courses.

Given the sheer size of the enterprise, the changes called for in the report will take time and require significant effort. Still, the Committee observes that the time is propitious for change, because there is substantial consensus on the goals for mathematics education and the processes to achieve them. The report concludes, "The national revitalization of mathematics is within our reach, if only we are prepared to make a serious intellectual and financial commitment to our children's and our nation's future."

The Mathematical Sciences Education Board (MSEB) plans a broad dissemination of the report and its accompanying *Action Plan* that will include mathematics department chairs at two and four-year college and university campuses and the leadership of the professional mathematical associations, societies, and organizations. Additional copies of the report will be sent to deans of colleges of science, the leadership of major scientific and higher education professional organizations, to policy makers, government agencies, and selected corporate officers in business and industry. Individual copies of the report are available for $7.95 from: National Academy Press, 2101 Constitution Avenue Northwest, Washington DC, 20418, (800) 624-6242.

The Committee on Mathematical Sciences in the Year 2000 (MS 2000) was appointed in 1988 by the National Research Council as a joint project of the MSEB and the Board on Mathematical Sciences (BOMS). The Committee was chaired originally by J. Fred Bucy, former Chief Executive Officer of Texas Instrument, Inc. Dr. William E. Kirwan, II, the president of the University of Maryland at College Park assumed the Committee leadership when Mr. Bucy was drawn away by other commitments. Bernard L. Madison was the project director through December of 1989; he was succeeded in the position by James A. Voytuk, who served through the end of the Committee's work in December 1990.

The publication of *Moving Beyond Myths* completed a ten-year analysis of the US mathematical science enterprise by boards and commissions of the NRC. These efforts included the work of the so-called David Committee on the status of US mathematical research, whose report *Renewing US Mathematics* was issued in 1984 and updated last year in *Renewing US Mathematics: A Plan for the 1990s*. Other reports resulting from NRC studies include *A Challenge of Numbers: People in the Mathematical Sciences* (1990) and *Everybody Counts: A Report to the Nation on the Future of Mathematics Education* (1989).

Michael H. Clapp is Associate Executive Director of the Mathematical Sciences Education Board (MSEB) and Director of its College and University Programs.

Colleges Urged to Make Radical Changes to Deal with National Crisis in Mathematics Education
A Report from National Research Council Says Students' Interest in Subject is at an All-Time Low

D.E. Blum, The Chronicle of Higher Education XXXVII:31 (April 17, 1991)

American universities may lead the world in mathematical research, but they must make radical changes in the way they teach mathematics to undergraduates, concludes a new report from the National Research Council.

Among the problems facing mathematics education are "casual" teaching, curricula lacking context, and "invisible" instructors, the report says. It urges mathematics faculty members to become as involved in the teaching and learning process in mathematics as they are in research and the creation of new knowledge.

Spate of Faculty Retirements

"We are asking that the quality of mathematics education rise to the level of mathematics research in this country," said William E. Kirwan, president of the University of Maryland at College Park, a mathematician by training. He headed the committee that prepared the report, *Moving Beyond Myths: Revitalizing Undergraduate*

Mathematics.

The report calls upon numerous constituencies – ranging from professors to legislators – to deal with what it calls a national crisis in mathematics education. Interest in mathematics is at an all-time low among entering college students, and mathematics professors are retiring faster than they can be replaced, according to the report. In addition, the report says, major segments of the nation's population – women and members of minority groups – are seriously underrepresented in mathematics.

The report urges universities to elevate the importance of undergraduate teaching of mathematics and to develop plans to recruit women and minority professors. It calls upon mathematics professors to abandon lectures for more effective teaching methods, and to introduce computers into the classroom.

Some Say Report May Go Too Far

While few in the discipline dispute that mathematics education at the collegiate level is sorely in need of improvement, some observers said the report might have gone too far.

William H. Jaco, executive director of the American Mathematical Society, an association of colleges and mathematicians, said he didn't see the crisis in mathematics education depicted by the report. He said the movement to improve mathematics education was instead driven by changing demographics and the need to incorporate computers into research and the classroom.

"I'm happy that the report and the issue of education reform are out on the table for discussion, but I am concerned about the provocative language of this report," Mr. Jaco said. "Saying research professors who are leaders in the field are taking a casual attitude to teaching is not only not necessarily true, but does not tend to build bridges and take a positive approach to having all the constituents of the mathematics community working together."

Latest in a Series of Reports

The report is the last in a series of reports on mathematics produced by three committees of the National Academies of Sciences and Engineering. It is the only one that deals specifically with mathematics in higher education.

Mathematics at the collegiate level deserves special attention because colleges and universities train teachers and provide the final training ground for people who take jobs in the business sector, Mr. Kirwan said.

The report concludes that teaching methods represent a major problem in mathematics education. Mathematics, it says, is often taught without context, so students see it as unrelated to other subjects or problems in the "real world." It is also frequently taught along a flawed model of teaching – blackboards lectures, template exercises, and isolated study – instead of through such methods as problem-solving, it says.

The report also concludes that those college mathemics professor who consider themselves primarily teachers rather than researchers are virtually invisible. They are confined to introductory courses taught most often in two-year colleges or "in the margins" of university departments of mathematics, it says.

The report blames many such problems on what it calls common misconceptions or myths about mathematics. One, it says, is that innate ability, not hard work, determines mathematics skills. Other misconceptions, it says, hold that women and members of certain ethnic groups are less capable of doing mathematics, or that most jobs require little mathematics.

Moving Beyond Myths: Revitalizing Undergraduate Mathematics is available for $7.95 plus shipping from National Academy Press, 2101 Constitution Avenue, Washington, D.C. 20418; (800) 624-6242. [It is reprinted in July/August AMS *Notices*. Ed.]

[The following page one lead for above article showed a picture of William E. Kirwan. Ed.]

Improving College Math: A Panel's Recommendations

◇ Elevate the importance of undergraduate teaching.

◇ Engage mathematics faculty in issues of teaching and research.

◇ Teach in a way that engages students.

◇ Achieve parity for women and minorities and the disabled.

◇ Establish effective career paths for college teaching.

◇ Broaden attitudes and value systems of the mathematics profession.

◇ Increase the number of students who succeed in college mathematics.

◇ Ensure sufficient numbers of school and college teachers.

◇ Elevate mathematics education to the same level as mathematical research.

◇ Link colleges and universities to school mathematics.

◇ Provide adequate resources for undergraduate mathematics.

Some Observations

G. Weiss, UME Trends 3:5 (December 1991)

Moving Beyond Myths: Revitalizing Undergraduate Mathematics is the latest of a series of publications sponsored by the National Research Council (NRC) designed to present an analysis of the total United States mathematical sciences enterprise. The principal reports in this series are the first *David Report* (published in 1984), its update, *Renewing U.S. Mathematics* (published in 1990), *A Challenge of Numbers*, and *Everybody Counts*. In addition to the considerable effort that went into the preparation of these publications, several national symposia and workshops were organized to consider the state of mathematics and mathematical education in our country. These topics were also the subject of discussions in committees of the AMS and MAA. Such activities involve a large expenditure in both time and money. I am convinced that this attention is needed; many things are wrong with mathematical education and the support of mathematics in the United States and much of this effort is providing guidelines for ameliorating this situation. However, I find this latest publication, *Moving Beyond Myths*, disappointing. My purpose here is to explain why I feel this way and to try to offer some constructive criticism of this document.

Moving Beyond Myths addresses the problems in mathematics education at the college and university level. The other reports and efforts described above are more focused on the status of the research establishment in the mathematical sciences (the David reports) and on the teaching of mathematics in the nation's schools. In the preface of *Moving Beyond Myths*, we are told that this report calls for a "sweeping change" in the teaching of mathematics at the undergraduate level. Shortly after receiving it, I went through this rather short document rather quickly and superficially. My first reaction was that "there was not much in it." I talked to some of my colleagues about the report and they expressed the same reaction. Since I have been asked to write this critique of *Moving Beyond Myths*, I have read it carefully and I still find myself with the same reaction.

As I indicated above, this is a short report. In its preface, it is stated that "two circumstances have combined to enable *Moving Beyond Myths* to be a shorter report than might be expected from a three year project": first, an overview of undergraduate education was presented as part of *Everybody Counts*; second, most of the supporting data were presented in *A Challenge of Numbers*. The latter "might properly be viewed as an appendix, just as *Everybody Counts* might be considered as the introduction to *Moving Beyond Myths*." Unfortunately, essentially all the readers of the latter will do precisely what I did at first: they will, at best, read through it quickly and find it wanting. Furthermore, almost all these readers will not have the other two reports easily available to use them as an introduction and as an appendix. Even if they did have them in hand, they would have difficulties in considering them so coupled with this last document.

I believe that much of what is wrong with mathematical education (at all levels) is due to factors that are beyond the control of those in the mathematical professions, school administrators, and those involved in the funding of both education and research in mathematics. There are forces in our society that have created an atmosphere that is inimical to the goals that have been enunciated in the above mentioned reports, symposia, and workshops. I do not want to analyze these forces here. However, they have created a significant group of students who are not very well motivated to study mathematics (and other subjects). They also contribute to discouraging women and certain minorities from entering a mathematical field. I believe that everyone agrees with this. I mention it here because it should be clearly kept in mind when we propose action that will, hopefully, result in an increase in the mathematically literate population. It is, perhaps, unfair to say that *Moving Beyond Myths* ignores these forces and gives the impression that a "sweeping change" in the way we educate undergraduates will achieved the goals this document announces. Nevertheless, this is a criticism voiced by many of my colleagues who have read this report. The material included between pages 23 and 32 of the document could be interpreted in this way.

There are very few concrete suggestions in the report.

For example, on several occasions we encounter criticism of the "traditional curriculum" and the "lecture-recitation" methods used in undergraduate teaching, but there is little indication of what alternatives can be considered. On pages 32 and 33, there is a short description of what is being done in SUNY Potsdam, apparently with great success. In this description we find the phrase: "Mathematics at SUNY Potsdam does not rely on novel use of technology or innovative curricula to attract students." I applaud the inclusion of the example (and will say more about it later), but I find it disturbing that little more is said about it and that an apparent contradiction is not addressed. Why does this mathematics department meet with such success using the traditional teaching methods that *Moving Beyond Myths* insists should be altered?

As stated above, there is little supportive evidence in the document that indicates how severe are problems involved in attracting more minorities and women into the "mathematical pipeline." Two pertinent illustrations are Figure 5 on page 19 and Figure 6 on page 20. What disturbed me about these two items is that they are somewhat outdated; they indicate that four hundred Ph.D. degrees were awarded to Americans in 1986. More recent data indicates that this number is now considerably higher. One of the "Myths" that is cited is that women and members of certain ethnic groups are less capable in mathematics. While I agree that this is a myth and does not reflect reality (this is one of the effects of the forces in our society that I alluded to above), I do not find it satisfactory that a rather absurd argument is given to show this: "Ample evidence shows such beliefs to be false. Experiences of countries such as Holland and Japan belie this myth." My immediate reaction upon reading this was to ask myself "how large are the ethnic minorities in these two countries?" I cite these relatively picayune items because I am not alone in noticing them and they do detract from the impact that such a report has.

I found the section entitled "An Action Plan" particularly wanting. It consists of a collection of independent sentences that cover seven pages. Many of these sentences express ill-defined goals; almost none represent a well-defined course of action that is to be taken. All this adds to the feeling of vagueness imparted by this report.

Many interesting and innovative programs and efforts connected with undergraduate mathematical teaching are being carried out in our nation. *Moving Beyond Myths* mentions a few (such as the one in SUNY at Potsdam). It would have been very useful to have more of these examples and an accompanying analysis in each case. The document makes several references to the value of computer use in undergraduate mathematics. On page 17, it asserts that "nothing in recent times has had as great an impact on mathematics as computers....Yet only in isolated experimental courses has the impact of computing on the practice of mathematics penetrated the undergraduate curriculum." It is my impression that there is much more involvement of the computer in undergraduate teaching than is indicated by this statement. This topic, however, is complicated. For example, many of the programs that involve the computer in calculus are very expensive (both in equipment and staff). Not every college can undertake some of these efforts, but there are some programs that are both stimulating and not very costly. It would have been valuable to have this report include a more thorough discussion of such programs and the feasibility of conducting them in institutions that have financial limitations. There are programs that have had varying degrees of success in attracting women and minorities into mathematics. Again, only a few are briefly mentioned in the document. A more in-depth analysis of what makes these programs work would have been highly desirable.

In the final analysis, I believe that *Moving Beyond Myths*, at best, serves as a first draft of a report that calls for a "sweeping change" in undergraduate teaching. In this role it has definite merits, but it should not be considered as a major document sponsored by the NRC.

Guido Weiss is at Washington University, St. Louis.

From the Executive Director
Undergraduate Mathematics

W. Jaco
Notices Amer. Math. Soc. 38:5 (May/June 1991)

The mathematics education reform movement is a topic of debate within the mathematical community, as well as in government circles and the popular press. Much of this debate concerns school mathematics, but, for a large segment of our community, reform begins much closer to home: undergraduate mathematics education. A recent report from the National Research Council (NRC) has underlined the importance of addressing the need for change at the undergraduate level, has identified many areas for reform, and has set forth an action plan for government, colleges and universities, and the mathematical community.

In 1988, two NRC boards, the Board on Mathematical Sciences and the Mathematical Sciences Education Board, appointed the Committee on the Mathematical Sciences in the Year 2000, or MS2000. The objective of MS2000 was to review the status of undergraduate mathematical sciences education in the U.S., develop a plan for revitalization of mathematics education at the nation's colleges and universities, and delineate responsibilities for the implementation of the plan. The reports *Everybody Counts - A Report to the Nation on the Future of Mathematics Education* (the executive summary appeared in the March 1989 issue of *Notices*, page 227) and *A Challenge of Numbers* (the executive summary of this report appeared in the May/June 1990 issue of *Notices*, page 547) were products of MS2000. Now, there is a new report which concludes the work of the MS2000 Committee. It is titled *Moving Beyond Myths: Revitalizing Undergraduate Mathematics*. The full text of this report will appear in the July/August 1991 issue of *Notices* (see News and Announcements in this issue of *Notices* for information on ordering copies of the report).

Moving Beyond Myths describes the need to revitalize undergraduate mathematics and demonstrates the challenge this task entails. Many of the problems of mathematics education at all levels are traced to myths about mathematics and the learning of mathematics, such as the idea that success in mathematics depends more on innate ability than on hard work. The report is likely to stir a good deal of debate within the community, for it contains a fairly serious indictment of the undergraduate mathematics education system and of the traditions and habits of mathematical sciences departments in colleges and universities across the nation. Some of the depictions are less than admiring with provocative titles such as "Wasted Breath", "Missing Context", "Casual Teaching", "Flawed Models", "Outmoded Values", and "Invisible Instructors". Nonetheless, the report goes right to the heart of many of the problems with undergraduate mathematics education and concludes with specific goals and an extensive action plan.

Crystallizing consensus and moving the community toward action is not an easy task. As the report itself states, the implementation of the action plan "will tax the creativity, commitment, adaptability, and energies of mathematical sciences faculty and departments, college-university administrations and trustees, professional societies, and federal and state governments." Reform in mathematics education must include, if not begin with, a revitalization of undergraduate mathematics and involves our community. The report specifically challenges professional societies in its action plan. The AMS cannot ignore this opportunity to make a difference.

January 1991 Symposium: Roles of Research Mathematics in Education Reform

N. Fisher, MER Newsletter 3:2 (Spring 1991)

The symposium on the Roles of Research Mathematicians in Education Reform, which was part of the special session on Mathematics and Education Reform organized by the MER Network, on Friday, January 18, 1991 at the San Francisco joint mathematics meetings, explored individual and organizational views of how and why researchers should be involved in education reform. In the first part of the program, five researchers –Andrew Gleason, Harvard University; Leon Henkin, University of California, Berkeley; Deborah Tepper Haimo, University of Missouri; William Jaco, American Mathematical Society; and William Thurston, Princeton University– shared their personal experiences in becoming involved in educational work. During the second part of the program, the roles of the mathematics community, universities and professional organizations in fostering and promoting more extensive and deeper involvement of mathematicians in education reform were considered by four panelists: Ray Shiflett, Mathematical Sciences Education Board; Judith Sunley, National Science Foundation; Marcia Sward, Mathematical Association of America; and Uri Treisman, The Charles A. Dana Center, University of California, Berkeley.

Individual Perspectives. In their personal accounts, the five invited speakers talked about how they became seriously interested in educational issues, the educational activities they have engaged in and what they have learned from these experiences, and they identified some crucial educational issues in the learning and teaching of mathematics. The initial venture into mathematics education, although decisive in starting a long term involvement, often was in response to an invitation to participate in educational discussions or to teach in a special program, rather than by one's own initiative. Over time, careful reflection about what mathematics education was and what it might be led to deeper commitment.

The individual stories of the speakers varied greatly in time and circumstances. Their experiences included: Gleason's working with young school children in the 1960s and posing mathematical problems using concrete models; Haimo's working with high school teachers to expand the teaching of applications of mathematics in high school; Henkin's supervising mathematics graduate students writing their theses in mathematics education; Jaco's recognizing, as a department chair, the frustration of some of his faculty and the missed opportunities caused by restricting a mathematician's role to the mathematics research model; and Thurston's developing and teaching "Geometry and the Imagination," an alternate course to calculus for Princeton undergraduate non-mathematics majors.

Significantly, by taking the risk of plunging into

new educational territory, these researchers felt they had learned a great deal about the the teaching and learning of mathematics. Gleason was impressed by how deeply first graders thought about the things they were doing, their enthusiasm in taking hold of any interesting problem presented to them, and their ability to solve and generalize problems that would usually be presented much later in the curriculum. Thurston found that undergraduates were much more capable than he had believed, when he observed them working in small groups discussing mathematics in their own terms with one another unhampered by their lack of fluency in mathematical language.

Envisioning Changes in Mathematics Education. Whether stated explicitly or implied, the motivation to try something new grew out of the dissatisfaction with the perceived narrow vision of prevailing mathematics education at all institutional levels. Henkin explained how the great dissatisfaction of Berkeley students with their undergraduate mathematics education led to recognizing the importance of elementary school experiences in laying the foundation for later learning. Henkin questioned why it is that students don't naturally come to understand mathematics starting with their early experiences, and why, indeed, so many students learn to misunderstand mathematics. Although current mathematics education stresses the vertical nature of mathematics, Thurston pointed out that the different layers of instruction are uncoordinated, and he suggested two approaches to consider: More attention should be given to the entire structure of mathematics education and more emphasis should be given to the horizontal, "bushy" nature of mathematics. Significantly, Gleason concluded that, although the content of mathematics is very valuable, the goal of mathematics education should be to teach our students to develop the habit of thinking about and analyzing whatever their situation is, and to help them gain confidence in their own abilities to understand mathematics by making the effort to think about what is before them.

"Changing the Culture": The Mathematics Community and Education Reform. The need for the leaders of the research community to support education reform was unequivocally affirmed by Sunley's remarks. "Whether you like it or not," Sunley stated, "the leaders of the mathematics community are the research leaders in the field. If we are going to talk about changing the culture, those are the people you have to involve. The leadership must be ready to move in this direction."

In his commentary, Jaco charged the mathematics community with the duty of promoting educational reform, from its marginal position, to a central and sustaining issue of the mathematics agenda, and he challenged the mathematics community to recognize that educational reform is the collective commitment of departments and the mathematical societies. The time in which a mathematics department could discharge its obligation to education reform by vaguely acknowledging

that "Oh yes, Professor X down the hall does mathematics education," has passed.

Jaco's remarks were underscored by several of the panelists, who envisioned expanded responsibilities of the mathematics community including its support for individual mathematicians engaged in education. If the mathematics community is serious about broadening the view of scholarship to include valuing creativity in the organization and exposition of knowledge and the dissemination of knowledge, then it is necessary that respect and rewards be given equally for teaching, research and scholarship, Shiflett stated.

Sward highlighted how the MAA has been moving to take on a dual role: its traditional role of serving its members and the new role of being an active agent for change. MAA has targeted involving minorities in mathematics and the role of mathematicians in the education of teachers as two prime issues on its reform agenda.

The need to develop the means to inform the mathematics profession about education reform and to develop ways for the mathematics profession to learn about schools and mathematics education at all levels– K-12, undergraduate and graduate was highlighted by Treisman. He further underscored an earlier statement of Thurston's of the need to recognize the two-way interaction between college/university mathematicians and precollege mathematics teachers. There is much to be learned in both directions.

Mathematics and Research in Undergraduate Mathematics Education

E. Dubinsky, MER Newsletter 5:2 (Spring 1993)

The discipline of mathematics is in the process of giving birth to a new field – **Undergraduate Mathematics Education (UME)**. Nurturing offspring is a hallowed tradition in our profession, especially in the twentieth century. In recent decades, Statistics, Applied Mathematics and Computer Science are well known examples of fields which got their start in mathematics departments and then went on to become disciplines in their own right. In some schools these fields branched out to become their own departments; in others, they remained as part of the mathematics departments; and in still others, everybody got together to form a department of mathematical sciences.

All of this is now happening with UME. Look through the position announcements in the Notices and see how many ads call for people who specialize in mathematics education. Come to an AMS/MAA meeting and see how many talks, sessions and panels are devoted to this new field. Read the old publications (and some new ones) and see how many articles are devoted to this area. It is happening. As with all kinds of births, this spawning of new fields by mathematics is not painless and

UME is no exception. We are having and will continue to have our skeptics, our controversies, and our failures. One purpose of this article is to try to answer some of the questions mathematicians are asking about this new field, and to say a little about the new **Joint Committee on Research in Undergraduate Mathematics Education.**

Why should we think about research in undergraduate mathematics education? I do not need to detail here the extent to which UME is not working. It is, perhaps enough to point out that reports on the inadequacy of our educational system have, for some time now, referred explicitly to K-14 as opposed to the earlier concern with only K-12 levels.

Another indication of the need for reform is NSF's decision to expand its support for undergraduate education from calculus to other courses and to include research in learning. This is not just a decision of a funding agency. The number and quality of proposals suggest that the field has been building up a demand for support of broad based curriculum reform with a strong learning research component.

The research in learning aspects of curriculum reform is essential. One of the most important things we have learned from five years of calculus reform efforts is the importance of developing new pedagogical approaches. It seems almost self-evident that we cannot figure out how to teach students effectively if we do not understand something about how these students learn. Such an understanding is not fully available in our profession today (or anywhere, for that matter) and research in undergraduate mathematics education (popularly known as RUME) is the only way we can develop it.

Why should RUME live in a mathematics department? I am going to try to answer this question, but the reader should realize that, having spent twenty five years as a research mathematician, I have always considered a mathematics department as my natural professional home. Now my scholarly work is entirely in research in undergraduate mathematics education. There is probably some degree of personal bias in my position on where research in undergraduate mathematics education should take place.

But the main argument is that a mathematics department is where the action is. It is where we must do our field work. It is where we can conduct pedagogical experiments to help develop our theories. It is where we can work with faculty and inculcate them with our new methods. Finally, it is where we can find graduate students, both to serve as potential recruits, but much more important, to learn to become teachers/researchers in their future positions as professors of mathematics in colleges of all kinds.

The other obvious candidate for a home for RUME is an education department. My argument against this is a little delicate. Anyone working in RUME must have a strong background in mathematics and education. It is unlikely that many students will become well versed in both at the same time so we must think in terms of learning first one, and then the other. It is extremely difficult and time consuming for a person who is trained as a mathematician to learn enough about education, epistemology, psychology, and research methodology to be an effective researcher in mathematics education. It is almost impossible to go in the other direction.

As a last argument, I would suggest that quality standards for scholarly work is at least as high and fair in mathematics departments as in any other area. I think that RUME could benefit greatly from living in such an environment.

What is CRUME doing? After MAA sponsored an ad hoc committee for two years, the AMS and MAA, with the participation of the National Council of Teachers of Mathematics (NCTM) and the American Mathematical Association of Two-Year Colleges (AMATYC), have established a joint Committee on Research in Undergraduate Mathematics Education (CRUME). The function of this committee is to foster and support activities in RUME.

The work of CRUME is varied. We sponsor speakers, research paper sessions, panels and mini-courses at professional meetings and other conferences. Most of the people who work with us are mathematicians who are moving into mathematics education. We are preparing a white paper on Mathematicians and Mathematics Education Research.

The audiences for our various activities are almost entirely mathematicians. As in the case with sessions run by MER, audiences at CRUME activities at professional meetings are quite large compared to audiences at other activities. This suggests that there is a need for what we are doing and an interest in the way we are doing it.

The biggest and most important activity of CRUME is to sponsor a series of annual volumes of research papers on undergraduate mathematics education. To be titled *Research in Collegiate Mathematics Education*, the volumes will be published by the Conference Board of Mathematical Sciences. It is considered to be a precursor for a journal in this area.

Our hope for this publication is that it become an outlet for the highest quality research papers in the field and that both the content and the writing style of the papers will make them extremely useful for teachers of mathematics at the post-secondary level.

Where can we go from here? As far as CRUME is concerned, the present need is to continue to build up the activities we are engaged in as a means of stimulating and supporting research in this area.

Taking a larger view, however, I would suggest we begin to think of ways to build bridges between the activities of MER and CRUME and other groups. For several years, I have been very interested in the idea of an institute or center for undergraduate mathematics education that would relate to curriculum development, research, preparation of college teachers, policy issues and other activities related to research in undergraduate mathematics education. Lately,. I have seen a number of indications that the time may be ripe for moving forward on such a project. Perhaps we should all get together and see how this might work.

Towards a Scholarship of Teaching

N. Fisher, MER Newsletter 5:2 (Spring 1993)

[Newsletter] Editor's note: This article is based on the address "Displaying Teaching to a Community of Peers" by Lee S. Shulman, Charles E. Ducommun Professor of Education, Stanford University, and President, National Academy of Education, at the Forum on Faculty Roles and Rewards sponsored by the American Association for Higher Education, on January 29-31, 1993 in San Antonio, TX.

If there is an imbalance between teaching and research in academic life, it is surely an imbalance in the value accorded each of these academic endeavors. Which has higher value, teaching or research, is in the eyes of the beholder. The public favors teaching, whereas the academic world, through its mechanisms of rewards and recognition (if not its rhetoric) favors research. To attack the teaching-research dichotomy, or more inclusively the teaching- scholarship dichotomy, Professor Shulman looks at teaching from the analytical/scholarly perspective. Not surprisingly, viewing teaching as scholarship is a powerful method for revealing new insights and strategies for change. Attention is given to Shulman's vocabulary and line of reasoning, as well as his conclusions.

Anticipating Misunderstandings. It was by examining an act of scholarship, or more precisely "what went wrong" in the scholarly process, that Shulman illustrated the inextricable relationship between teaching and scholarship. A colleague had called Shulman's attention to an article in the American Educational Research Journal in which the author had misrepresented Shulman's thesis of the use of cases in teaching. In earlier publications, Shulman had argued that knowledge of teaching was not only a knowledge of theories and maxims, but if we truly understand what great teachers know we would find that they have case knowledge, i.e., knowledge of particulars. Case knowledge, rather than being anecdotal knowledge, is viewed as knowledge of a case of something, an instance of a larger class of particulars that share certain distinctive properties. Case knowledge, in which the instances are organized in some fundamental way, has a theoretical base. But in the article in question, the author asserted that all cases were themselves cases of theory. In fact, Shulman had suggested that there are three kinds of cases: prototypes, which exemplify theoretical principles; precedents, which exemplify

principles of practice; and parables, which teach moral or ethical models.

What caused this miscarraige of scholarship? Shulman concluded that the problem was his own; as a scholar, he had done a bad job of teaching. A basic property of good teaching should be to anticipate the many ways in which the community one is addressing might possibly misconstrue what the teacher/scholar has in mind, and to build into the pedagogy correcting mechanisms. "Scholarship is unconsummated until it is understood by others in the community," Shulman concludes. Teaching, as the communication of knowledge, is integral to scholarship.

Teaching as Community Property. But there does exist a dichotomy between teaching and scholarship. If the dichotomy is not intrinsic to these pursuits, the explanation must lie with external factors. Pointing to the organizational distinctions in the ways teaching and scholarship are carried out, Shulman not only provides an analysis for the current situation but lays a foundation for investigating models to join teaching and scholarship.

Shulman recalled his keen emotional response on being "welcomed to the community of scholars." His vision of the scholar was of a solitary individual, laboring quietly and obscurely. In "splendid solitude" the scholar pursued his or her investigations. It was as a teacher, in Shulman's vision, that the academician entered the social order, interacted with others and became a member of the community.

"The realities are just the opposite," Shulman learned. We experience isolation as teachers. It is in the classroom that we do things about which we almost never have the opportunity to talk about with others; we close the door and experience pedagogical solitude.

In contrast, as scholars we are members of active communities. We converse, evaluate, and gather with peers from other universities to exchange methods, findings and excuses.

Teaching is not valued in the academy because the way in which we treat teaching removes it from the community of scholars. It is not that universities in principle diminish the importance of teaching nor is it in principle that investigation is seen as having greater weight than teaching; rather we value those aspects of our lives that can become community property.

Redefining the Process of Teaching. Shulman would have us consider three principles for moving towards a balance in teaching and scholarship within the community of scholars: teaching needs to be identified with the disciplinary community; teaching needs to be represented by visual artifacts; and, as part of the disciplinary culture, the evaluation of teaching needs to include peer review.

Disciplines are the basis for our intellectual communities; disciplined based scholarship is valued more than forms of scholarship that are non-discipline (note that non-discipline does not denote interdisciplinary). But look at how we support and evaluate teaching. We have universitywide centers for the teaching of learning; student evaluation forms are uniform, as though the teaching of civil engineering and the teaching of Chaucer were the same activity.

The message is clear: Teaching is generic. Teaching is not within your discipline, it is not essential to what you do, it is something else. Teaching is technical, a performance, something on top of what you really do. Bringing teaching into the community of scholars requires that the review, examination, and support of teaching become the responsibility of the disciplinary community.

Continuing this line of argument, to put responsibility for quality control of teaching at the disciplinary level, teaching has to become a public act. It will be necessary for peers to be knowledgable about one another's teaching. And the community of peers, as with scholarship, is construed as the professional disciplinary community outside one's own university as well as one's departmental colleagues. Revealing teaching, making visible what is otherwise invisible so that it can be shared, discussed and critiqued is the challenge.

Building a Culture of Teaching. By changing practices, we should anticipate changing beliefs. Taking a leaf from the work on the assessment of elementary and secondary teachers, Shulman promotes the idea of "consequential validity." Not only should we be judging teaching effectiveness, but we should be using procedures that help teachers teach better. Using a variety of procedures, candidates for review should be able to make a case for their teaching worth, as they are expected to do for their scholarship. Similarly, requiring a candidate to present a teaching colloquium as well as a colloquium in the discipline in which he or she may include a scholarly analysis of a course would go way beyond changing the recruitment process. Faculty would have to discuss education seriously, openly, and systematically as a central part of their professional lives. To create a culture of teaching we need to aim at the fundamental ways in which we live and think.

Where We Are. Where we are now is that we've "kind of got the words but we're nowhere near knowing the music." The search is not simply to improve the precision of evaluation. There needs to be a commitment to active, vigorous, and difficult experimentation. This is not a call for radical, new initiatives, but a call to reassert the view of education as a perpetual experiment whose methods should be changed to meet either new conditions or better insight into prevailing conditions. "[If] we feel that the spirit of experimentation and inquiry is an appropriate value to place on those things we study, then we can hardly exclude ourselves and our own work from the same moral obligation."

To Enhance Prestige of Teaching, Faculty Members Urged to Make Pedagogy Focus of Scholarly Debate

B.T. Watkins, The Chronicle of Higher Education
XXXVII:9 (October 31, 1990)

For teaching to gain prestige in higher education, faculty members must make pedagogy a subject of scholarly debate. And they must focus that debate on the content of the disciplines, an educational researcher said at a conference here.

"What colleges and universities are about is scholarship," said Lee S. Shulman, a professor of education at Stanford University. Teaching will be considered a scholarly activity only when professors develop "a conception of pedagogy that is very tightly coupled to scholarship in the disciplines themselves."

For example, Mr. Shulman said, "when we think about the pedagogy of mathematics, we have to be thinking at least as deeply about the mathematics as the pedagogy. And similarly for history and for all the other disciplines."

"As long as the only things that a faculty has in common are parking, football, and anxiety about what's happening to TIAA-CREF, that's what the discourse is going to be around. That's what's shared."

Mr. Shulman added: "If we want a discourse on pedagogy, we have to make the pedagogy worthy of conversation."

A New Concept of Scholarship

Mr. Shulman presented his views at a national conferences called "The Scholarship of Teaching," arranged by Iona College to give administrators and faculty members an opportunity to consider some of the ideas in a forthcoming report from the Carnegie Foundation for the Advancement of Teaching. The foundation made preliminary copies of the document available to speakers before the conference.

The report, "The Professoriate: a Community of Scholars," proposes a new concept of scholarship that would give greater recognition to teaching. It is expected to be released before the end of the year.

Ernest L. Boyer, president of the foundation and co-author of the report, told educators at the conference that "the task before us now is to take a step back and get away from the tired old teaching-versus- research debate."

For a long time, he said, "we have undervalued the importance of conveying knowledge and generating a new generation of scholars through the powerful and demanding task of teaching."

Mr. Boyer said, "The time has come for us to inquire much more carefully into the nature of pedagogy. It's the most difficult and perhaps the most essential work in developing future scholars."

'Not an Exciting Topic'

"At the moment, teaching is not an exciting, intellectual topic to bandy about in a faculty lounge," said Russell Edgerton, president of the American Association for Higher Education, because "the image of teaching is teaching as technique."

Discourse about teaching methods – how to deliver a lecture or lead a discussion group, for example – is not "very rich and deep," he said.

On many campuses, the speakers said, faculty members already have opportunities for scholarly discussion.

When they develop a core curriculum, for example, "faculty members have to make some tough choices about what kinds of things they want students to read and know," said Mr. Shulman.

The Case-Study Method

Creating a core "gives the faculty something to talk about," he said. "It creates the curricular conditions for members of a community to commune about their pedagogy."

The case-study method of teaching provides much the same opportunity.

When professors who use the same case studies get together and one says, for example, "You'll never believe what my group did with 'People Express' [airline] this week," Mr. Shulman explains, "you have a community that can carry on a discourse around pedagogy."

Mr. Edgerton said professors could consult with each other about teaching materials the same way they do about research proposals.

A professor would never send a grant proposal to a funding agency without asking several colleagues for their opinions, he said, but "a lot of faculty members do not think of sharing the syllabus of a new course with their colleagues."

K. Patricia Cross, a professor of education at the University of California at Berkeley, proposed that professor use "learners, which we all have in common," as a point of departure for both research and scholarly discussions.

"The intellectual challenge of teaching lies in the opportunity for individual teachers to observe the impact of their teaching on students' learning," she said. "Yet, most of us don't use our classrooms as laboratories for the study of learning."

Ms. Cross said: "We talk about having to keep ourselves intellectually alive through doing research in our disciplines, while the most fascinating intellectual challenge of all sits before us every day."

She said. "The more we know about learning, the more intellectually challenging teaching will become."

Everybody Counts — Even Our Ablest

A. E. Ross, UME Trends 1:6 (January 1990)

"Be doers of the Word And not hearers only Thus deceiving Your own selves."

— Epistle of James 1:22

Quo Vadis

The policymakers and many of our colleagues are almost exclusively concerned with the frightening prospect of a large proportion of our citizenry not only unemployed but unemployable. To quote the *Economist*, "If you think education is expensive, try ignorance". As important as this concern is for our nation (sadly enough we have not been making much progress here!), to neglect the nurturing of young talent can only lead to economic and political disaster.

The narrow outlook of orthodox egalitarians in the foundations and in the government, combined with the shortsightedness of industry (cf. the MIT study, *Made in America*) drag down efforts to cull out talent. They vigorously insist that the "elite" are the affluent rather than the shock troops of exploration and intellectual development.

Although there exist vivid instances of late flowering of talent, by and large intellectual awakening of creative people can be traced to early influences. All too often early awakening of talent brings the individual into conflict with the existing educational practice (Pasteur, Pierre Curie, etc.).

The eloquent appeal of a distinguished scientist, Freeman Dyson, for greater courage in probing for new ideas will not bear fruit unless a habit of adventure is acquired early in life.

A Treasure Hunt

We justifiably refer to our creative people as the "elite". The well being of every profession depends upon its capacity for renewal through the discovery and development of new talent and upon attracting this new talent to the profession in order to enhance its capacity for creativity. Dynamics of exploration today compels us to be concerned with the development of inquisitiveness in the very young.

Student indoctrination through (teacher proof!) expositions, however eloquent and complete, is not what is needed. This practice has been ineffective. Teachers as well as students should be given an opportunity to explore and improvise.

Our problems have multiplied and have grown to such complexity that capacity for creativity has become a vital national asset.

Despair vs. Hope

Do we have an educative infrastructure to assist us? The tragic reality is that we do not! "Precollege education is nothing short of scandalous," said Dr. D. Allan Bromley at his Senate confirmation hearings as the new science advisor to the President. The sad secondary school reality affects our practices in college.

There is much ancient wisdom which should find its way into the classroom. The teacher is the individual in a strategic position to bring this about. Today more than ever, a dedicated teacher, to be effective, must understand more mathematics and science than is contained in the traditional prescriptions for classroom content. Ultimately, what teachers can (and should) learn in a challenging program of studies becomes pertinent in their own classroom. Teachers' own involvement in exploration leads to a new outlook on the nature of fruitful interaction between teachers and their students.

The need to explore and exploit our native resources of talent is great. It would be politically and socially untenable to deny our own able youth the kinds of opportunities for growth which make our accomplished newcomers from abroad so very valuable. Any genuine search for talent must be carried out with the broadest possible social and economic base. The springs of talent are widely dispersed and are no respecters of geographic, social or economic boundaries.

We would like to think that a growing grassroots involvement of our concerned colleagues may give us hope for a better future.

The dedicated activities in San Antonio of Manuel Berriozabal show vividly (cf. *UME Trends*, May, 1989) that one can achieve excellence in education programs for minorities. The special sessions (organized by Keynes, Fisher and Wagreich) of the Mathematics and Education Reform Network at the AMS meeting in January 1990 bring together accounts of varied projects ranging in their concerns through students, teachers, and curriculum content. Also at AMS in January 1990, a session on Humanistic Mathematics (organized by Alvin White) discusses cultural aspects of mathematics. One should mention that the Mathematical Association of America, American Mathematical Society and American Association for the Advancement of Science all contribute to the support of the American Regions Mathematics League (ARML). This league of secondary schools encourages deep student involvement. Each member high school develops an exploration seminar under the guidance of a master teacher. Once a year, ARML sponsors an imaginative national competition. All this activity is developing a way of collaboration which will facilitate the creation of an elite among all students including underrepresented groups.

Contrary to the all too common justification ("they will make their way") of the neglect of able youngsters, we must realize that the value added by what we can do for them enhances the quality of their lives and contributes significantly to the welfare of our nation. Thus, everybody counts – even our ablest.

Professor Arnold E. Ross, formerly chairman of the mathematics departments at Notre Dame and Ohio State Universities, has been directing programs for mathematically gifted youngsters for over fifty years.

CBMS To Publish Research in Undergraduate Mathematics Education

MAA FOCUS 12:5 (October 1992)

On the recommendation of The Mathematical Association of America (MAA)-hosted conference, Communicating Among Communities, (see MAA's FOCUS, February 1992 or UME Trends,March, 1992), the Conference Board of the Mathematical Sciences (CBMS) has agreed to publish a number of annual volumes describing the state of the art in, and entitled Research in Collegiate Mathematics Education. The volumes, which will appear in the CBMS series, Issues in Mathematics Education, will be co-edited by Ed Dubinsky (Purdue University), James Kaput (University of Massachusetts, Dartmouth), and Alan Schoenfeld (University of California, Berkeley). The MAA's Committee on Research in Undergraduate Mathematics Education (CRUME), augmented to include individuals from the various communities represented by CBMS, will serve as the Editorial Advisory Board.

It is expected that the first volume will appear in 1994. More specific guidelines for authors will be available by January 1993 and will be disseminated widely throughout the mathematical community. While some of the papers appearing in these volumes will be solicited, others are expected to be contributed by researchers. Authors wishing to submit papers should contact one of the three editors.

The following remarks describe goals and rationale in more detail.

The past half dozen years have witnessed astonishing changes in (a) the intellectual community's understanding of the importance of collegiate mathematics education, (b) the development of a community of researchers who have made a firm commitment to doing research on issues of mathematics education at the college level, and (c) the mathematical community's willingness to address fundamental issues of undergraduate education in a serious way.

The best one-line summary of our needs is given in *Everybody Counts*:

> "Reform of undergraduate mathematics is the key to revitalizing mathematics education." (p. 40).

Undergraduate mathematics education affects all aspects of mathematics preparation, for it is in collegiate mathematics classes that the nation's schoolteachers receive their mathematical preparation (and thus the view of mathematics to which their students will be exposed), and it is in those classes that the nation's scientific and mathematical elite receive their fundamental grounding in mathematics.

In the late 1980s, two simultaneous developments resulted in the creation of a strong and growing collection of producers and consumers of research in mathematics education. The first is that a combination of mathematicians and educators have become interested in issues of undergraduate instruction, bringing to those issues the necessary synthesis of mathematics knowledge and increasingly sophisticated research skills. In recognition of the existence and importance of the research community, the MAA has an active Committee on Research in Undergraduate Mathematics Education. It is anticipated that this committee will soon become a joint AMS/MAA committee, with ties to the National Council of Teachers of Mathematics and the American Association of Two-Year Colleges.

The second development is a growing recognition by the mathematical community that fundamental curriculum reform can and should be informed by careful, sustained research into learning and teaching, as well as other factors involving undergraduate mathematics education. This recognition is an outgrowth of mathematicians' greatly enhanced interest in matters educational. It is also one consequence of the development of the field of research in collegiate mathematics education.

As little as ten years ago the thought of education being the topic of serious attention at the Joint Mathematics Meetings would have been almost laughable. But the 1990 meetings featured an invited main presentation on "Teaching Undergraduate Mathematics: Insights from Education Research;" subsequent meetings included an AMS panel on educational issues; and in the past two years' meetings, contributed paper sessions on Research in Undergraduate Mathematics Education (in addition to those sponsored by Mathematicians and Educational Reform, and other educationally related sessions) occupied a significant part of the program and were very heavily attended.

There is now a substantial community of mathematicians eager to know of and use the results of high quality inquiry into collegiate mathematics education. The existence of UME Trends similarly points to consumer demand — but it doesn't offer a scholarly outlet for the research papers themselves. Indeed, the absence of a central print locus for research in undergraduate mathematics education has had a doubly constraining effect. On the one hand, there has been no straightforward way to consistently bring high quality work on collegiate mathematics education to the attention of the mathematical community. On the other hand, the absence of a clear dissemination mechanism has stifled the growth of the research community. Its presence, however, could have a strong catalytic effect.

In recognition of these problems and opportunities, the MAA conference made the following recommendation:

Recommendation 1: The Mathematical Association of America (MAA) and the American Mathematical Society (AMS), in cooperation with the National Council of Teachers of Mathematics (NCTM), should plan a series of annual special volumes presenting exemplary research

papers in collegiate mathematics education. These volumes would serve as precursors to the establishing of a journal.

CBMS has now agreed to publish these volumes in the series noted earlier.

Manuscript Solicitation and Review. Initially, the editors and editorial board will solicit manuscripts designed to focus on major issues and showcase research of the highest quality and significance. Such work may include integrative summaries of what is known in areas of critical importance (e.g. calculus reform, linear algebra, gender and minority issues, the uses of technology in instruction, functions and reasoning about them); they may include discussions of methodological concerns; they may include exemplary individual studies exploring aspects of mathematical thinking or instruction at the collegiate level. Of course, the news of the volumes' impending existence will be widely announced, and manuscripts will be actively solicited through the announcements. All submissions, whether they have been solicited by the editors or not, will receive a minimum of three independent reviews, and final decisions will be made in conference by all three editors (at least one of whom will have been one of the reviewers for the manuscript under discussion).

Contact one of the editors if you would like more information about the volumes or about submitting articles.

The International Congress on Mathematical Education: Its Roots

E.L. Poiani, MAA FOCUS 12:2 (April 1992)

The roots of the International Congresses on Mathematical Education, more popularly known as "ICMEs," trace back to a paper by the group of American commissioners who participated in the first International Commission on the Teaching of Mathematics. It reads as follows:

"The preparation of the reports [on the teaching of mathematics] calls for a comprehensive survey of our educational system in general and of the work in mathematics in detail; for a sketch of the unparalleled activities of recent decades in the development of existent institutions and in the genesis of new ones; [and] for an account... [of] the reforms that are still under consideration... Work in mathematics must be regarded and interpreted in the light of its environment, and our reports should furnish the reader of other nations with information respecting our educational system and conditions analogous to that which we shall expect from them. ...

America is unique in the liberty left to individual initiative in matters of education, and in the absence of authoritative central legislation and supervision. It is desirable, therefore, that the reports describe clearly the practical working of this freedom and its effect, good and bad, upon our progress in general and in mathematical education."

The goals expressed by these American commissioners ring familiar, although they were written in New York city over eighty years ago (26 and 27 March 1909), just one year after the Fourth International Congress of Mathematicians met in Rome in April 1908 and created the first "International Commission on the Teaching of Mathematics."

Historically, the ICMEs evolved in the following stages:

◇ In 1905, David Eugene Smith, an American mathematics educator, first proposed, in writing, that there should be an international group to look at mathematics instruction and curriculum.

◇ The International Commission on the Teaching of Mathematics, created in 1908, was headed by Dr. Felix Klein who remained its president until his death in 1925.

◇ The International Commission met in fall 1908 with delegates from thirty-three countries and concentrated on a survey of teaching practices. Two world wars interrupted its progress, but the Commission reported on mathematics education at the 1932 and 1936 International Congresses of Mathematicians (ICMs).

◇ In 1952, the International Commission on Mathematical Instruction (ICMI) became a commission of the newly created International Mathematical Union (IMU), with twenty-seven countries forming their own national committees. The IMU belongs to the International Council of Scientific Unions.

In the US, the National Academy of Sciences (NAS) serves as the national member of ICSU. The US Commission is called the United States Commission on Mathematical Instruction (USCMI) and began in 1952. Today, more than sixty countries have national committees.

◇ The Chair of the USCMI sits on the General Assembly of the ICMI. The Assembly convenes at each ICME.

◇ USCMI composition has a broad national base of research and education with two members each drawn from the MAA, the National Council of Teachers of Mathematics, (NCTM), and the United States National Committee on Mathematicians (USNCM). USNCM is the American liaison for the ICMs.

◇ Rifts have existed between mathematics researchers and educators. Reports on mathematics education were tucked into ICM programs, but not ordinarily given much attention.

◇ The need for international congresses focuses specifically on mathematics education became evident, leading to the first ICME in 1969 at Lyons with six hundred in attendance. The other ICMEs with par-

ticipation numbers have been as follows:

   ICME-2 (1972): Exeter (1,400)

   ICME-3 (1976): Karlsruhe (1,800)

   ICME-4 (1980): Berkeley (2,000)

   ICME-5 (1984): Adelaide (2,000)

   ICME-6 (1988): Budapest (2,400)

◇ Quadrennial congresses are now held regularly. The next, ICME-7, will be held 17–23 August 1992, at the Universite Laval, Québec, Canada, 3,000 participants are expected.

Sometimes the only international news on mathematics we read is comparisons of performance on International Assessment Tests. ICMEs go far beyond this level and offer special benefits to the participants and those with whom they interact. ICME-6 provided many networking opportunities for the mathematical community, constructing bridges within a beautiful city of bridges. ICME-7 will build on the previous exchanges and look at new developments in mathematical education. For example, the dynamic changes caused by the dissolution of the Soviet Union should have an impact on how and what is taught in the mathematics classroom.

Eileen L. Poiani is a former Chair of the United States Commission on Mathematical Instruction (USCMI) and a professor of mathematics at Saint Peter's College.

The Japanese University Entrance Examinations in Mathematics

R. Askey, MAA FOCUS 13:1 (February 1993)

The December issue of FOCUS included a special supplement on the Japanese University Entrance Examinations. In this month's Personal Opinion column, Professor Richard Askey of the University of Wisconsin at Madison, considers some of the issues raised by the appearance of that document.

The National Educational Goals include having U.S. students first in the world in mathematics achievement by 2000. I hope all of us know this is impossible, although the mathematical community has not said so publicly as far as I know. However, impossible or not, it is necessary to try to improve our mathematics education, and it is useful to look at the rest of the world to see what students can learn.

Japan has an excellent educational system, and information on it at most levels is now available in English. Preschool is very important for socializing the students for school. See Peak [2]. Elementary schools are the backbone of their success. See Stevenson and Stigler [4] for an account which is slanted toward mathematics education. There is an excellent video [3] which shows what goes on in some Japanese and Taiwanese elementary school classrooms. At the lower secondary school level it is now possible to look at translations of textbooks. See [1].

The exams published in [5] provide an opportunity to see what students have learned.

First, there is very little which is learned by rote. That is already true in elementary school, as one can see in the video [3]. It continues to be true as seen in these exams.

Second, when our SAT is compared to the Japanese UECE exams we see part of our problem. We ask one step problems, they ask many step ones. While either type of question will probably give a similar ranking to a large group of students, the message being sent about how well one needs to learn something is completely different.

Third, a large part of our calculus reform movement has decided that the algebra skills of our students are so poor that they get in the way of students learning the ideas of calculus, and so have looked for ways around this. Seeing what the Japanese students can do in the way of algebraic calculations makes it clear that most students can be taught this. I worry very much about the message we are sending back to schools about the lack of importance of technique.

The way in which answers are given in the UECE exams is very interesting, and is so much better than the way we give answers in multiple choice exams that I felt stupid for not having thought of this possibility for giving machine graded exams. I hope it will be adopted both for national exams and for placements tests when it is necessary to grade a large number of exams in a very short time. The care with which the UECE exams are analyzed after they are given is impressive, and we should do something similar. Here the Japanese have an advantage over us, since they have a real national curriculum rather than the defacto one we have, and so are better able to spot places where the essential ideas were not learned by enough students, and can feed this back into future texts. This would be much harder for us to do, but we do not even try. Some readers will comment on the absence of many applied problems. These exist, but are given in the science exams. When all of the students have a good mathematical background, science courses can be more mathematical than ours are.

Japanese cram schools exist, and as you can see from the comments, the Japanese educational establishment does not like them. I was pleasantly pleased by the response from students when they were asked if they thought it was sufficient to have just studied their texts to prepare for the UECE exams. Even a majority of the humanities students said yes. The exams given by individual universities are in general harder, and here some extra preparation is probably useful. There are high school mathematics magazines completely unlike any we have. The best is monthly, almost 90 pages per issue, and in addition to many problems and solutions,

has articles about more advanced mathematics. A colleague, H. Terao, said he learned about groups and some of their properties from reading one of these magazines when he was in high school. We need something similar here to supplement "Quantum", which has a different focus.

The individual exam which surprised me most was the one for the Education Division of Shiga University. Shiga is one of the prefectures, so the students taking this exam are similar to the students studying education at a school like the University of Wisconsin. After having taught one of our two courses for elementary school teachers, I have a "modest proposal". Our elementary school teachers will teach mathematics every day, so they should know as much mathematics when they graduate from college as future elementary school teachers in Japan know when they graduate from high school. At present the best of the 24 students I had last semester should be about ready to start the Japanese 9th grade after they take our second course for teachers. This goes a long way toward explaining where some of our problems are. This is our responsibility, for we teach these teachers. Our texts are also part of our problem. They have too many worked examples, home work problems are just like the examples in the book, and there are too few problems where the students have to think. Here some of the reform efforts are good, but I am frightened by comments like those in an interview in the last newsletter from the Harvard Calculus Consortium. After saying that students' manipulation skills have become much weaker, Anthony Phillips says; "And the CCH curriculum makes a great virtue out of this necessity. By eliminating some of the symbolic manipulation from calculus, they were able to make the course more accessible to students." There are important ideas behind successful symbolic manipulation, and when this is not realized and taught, we cheat our students. The Japanese do this, we must also.

[1] Japanese Grade 7 Mathematics, (also grade 8 and grade 9), The Univ. of Chicago School Mathematics Project, 1992.

[2] L. Peak, Learning to Go to School in Japan: The Transition from Home to Preschool Life, Univ. of California Press, 1992.

[3] The Polished Stones, video available from Center for Human Growth and Development, Univ. of Michigan, 300 N. Ingalls, Ann Arbor, MI 48109

[4] H. Stevenson and J. Stigler, The Learning Gap: Why Our Schools Are Failing and What We Can Learn from Japanese and Chinese Education, Summit Books, New York, 1992.

[5] Ling-Erl Eileen T. Wu, Japanese University Entrance Examination Problems in Mathematics, Math. Assoc. of Amer. 1992.

Richard Askey is a professor at the University of Wisconsin–Madison. The opinions expressed in the above article are those of the author, and do not nec-

essarily represent the views of the MAA.

Mathematics Education: A Case for In-Context Placement Testing

M. Hassett, F. Downs, J. Jenkins
AMATYC Review 14:1 (Fall 1992)

Introduction

Until 1988, placement testing for entry level mathematics courses at both Arizona State University (ASU) and the Prescott, Arizona campus of Embry-Riddle Aeronautical University (ERAU) was mandatory and followed a model which is widely used in post-secondary institutions. A multiple choice test was given to all incoming students to assess their background in mathematics. It was furnished by an outside agency which requested that the questions not be released to the public and was not written to correspond directly to any particular college course. This type of placement testing will be referred to as "out-of-context" placement, since it is designed to test prerequisites entirely outside of the course which the student wishes to take. In 1988, the authors independently became dissatisfied with the results of their placement programs and instituted new placement programs which turned out to have essentially the same structure.

In the new placement programs, students are advised to select courses on the basis of prior mathematics coursework at either pre-college or college levels. This advice is intended to place each student in the highest level mathematics course in which he or she should be successful while at the same time advancing the student's academic program. All entry level courses (from algebra through calculus) begin with a review of important prerequisite material. Students are tested over this material at the end of the first few days of class. Students scoring poorly are advised to drop to the prerequisite courses during a special drop-back period. This is an "in-context" placement system.

In independent trials with this in-context placement structure at each institution, an improvement in student attitudes about the placement process has been observed. There has been an increase in the number of students electing to start at higher levels with no decline in success rates. In fact, most courses have improved success rates. The relative merits of in-context and out-of-context placement will be discussed in detail in this paper.

Problems Arising From Out-of-Context Placement Testing

Theoretical Problems. Research in statistics and educational psychology has shown that the correlation coefficient between a placement test and a course grade should be .80 or higher if use of the test is to have a significant effect on success rates. When the correlation is lower, use of the test is not likely to result in a significant improvement in the success rate. In fact, an unacceptable number of students may be misadvised by their placement scores. Hassett & Smith (1983) summarized this research and applied it to mathematics placement.

Other studies have also concluded that many placement tests do not reach the desired correlation of .80 with course grade. Noble & Sawyer (1988) state:

> Numerous studies have examined the relationships between admissions and placement test scores and specific course grade. . . . The mathematics validity studies comprised a larger portion of the research on predicting specific course grades. A variety of predictors were used, including ACT subtests and composite scores; SAT-V, SAT-M,, and SAT-Total scores; high school rank; and scores on specifically developed mathematics placement tests. The correlation coefficients ranged from .04 to .75. (p. 3)

The inaccuracy of out-of-context placement for many students is also apparent in a study done at Bucks County Community College in 1986 and 1987. Hoelzle (1987) showed that 47% of the students who placed themselves above the course predicted by the placement test were successful.

In 1987, both ERAU and ASU used the MAA placement tests. The correlations with course grades ranged from .30 to .50. Correlations using other available standard instruments (ACT math, SAT-M) were in the same range or lower. The placement instruments available to us did not correlate well enough with course grades to be used as the main predictors of success.

Practical Problems. Advisors believed that placement test scores were extremely valid. Thus they were hesitant to permit students to register for courses other then the ones indicated by the placement test. This tended to make advising superficial. Since some advisors were unfamiliar with the course content of mathematics courses, they accepted the results of the placement test as correct even when the student protests were supported by sufficient preparatory coursework.

Most of the students who took the out-of-context mathematics placement tests were not enrolled in a mathematics course when they took the test. Moreover, the importance of the test was often not made clear to the students, so that they made no effort to review any material. As a consequence, their mathematics knowledge and skills were at an ebb and they tended to score lower than they would have scored during the school year.

The combination of advisor ignorance and student rustiness forced many students to register for courses they had already passed elsewhere. When those courses were remedial, the students were delayed a semester or more in starting required courses. They also adopted poor work habits that usually led to low or failing course grades. Instructors in remedial classes complained of poor attendance and poor attitudes by students who said they already knew the course material.

Many students became hostile. They had good reason. They had passed the course at another institution, but were being required to repeat it. The placement test was perceived as a barrier to registration, not an aid. A great amount of office staff time was spent with students complaining about the system and filling out petitions to override placement scores.

The placement test was acting as a filter instead of a pump in the mathematics pipeline. The university system was placing students who had taken four or more years of college preparatory mathematics in high school into a precalculus curriculum. Consequently, many competent high school students were being turned into mediocre college students in one short semester. Furthermore, the placement process was not meeting its goal of improving success rates in the targeted classes.

There was also an adverse financial effect. University funds that could have been used on instructional assistance were being spent on the mandated placement testing program. Of course, out-of-context testing does make an advisor's job easier. Advisors can process a large number of student students in a relatively short amount of time because they have a relatively short amount of time because they have a template to follow. Such "placement procedures tend to be used for their practical advantages, such as their greater ease of operation, or their greater ease of explanation to staff and students" (Noble & Sawyer, 1988, p.2).

Advantages of In-Context Placement Testing

In 1988, both ASU and ERAU eliminated compulsory use of out-of-context placement testing and began to use in-context testing. Data analysis of the effects of their change in approach will be discussed in the next section, although additional analysis remains to be done. Analysis notwithstanding, the practical advantages of this new system were immediately obvious.

Advising Became More Thorough. At Arizona State advisors received new materials on how to look at a students' record in mathematics. A self-advisement flow chart was created that students and advisors could use to guide the choice of a mathematics course based upon a student's background and past performance in mathematics courses. Seminars for advisors stressed the importance of focusing on each student's mathematics preparation rather than on a single placement test score. Without a trivial algorithm advisors were forced to worry about the quality of their advice. The University Testing Center still provided the MAA placement tests as an optional support vehicle and campus computer sites made available the self-advising computer program *Are you Ready for Calculus?* This is a program developed by Dr. David Lovelock of the University of Arizona.[1]

At Embry-Riddle, advisors were given more training in mathematics placement. The time usually spent during student orientation for placement testing was used for mathematics orientation. This was directed by a faculty member from the mathematics department. Students were allowed to ask questions and they were given outlines for the various courses open to them. They were encouraged to sign up for the highest level course for which they felt they had a chance to be successful. Accordingly, students who had done well in a year of high school calculus, but who had not taken the AP Calculus Examination, were allowed to register for second-semester calculus.

On Both Campuses Student Hostility Disappeared. Teachers and office staff no longer have to contend with hostile students who have covered the material before. Students are no longer being told that they cannot enroll in a mathematics course because they have "failed" the placement test. Instead they are told that the Mathematics Department will help them decide if the course they have enrolled in appears to be too difficult, but will leave the actual decision up to each student.

At ASU students are given two weeks to make their decision. The first week's assignments are based on a review of essential prerequisite material, and a test on this material is given on Thursday or Friday. Students who score below 70% are advised to drop to a lower level course. This can be accomplished through the end of the second week of the semester by special arrangement with the registration office.

At ERAU students have the first three weeks to decide if they wish to transfer. This is an extension beyond the normal drop/add period, but because results have shown significant positive benefits the administration has allowed this arrangement to continue.

Enrollment in Remedial Courses Decreased at Both Schools. In general, students enrolling in remedial courses do so because of a rational self-assessment of their needs. The experience of the course instructors through several semesters bears out the validity of this statement. The teachers of these courses have indicated that attendance has improved and that student attitudes are better. As was hoped, success rates in introductory classes have not deteriorated. With the absence of compulsory placement testing student elect to start out in a higher level course, and they do so with less risk of failure.

Results of Preliminary Data Analysis

The goal of in-context placement is to put each student into the course most likely to lead to success. As the new system began to operate, trends in enrollment and success rates were observed and data were collected. This section contains analysis in three areas: predictive

---

[1] This software, along with similar "readiness" software, is in the public domain. It can be obtained for nominal charge to cover costs of the computer disk, handling and mailing. Address requests for information to Mathematics Department, University of Arizona, Tucson, AZ 85721.

validity, enrollment trends, and success rates.

Predictive Validity. An earlier study (1983) at ASU had indicated a correlation of .80 between course grades and grades on the first week review test in College Algebra. Similarly, a correlation of .85 between course grades and grades on the first week review test in Calculus I was observed at ERAU in 1987. These correlations fell within the desired range and were used as justification for using the first test as a placement criterion. No revalidation was planned under the current structure since the policy of encouraging students with low scores to leave the course makes it impossible to see how they would do if they stayed.

However, one new instructor at ASU, teaching 490 students in three large sections, failed to inform students with low grades that they could drop to a prerequisite course. The grade results from these sections provide data which allow us to study the effectiveness of the review test as a predictor. This data indicate that the first test was quite useful as a placement advisory tool:

Table 1. Intermediate Algebra: Fall 1989 Course Grade vs Score on First Week Test in Numbers of Students at ASU

		Course Grade		
		ABC	DEW	Totals
1st	ABC	180	108	288
Week	DE	32	170	202
	Totals	212	278	490

(a) It is clear that students with low first scores were at risk. Of the students who were unsuccessful, i.e., D or E, on the first test, 84% (170/202) failed to achieve a satisfactory course grade. In contrast, 63% (180/288) of those students who received a satisfactory grade on the first test passed the course with an A, B or C.

(b) By using the first test score as a predictor, it is possible to identify an at-risk group whose actual success rate in the course was only 16% (32/202). If this entire group, i.e., 202 students, were excluded, the success rate would be raised from 43% (212/490) to 63% (180/288).

(c) If the first test had been used for placement, only 15% (32/212) of the students who passed the course would have been excluded while 61% (170/278) of those who eventually failed the course would have been excluded. The key objective of a placement test is to identify students who are at risk of not succeeding in the course. the major error would have been the exclusion of students who in fact could satisfactorily pass the course. In identifying 61% of the students destined to fail and only 15% of the students appearing to be at risk but who actually achieved success in the course, the value of this in-context procedure is apparent.

Enrollment Trends and Success Rates. Both ASU and ERAU have observed decreased enrollment in re-

medial courses and increased enrollment in standard required courses. In looking at the engineering and computer science freshmen at ERAU over a four year period, we note that enrollment in Calculus I has continued to rise. These students take Calculus I as their first mathematics course that counts toward the degree. In 1987, 51% of these students registered for College Algebra based on the results of an out-of-context placement test. In 1988, the in-context test was first used as a mandatory placement vehicle and all students who failed this test were automatically dropped back to College Algebra (Jenkins, 1989). In 1989 and 1990 the drop-back after the first test became advisory rather than mandatory. Table 2 shows the placement and success rates for these four years.

Table 2. Placement and Success Rates for Freshmen Engineering and Computer Science Students at ERAU

	College Algebra Place (ABC)	Calculus I Place (ABC)	Calculus II Place (ABC)
1987	51%(88%)	42%(60%)	7%(75%)
1988	27%(65%)	69%(67%)	4%(67%)
1989	17%(52%)	79%(77%)	4%(86%)
1990	14%(71%)	80%(60%)	6%(75%)

Has the change in placement policies favorably affected the success rates at ASU and ERAU? It certainly has not lowered success rates at either school. At Embry-Riddle the success rate, i.e., A, B, or C, in Calculus I fluctuated somewhat, but the enrollment increased dramatically. What is significantly more important is the rise in the percentage of students who achieved success in their first required math course, i.e., Calculus I, in their degree program during their first semester in school. This can be seen from Table 2. Consider all of the freshmen in engineering or computer science at ERAU to be the "calculus pool." In 1987, 30% $[(.42)(.60) + (.07)(.75)]$ of this pool was successful in a calculus course in the first semester. The remaining 70% of the pool either took College Algebra or received a D, E or W in calculus. In 1988, the first year of in-context placement, 49% of the calculus pool students were successful in calculus in their first semester. In 1989 the success rate took another jump to 64% before tumbling back to 53% in 1990. Even this last figure represents a remarkable improvement over the 1987 rate of 30%.

Similarly at Arizona State University the percentage of College Algebra students earning A, B, or C has increased from 50% to 65% since the change in placement testing. The A, B, C rate in beginning calculus remained at the same level of 48% until the Fall '90 semester, when it rose to 57%. While there have been changes in personnel, and, consequently, instruction, in these math courses over the time in question, the dire predictions by some faculty of large scale failures without out-of-context placement testing have not occurred.

It seems clear that in-context placement pumps students into the major in accordance with their original career choices, whereas the mandatory out-of-context placement filters them out, lowers self confidence, and causes unnecessary changes in their career choices. A memorable, and tragic, example of a freshman student at ERAU illustrates this point. In fact, it was primarily this case history that provided impetus to the author to instigate an in-context placement program.

The student had five years of college preparatory mathematics in high school. However, she did poorly on the mandatory out-of-context placement test and was forced to take College Algebra. Part of the reason for her low score could have been that some of the questions on the placement test covered material she had not seen in over two years. Realizing that she knew the mathematics of College Algebra, bored and lacking challenge, she rarely studied and skipped several classes. Consequently she earned a course grade of B, instead of the A she should have had. The next semester the same pattern with the same result was repeated in Calculus I. The following fall semester she did poorly in Calculus II. She had finally run into some new material and had forgotten how to study mathematics. What had the system done? Through the use of an inept placement process it had taken a highly competent high school student and turned her into a mediocre college student.

Summary

An effective placement system is important to helping students and faculty achieve educational goals. Our experience at ASU and at ERAU has led us to believe that reliance on a single out-of-context placement examination is not conducive to achieving the desired degree of excellence, and may even become a harmful filter that keeps students out of the course most beneficial to them. By contrast, in-context testing shows that more students with prior success in mathematics prove to be well prepared if they are given some course time for serious review. We believe that an in-context system involving continuing positive interaction between students and faculty in the classroom is the more effective placement process.

Matt is Associate Chair for Undergraduate Mathematics at ASU. He has taught there since he earned a PhD from Rutgers in 1966. Floyd is Director of Undergraduate Mathematics at ASU. He has an AB form Harvard and an MA from Columbia. John has taught at ERAU since 1978. He has a BA from the University of Tennessee at Chattanooga and an MAT from the University of Florida.

Arizona State University Tempe, AZ 85287-1804

Embry-Riddle Aeronautical University Prescott, AZ 86301

References

Hassett, M.J., & Smith, H.A. (1983). A note on the limitations of predictive tests in reducing failure rates, *Placement Test Newsletter.* Mathematical Association of America, 7,1.

Hoelzle, L. (1987, November). *Bucks assessment testing program.* Paper presented at the 13th annual meeting of the American Mathematical Association of Two-Year Colleges, Kansas City, MO.

Jenkins, J. (1989). *Analysis of placement practices for calculus in post-secondary education.* Final Report, Exxon Education Foundation Grant. (ERIC Document Reproduction Service No. ED 313 084).

Noble, J.P., & Sawyer, R. (1988). *Predicting grades in specific college freshman courses from ACT test scores and self-reported high school grades* (ACT Research Report Series 87-20). Iowa City, IA: American College Testing Program.

Committee on Testing

MAA FOCUS 12:3 (July 1992)

The mission of the Committee on Testing (COT) involves mathematics testing and assessment for grades eleven and twelve and the undergraduate level. It also maintains and routinely improves the MAAs Placement Testing (PT) Program. In addition, COT develops position statements on issues in mathematics testing and assessment (e.g., the use of calculators on placement tests); initiates and oversees externally funded projects relevant to its mission; and cooperates with other MAA committees with related missions (e.g., the Subcommittee on Assessment of the Committee on the Undergraduate Program in Mathematics (CUPM)).

MAA Placement Testing (PT) Program. Since the PT Programs inception in 1977, COT has overseen its activities. In March 1992, more than five hundred post-secondary institutions subscribed to the PT Program. In 1986, the PT Program test packet included six college-level placement tests and two prognostic tests targeted at high schools. During the ensuing years, the PT Program test packet expanded and now contains fourteen tests including calculator-based versions of its Arithmetic and Skills, Basic Algebra, Algebra, and Calculus Readiness tests. The Committee is also developing a new Basic Algebra test and three calculator-based prognostic tests. Two of these calculator-based prognostic tests will require students to use a graphing calculator and the third test, the Calculator-Based Advanced High School test will, for the first time, examine high school juniors enrolled in calculator-based precalculus courses. In 1992, the Committee will begin development of calculator-based versions of the Advanced Algebra and Trigonometry and Elementary Functions tests.

Funded Projects. COT continues to oversee three externally funded projects for the MAA: the Calculator-Based Placement Test Program (CBPTP) Project, the Computer-Generated Placement Test (CGPT) Project, and Teaching Mathematics with Calculators: A National Workshop (TMC).

Grants from Texas Instruments Incorporated fund the CBPTP Project. This project, initiated in late 1986, develops calculator-based placement tests and prognostic tests for the PT Program test packet. When the Committee completes this project in 1993, the PT program will feature six calculator-based placement tests and three calculator-based prognostic tests.

The Fund for the Improvement of Postsecondary Education (FIPSE) funds the CGPT Project. This Project uses the item-generating functions COT developed to produce software that, with monitoring, will generate PT Program tests. During these production years, this Project has also conducted research on the statistical parallelism of PT Program test items and has adjusted the parameters associated with each item-generating function to produce the desired parallelism. Furthermore, the software facilitates customization of PT Program tests, and COT will soon offer PT Program subscribers an opportunity, on a trial basis, to order custommade placement tests.

Both the MAA and the National Council of Teachers of Mathematics (NCTM) administer the Teaching Mathematics with Calculators: A National Workshop; both the National Science Foundation (NSF) and Texas Instruments Incorporated provide financial support. This Workshop prepares middle and secondary mathematics faculties in the Mesquite and Fort Worth, Texas school districts to use calculators effectively in their classes. In addition, the workshop has developed and released two instructional packages containing a videotape and printed materials designed to aid teachers from other school districts implement the use of calculators. During 1992–1993, the TMC Workshop will conduct summer institutes in both school districts and develop two additional instructional packages.

Outreach and Training. COT members continue to address mathematics and assessment testing issues at national meetings; during the past twelve months, COT members have delivered talks at the annual meetings of the National Association of Developmental Education (NADE), the National Council of Teachers of Mathematics (NCTM), the International Conference on Technology in Collegiate Mathematics, and at the MAAs own 1992 Annual Meeting in Baltimore, Maryland. COT continues to offer its successful minicourse on placement testing at the Associations annual meetings. It also offered a similar minicourse at the annual meeting of NADE.

To receive an MAA Placement Testing (PT) Program informational packet, contact: Hanta V. Ralay, MAA Placement Test Coordinator, The Mathematical Association of America, 1529 Eighteenth Street Northwest, Washington, DC 20036-1385; (202) 387-5200; maa@athena.umd.edu. Fax: (202) 265-2384.

For additional information on COT activities and projects, contact: John G. Harvey, Chair, MAA Committee on Testing (COT), Department of Mathematics, University of Wisconsin at Madison, 480 Lincoln Drive, Madison, Wisconsin 53706-1388; (608) 262-3746; harvey@math.wisc.edu. Fax: (608) 238-4477.

The Major in the Mathematical Sciences
B. A. Case, MAA FOCUS 11:3 (June/July 1991)

Copies of the full report are available from the Mathematical Association of America by calling 800-331-1622 or 202-387-5200.

Approximately once a decade CUPM issues recommendations concerning the undergraduate major. In January 1991 CUPM endorsed a report from its Subcommittee on the Major (SUM) that updates the more lengthy 1981 *Recommendations for a General Mathematical Sciences Program* (which was reprinted as the first six chapters of *Reshaping College Mathematics*, MAA Notes No. 13, 1989). The new CUPM report emphasizes a unified (but not uniform) major in the mathematical sciences which supports various concentrations or tracks.

Although forward looking, these recommendations are anchored firmly in the reality of current practice and owe much to the 1981 CUPM *Recommendations.* The first five of nine tenets of program philosophy are similar to those in the 1981 report, sharpened to encourage independent mathematical learning and written and oral communication of mathematics. Four added tenets deal with availability of options, increased advising responsibilities, effects and applications of technology, and easing transitions between levels of study. A unified structure, presented as a tool for fashioning departmental course requirements, encourages broad course choices. Issues of teaching and advising are natural adjuncts to the discussion.

Philosophy

The report describes a curricular *structure* with fixed *components* within which there is considerable latitude in specific course choices. Combined with specialized curriculum concentrations or *tracks* within the major, this structure provides flexibility and utility. The structure involves both specific courses (e.g., "linear algebra"), and more general experiences (e.g., "sequential learning") derived through those courses. By making appropriate choices within components, students can obtain a strong major for prospective secondary teaching or for graduate school preparation. Although most of the tracks may be considered as options within applied mathematics, the needs of a student desiring a background in classical pure mathematics can be met by careful course selections. Every applied mathematics track should include fundamental components of "pure" mathematics, just as each pure mathematics student should gain experience with applied and computational mathematics.

The component structure with tracks is typical of the pattern of many of today's undergraduate mathematical sciences departments in that it allows many curricular choices. Track systems lead students to make lifetime choices with only minimal knowledge of their ramifications. Consequently, departments must accept advising responsibilities significantly greater than what was necessary in earlier, simpler days when all mathematics majors took essentially the same sequence of courses. Students need regular and frequent departmental advice to explore the many intellectual and career options opened by a mathematical sciences major. Advisors should pay particular attention to the need to retain capable undergraduates in the mathematical sciences pipeline, with special emphasis on the needs of underrepresented groups.

Components and Tracks

Seven components form the structure of the mathematical sciences major:

A. Calculus (with differential equations)

B. Linear algebra

C. Probability and statistics

D. Proof-based courses

E. An in-depth experience in mathematics

F. Applications and connections (e.g., computer science, physics)

G. Track courses, departmental requirements, and electives

The first six components normally require nine or ten courses (typically 3 or 4 semester-hours each), at least seven of which would be taught by the major department. The seventh component highlights the common practice in many colleges of offering formal clusters of courses from which students majoring in the mathematical sciences may choose; these tracks may be offered in one or several mathematical sciences departments.

The report includes information about typical tracks, including actuarial mathematics, administrative or management science, computational and applied analysis, computer analysis, data or systems analysis, discrete mathematics, operations research, pure mathematics, scientific computing, statistics (applied), teaching (secondary). In some, specificity of the structural component courses may be desirable. For others, many courses may be required (either outside or inside the mathematical sciences) to complement the structural components. Note that no track is specifically labeled as preparatory for graduate study: with appropriate choices for the structural components, most tracks could provide sound preparation.

Completing the Major

In addition to courses and components, the mathematical sciences major should also involve a variety of other types of experiences and activities that are, in some cases, "co-curricular." Several supportive activities are specifically cited in the CUPM report as contributing to students' self-confidence and ability to work with others:

Integrative experiences. Every major should be encouraged to think about the mathematical sciences as a whole. Common means include senior independent projects, research or scholarly investigations, problem-oriented senior seminars, independent study that includes writing about some area of mathematics, seminars in which students present accessible journal articles, or an undergraduate colloquium series.

Communication and team learning. Major programs should prepare students to communicate mathematics, both orally and in writing. Communication skills are honed when upper-division students assist freshmen and sophomores, tutor high school students, or work in teams to investigate mathematical problems of some complexity.

Independent mathematical learning. Whether a mathematics graduate enters a mathematical sciences career immediately, goes directly to graduate school, or enters another career path, the student will need to function as an independent learner. Independent study projects and teaching strategies that increase the numbers of student presentations encourage independent learning.

Structured activities. Out-of-class student activities have a long tradition on many campuses, including mathematical honorary societies, MAA Student Chapters, local club activities, and honors programs. Other activities can provide valuable broadening opportunities in teamwork and independent learning; these include preparations for mathematical contests, undergraduate research experiences, and internships and cooperative education.

Implications

The statements of philosophy in the report embody educational principles that can lead to an enriching educational experience and the recommended program structure provides a flexible vehicle for fulfilling those principles. One underlying tenet, however, transcends the particular form of curriculum implementation: It is only by requiring substantive achievement of our students that we will be able to produce the sort of quantitatively expert individuals who are going to be the mainstay of the discipline and of society for the next century.

A Forum to present and discuss these recommendations will be held at the Orono meeting in August 1991.

Bettye Anne Case is at Florida State University and served as Chair of the CUPM Subcommittee on the Major which prepared the report.

Undergraduates Experience Research Firsthand

SIAM News 23:4 (July 1990)

What does it mean to be a mathematician?

The National Science Foundation's Research Experience for Undergraduates (REU) program helps talented undergraduate students answer this question.

The REU program, established in 1987, provides undergraduates with opportunities for active participation in mathematics, science, and engineering research. Exposing talented students to the excitement of research when they are making career choices may encourage these students to continue their science and mathematics studies after graduation. Currently NSF budgets approximately $14 million annually for the program.

Approximately $500,000 of each years total is used to fund 15 selected mathematical research sites around the country. Between six and 12 students participate at each site, for eight to 10 weeks of intense research, usually in the summer. The program configuration varies from site to site. A student might contract with a faculty member to work on a specific project, a group of students might work together on a project, or the students in a group may work on individual but related problems.

Among the areas of mathematics represented in the program are combinatorics, graph theory, dynamical systems, group theory, matrix analysis, and game theory. Topics are selected for their accessibility to students at the undergraduate level. The student researchers vary from very talented freshmen to seniors, although it is mostly sophomores and juniors who participate in this highly career-oriented program. The students live on campus for the duration of the program, often sharing apartments or dormitory rooms.

Students are encouraged to write up their results in a form suitable for journal submission, and rewrites and revisions can continue throughout the year. Students are then encouraged to attend regional or national meetings to talk about their results.

A Summer of Math on Lake Superior

Every year since 1977, with one exception, Joseph Gallian, a professor of mathematics at the University of Minnesota, Duluth, has conducted a mathematics research program for undergraduates. NSF has funded the program for 10 years, and since 1987 the Duluth site has been part of the REU program. Gallian's program runs for 10 weeks in the summer. The number of participants has varied over the years from two to six. Selection criteria include letters of recommendation, Putnam competition performance, home school reputation, and responses to questions on the application.

The Duluth program is loose and informal. The students live on campus, sharing three-bedroom apartments. Each student is assigned his/her own problem but is encouraged to discuss the work with the other program participants. The atmosphere is cooperative rather than competitive. In addition to the undergraduate researchers, a few past participants in the Duluth program return as research advisers. David Witte first participated in the Duluth program in 1979 as a student; he returned as a research adviser each summer for the next eight years.

Each week the students present talks on their work of the previous week. Occasional guest speakers present colloquia. As a group the students and Gallian lunch together a few times a week, and Gallian is available to meet with the students individually. The last week the students present one-hour colloquia on their work. One of the main goals of the Duluth program is for each participant to publish the results of his/her work. With this goal in mind, Gallian typically chooses recently posed problems that have a short literature and for which at least partial results are probable. Gallian starts the summer with twice as many problems as students, and considers the students' backgrounds when assigning problems. Sometimes a problem turns out to be too hard for a student – in that case a change can be made. Gallian does not see this experience as a job for the students and so considers fun an important part of the summer. He organizes many "field trips" for the students including kayaking, picnicking, softball games, and visits to Lake Superior and area parks, and the math department at Duluth provides a car for the students' use on weekends. Mount Holyoke Program Mimics

Research Institute

In 1988 the Department of Mathematics at Mount Holyoke College established an undergraduate research site with partial support from the REU program. The Mount Holyoke program runs for nine weeks, from June to July, under the direction of Donal O'Shea. The other principal investigators are Janice Gifford, Alan Durfee, Mark Peterson, and Lester Seneshal. Twelve students participate each year, working in groups of four. Under the supervision of a faculty member, each group carries out a separate project. The students come from various schools in the New England area.

The premise of the program at MHC is "that exploitation of the computational potential of workstations tied to minicomputers offered a good possibility for engaging undergraduates in mathematical research in areas usually considered to be beyond the scope of undergraduates." The organizers of the program feel that through the project they can utilize the advantages of working with undergraduates, who need not be as results-oriented as graduate students and can be assigned "high-risk/big-payoff" questions.

Each faculty member selects a topic – in pure mathematics, statistics, or applied mathematics. The participants are selected with a balance of talents in mind, and at least one student in each group has extensive computer experience. The program was designed to evoke the atmosphere of a typical scientific research institute. Each group is assigned a large room with chairs, chalkboards, a conference table, and several DEC and Sun workstations and a terminal networked to a mainframe. All three groups share a common room with a blackboard, coffee-maker and refrigerator. Each group begins its day by meeting with its faculty adviser, where a plan is made to divide the day's work, and other meetings and reports are scheduled. Each group meets again in the afternoon, and the day ends with afternoon tea. The students give formal reports each Tuesday at the program's colloquium, and guest speakers give talks on Thursdays. The last week of the program is devoted to the preparation of final project reports.

Because of the group nature of this project, the issue of differing levels of preparation and ability arose. The organizers of the MHC program decided "that differences in ability are a part of life and the sooner one finds out about them the better." With regard to differences in preparation, each of the three problems involved a large amount of mathematics so that everyone, including the faculty, needed to learn a great deal. The students were very cooperative and even exploited the differences in preparation – they taught each other the necessary information. According to O'Shea, "The overwhelming positive feature of the MHC model is that the amount of mathematics and computer science each student and faculty member learned was truly impressive."

Problems Beyond the Math

Followthrough of the students' work – the writing, editing, and revisions necessary for publication – can be the most difficult part of these programs, according to Gallian. By the final week few students have their work ready for publication. In most cases, contact between the students and faculty continues throughout the school year. The adequacy of financial support is a consideration for program leaders. Gallian feels that the NSF-recommended stipend levels are low, especially given the fact the students who qualify for research programs often have other, more lucrative, summer job opportunities. The faculty involved in these programs must also make financial sacrifices, according to Gallian. Four weeks' salary is the maximum the NSF grant has ever paid him, he says.

According to John Ryff, REU coordinator at NSF, the sacrifices imposed on faculty advisers by the award limitations are recognized at NSF. He says, "An increase in the average [grant] to $5,000 per student is expected. This should permit greater budget flexibility."

Even with financial support, a substantial time commitment may be demanded of the faculty involved in these programs. Thus, it is imperative that the projects directly benefit the faculty members' research.

Despite these problems, Gallian and O'Shea remain enthusiastic about working with undergraduates outside the regular academic curriculum. Ryff has visited a number of sites and considers the program "a smashing success." Patrick Headley, a participant in the Duluth program, gives the student perspective:

I made the right choice in going to Duluth, and I would advise other undergraduates to enter similar programs. I have a much better understanding of what it means to do mathematics – deciding what is worth trying, applying ideas persistently, and communicating those ideas to others. . . . Many undergraduate mathematics majors enjoy mathematics but are not sure what it means to be a mathematician. If these students are given the opportunity to do original work, they will be better students and may stay in mathematics when otherwise they might not.

For more information, contact John V. Ryff, REU Coordinator, National Science Foundation, Division of Mathematical Sciences, 1800 G Street, NW, Washington, DC 20550; (202) 357-3455; jryff@nsf.gov.

AMATYC Aims to set Standards for the First Two Years

A. Selden and J. Selden
UME Trends 5:1 (March 1993)

Professional standards were a big concern for the approximately one thousand attendees at the eighteenth annual conference of the American Mathematical Association of Two-Year Colleges in Indianapolis last November.

Currently a volunteer organization with no headquarters or permanent staff, AMATYC is well aware of the fact that two-year colleges teach over half the students taking lower-division mathematics and thinks the time has come for sweeping organizational change. Supported by a grant from the Exxon Educational Foundation, AMATYC engaged in a year-long planning effort resulting in a strategic plan with twelve main goals and numerous lesser objectives. This was approved by its contentious delegate assembly, whose energetic debate used different microphones for those speaking for or against an issue.

AMATYC's mission statement contains a call for the development and implementation of curricular, pedagogical, assessment, and professional standards for two-year college mathematics education. AMATYC would like graduate mathematics departments to establish master's and doctoral programs specifically for preparing two-year college faculty consistent with guidelines passed at this conference. Prepared by the Education Committee's Qualifications Subcommittee, chaired by Greg Foley of Sam Houston State, these *Guidelines for the Academic Preparation of Mathematics Faculty at Two-Year Colleges* have been under discussion by the membership for six years. They recommend that two-year college mathematics instructors should begin their careers with at least a master's degree in mathematics or a related field with at least 30 semester hours of graduate-level mathematics, with courses chosen from such areas as discrete mathematics, computer science, modeling, probability, statistics, combinatorics, as well as linear and abstract algebra and real and complex analysis.

New hires should have teaching experience at the secondary or collegiate level, which could be fulfilled through a program of supervised teaching as a graduate student. In addition, they should have coursework in pedagogy such as the psychology of learning mathematics; using calculators and computers to enhance mathematics instruction; measurement, evaluation, and testing; and a college mathematics teaching seminar, an outline of which was given. Hiring committees should consist primarily of full-time two-year college mathematics faculty who are in a position to evaluate candidates' credentials. Individuals trained in other disciplines should not be permitted to teach mathematics unless they have sufficient mathematical training as well. Overall, the emphasis was on establishing a cadre of well-trained, professional mathematics teachers, whose education was somewhat different from that of the traditional research-oriented PhD.

In addition to considering the preparation of two-year college faculty, the strategic plan calls for reform of content and pedagogy, including the appropriate use of assessment, technology, and a variety of instructional strategies, in all mathematics courses in the first two years of college. To this end, AMATYC, working with the assistance of MSEB, is beginning a five-phase initiative called *Curriculum and Pedagogy Reform of Mathematics for the First Two Years of College* (CPR-MATYC). As explained by AMATYC President Karen Sharp of C.S. Mott Community College, a proposal has been submitted to NSF calling for the development of a "Standards document" for lower-division mathematics. This would be widely disseminated to achieve a consensus. Subsequently, a national convocation for formal presentation to the mathematics community would be held, followed by regional workshops. A "phase zero" steering committee with broad representation, including two members representing the *NCTM Standards*, begins work early in 1993.

Other goals refer to adjunct faculty, recruitment of minority and women students, formation of collaboratives among secondary schools, two-year colleges, and four-year colleges, establishment of links with policy makers, appointment of two-year college mathematics educators to MSEB and NSF, and professional development and support for members. Noticeably lacking in the document was any reference to MAA or AMS. Many participants voiced the opinion that two-year colleges are the place where "real teaching" takes place, causing others to come to the defense of four-year colleges in this regard.

The program had four-hour calculator minicourses, as well as two-hour workshops on study skills, math anxiety, bridging the gap from arithmetic to algebra, developmental algebra, alternate forms of assessment, cooperative learning, proposal writing, and computer software varying from `PCSolve` to `CONVERGE`. Macintosh, IBM,

Casio, and Texas Instruments made product presentations. There were hour-long papers of varying quality of topics ranging from retention to earth algebra, exhibits, and a poster session, as well as invited speakers at three general sessions, one a very pleasant sit-down breakfast. CRUME sponsored a panel on implications of research in mathematics education for two-year colleges.

Name tags emphasized first names and featured ribbons for first-time attendees and presenters, reminding one wag of "prize sows." A pleasant social evening of gaming with play money was provided at Casino de Fantasia. The conference site selection process was debated by the delegate assembly as many vied for the honor of hosting future meetings. Next year's AMATYC Conference will be in Boston, November 18–21, 1993.

New MAA Committee
Seeks Advice on Advising

D. Lutzer, UME Trends 4:6 (January 1993)

College is more complicated these days. Once, a mathematics major meant three years of required courses with a few electives in the senior year. Advising was easier then; checking off courses listed in a college catalog might have been enough.

Mathematics students today face a spectrum of early choices. In some mathematical sciences departments, students as early as the sophomore year need to decide between an actuarial mathematics program, a biostatistics option, a classical pure mathematics track, a discrete mathematics focus, and so on. Lifetime choices are being made by students early in their undergraduate years, and students are understandably worried. Alumni surveys in all majors show that students want to turn to faculty mentors through the advising process, but that they are not satisfied with the results. At many colleges and universities, alumni report being less satisfied with advising than with any other aspect of their undergraduate careers. More systematic data indicate that concerns about advising are cited by over half of science, engineering, and mathematics students.

Academia realizes that advising is a problem and is beginning to understand what needs to be involved in the solution. The Association of American Colleges wrote: Advising must be more than monitoring students to ensure that they are making satisfactory progress, and it must be more than suggesting choices among possible options. Quality advising includes discussing goals and expectations: of the program, of the institution, and of the student. It includes discussing opportunities in the field and strategies for achieving students goals both during and after their program of study. It includes discussing the relationship among courses in the program and between the program and general education.

The mathematical community, too, sees the special

importance of advising. The most recent CUPM report lists careful and sustained individualized advising as central to todays mathematical sciences curriculum. CUPM writes that the diversity of mathematical program options available today imposes advising responsibilities of a new order of magnitude upon departments. Advising todays mathematics majors must go far beyond monitoring student progress in completing a list of required courses; indeed, todays automated degree audit systems can probably do a better job in that area than most human advisors.

A recent MAA publication by the Committee on the Teaching of Undergraduate Mathematics, *A Source Book for College Mathematics Teaching*, argued that the advising program in a mathematical sciences department should include much more than the traditional meetings between a faculty member and a mathematics major to choose next terms courses. It praised Ohio State Universitys involvement in advising thousands of Ohios pre-college students about the mathematical expectations of various colleges and majors in the state. It focussed on the special need for careful placement of freshmen and transfer students and for articulation agreements between regional two- and four-year colleges. When it comes to advising mathematics majors, CTUM made it clear that advising should assist students in long term planning, whether for careers or for graduate school. Other articles have stressed advising as particularly important in meeting the needs of women and minority mathematics students (see *UME Trends*, March 1991).

To focus additional attention on the advising role of mathematics faculty, the MAA recently created an ad hoc committee to consider the advising of undergraduate mathematics students, freshmen to seniors. The committee is particularly interested in examples of successful advising practices around the nation and in career advising materials created by local departments to supplement the brochures published by various mathematics professional societies. If your department has a particularly effective or innovative advising program, the Advising Committee would like to hear details. *You may contact the committee by calling its chair, David Lutzer at the College of William and Mary (804) 221-2470, or by sending e-mail messages to the committee at adviscom@wmvml.bitnet.*

Dave Lutzer is at the College of William and Mary. He is chair of the MAA Committee on Advising.

Crisis in Math Advising

S.Z. Keith and G.A. Earles
UME Trends 3:1 (March 1991)

The need to bring more students into the mathematical pipeline has received substantial national attention. Included in the reform work are many programs to advise K12 students of the importance of mathematics in their future. On the other hand, less attention has been given to the issue of advising students at the university portion of the pipeline, in spite of high attrition rates at that level. Curriculum enhancement and improved instruction are obviously important goals in the reform of mathematics teaching. But these programs may be dead-ends without excellent advising. Unfortunately, many colleges seem to shrug off responsibility on the subject, and students continually report their disappointment with advising programs. As educators, we owe it to students to provide them with some sense of the future, and as a beginning, the issue of better advising needs to achieve more visibility in the mathematics community.

Characteristics of the Problem

The majority of women and minorities now receive their education at public colleges. These students, together with the broader economic base of white males, are probably working while in school and have strong vocational goals. Most have never contemplated graduate school, where the costsincluding lost income-may be a formidable barrier. These newcomers to higher education demand more than the standard advice they are liable to receive: that mathematics is a good background, but that they should be prepared to retrain. Many want answers to immediate, job-specific questions, such as what types of jobs are available within their state, what types of majors, minors, or courses of study make them more hireable?

Without more specific advice, many good students are likely to drop out of the mathematics major. And while we need to "pitch" to students the usefulness of mathematics, we cannot afford to create false illusions about their future, either. Excellent advising enables students to develop and pursue realistic expectations. Furthermore, for us to ignore the job-specific needs of these students will cause us to risk being a "weeding-out" system at a time when declining enrollments suggest this attitude is no longer supportable.

Unfortunately, departments and placement offices are not well equipped to collect the data necessary to interpret the complex world facing a mathematics major. In fact, the problems in collecting data relevant to advising are at least three-dimensional in scope: we need longitudinal studies, regional studies, and studies pertaining to advising at different stratifications of education. And bluntly, it is becoming difficult to interpret the hireability of the mathematician at any of these levels, as contradictions in the data appear to surface.

On the one hand the Joint Policy Board for Mathematics reports the need to double the number of mathematics Ph.Ds we produce. But will there be jobs for these specialized individuals if we produce them? Some mathematics Ph.Ds are currently without jobs, for reasons not clearly understood. And even if we anticipate an attrition in college faculty in the next ten years, can we rely on colleges not to increase their student/faculty ratios?

Suggestions for Improved advising

1. Mathematics professional organizations should collect information pertinent to the issue of advising students into jobs and into graduate schools. This information should be consolidated with that which is available from industry and government and made available to the academic community on a regular basis. We may need a base of questions that can be more broadly interpreted, such as that provided by attitude surveys.

2. Mathematics meetings should give courage to the issues of advising in debates, panels, interviews, and minicourses that feature model programs. Characteristics of model advising programs should be formulated.

3. In an ongoing discourse on the role of mathematics in our future, the mathematics organizations should help us widen channels of communication with other scientific disciplines.

4. Colleges and universities should be encouraged by the mathematics organizations to select excellent, sensitive advisers and provide comprehensive advising programs to all students, from entering freshmen and graduating seniors to graduate students. Furthermore, administrations should give professional credit to the individuals who implement these programs.

Sandra Keith and Gail Earles are at St. Cloud State University.

On the Far Side of Curriculum Reform

C. R. Curjel, UME Trends 2:3 (August 1990)

Traditionally, "working on a course" means for most of us "crystallizing and organizing the subject matter" and "creating classroom material such as texts and software." As a rule, questions on other aspects of a course are less prominently discussed. I call these aspects "course format." By this I mean the totality of

⬦ the institutional arrangements such as class size, class hours, credits, admission, registration, course evaluation policies;

⬦ the course organization such as syllabus, homework, tests, grading policy;

⬦ the instructor's view on issues such as her/his role in the classroom, student-instructor and student-student interactions, the students' learning processin short, the views that underlie the social organization, so to speak, of the course;

⬦ the students' assumptions on the kind and amount of work a course will require.

Among the many reasons for our reluctance to question and discuss course format I see the following:

⬦ the institutional arrangements seem to be inviolable;

⬦ it is far more efficient to continue last quarter's course organization and, in general, to base one's teaching on one's own experiences as a student;

⬦ as mathematicians we embrace the belief that our field, as the ultimate science, is especially objective, and that personal and social matters should not, and therefore do not, affect our work;

⬦ we are afraid of being dragged into "educational" issues or into areas of motivation, feelings etc., and of wandering away from mathematics;

⬦ teaching is an intensely personal matterfor example, witness how deeply triumphs and failures in the classroom affect us. To have a detached view of one's teaching amounts to questioning oneself, an unpopular activity.

I notice that much of the recent work done to improve undergraduate mathematics instruction bears on *subject matter reorganization* and *creation of new course material*, two absolutely necessary activities if we want to make a change. My instincts tell me, however, that similar efforts in rethinking the *course format* are called for if the reorganized subject matter and the new course material are to have a lasting effect on students. The course format shapes the everyday life of students and instructors as much as anything else.

Now what about the forces that seem to militate against a serious discussion of course format? Here are my reactions to some of them.

It *is* possible for mathematicians to work on course format without getting into educational quicksand. Between 1973 and 1978 a group of faculty at the University of Washington, with G. S. Monk as the driving and guiding force, took over an existing course, reorganized the subject matter, wrote new course material, radically changed the course format, and successfully taught the course for more than ten years. What kept us going was a *dogged insistence that any discussion of course format has to take place in the context of specific mathematical material of a specific course*. In this view, a course is considered as an organism in which every single aspect is related to the overall structure, and vice versa. For example, the dynamics of precalculus material is different from that of calculus material, and there is no universal, "correct" course format applicable to both.

Another telling example of a serious discussion of course format can be found in Brown-Porta-Uhl's *Mathematica Notebooks* project. The authors report that at one point the reality of the computer collided with the traditional lecture format. This in turn led them to reflect on lectures, to assign them a new function, and finally to a sweeping redefinition of the lecturer's role. These course format items are an integral part of the project, not just peripheral details. A truly new course format is being proposed.

As for the institutional arrangements, they have an air of stability because they are held in place by administrators whose power is visible and accepted. Most faculty members shy away from meeting administrators head on because their interests lie in different lines of workafter all, they have not been selected for administrative street savvy. Thus, factors such as admission standards, course requirements, enforcement of prerequisites, student expectations about grades, public views, and pressure on curricula are rarely questioned in spite of the enormous influence they have on teaching any course. It is unclear at best how a restructured curriculum and a renewed sense of mission will fare in an institutional framework that is built on assumptions held over from the times during which our current crisis developed. If the reform wave dissipates in the face of administrative traditions then it was not worth much to begin with.

Other obstaclesthe myth of the objectivity of the instructional process, the self-propellency of one's past, the difficult task of looking in a detached way at our deeply ingrained beliefs on teachingmust be addressed by each of us on our own. What we can do publicly, however, is to acknowledge that there is such an issue as "course format" and to question traditional course formats as much as we have come to question traditional syllabi and textbooks.

Caspar Curjel is on the faculty of the University of Washington, Seattle.

MAA Efforts in Assessment

B.L. Madison, UME Trends 4:3 (August 1992)

Attitudes about programs of assessment of student learning outcomes vary widely among MAA members. Some believe assessment programs are positive, some are very skeptical about the value of assessment, and many are uncertain about the meaning and purposes of assessment programs. This is the feedback to the CUPM Subcommittee on the Assessment of the Undergraduate Major (SAM) during the past eighteen months of committee activity. Although SAM members have been aware of this ambivalence toward assessment, plans for at least a partial MAA response to the national assessment movement are proceeding.

Responding to requests from departments for help, SAM has established a task force to draw up a set of guidelines to use in structuring a program of assessment. Most of the requesting departments are facing mandates to establish student learning outcomes assessment as part of program, departmental, or faculty evaluation, or more simply to use actual student learning as a measure of success.

The Guidelines task force was to begin work in Spring 1992 but has been delayed by funding proposal development and the resulting discussions about an appropriate position toward assessment by the MAA. The discussions are driven mainly by skepticism about the effectiveness of assessment of student learning and the fact that most assessment is being mandated from outside of mathematics department faculties. The issue for some MAA members: Do we help even though we do not believe? Does the establishment of guidelines for structuring an assessment program amount to endorsement of the program? The position about which SAM activities has revolved is that we need information and guidelines to make assessment programs more constructive. The attitudes of SAM members vary widely too, and SAM has taken no position on the advisability of assessment programs.

SAM has focused its efforts on gathering and disseminating information in planning an MAA response to the national assessment movement. The task is easier because of a foundation built by other MAA projects: (1) *The Undergraduate Major in the Mathematical Sciences*, a 1991 CUPM report; (2) *Challenges for College Mathematics: An Agenda for the Next Decade*, a report by a joint task force of the MAA and the Association of American Colleges; and (3) *Assessment of Undergraduate Mathematics*, a report of an MAA Curriculum Action Project focus group. All three of these, and six other chapters, constitute the MAA Notes, Number 22, *Heeding the Call for Change*, published in 1992. The first two reports cited above form a basis for setting learning goals for the undergraduate major, the first step in establishing an assessment program. The second and most difficult step is measuring the extent to which learning goals are being met. The third step, which is supposed to be the payoff and which worries skeptics most, is the use of and response to the data gathered in Step Two.

On the basis of committee deliberations and responses to requests for information about departmental assessment programs, SAM has concluded the following:

◇ Many departments are working on setting up assessment programs, mostly mandated from outside the department.

◇ There are very few assessment programs with any evaluated experience; hence, there are few exemplary programs.

◇ Among academic disciplines, mathematics is well positioned in that it has developed a rather substantial intellectual basis for setting goals for student learning.

◇ A successful assessment program will probably require multiple assessment instruments, almost none of which are standardized; hence, alone, standardized tests will be deficient.

The first and second conclusions above appear to summarize the general state of affairs in the assessment of mathematics departments: many are just beginning and few have experience. SAM is sponsoring a contributed papers session at the Mathematics Meeting in San Antonio in January 1993 to provide opportunities for presentation of work on assessment. The organizers are Charles F. Peltier, Saint Marys College, and James W. Stepp, University of Houston.

In June, 1992, SAM arranged two sessions at the Seventh Assessment Conference sponsored by the American Association for Higher Education. This annual conference had as one of its 1992 special emphases assessment in the discipline and the major. One session, entitled Lessons Learned in Assessing Collegiate Mathematics, featured presentations by Lanny Morley of Northeast Missouri State University and Richard D. West of the United States Military Academy. The other session was an overview of the SAM guidelines project by Bruce Peterson of Middlebury College and the author.

SAM is particularly interested in information about departments experiences with assessment of the major or other curricular blocks of mathematics courses. Send such information to *SAM Chair Bernard L. Madison, Old Main 525, University of Arkansas, Fayetteville, AR 72701, BMADISON@UAFSYSB.UARK.EDU, phone (501) 575-4804.*

The Lower/Upper Division Gap

S.V. Keny, UME Trends 3:1 (March 1991)

At Whittier for the last three semesters I have run a seminar designed for math majors to help them in their transition from more computation-oriented classes to more abstract classes. This transition becomes very difficult for students who have never been exposed to the various techniques in logic.

Very often the students see these techniques for the first time in a class like Advanced Calculus. The new results and their proofs become an overwhelming experience for the students and a very frustrating experience for the instructor. So last year I started a sophomore-level student-run seminar on logic.

I am sure this is not the most original approach in teaching. For example in UME TRENDS, August 1989, Roger H. Marty discussed a similar course. Just like Marty, I emphasize the logical techniques. Very often the mathematical results discussed are the ones the students are already familiar with, and this helps them to concentrate more on the reasoning aspect.

But there are two aspects of this course that are different from those of Martys:

1. First, this is entirely student run. On the first day of the semester I assign different chapters and sections thereof to all the students and give them the lecture schedule. Of course I meet with the students individually prior to their lecture and help them if needed. I also emphasize writing of the proofs. But the major responsibility lies with the students to make sure that they are prepared with their material before their lecture. This also teaches them to read math books. I do not teach the techniques of oral presentations. Often, the lecture turns into a constructive discussion session and the students come up with some nice examples.

2. Second, I try to relate this seminar to my other upper division math classes. When I prove a theorem, I point out to the students precisely the technique I use in the proof and refer them to the corresponding result from the seminar. This makes them appreciate the seminar even more.

This seminar is designed around the book titled Bridge to Abstract Mathematics (Mathematical Proof and Structures) by Ronald P. Morash, as a one unit seminar every semester. Math majors are required to take it twice in a sequence. The students earn the one unit based on their lecture, home works, class participation, and attendance.

I have run this seminar for the last three semesters, and most of the students have told me that they enjoyed the teaching part very much. Of course initially they had a lot of difficulty preparing their lectures, but this improved significantly towards the end of the semester. I feel that preparing and delivering lectures to their colleagues helped them develop their self confidence about being able to handle an abstract math class.

Since this is only the second year that I am running this seminar, I do not have any scientific results to point towards its effectiveness. But having seen many students in upper division math classes before the seminar and comparing them to just one group from my last years seminar, I see a tremendous potential for the seminar in helping students in their major.

Sharad Keny is at Whittier College.

The Calculus Turmoil

P.R. Halmos, FOCUS 10:6 (November/December 1990)

What's all this calculus turmoil about? Have all mathematicians except me taken leave of their senses? Surely it isn't just a racket – a way of getting money from the NSF? But if it is not, then why are otherwise honest and intelligent people screaming about it? Will somebody please explain to me what's going on?

The two main symptoms (or are they the causes?) of this *new* disease are that lots of students *flunk* calculus and that most *textbooks* are bad.

Well, let's see. "More than a third of the 600,000 or so people who study calculus in college every year fail to complete the course." (Peter D. Lax, "Calculus Reform A Modest Proposal," UME *Trends*, 2 (Number 2, May 1990). For a different view, see Leonard Gillman's "Two Proposals for Calculus," *FOCUS*, 7 (November 4, 1987).) I am prepared to believe that, and I am neither surprised nor horrified. Teachers have been discussing failure rates with one another ever since the concept was invented, and 40% is not unusual among the numbers that they bandy about. The failure rate varies a lot, of course – it depends on the quality of the university in question as a whole, and, derivatively, on the quality of the teachers and the students at the university. A part of the reason is the reason that students take calculus in the first place. The word has a mystique; it's a Mount Everest to climb. It is an essential part of the tool kit of many who wish to study science, and science is a glamorous and well-paid(?) and highly respected profession. Students who take calculus "because it's there," or for the glamor, the money, or the respect, are taking it for the wrong reason, and their failure is neither surprising nor worrisome.

What is the failure rate for people taking piano lessons? I don't know, but I could make up some statistics as well as the next guy, and I wouldn't be surprised if the true statistics pointed to a situation as "lamentable" as the one in mathematics.

As for how the books are bad: they contain "inert knowledge," including "integration techniques that work only in specially rigged cases," "fake applications," and "tedious calculations."

Are those comments new? And, new or not, are they correct? Yes, they are correct, and no, they are not new: they have been made for many decades (centuries?), and they point to something bad. To something that is all bad? I am not sure.

Very few people advocate epsilons and deltas in elementary calculus, and, for sure, I am not one of them, I confess that when it comes to arithmetic I am still in favor of the multiplication table being drummed into innocent heads (at least through 9×9), together with some of the elementary and almost universally needed techniques of estimating orders of magnitude. (If each of 1,000 students writes 10 themes a week for each of the 30 weeks of the academic year, and if, on average, the labor of grading each theme costs 20 cents, then we should be able to decide whether the money we need from the administration is \$6,000, \$60,000, or \$600,000.) Similarly, in calculus, I am still in favor of teaching the classical basic techniques of the differential and integral calculus. I want a student who sees x^3 or sin x to snap back $3x^2$ or cos x, and I want a student who sees $\int [dx/(1+x^2)]$ and $\int [dx/(1-x^2)]$ to know the difference between them and to be able to come up with the indefinite integral in both cases, even if a pocket calculator can do it more quickly. It is true that much of the classical drill (that I for one went through in the 1930s) can be left to the button pushers – pocket calculators do very well what for a long time we haven't been wanting to do at all. But I am still in favor of teaching some (many?) rigged integration techniques and tedious calculations – they are as rigged and as tedious as finger exercises for the pianist, and just as indispensable.

Is computing to be "an integral part" of calculus? (I can't help wondering whether that's an intentional pun.) Yes, of course, that's just what I've been saying. By computing, to be sure, I mean what calculus teachers have always meant – computing with rational, trigonometric, exponential, and logarithmic functions, and getting used

to their behavior and their relations to each other. I do not mean computing in the sense that computing machines are good for – horrors, no!

The greatest "mathematical" invention of the last few centuries is not electronic but notational – I refer to decimals and their relation to arithmetic. An electronic typewriter that can crunch numbers is useful – but I don't want to teach spelling in courses in literature, decimal calculations in courses on number theory, and computing techniques in courses on calculus. I like spelling and decimals and computers – but they should be kept in their places.

What then should a calculus course emphasize, if not integral rigging and number crunching? Qualitative understanding of functions? – asymptotics and approximation as implied by the differential equations they satisfy? Fine – sure – why not? Substantial applications – in engineering, in economics, in psychology? – yes, maybe – but our focus should always remain at the mathematical center. We, teachers of calculus, are not usually practicing engineers, economists, or psychologists, and just as we might worry about the competence of those people to teach our stuff, they have the full right to worry about our encroaching on their expertise. But I don't really thing there is anything serious to worry about as far as the content is concerned. Most calculus courses (certainly the ones I took, the ones I taught, and the ones I hear colleagues tell about) are pretty much free-wheeling – they emphasize the mechanical techniques to the extent that drill is necessary (or a little too much, or not quite enough, but more or less to the right extent), and they contain the illustrations and applications that the teacher is competent to explain and the students are ready to receive.

Do bad books contribute to making the courses bad? Yes and no – mainly no. At the calculus level, in my experience, students simply don't read the book – most of them cannot and none of them has to. They cannot because they just don't know how to read – the formal language of the text is repulsive to them, and boring, and difficult – most of our class time has to be spent in repeating the book in the vernacular. And they don't have to because most of our class time is spent in repeating the book in the vernacular.

A famous (infamous?), old (ancient?), bad (dreadful?) calculus book is Granville, Smith, and Longley – and its badness is irrelevant to using it as a text. It tells some lies, yes, but who reads them?! What matters is that it has a table of contents, to which you can add and from which you can delete – and it has a large collection of numbered problems that you can assign for homework. Being fearful of a largely imaginary and hugely exaggerated disease allegedly caused by the enforced use of uniform texts is a rather silly phobia – any calculus teacher who has taught the subject once can use any book and do as well with it as with any other.

Yes, there are good books and they have been with

us perhaps even longer than the bad ones that water them down – Landau, Courant, and Apostol may have lived in vain, but they did live. No, good books don't catch on – and a large part of the reason is that our calculus students are not ready for them. Most students are more than intelligent enough to cope, but they've been held back – pushed back! – since their early years in grade school and high school, and they arrive at the university not just untrained or badly trained but almost negatively trained. Some of them in their turn become poorly trained grade school teachers and high school teachers and go on blissfully to give their students training at least as bad as they had received.

The results of the bad training are visible everywhere. It shocked me to find, as I was teaching a course on analytic functions that my students (all of whom had credit in a course called "advanced calculus") not only did not know the sum of a geometric series but maintained that they had never even heard of it – that's bad college calculus. But, in the same course, I was even more shocked to learn that they knew neither the factor theorem (the relation between the roots of a polynomial equation and the factors of the polynomial) nor its name – once again they had never even heard of it – and that's bad high school algebra. And, incidentally, and strongly and relevantly, the same sort of ignorance is visible in other subjects, notably in the use of language. When I require that the solutions of problems be communicated to me not just by scribbled equations but by sentences that begin with a capital letter, end with a period, and have a reasonable sprinkling of commas between, my students groan and tremble – they don't believe me, they think I am unreasonable, and they simple don't (can't) do it.

Did all this start "in the fifties" – is it all a new disease caused largely by "the vast majority of research mathematics?" Balderdash! The problem is not new. Practicing mathematicians have known about it long before the 1930's, when I began to dip my toes into the mathematical ocean, and they will keep talking about it, I predict, long after we are all gone. Old people always say that the world is going to hell in a handbasket, and I'm ready to subscribe to that tenet – things might always have been bad, but maybe, maybe, they are getting worse.

Students know less and less about language (because they are taught less and less), and they know less and less about calculus (because they are given less and less preparation, and, in fact, progressively worse preparation, in high school). Is calculus the only mathematical sufferer? No – absolutely not – and I suppose that the reason calculus is usually singled out is that it's a huge industry – it is the bread and butter of many of us. But precalculus (multiply and divide polynomials and sines and cosines) and postcalculus (add vectors and multiply matrices) are in equally bad shape.

Is the message coming through? I am saying that there is a lot of unjustified turmoil, but I am not saying that there is nothing wrong. I am saying that the calculus

turmoil is rather silly – it is both misplaced and exaggerated – but there is something wrong, and what's wrong is much broader and deeper and more serious than calculus students who flunk and textbooks that are dull.

Are most human beings too stupid to learn calculus and grammar? No, the trouble is not genetic – it is, for want of a better word, sociological. In Japan, we are told, students learn calculus, and they are the same kind of human beings as the ones in Manhattan, New York or Manhattan, Kansas. But in Japan there seems to be an almost universal respect for learning, a social background from which children absorb the feeling that intellectual effort can be pleasant and rewarded and applauded – an atmosphere in which scholarship is encouraged as much as (more than?) athletics and commerce, a social structure in which professors are as good as (better than?) baseball pitchers and airplane pilots. Such respect, such a social background and atmosphere, have been eroding in the US for most of the twentieth century – since, as a rough approximation, the time of John Dewey, my favorite villain, one of whose effects appears to have been to cause people to sneer at abstractions and admire the "practical". That, I think is why the US is at or near the bottom of many school achievement lists in which the countries of the world are compared – that is why we have a "calculus crisis" but Japan does not.

Yes, there is a disease, but calculus is neither its cause nor its main symptom. We mathematicians can do our small bit to cure it, but not by rewriting calculus books. All that we can do, all that we are professionally able to do, is to insist on raising the quality of primary and secondary education by establishing and maintaining a high quality in college courses, by insisting on and strictly enforcing severe prerequisites, and by encouraging and properly training prospective grade school and high school teachers. That we can do, and I hope we will.

Turmoil in Teaching
A Response to Paul Halmos

P. Hilton and J. Pedersen
UME Trends 3:1 (March 1991)

Our friend Paul Halmos, in his stimulating article "The Calculus Turmoil," which appeared in the November/December 1990 issue of FOCUS, has performed a characteristically neat sleight of hand! His opening paragraph appears to dismiss the general concern over our poor success in teaching the calculus as some sort of professional aberration on our part. ("Have all mathematicians except me taken leave of their senses?") As we continue to read, we are told that the quality of textbooks does not matter because, in any case, "students don't read the book." But as we read yet further we find that what he is saying *quite explicitly* is that the situation is really far worse than is generally claimed. "Yes, there is a disease, but calculus is neither its cause nor its main symptom." The disease, Halmos perceives, infects our whole educational system, our whole society; it penetrates deeply and broadly, and it is not a recent phenomenon. So there is no point in trying to treat just one relatively unimportant symptom we must attack the disease at its root by insisting "on raising the quality of primary and secondary education by establishing and maintaining a high quality in college courses. . . ."

With Halmos's argument that the problem goes way beyond the teaching of calculus we have no quarrel whatsoever; but we do believe that the effective teaching of calculus is of special importance today and that our woeful failure is a particularly ill omen for our society. We also believe that the textbook situation is significantly bad, for reasons which we will give. Moreover, while we unreservedly accept Halmos's diagnosis of the nature of the disease, we do not believe that we university teachers would be justified morally in confining the blame to the inadequacies of precollege teaching, nor are we comfortable with a recommended treatment consisting *exclusively* of "high quality college courses," when the patient is generally insensitive to, or may even reject, high quality mathematics. For example, it is vitally necessary that university mathematicians assume their responsibility for the quality of precollege mathematics education and the proper preparation of precollege mathematics teachers; it is just as necessary that all mathematicians recognize the merit of such activity.

Let us elaborate on some of these points.

In the past the learning of mathematics at the university and the passing of tests simply constituted a set of obstacles in the student's path to achieving the qualifications needed for certain types of employment. What the student learned, and the manner of learning it (active or passive, by the exercise of intelligence or memory), mattered scarcely at all, because, for most students, there was little reason to expect that they would, in their lives and careers, actually use any of the mathematical skills they were certified as having acquired. This situation has now changed; the capacity for quantitive reasoning is now a desideratum for most responsible positions and for the exercise of informed judgment on the complex problems besetting contemporary society in the first world. Moreover, the culture of that society is characterized by change, both continuous and discontinuous, even more than it is characterized by advanced technology, so that today's and tomorrow's citizens must be prepared not only for a full comprehension of the role of the computer but also, and just as importantly, for understanding the mathematics of change. Thus the calculus is not just a subject suitable for its power to discriminate in tests; it is, first and foremost, the area of mathematics relevant to the student's world. Let us be quite clear that the Japanese have not simply decided to master the calculus because it appears on their tests! All those nations whose leading citizens today conspicuously value the advantages of good educationunfortunately, that excludes the

United Statesoutshine us in mathematical performance at all levels including, of course, the study of calculus.

Moreover, if the students are subsequently to use what they have ostensibly learned, then they should have learned with understanding and imagination and not merely by developing their pattern-recognition and memory skills. For material acquired without genuine understanding is not really usable, and the skill acquired is only temporary and does not bring with it the confidence needed for successful and enthusiastic exploitation. Thus cookbook courses from books of mathematical recipes serve no useful purpose, but this is what most students wantand get. Halmos cites many examples of the unfortunate effects of "bad training." We would go further and assert that a fundamental problem in this country is that training is confused with education and substituted for itand training that does not lead to any real understanding is always bad education.

There is, further, a real difficulty about improving the textbooks, but we have drawn attention elsewhere. The textbook publishers are naturally interested in making a profit and this leads them, as Murray Protter has so cogently observed in a recent Mathematical Intelligencer article, to play for safety in publishing calculus texts that conform to the standard pattern, characterized by very broadbut not very deepcoverage of topics and a mass of stereotyped exercises. Thus texts that depart significantly from this format are less likely to command support from publishers. Put in a nutshell, one might say that the only texts likely to be published by any of the larger publishers are probably unnecessary because they closely resemble those texts already available; and mathematicians struggling to bring about essential improvements are having to resort to the practice of 'desktop publishing'.

Halmos argues that students are studying mathematics in excessive numbers because "science is a glamorous and well-paid (?) and highly respected profession" and that, given these swollen numbers, it is to be expected that we will have a high failure rate (he quotes Peter Lax's figure of more than a third of the 600,000 students of calculus failing to complete the course). Now it is perfectly true that we are teaching mathematics in our core program or in service courses to more and more students, but the reason resides in the fact that so many disciplines now require some minimal mathematics qualification for their majors. It is simply not true that large numbers of our young people are attracted to a career in science; it is generally true that today's students are attracted to those studies that they expect will lead to lucrative careers, business, and law courses. Such students are not likely to distinguish themselves by outstanding performances in calculus courses.

Thus while we must try to maintain the integrity of our courses, we have a real difficulty in that so many of our students are learning mathematics reluctantly if they are learning it at all. (It is our experience that even students majoring in computer science do not take kindly to being obliged to take 'quality' mathematics courses—their want of interest in mathematics leads one to wonder whether they are even interested in computer science, or whether they merely see the computer science track as a promising avenue to job security and affluence.) In our judgment, Halmos's call for the maintenance of standards in our courses must be tempered with a realistic assessment of what our students are capable of learning and the pace at which they can progress. In many Canadian universities the drop-out rate in first year courses is unacceptably high. We in this country have a tendency to mask the true failure rate by a policy of indulgent grading and the setting of undemanding tests involving questions of a familiar and predictable pattern. How many of us have employed the D grade as an act of kindness in the face of an F performance?

But let us not lose sight of Halmos's main point—or what we choose to regard as his main point. There is a malaise afflicting American education, and it is a malaise which will afflict any acquisitive, materialistic society. Such societies do not value education itself highly, since education—as distinct from qualification certification—is not necessary for the individual's expectation of enjoying high living standards. Amidst the euphoria engendered by the great tide of freedom that has swept through Eastern Europe, there are those who are more perceptive and see a general weakening in cultural and moral values accompanying the headlong rush for television sets, refrigerators, and Mercedes. It is not unduly pessimistic to predict that the triumph of the market economy will see a decline in educational standards and in the respect accorded the (now relatively poorly-paid) teacher. After all, Margaret Thatcher, whose political philosophy epitomized the entrepreneurial society, was refused an honorary degree by Oxford University on the grounds that she was hostile to education.

Thus we must, in a very important sense, agree with Paul Halmos the problem is more serious than just our students' poor receptivity to the ideas of the calculus. We are facing a complex social system that militates against any form of successful education. Our materialistic society places little value on education; commercial television encourages a popular appetite for instant gratification; and our children's role models, our society's successes, are certainly not characterized by a dedication to education, nor by occupying positions open only to the well educated.

As mathematicians and teachers we must do what we can, but our relative failure in teaching the calculus, while an important symptom of the problem, of vital importance to the maintenance of our eminence as an industrial society, is still only a symptom, not, as Paul Halmos points out, itself the disease.

Peter Hilton is at SUNY Binghamton, and Jean Pedersen is at Santa Clara University.

Opinion: What's Better About Reformed Calculus?

D.A. Smith, UME Trends 3:2 (May 1991)

The first good thing to come out of the calculus reform movement is the movement itself. It is a movement: witness the CRAFTY volume [MAA Notes No. 17] with information on more than 70 projects that were already under way a year ago. Consider the crowded sessions—panels, featured speakers, contributed papers, minicourses, poster session—at national and regional MAA meetings on the subjects of calculus, writing, use of computers, and education. And the level of interest is still accelerating. Those of us with the head starts provided by NSF and other funding are receiving a steady stream of invitations to visit other campuses and help other faculties get started with reform of their own programs. My experience with such visits suggests that many mathematics departments and the departments they serve are deeply interested in and committed to making substantial changes. Contrary to the beliefs of many mathematicians who haven't bothered to check, their administrations are also interested, committed, and ready to find the necessary resources.

The most positive thing I see happening with the movement is a diversity of approaches, fueled by widely divergent ideas and circumstances, converging to a common set of goals—but not toward a common set of materials. In a few years time we will have a constructive proof that there are many ways, affordable and transportable ways, to teach calculus effectively, and we need never again settle into the rut of offering the same course over and over, year in and year out. In the remainder of this brief essay, I want to describe the components of the emerging consensus, roughly grouped under the headings of Content, Environment, Students, and Instructors.

Content

Everyone agrees that students must constantly experience calculus concepts in a rich interplay of symbolic, numerical, and graphical forms—what the Harvard group led by Deborah Hughes Hallett and Andy Gleason calls the "three-fold way." Most agree that there must also be a steady interplay between the discrete and the continuous—in phenomena being studied, in the models of those phenomena, and in the methods for dealing with those models.

Every calculus reform program is focused on developing thinking skills and conceptual understanding—on eliminating the possibility of students being certified as having learned calculus when all they have demonstrated is modest proficiency at memorizing formulas and manipulating symbols. Most reformers recognize the importance of communication and verbalization of thought processes, especially through writing, which is a powerful tool both for student learning and for faculty observation and measurement of that learning.

There is a popular misconception that calculus reform

means "using computers (or calculators)" in calculus–in fact, most of the reform projects do use computers or calculators in some way. But some (for example, those at New Mexico State University and Ithaca College) started out with no emphasis on technology. The emerging consensus recognizes the importance of using "appropriate technology," which means different things on different campuses, depending on available resources, the particular focus of the course, and many other factors. The use of computers and calculators in reformed courses is not just a trendy thing to do; these are the tools that enable us to break out of the mindless symbol-pushing mode and allow students to experience the interplays of the graphical, numerical, and symbolic and of the discrete and continuous.

Finally, while we are all still looking at the content of calculus as what we had always hoped the traditional course would become, we see very real changes occurring in the syllabus. First, there is the recognition that there need not be just one syllabus. Second, there is the recognition that the course must not be measured out in 50-minute sound bites and 4-minute exercises. Beyond that, we see a lot of different things happening, but with a number of common themes. For example, many are discovering the value of real-world problems, not as afterthoughts, but as up- front motivators, as "hooks" to capture the interest of students who have (or think they have) no interest in mathematics for its own sake. (That describes somewhere between 95 percent and 99.9 percent of all beginning calculus students on almost all of the campuses I know about. St. Olaf College might be an exception.)

Environment

Sally Berenson of North Carolina State University puts it this way: "Telling is not teaching, and listening is not learning." Every reformer either knows or quickly discovers the need to focus on learning and on creating an environment in which learning can take place. For some, the physical location is the traditional classroom; for some, a laboratory; for some, both. In most projects, "lecture" is being largely replaced (totally, in the case of Purdue) by multiple classroom and/or lab activities. The operative word here is activity – active involvement of the student-learners replacing passive reception of "the word" from the all-knowing lecturer. These activities often involve discovery, experimentation, and open-ended problems.

They also often involve teamwork; students can learn a lot from and with each other that they would never learn working in isolation or listening to a lecture. However, cooperative learning environments also raise issues on which we do not yet see a consensus emerging: issues of individual vs. corporate responsibility. Our students have been programmed to compete with each other, to see other students (rather than ignorance) as the enemy. And we are conditioned to using that sense of competition as a motivator and to seeing cooperation as "cheating." So how do we get them to share their learning

experiences? And how do we evaluate their efforts when they do?

These are difficult questions to answer, but they are not impossible. Consider this: Our salaries are paid by our students (or their families) through tuition and taxes. We are not paid by the medical, engineering, and business schools who want us to screen their applicants. Thus, we must assign the highest priority to learning (our students' best interests, whether they know it or not) and relegate evaluation and certification to distinctly inferior roles. What we know, and what we are learning, about learning tells us that teaching to tests is counterproductive. And now that we have a wide range of other student activities to observe–lab reports, project reports, worksheets, group interactions–we are free to de-emphasize tests as the primary focus of the course. I think we will see a consensus on that point, even if we never see one on just how important tests are.

Students

We hear the same things from a lot of students in a lot of experimental courses around the country: "This isn't what we did in high school." That's meant to be a criticism, but I take it as high praise. "You're making us work too hard." Same comment.

At the Ohio State Technology Conference, Deborah Hughes Hallett described how scary it is for students to be thrust into a totally new environment with totally new demands, when they were secure in the knowledge of what "math" is and how it works–even if they saw themselves as people who could never "do math" very well. But once we get them through that scary period (no mean feat–more about that below), our students find they are having fun, and we see them learning and doing things in calculus that students never did before, such as writing intelligent papers about solutions of real problems. Furthermore, they are finding meaning in what they are doing (for example, in the modeling and interpretation phases of problem solving), contradicting their prior beliefs that math has nothing to do with anything else and that it has no intrinsic meaning either.

Instructors

We're just like the students: Changing the way we have always "done math" in the classroom is scary. But once we get past that, we are having fun, working harder than we ever did before, and finding meaning in what we are doing. Our colleagues see that, and some of them are eager to get a piece of the action.

There are a lot of metaphors for the instructor's role in a reformed calculus course. Steve Hilbert of Ithaca College describes calculus reforms as "a decathlon, not an event." Tom Dick (Oregon State University) speaks of changing his role "from a sage on the stage to a guide on the side." Alvin Kay (Texas A & I University) frequently finds himself in the role of drill sergeant. And many people have described what we do as "coaching." Keith Schwingendorf (Purdue) says that he gets much more respect at cocktail parties now; when people ask him what he does for a living, he answers, "I coach at Purdue."

I have another metaphor that is deeply personal and that relates especially to the scary, but critical, time of forcing students out of their traditional beliefs and habits and getting them to function in the brave new world of reasoning. I stumbled on this over a year ago in one of those dinnertime conversations with my wife that starts with "How was your day?" It suddenly dawned on me, as we were comparing notes on the day's activities, that she and I are in the same line of work. She's a hospice nurse.

In the course of caring for terminally ill patients, helping them to die with minimal pain and maximal dignity, my wife is constantly dealing with the stages of death and dying identified by Elizabeth Kbler-Ross; denial, anger, bargaining, depression, and finally acceptance, often a joyful acceptance. Not everyone goes through all the stages or in the same order, and some never make it to acceptance. But this is what I was seeing in my students, and it was clear that I had to be their nurse as they ended one kind of life and started another.

On reflection, we all do a lot of dying in our lifetimes–graduation, marriage, divorce, employment, unemployment, mid-life crisis, and so on. For all of our death experiences, except possibly death of the physical body, we know there is another life awaiting us, but we still find these experiences scary. When we change the rules of the educational game, we impose on our students a termination of their way of life–and they are not at all sure about that afterlife. Thus, we have to recognize and help them through their stages of dying, easing their pain and respecting their dignity, in the hope of their reaching joyful acceptance and new intellectual life.

This article is based on remarks presented at the third annual meeting of principal investigators on NSF calculus grants, Columbus, Ohio, November 11, 1990.

Innovation Need Not Be Expensive: A Pilot Project in Calculus Reform at Butler University

J. H. Morrel, UME Trends 1:4 (October 1989)

Butler University is a small, private school, grounded in the liberal arts and offering professional and preprofessional programs. Due to the requirements of various majors, approximately 40% of all Butler students enroll in calculus. Despite a modest level of high technology equipment, Butler is, in some ways, well suited to implement innovative changes in calculus instruction. Mathematics classes are limited to 40, students are relatively well-prepared, and nearly all calculus is taught by full-time faculty. Based on the project described below, I submit that it is not necessary to have a large investment in technology or to wait for one in order to improve

calculus instruction and learning.

A grant from Lilly Endowment, Inc. in 1988 allowed me to revise the first semester of calculus. Underlying this project is the notion that the most important objectives for students to achieve in beginning calculus are intuitive understandings of change vs. stasis, instantaneous behavior vs. behavior on the average, and approximation techniques vs. exact ones. I also believe that the most important skill to be acquired in calculus is problem-solving. These ideas, and the belief that actively engaged students learn more easily than do passive ones, comprise the framework for my calculus revision.

From the MAA Notes *Toward a Lean and Lively Calculus* and *Calculus for a New Century*, I selected those suggestions which were consistent with my philosophy and which could be implemented without a large infusion of money.

The objectives of the pilot course taught in the fall of 1988 were to insure that students

◊ write coherent mathematical arguments or solutions,

◊ read and understand mathematics on their own,

◊ apply problem-solving skills in a broad range of problem situations,

◊ exhibit thorough understanding of the basic concepts enumerated above,

◊ perform a wide range of computational skills, both by hand and by machine, and

◊ adopt an enthusiastic attitude toward mathematics.

In my lectures, I replaced proofs with geometrically intuitive explanations. I lectured only about half the time in a 5-hour course. Since all required material could not be covered in lecture, the students had to dig through the text, helping meet one of the other objectives. New ideas were often introduced and reinforced through group problem-solving or "brainstorming" sessions, using hand-outs and dialog between the students and me to work through definitions and applications of new ideas.

The composition of my tests also changed dramatically. I included more conceptual problems, more "word" and "non-routine" problems. Previously, I would not have put more than 2-3 word problems on an hour test, and non-routine problems were present only as bonus questions. Last semester, half the problems on the final examination were of those types.

The problem-solving emphasis reflects my belief that elementary calculus is most effectively taught as an introduction to applied mathematics. I gave two lectures and a hand-out on Polya's ideas for problem-solving, and every problem was examined within this framework. Two days a week were set aside for in-class activities. Sometimes I separated the class into small groups and handed out problems to be attacked by the groups while I roamed around as advisor. On other occasions, we worked as a "committee of the whole", while I recorded our solution at the chalkboard. The students often worked, in groups, theoretical or non-routine problems which most of them would never have attempted alone. In addition to these in-class activities, I distributed five problem sets. These sets, independent of a particular section of the text, included multi-step and open-ended problems, some requiring library research, and a few requiring computer use. The students had three weeks to work on each set, were allowed to work together (with acknowledgement), and were required to write solutions clearly and coherently.

I used calculator and computer technology to some degree, recommending the CASIO FX-7000G graphing calculator, and installed True Basic's CALCULUS 3.0 on the 8 departmental microcomputers. About two-thirds of the class purchased the calculators and several class sessions involved their use. Although the University has no electronic classroom, I used a microcomputer equipped with a "PC Presenter" and an overhead projector in class. Several problems from the problem sets required the use of CALCULUS 3.0, and many students used it to check their routine homework, all of which was graded and returned.

The results were encouraging. Over the past five years, calculus students have averaged 67.5% on my comprehensive final; this class averaged 74.5% on an arguably more difficult examination. Eight students actually improved (as measured by test scores) as the semester progressed. The course had a 15% failure-to-finish rate and a 7% 'D' rate. Previously, the same rates, measured over the past five years in my calculus classes, were 25% and 15%, respectively. Finally, despite the fact that the class met at 2:00p.m., attendance improved.

The class was very lively. The students were enthusiastic (even on Fridays), and a sense of comaraderie developed, due in large part to the group sessions, which also bolstered the students' confidence in their ability to solve difficult problems. The graphing calculator, useful for locating roots numerically, graphing functions quickly and seeing the relationship between the graph of a function and its derivative, had a significant impact on the level of mathematical intuition developed by the students.

Many students enroll in calculus at institutions which cannot afford large investments in computer hardware. These institutions can still be leaders in calculus innovationinnovation which does not require such an investment, i.e., the type described above. While larger institutions may show us how to integrate technology into instruction on a large scale, we, at colleges and universities like Butler, can lead the way in curricular innovations and classroom techniques.

Judi Morrel is at Butler University.

Send Money: Is Calculus Reform a Mistake?

S. A. Garfunkel, UME Trends 1:5 (December 1989)

Things are bad ... no, terrible. The number of U.S. math Ph.D.'s is shrinking. U.S. students perform at the bottom of international assessments. We are losing our competitive edge – Send Money!

Reform is in the air. The mathematics community is mobilizing – initiatives, frameworks, standards, strands and threads – reports and proposals are everywhere we turn.

There is much that it is encouraging in all this. Certainly math education is in need of reform. Certainly it is exciting to see members of our diverse community, from all levels of the educational system talking and working together, to focus the attention of the general public on these problems.

But what if we win? What would happen if tomorrow the government said to the math community, "You're right. This is a national disaster. Here is all the money you want. No more studies, no proposals, you're the experts ... Fix it." What would we do?

In truth, I find this prospect quite frightening. To a large extent, this is due to the expectations game. Americans are not a patient people. We would have to give progress reports. "You've been at this for two years now with X billion dollars – why hasn't understanding of calculus improved? Why are we still graduating 60% foreign math Ph.D.'s?" The problem is clear. If we raise expectations, the public will look for quick solutions. And the only quick fixes are antithetical to real reform. Certainly we can drill students to perform better on standardized tests. Certainly we can pay for more Ph.D.'s – buy some potential physicists, chemists and engineers. But these are not solutions – no matter how attractive they might seem.

The truth is that educational reform must be measured in decades (or more realistically, quarter centuries). And this is a message we must shout from the rooftops – whether it is good for next year's NSF budget or not.

All right, but since we're being hypothetical, let's imagine that we win this one too. We're given the grace of both time and money. Now what?

I honestly believe "we" don't have very clear ideas. The national reports speak to broad themes – problem-solving, problem-posing, activity-based, doing mathematics in context, applications, technology, etc. But what should we actually do?

To begin with, let's look at where we are. What, if anything, do we do well? I personally believe that *the best are getting better.* Those talented and motivated students who stay in mathematics (at both high school and college) have a broader and deeper grasp of the subject. All we need do is look at Westinghouse Projects, the Putnam Examination and the Mathematical Contest in Modeling. The creation of new special schools around the country has clearly had a positive effect. If we look at the curriculum of these schools we see courses in mathematical modeling, discrete math, calculus, probability and statistics. We are giving our talented and motivated students *more* (contrast this with the 'good old days' when more meant only more calculus). But if this is true (and I firmly believe that it is), why all the fuss?

Well, one of the most quoted statistics is that the half-life of a math student equals one-year. That is, from the ninth grade on, we lose half of the students who have an interest in mathematics <u>each</u> <u>year</u>. Here, I believe is something we can fix. But to fix it we need to seriously change the way we look at the world! Or at least the world of math education.

First, some axioms, or if you will, articles of faith.

◇ Math education does not only mean the education of mathematicians.

◇ Math education is like sex education, (you don't tell students everything you know).

◇ There is no magic linear sequence of math courses which must be taken in order.

◇ Calculus in college and algebra in high school are not mythical subjects which every person must learn.

◇ We teach too many techniques and not enough ideas.

We need to focus our attention on the average student. There are more of them. Their attitudes become their children's attitudes. They will be the work force in the new technological society. They will pay our salaries. Their education is what we're paid to provide. (They also are our school teachers – look at the SAT scores.)

There is an important assumption underlying all this – a possibly controversial position I should make explicit. *I believe we should gear our efforts to the average students, and when we identify brighter or more interested students we should move them into a different curriculum.*

At present, (leaving aside the tracking system) we teach a curriculum designed for the future math major and let the other students fall by the wayside. Our notion of educational democracy simply provides most students with an equal opportunity to fail.

So, now we have money, we have time and we have the average student in front of us for an hour or so daily. What are our goals?

If we look at a set of goals for math education, we need to look for *end* goals. What do we want students to know and understand when they are finished taking math courses. Here we might look at three kinds of goals: attitudinal, conceptual and skill oriented.

With respect to attitudes we want students to have a feel for mathematics, to understand something of what it is as a discipline and how it relates to the context of human endeavor – both historically and personally. We want students to enjoy the subject, to have a sense of its beauty, utility and fun.

Conceptually, there are very large and basic notions we hope to impart. I won't attempt to make an exhaustive list, and mine would certainly differ from yours, but here is a sense of what I mean: approximation, function, cardinality, proof, chance, symmetry, mathematical model are a few.

Skill objectives are more difficult to formulate because there are so few. Remember these are end goals. Things we feel need to be there ten and twenty years out from school. Basic calculation certainly, graph interpretation, modeling skills, data analysis, visual thinking. It gets harder and harder to justify specifics – especially now that computers can get an "A" on a calculus final in 15 minutes.

The truth is that the conceptual and attitudinal goals are what we really care about. And in a real sense there is a tremendous amount of freedom in producing a curriculum which gets us there. You can teach the nature of mathematical modeling as easily by using graph theory to model garbage collection on city streets as calculus to model Newtonian mechanics. We don't have to teach Euclid to give students experiences with proof. And applications can range from compact discs to TV scheduling, to building your own bookcase. The real point is that once we consider the education of the average student, (not the average mathematician) and set some reasonable conceptual goals, we have a wealth of curricula to choose from.

Now things are really getting scary. Viewed, from this context, there are almost no rules or guidelines. This much freedom brings an awesome amount of personal responsibility. It is the mathematical equivalent of what many Russians are going through – moving from a controlled social and economic environment to the uncertainty of Glastnost and Perestroika. The new orthodoxy is experimentation and, to mix Marxist metaphors – it is time to let a thousand flowers bloom.

Having said all this – I am, after all, a curriculum developer, and I have very definite ideas about curricular change. I'm worried that we may be charging off in the wrong direction.

The problem with the NSF Calculus initiative, for example, is not that calculus reform is wrong-headed – light is preferable to heavy, lively is preferable to dull. The problem is that this initiative has framed the debate in a way which begs the real question. Instead of honing a specific course as though our offerings were set in stone, we should be questioning the entire first two years' sequence. Calculus reform, if interpreted as creating a new calculus course for all first year undergraduates, simply put, must fail.

Consider the undergraduate curriculum. Remember we're still talking about the average student – that's where our problems and our concern are. In just the past year or so, there has been an enormous amount of attention paid to reform of the calculus. Nationally, failure and drop-out rates approximate 50%. This really shouldn't be a shock. Almost 1 million students a year take calculus. Our average student is very likely to be poorly prepared – calculus is hard. But more important, it is frequently the wrong course for the wrong students at the wrong time. If you enter college as a freshman, you can take the Pysch 101-102, Bio 101-102, Chemistry 101-102, English 101-102, and Calculus. Whatever happened to Math 101-102?

Of course we lose math majors! Of course we have large drop-out and failure rates. We take generally poorly prepared students and teach a rather specialized, (and extremely *algebraic*) branch of mathematics. *Every* other department tries to present a broad view of their discipline – showing students a tremendous range of ideas – showcasing the important directions of their field.

The reasons for their approach are compelling. If students stop after the first courses, they have a balanced view of the subject, a view the experts in the field have chosen. If they go on, they have a common knowledge base and a sense of subfields in which they might specialize. What have we got? Disheartened students who learn a year of calculus with little sense of what mathematics is, and less reason to continue. We need Math 101-102. We can explain the breadth of the ideas of our subject – for those who stop and for those who go on. In particular, for those who choose calculus after this introduction, we will have a common base from which to start.

The service course argument – physicists and engineers must know derivatives by week 3 – is specious and pernicious. First and most importantly, we have to take responsibility for our own offerings. We cannot bemoan the loss of majors on the one hand and let other departments dictate our curricula on the other. If we are content with the FTE's we receive as a service and remedial department we should teach these courses and shut up. But we don't have to. Other departments can design their syllabi around ours. They do anyway. Physicists and engineers teach the two weeks of calculus they feel they need – independent of what we teach. The mathematics curriculum is our domain, pure and simple.

Our present approach is to teach specific skills in specific areas and wait for a synthesis to happen (an ah-ha experience). This only works for a small percentage of students. Average students stop too soon and are left with half-remembered algorithms and a bad taste in their mouths. We can't afford it – we're paying too high a price.

If we present an honest overview of mathematics, then in effect we can treat every other math course as though it were the last course a student might take – the data show that this is in fact true for half our students. Math courses then become not prerequisites for the next course, but prerequisites for life.

The reform of mathematics education cannot and should not produce a new orthodoxy. As strongly as I believe in teaching mathematics through applications, I do not want applications to replace concepts in some new 'new math'. The truth is that different students learn mathematics (as any discipline) differently. There is no "correct" way to teach our subject. The challenge to all of us working in the field is to present teachers with as many and as flexible tools as possible, along with some ideas about the pedagogical principles governing their use.

Reform takes place in the classroom, not in proposals, or research papers, or newsletter articles. We must provide classroom teachers with the materials, the time, and the freedom to find what works. For our students we must tear down the artificial barriers to their understanding and appreciation of what mathematics is and how it effects their lives.

I began by asking what would we do if we were to win. I did that because I believe we will. We *will* be given federal money (and some time), to work on the problems of math education. I believe that if we set reasonable expectations and work honestly with the real students in front of us, we can make a significant impact – and have a hell of a good time doing it.

We won't save the world (we might not even reduce calculus attrition), but we might make our own experiences and those of our students a bit more meaningful – and that would really be winning.

Solomon Garfunkel is Executive Director of COMAP.

Confessions of a Calculus Cop-Out

P.J. Davis, SIAM News 25:5 (September 1992)

I am absolutely disgusted with myself. I have been teaching a summer course in freshman calculus, and I have been doing it pretty much the same way that it would have been done 50 or 100 years ago. Has nothing really changed in the interim? Twenty-five years ago, in the days when the IBM 650 with all its punched cards reigned supreme, I incurred the irritation of numerous colleagues by suggesting that it was time calculus joined the 20th century. A year later, I went around the country preaching the virtues of "The Computer Calculus" in a AAAS Chautauqua series. Since that time I have not taught calculus at all.

How has my current conservatism come about? Have I lost that "old time religion"? The truth of the matter is extremely simple: When I was asked to teach the course, I just followed the current practice of my mathematics department (and many mathematics departments). I perceived that this practice would be the path of least resistance for me. There would be no irritated chairman to question my personal selection of materials. There would be no parents who preferred not to have their children turned into educational guinea pigs. There would be no frustrations due to computer ombudsmen who are always on the phone when you need them most. There would be no evening labs to learn a powerful language. (Metatheorem: the more a language does for you, the less friendly it is.) In short, I became — only for the time being, I hope — what could be called a mathematical couch potato.

I used a standard syllabus — no change in a hundred years. I used a popular textbook — a number of changes: polychromatic layouts, beautiful computer graphics, lots and lots of baby real variable theory (ignored), a nod, here and there, to numerical methods (ignored), occasional historical remarks going beyond a name and a date (ignored) — and all this spread over 900 pages of a book sufficiently elephantine to induce a hernia in anyone not in top shape.

I had an ideal class of 15 bright students. I lectured to them. (Why lecture? Can't they read the book?) I found that getting them to ask questions was worse than pulling teeth. I found that they wanted to know only what was required for the test and what wasn't. I assigned the usual sort of problems with no computer work of any sort, and I got back the usual sort of homework papers, some neat, many scribbled, and displaying the usual amount of copying.

You can't fight something with nothing, as they say, so what are the alternatives? The alternatives are contained in the reports of hundreds (it would seem) of conferences, pilot programs, newsletters, new textbooks, transparency acetates, computer programs, floppies, videos, multimedia constructions. I have heard it said that if you want to have buckets of grant money dumped on your shoulders, put in an application to try out a new wrinkle and call it an experimental course in first-year calculus. I can imagine that some enterprising grant consultant has produced a form, much like a form for leasing a house or for obtaining a no-fault divorce, in which the applicant simply fills in the personal wrinkle in a few lines while the remaining 98% of the form is the usual boilerplate dealing with background motivation, organization, funding, validation, follow-ups, standard deviations, coffee breaks, etc.

It is not my intent, nor do I have the knowledge, to summarize all that is around and available. Much of it has to do with the manner in which the computer is to interact with the material of calculus. But this is by no means the only issue. Many of the issues cut across computer usage. For example, there are the "special-interest-group calculi" directed toward statisticians, numerical analysts, psychologists, economists, or quantum electrodynamicists. There is calculus for believers in nonstandard analysis. There is calculus for liberal arts students, for poets. For all I know, there may now be calculus books for one-armed paperhangers. There is the Great Books approach to calculus, and there are enthno- or paleo-calculus books, anti-eurocentric calculus books that tell the student how π was computed to 15 places in 15th-century India. Should first-year calculus now be de-Soviet-Unionized to accommodate all these separate constituencies?

To what end do you teach, and, as a consequence, what do you teach and how do you teach it? Do you teach for economic competitiveness? To reach for the gold in the Math Olympiads? (Who sets the problems in the Olympiads, and why are they set as they are?) Do you teach to display a subject that contains some of the finest products of the history of ideas? To convince the unconvinceable that the subject really is fun? To show that mathematics is a vital part of the conceptual infrastructure between calculus and society? Do you teach for political correctness?

Do you think that calculus should be taught à la Euclid, as an axiomatic and deductive doctrine? That the mean value theorem is critical? That the students should be able to derive the formulas for the radial and transverse accelerations of a particle?

Do you want to employ the Socratic method of the Moore school of "Texas topologists" in a conversation pit? Do you think that Piaget and his contemporary admirers have anything to contribute? Do you teach for intuitive understanding? Do you want to install the project system, using realistic word problems and models to motivate learning? Or do you want to trudge along traditionally with the old concepts, algorithms, and the two famous ships, one travelling north and the other east?

What is the roll of drill? Is there a difference between drilling with pencil and paper in the old-fashioned way and drilling on Mathematica or Maple or Matlab? Or is it just boredom and rote at one level versus boredom at another level? What is the role of memory? Should everything the author has kindly highlighted be committed to memory? How much of the big, fat users' manual is really necessary?

How and for what do you test? Do you test for "plug and chug" manipulation or for fundamental concepts? Have you identified the fundamental concepts, and if so, what gives you confidence that they will be thought fundamental next year? If you ask your students to write an essay, won't they merely reproduce what has been highlighted?

How is teacher training to foster a new scheme to be installed? Is is true that a computer package with hypertext features will replace human teachers, many of whom, in any case, are unqualified mathematically and educationally, are bored stiff, or are researchers who find no professional status or satisfaction in "mere teaching"? Should a hypertext system used in Florida be in English or in Spanish? Will a hypertext system, by virtue of its own elephantine possibilities, create hernia of the mind with information overdosing?

Will a high-tech, cutting-edge, state-of-the-art system save the universities or the high schools oodles of money, or will it merely transfer three times the amount into the pockets of the software producers and vendors? Will the teachers in a future super-high-tech age be mere computer systems installits and fixits? Is the whole traditional teaching profession an endangered species? If so, can a sense of what is important to do, a burning passion for learning, knowing, thinking, be instilled by a machine?

The number of questions that can be raised about first-year calculus is endless, and the answers that I have seen to some of them do not convince me one way or another. A verse of Rudyard Kipling describes my feeling about the myriad possibilities that abound:

There are nine and sixty ways

Of constructing tribal lays

And every single one of them is right.

Exposure to first-year calculus is one of the tribal rites of the higher education community, and as with all rituals, it can engender very strong feelings as to the right way it must be performed.

Go back now to my opening paragraph. How would I, personally, without self-loathing, teach first-year calculus? Confronted with all the attractive opportunities, which would I select? Frankly, I don't know, and perhaps that answer explains my course of least action. And I have the feeling that the country, having survived the disastrous era when calculus went abstract and theorematic, now keeps calculus firmly in the 19th century for want of a clear and general perception of the correct road to take. Mathematics in general and calculus in particular have been overwhelmed by a multiplicity of new demands and by the availability of new potentialities — even while the financial rewards for mathematical expertise have been going down the drain.

I know this much, though. If I am asked once again to teach first- year calculus, I will try to do other than I did, even if it means that, like Stephen Leacock's hero, I jump on my mathematical horse and ride off in all directions simultaneously.

Philip J. Davis is a professor in the Division of Applied Mathematics at Brown University.

Discovering Calculus through Student Projects

D. Pengelley, UME Trends 3:3 (August 1991)

How can we, as teachers, create and use substantial student investigations or projects that guide students into discovering for themselves important ideas and applications of calculus and differential equations? This was the question focusing the attention of twenty-five participants who attended an NSF-funded conference in March at New Mexico State University.

Participants arrived with examples showcasing their own use of projects, perhaps designed to challenge a student with experiment, conjecture, and then proof, or possibly to teach a fundamental concept through discov-

ery. They also brought rough ideas for student projects, which were then discussed and honed in small working groups on the second day. This in turn stimulated discussion of pedagogical issues related to student projects. Examples of project ideas worked on included:

1. Discovering various central results in calculus: for instance, analyzing and designing the smoothness of a cruise control mechanism to reveal the epsilon-delta nature of continuity, from Steve Hilbert and Diane Schwartz at Ithaca College.

2. Taylor approximations in two variables: intuitive aspects emanating from the graphics software of Bev West and Bjorn Felsager at Cornell, melded (via discussion of how lava flows) with theoretical approaches by Marcus Cohen of New Mexico State.

3. Calculus behind physical experiments and models: why do the tops on McDonald's pie packages snap open and closed, and why do the snap bracelets popular with school children "snap"? stimulated by Elgin Johnston of Iowa State.

4. Computer simulation of differential equations raising and answering questions: funnel/antifunnel behavior of $\frac{dx}{dt} = x^2 - t$, led by Bev West and by David Lovelock from Arizona.

This concrete work raised many pedagogical issues such as,

a. Teaching students when to use and not use calculators and computers and that these machines are often not to be trusted (Dick Metzler from University of New Mexico, and David Lovelock).

b. Group work and other aspects of a good project assignment: assigned tasks for individuals, group versus individual reports and presentations, evaluation, presubmissions or two-part assignments, length of projects, an all-projects course, connections to in-class work and labs, making projects so enticing that students can't put them down (contributions from Keith Schwingendorf of Purdue, Dave Arterburn of New Mexico Tech, Aaron Stucker of U.S. Naval Academy, Brian Farr of Worcester Polytechnic, and many others).

Final discussion centered on our goals for students. We want to foster their abilities to think independently, synthesize different approaches, solve multistep problems, and ask whether an answer is reasonable; to be able to read and write mathematics and to talk about and analyze mathematics with others to learn by exploration, learn to model the real world, and indeed to learn how to learn. Ambitious goals perhaps, but the most exciting thing is that our diverse experiences with assigning projects to students is convincing us that these goals are achievable.

David Pengelley is at New Mexico State University.

Mathematics Outside of Mathematics Departments

S.A. Garfunkel and G.S. Young,
Notices Amer. Math. Soc. 37:7 (September 1990)

The following article is based on a survey commissioned by the Exxon Education Foundation. Solomon A. Garfunkel is Executive Director of the Consortium for Mathematics and Its Applications (COMAP) in Arlington, Massachusetts. Gail S. Young is professor of mathematics at Columbia Teacher's College. The authors wish to express their deep appreciation To Dr. Richard F. Link, who prepared the sampling plan, oversaw data analysis, and kept two mathematicians statistically on track.

One could view the present state of American mathematics as one of brilliant success or extreme crisis. Mathematics research is more fruitful than ever before in the history of the discipline. Mathematical methods are making real contributions in more and more disciplines, and many new fields are coming into being. Undergraduate mathematics majors who have learned some computer science can get jobs with starting salaries comparable to new mathematics Ph.D.s. New mathematics doctorates are in great demand and, in academia, can get starting salaries comparable to associate professors in the humanities.

On the other hand, the number of mathematics Ph.D.s awarded each year has dropped steadily for more than a decade, and more than half the new doctorates are going to foreign nationals. The number of bachelor's degrees in mathematics, after more than doubling between 1960 and 1970, dropped during the 1980s to the 1960 level. Calculus-level enrollments more than trebled in the fifteen years following 1960. In 1985 calculus accounted for 39% of the mathematics enrollments, whereas remedial and precalculus courses accounted for 52% of the enrollments, advanced courses comprised just 9% (approximately 147,000).

This last fact is especially curious. Consider that, between 1960 and 1983, engineering bachelor's degrees rose from 38,000 to 72,000; biology degrees from 18,000 to 44,000; social science degrees from 16,000 to 25,000. Mathematics has become increasingly important in all of these fields, so one would assume that enrollments in advanced mathematics would have risen in that period. Where are the students of advanced mathematics? This was the question motivating this study, undertaken in 1988. What we found is that a rise in advanced courses seems to be occurring, but it's not occurring in mathematics departments.

The Study. Anecdotal information and even a cursory look at college and university catalogs indicated that the content of many courses in departments (other than mathematics!) consists entirely or mainly of advanced mathematics. We found such courses not only in the expected places — departments of physics, engineering, computer science — but also in such departments as

economics, political science, biology, and management. Therefore, the fact that enrollments in advanced mathematics courses have not risen may reflect an increase in enrollment in mathematics courses being taught outside mathematics departments. This study was prepared to test his hypothesis and to provide a baseline for future studies.

We sampled 425 schools according to a statistically-based sampling plan (details are provided in the appendix). Course catalogs were used to select the non-mathematics department courses which appeared to contain significant mathematical content and which used mathematics at or above the level of calculus. Statistics departments were specifically excluded from the study. In cases where it was difficult to assess the mathematical content of a particular course, we telephoned department chairs to get their appraisal. Once the courses were selected, we obtained enrollment data from the institutions (in addition to other information about the courses).

Out study determined that, each year, there are over 170,000 enrollments in advanced mathematics courses being taught outside mathematics departments. These data are presented in the tables at the end of this article, which also sort the numbers according to school, size, type (private or public, university or college), and region.

Engineering departments recorded the largest number of enrollments (83,854) with significant mathematics content. Of the courses represented, 86% developed the mathematical content before it was applied; this was especially true of electrical and chemical engineering courses. Among the mathematical prerequisites mentioned for engineering courses were: calculus, linear algebra, discrete mathematics, and modeling. The term *mathematical maturity* was used in several responses. (Interestingly, engineering courses also had the largest average class size: 113 students.)

Next in order of enrollments were business courses. Calculus, linear algebra, and statistics were common prerequisites, with "the mathematics involved in operations research" mentioned quite often. The survey found 61 courses in "business forecasting", which seems to employ large amounts of probability and modeling; these courses are probably required, because they had large enrollments.

The sciences use a great deal of mathematics, and science instructors seem to be expending a great deal of time teaching mathematics before they actually use it in their courses. Courses, such as Mathematical Methods of Physics or Numerical Methods in Biology, specifically train students in the mathematical applications of a scientific discipline. Many of these courses seem to be required. Another interesting development is the prevalence of probability being taught in genetics classes, despite the fact that it is a prerequisite for almost every one of these courses.

Agriculture and the social sciences also recorded

courses with advanced mathematics. Many courses stress the need for *mathematical maturity*, with calculus being the formal prerequisite. Discrete mathematics and probability are also common prerequisites in these areas. Our deletion of statistics courses from the survey contributes to the small size of the listed enrollments for social science courses. For example, a course entitled "Psychology Statistics" is present in over 50% of the departments surveyed.

Department Responses. We sent a preliminary draft of the first sections of this report to the chairpersons of responding departments, partly as a courtesy, but also in the hope that many would respond to the request in our cover letter that they tell us why they think these courses exist and whether their campuses know that so much mathematics is being taught outside the mathematics departments. We sent out 714 drafts, and have received 292 replies (a response rate of 41%). This is quite good considering the lack of pressure in our letter, the fact that we are clearly not in their fields, and the lack of follow-up.

In fact, many of the replies are quite long and detailed, showing considerable thought and concern. The nature of the comment does not depend on the size or status of the university of college. Although some of the comments are vitriolic, many appreciate some of the problems of mathematics departments, and some regard the present state as inevitable and even desirable. But overall the comments are critical, and very few of the respondents are really happy with the mathematics departments at their institutions.

The comments can be classified roughly into five categories:

(1) The mathematics faculty does not know or appreciate applications.

(This perception is held even in the case of departments which we believe are strongly involved in applied mathematics.) Some typical quotations:

◇ "[There is an] inability of mathematicians to come to grips with a difference between 'pure' and 'applied' mathematics. Mathematicians of the 'pure' strain look down their noses at the other strains. Thus science and engineering departments feel that their students will not be adequately served by math departments. The math departments are unconcerned because they view this as 'bogus' math which they have no interest in offering anyway..."

◇ "Modern technical fields, with their complex applications of math/technology to real-world problems, have requirements that can't easily be met by isolated, largely theoretical, overly generalized presentations that are not (usually) presented as being useful or interesting to a practical person. The math department courses 'turn off' most of our kids and it's up to us back in the professional departments to turn their enthusiasm back [on] again."

◇ "Mathematics departments have become so abstractly oriented that their courses are not given any applied content ... Most math Ph.D.'s never take an applied math course, so why should they teach it? Our engineering school tried to convince our math department to teach some applied courses but eventually gave up."

(2) Mathematics faculty teach mathematics as an art with full abstraction, not as a tool.

◇ "The content of most math courses focuses on theoretical development. This is not 'bad' per se but leaves most students wondering about the 'what, when, where, and why' of applications ... seldom addressed in traditional math courses."

◇ "Applied departments use math as a tool. An individual topic is analogous to a hammer perhaps. They wish to 'hammer' with it. On the other hand, math departments often become more interested in its description and generalization of the 'hammer' itself."

◇ "Engineers find math to be a need, not a love. Mathematicians. . .are out of touch with the real world and are more like mathematical artists than real world scientists."

(3) Topics span too many mathematics courses.

◇ "There is not room ... for every student to take the separate courses in differential equations (ordinary and partial), vector and tensor analysis, complex variables, Fourier series, probability ... all these are covered in our one-year course by omitting the detailed proofs and generalization ..."

◇ "Other reasons [for such courses] include the claim that the mathematics needed cannot usually be found in one single mathematics course. Bits and pieces of several courses are wound together."

(4) The mathematics departments have not kept up with new applied mathematics.

◇ "Mathematics departments generally do a lousy job with mathematics their faculty has no training for. Specifically, Shannon's information theory; automata; transformation geometry; graph theory, particularly as employed in practice; algebraic coding theory; polynomial rings and finite fields; computer ability; and, one suspects, probability and statistics."

(5) Mathematics courses do not give students the knowledge or the mathematical maturity for further work.

◇ "An attempt by mathematicians to do 'something for everyone' in basic calculus courses rather than concentrating on a generic mix of basic skills and concepts. Engineering has further suffered by the loss of more preparatory subjects such as analytic geometry."

◇ "I cannot take it for granted that [students from calculus] are able to use their mathematical skills in problem solving. What appears to be. . .lacking is the ability to formulate a problem quantitatively and then to solve it using the tools they learned in their calculus course."

We are reporting these opinions as ones widely held in other fields, and we are not endorsing or contradicting them. They are, however, very widely held by our respondents, and therefore, must be considered.

Conclusions. As this is to our knowledge a baseline survey, it is difficult to measure any trends. However, anecdotal evidence indicates that there have been substantive increases in enrollments in the courses surveyed. It is certainly clear from the survey that there are more enrollments in advanced work in mathematics outside of mathematics department than within (approximately 173,000 as compared to 147,000). It is unclear whether this situation is fully understood at individual schools (and in national assessments).

Our follow-up questionnaire to non-mathematics departments indicates that these departments believe that the substance of the survey is well-known and understood on their campuses. In fact, it is clear from their responses that the department believe that mathematics faculty are responsible for and content with this state of affairs.

In a preliminary form of this report we concluded with, "Finally, we should like to stress that this report makes no value judgments on the present state of affairs. It may or may not be healthy for undergraduate instruction in mathematics to be distributed among many departments."

However, on a personal note we find both the survey results and the attitudes of non-mathematics departments deeply disturbing. There is an indicated residue of ill-feeling toward mathematics departments. Worse yet, mathematics faculty and curricula are often seen as at best irrelevant and at worst counterproductive. Moreover, respondent after respondent expresses the belief that mathematics departments are unconcerned about issues of course offerings.

Perhaps they are correct. We have shared the results of this survey with a number of mathematicians and mathematics educators and seen no evidence of shock, dismay, or surprise. We are anxious to get reaction from the broader mathematics community.

This survey was commissioned by the Exxon Education Foundation. The purpose of this and several other programs funded by the Foundation was to assess the state and health of undergraduate mathematics education. Moreover these projects were designed to point the way toward needed reform. Given these results, what, if anything, should be done?

We would like to ask that question of the AMS membership. Are these results surprising? Are they troublesome? What, if anything, should they spur us to do? At the very least, we hope to begin a public debate. Please write to us at: COMAP, 60 Lowell Street, Arlington, MA 02174. Your ideas and suggestions are important. They can have a direct effect on programs designed to

improve the quality of undergraduate instruction and on the health of our profession.

Why Dont Most Engineers Use Undergraduate Mathematics in Their Professional Work?

R.W. Pearson, UME Trends 3:4 (October 1991)

I pose the question:

Why do 50% (probably closer to 70%) of engineering and science practioners seldom, if ever, use mathematics above the elementary algebra/trigonometry level in their daily practice?

(Editor's Note: See the article, The Occupational Displacement of Mathematical Scientists in Commerce, UME TRENDS, August, 1991, for background on this issue.)

Based on my fifty-four years of experiences as a design engineer, as an engineering manager, as a member of management assigned to help alleviate engineer-shop design and manufacturing problems, as a product cost and reliability analyst, as a corporate executive, and as an undergraduate mathematics instructor, I sincerely invite mathematicians to look more deeply into the end use of their product in the "real" world.

In this spirit I ask:

Is there something drastically wrong with our method of teaching mathematics to undergraduates which might result in the cause behind the first question above?

My work has brought me into contact with *thousands* of engineers, but at this moment, I cannot recall, on average, more than three out of ten who were versed enough in calculus and ordinary differential equations to use either in their daily work. It certainly would have been better for technology if they had.

Over the years I have wondered:

Is it because the mathematics courses were (and are) required by some "education authority," so the student reasons that "the sooner I get them behind me, the sooner I can get on with *real* engineering stuff?"

Or, did the mathematics professors "throw" the stuff at the "lowly engineering" students because it was part of the curriculum? (Gosh, how I hate to think this could be so!)

Or, were the examinations so easy that little was really accomplished?

Or, did the professor do such an excellent job of covering the subject at hand that the students failed to grasp the important intent of the presentation?

I personally have taken math courses where I experienced each of the above situations.

Moreover, there were/are extremely competent engineers/scientists who never had a course in calculus. Edison was one. I have known four engineers who have been granted many United States Patents. In fact, two of them never went to college, one of whom was chief electrical engineer of a world renowned electrical manufacturing company. Another had only a degree in political science! I cite these examples solely to indicate that although higher math is certainly a great advantage in many areas, it is not the only necessity for technological growth, and it seems quite apparent to many engineers that they don't need such math for their livelihood.

Many engineers can get along very well in the workplace without knowing "higher math" by having their mathematical needs answered by a research and development specialist, a mathematician, or by looking up the formula in a handbook. For these engineers, it is easier *not* to go back to the math textbook and plow through the "if and only if's" and other conditions on the use of given equation, which, from a managerial viewpoint, is a waste of time and money, if a person versed in mathematics is available.

I relate the cursory use of "higher math" in engineering to the quality of the usual type of course competence examination questions which state:

"Given the set of circumstances, develop the equation needed, then find the answer."

This is closed indeed. We need more to insure that the student really understands all of the nuances overlooked or forgotten in the statement and solution of that problem. Further, the student's tendency to jump toward a given type of equation may have been just rote, because *that equation* was discussed in class several sessions ago. Does one routinely ask the question:

"Are there several methods of solution? If so, which is the one that tells me the most about the limits, or other nuances, of that solution?"

As I think of the exceptional reasoning power of our great mathematicians, which has created the theorems and equations that have permitted the elevation of engineering/science to its present state, I stand in awe. Why do practitioners in these fields not use this knowledge in their daily work? Is the logic so deep that it scares the average engineer/scientist away from its use? Are the necessary mathematical manipulations overwhelming the memory capabilities of the practitioner? Are we trying to cover so much in our class presentations that the student dreads the subject?

By all means the *mathematician* must continue to create the logic and proofs in her or his work, but is it *absolutely necessary* for the ordinary practitioner to be confounded with such deep thoughts? Are we covering too many parts of calculus and differential equations in our elementary texts? Would we get more generalized use of these studies if we stressed understanding of those topics and equations most often used by practitioners? We would need some sort of poll to determine this.

Some mathematicians in the calculus reform movement have polled the practitioners for use-information of this nature. So why do we continue to teach what the mathematics professors *think* the engineer/scientist needs, as contrasted with what is *actually* needed in industry and commerce?

So, Dear Mathematicians, we Practitioners need your help — to bring our engineers up to mathematical speed.

Robert Pearson is Chairman of the Board of P-W Industries, Inc.

Can We Modernize What We Teach?

W. F. Lucas, UME Trends 1:5 (December 1989)

Two things fill my mind with never ceasing awe: the starry heavens above me and the ethical law within me.
Immanuel Kant

Mathematics education is in very serious trouble. Some of the major reasons for this failure are: (1) a lack of sufficient resources, both financial and human, with bleak prospects for the near future; (2) a variety of societally related shortcomings as illustrated by students' short attention spans and unwillingness to pursue intellectual challenges; and (3) a very bad "image problem" which is of our own making. I consider (3) to be at least as serious as (1) and (2), and the problem that should be the least expensive and easiest to eliminate, if we only direct our minds and efforts towards it. I begin with three popular beliefs about mathematics and how it is taught. These negative attitudes persist despite the enormous breadth of new, exciting, and highly teachable topics from contemporary mathematics which could so easily alter these views. I conclude with some suggestions on how to begin to solve this image problem by modernizing what we teach.

Some Unfortunate News

First, mathematics is viewed by our students and the public as mostly *old stuff*. Important? Yes! But almost all out of the distant past. A rather "completed subject." The "Latin of the sciences," rather than a vital modern science. (To many, the "distant past" means any time before they were born.) High school students know that physics and biology are alive and growing, and a 40 year old course would really be obsolete even though it may have much in common with one today. A 40 year old algebra course would be fine today, and an 80 year old one may well be better. There is nothing wrong with studying the classics, but you would think that a subject which is doubling its activity every decade would have more new to offer at *all* levels.

Second, there is an unnecessary shortage of real excitement, true enjoyment, and interesting challenges in our math classes. Outside of school, people routinely choose to play games, solve puzzles, schedule activities, and select from alternatives. They frequently engage in considerable quantitative thinking, but rarely associate this with mathematics. Students tend to tune out real thinking and turn on the more rote and algorithmic parts of their brains as they enter the classroom. Teachers, on the other hand, can employ a variety of techniques to maintain attention and increase interest. These approaches include group interaction, team projects, experimentation, modeling, discovery methods, and forms of problem solving. There exists today many new math subjects which could provide much more suitable and rewarding topics for these or other modes of teaching.

Third, precollege math is probably the most profound case of intellectual *delayed gratification.* Students hear over and over again how important math is and how they will need it in the future. A distant future most will never achieve. Math remains the butt of jokes in this regard. We invite in famous people from the physical and life sciences to testify to the great importance of mathematics in their work. Nevertheless, students lack conviction. This situation is so terribly unnecessary in light of the many new mathematical subjects that have elementary topics which could really convincingly demonstrate the importance of math to school children's *everyday* life. Most of us know from personal experience that just a little bit of fun or applicability can go a very long way in terms of personal reward.

New Directions using Mathematics

Continuous math has, for very good reason, dominated mathematics for three centuries. To prepare students for calculus without delay has been an aim of almost everything taken before it. For most students, however, the role of calculus is that of a service course. It prepares them to study the physical sciences and traditional engineering. This situation persists even though few of our students will ever reach calculus. Many of the other goals of precollege math could surely be met as well, or better, with alternate topics or strands. Many recent, interesting, and immediately relevant concepts could achieve the same ends regarding quantitative reasoning. Should we continue to "prohibit" people from gaining a serious appreciation for modern (as well as much of classical) mathematics until they have mastered the "high hurdles," i.e., been through this rite of passage called "calculus"? Many will be left far behind, because they failed to clear even the "low hurdles" of high school algebra.

Megatrends in applied math in recent decades include its use to analyze very large scale systems involving various constraints and human design; as well as to study humans themselves: their values, behavior, interactions, institutions, and decision making. Computers allow us to solve many problems we shied away from in the past. Many of these new applications have not been dominated by continuous math. They have often stimulated theoretical advances in discrete math, and the ongoing synthesis of these is is turn rapidly creating an exciting new modern algebra. On the other hand, many new developments in pure math, such as deep insights into the nature of higher-dimensional geometry, may well find major uses involving algorithms and complexity in some of the more recently mathematized areas, which are inherently multivariate (but finite). In any event, it is evident that modern uses of mathematics in the behavioral, social, managerial, operational and decisional sciences has provided a plethora of new materials which could, if we were to pursue it, have truly profound impact on math education at every level. Students today can be seriously motivated to study math (pure or applied) for many other good reasons in addition to understanding the beautiful structure and workings of the physical world and its technology.

Some Suggestions

We must alter the popular image of mathematics as a dead field and of serious interest only to a few experts. The wealth of new ideas in mathematics today can convincingly demonstrate that it is a modern, exciting, and useful subject. We should give our students an appreciation for twentieth century mathematics before the century ends! The rewards could be great if we would only "risk" seriously altering a bit of our curriculum. (Mere reports asserting that we are doing so are not sufficient.) Such changes would revitalize our teachers as well. Several other current concerns, like writing in math class and humanistic math networks, would also be more successful in this new mode.

To teach more recent developments in mathematics and its applications, one must, of course, know something about them. Our leaders, societies, and funding sources must assist in making new advances accessible. Our record to date has not been outstanding. We must eliminate various elitist attitudes and erroneous misconceptions such as: real researchers don't do education; some types of math are better than others; some directions of application (e.g., physical science) are better than others (e.g., management); beauty and utility are somehow in conflict (The Hardy Syndrome). These must be replaced by a new code – a code that includes the assertion that no mathematician has a right to not contribute to the educational system that made her or his career possible. We could learn much from the chemists in this regard.

There are three places where we could begin to introduce more new, motivational, and immediately applicable topics into our curriculum which would minimize significant change in our current habits and practices. (i) In the first-year college course where students are merely seeking science or math credits with no plans for continuing in such fields. (ii) In senior high school where many should take a concurrent or alternate course to "analysis" or AP calculus. Some students will not, or should not, immediately take the serious precalculus and calculus courses offered in many high schools; but they also should not interrupt or terminate their study of mathematics. Many topics from discrete math, statistics, modeling, computers in math, as well as entirely fresh ideas could be incorporated to enrich current courses and to design entirely new ones. (iii) In middle school where there is enormous repetition and slack in current offerings. I believe that quick success in such places would cause rapid spread of these ideas to other levels.

In summary, there is much in more recently discovered mathematics (including the general direction of Kant's second awe) which could go a long way towards modernizing our image and improving student motivation, if only some leadership in this direction were to emerge.

Bill Lucas is Chairman of the Mathematics Department at The Claremont Graduate School and an Avery Fellow to Harvey Mudd College.

Let Us Teach Exploration

G.D. Foley, AMATYC Review 10:1 (Fall 1988)

Preparation of this article was supported in part by grants form the British Petroleum (America), the National Science Foundation, the Ohio Board of Regents, and The Ohio State University. The opinions and conclusions herein are those of the author and do not necessarily reflect the views of these funding agencies.

I thank the colleagues and associates who offered suggestions during the preparation of this article.

Polya said, "Let us teach guessing" (1950). This article attempts to expand on this worthwhile theme. It advocates our helping students view mathematics as a lively, experimental science in which they learn through the exploration of ideas and interrelationships. It advocates moving from a computational focus to a conceptual focus by exploiting modern technology to avoid needless computations and by using the time saved to address important issues. In this way, perhaps we can make mathematics fun and satisfying and foster in our students a spirit of adventure and a thirst for knowledge.

Exploration in mathematics is not new. Mathematics began through observation. Early humans developed their mathematics through abstraction of natural forms and recognition of environmental patterns. Abstracting from specific instances to general properties and recognizing patterns still have an important place in mathematics today. Trial and error, guess and check, conjecture and proof are all a part of creating mathematics. An exploratory approach to teaching and learning recognizes this.

Modern psychology tells us that individuals learn through self-construction of ideas and inter-relationships (see Bodner, 1986). Each person builds her or his own mathematics. The teacher acts as a guide during this building process. An environment of experimentation facilitates learning via construction, and calculators and computers can serve as tools for this mathematical experimentation. Exploration is in keeping with this empirical, constructivist view of mathematics teaching and learning.

Exploration, Problem Solving, and Thinking. Exploration as an approach to mathematics teaching and learning is closely related to two currently advocated educational themes: problem solving and higher order thinking. In 1980, the National Council of Teachers of Mathematics proclaimed that "problem solving [should] be the focus of school mathematics in the 1980s" (NCTM, p. 1). Since then, problem solving has become the focus of much mathematics education literature (see Suydam, 1987). Thinking, especially higher order thinking, which many educators claim should be a major focus of the curriculum, has been a pervasive theme in recent general educational literature (see Patterson & Smith, 1986). The view of mathematics education as exploration captures the spirit and the intent of problem solving and higher-order thinking.

Exploration is related to problem solving. The term exploration suggest the sustained investigation of mathematical concepts and interrelationships. working toward the solution of a challenging problem can provide an environment for exploration, but problem solving and exploration re not synonymous. Exploration is often called for at various stages in the problem solving process - understanding the problem, devising a plan, carrying out the plan, and looking back (see Polya, 1957) - and does embrace many of the commonly listed but infrequently taught problem solving techniques: guess, check, and revise; work backwards; search for patterns; conjecture and try to prove or disprove; draw a graph; make a table; and solve a related problem. In other words, much of what we do when solving problems , especially complicated ones, is exploration. Problem solving, though intended to communicate a much deeper and much richer activity, is often interpreted as answer finding, and students end up working exercises rather than solving problems. To some, problem solving means doing story problems, but story problems too can become automatic, drill exercises, and many interesting problems are no based on stories.

Exploration is reached to higher order thinking. Higher-order thinking may evoke an image of a human figure leaning forward with chin on fist: and image of contemplation without investigation. But, according to *Webster's Ninth New Collegiate Dictionary*, to explore means "to investigate, study, or analyze" (1986, p. 438). Patterson and Smith (1986) define higher-order thinking as occurring "when a person is engaged in active and sustained cognitive effort directed at solving a complex problem and when the person makes effective use of prior knowledge and experience in addressing the problem [italics omitted]" (p.82). This definition conveys fairly well the notion of exploration used in this paper, but exploration has an added shade of meaning that is not emphasized by the definition. Exploration involves making use of new, uncharted (for the student) territory. The explorer not only makes use of prior knowledge and experience but also constructs new knowledge and fits in into an existing structure. Exploration involves *learning*, not just thinking.

Exploration conveys more than either problem solving or higher order thinking alone and is less likely to be misinterpreted. Perhaps this is why the word explore has such a prominent place in NCTM's *Curriculum and Evaluation Standard for School Mathematics*:

> To provide all students an opportunity to learn the mathematics they will need, the emphasis and topics of the present curriculums should be altered. More importantly, methods of instruction need to emphasize exploring, investigating, reasoning, and communicating on the part of all students... Classrooms should be places where interesting problems are explored using important mathematical idea. For example, in various classrooms one could expect to see students recording measurements of real objects, collecting information and describing their properties using statistics, and exploring the properties of a function by examining its graph. (Commission on Standards for School Mathematics. 1987 p.1)

This quote comes from a working draft of the *Standards*, but based on individual comments, regional meetings, and open discussions at the 1988 NCTM meeting in Chicago, there seems to be a consensus that this draft does provide an appropriate framework for the mathematics that should be taught on schools, and the final form of the *Standards* will likely be very similar. Exploration, as a major theme of the *Standards*, is linked to other themes: problem solving, reasoning, communication, and access to and appropriate uses of technology.

How Can Exploration Be Taught? Mathematics is sequential, and with few exceptions, there is a next course for which we always seem to be preparing our student. So the first step is to take a hard look at what content students really need in order to go ion. We must ask and answer, What are the central concepts and critical supporting ideas and interrelationships? Douglas (1986) and Steen (1988) claim that, at least in calculus, there is some fat we can trim so that we can focus our students' attention on the meat of the subject. Calculators and computers speed computations and graphics, and thus both free time for the exploration of concepts and make new sorts of exploration possible. Once topical emphases and curricular themes are established, exploration can begin. The following list provides some starting points for creating an environment in which mathematical exploration can take place:

1. *Modeling Exploratory Behavior:*Instructors must exhibit the kind of approach to mathematics they wish their students to follow. Planning, looking ahead, midstream checking, and looking back are all appropriate mathematical behavior. Pedersen and Polya (1984) give several excellent examples of problems that can be approached in many ways; students need to see their instructors investigating alternatives if they are to do it themselves.

2. *Classroom Demonstrations:*Demonstrations can be memorable. Flipping graphs on overheads to demonstrate how to obtain the graph of an inverse, shining conical beams of light on planar walls and ceilings of darkened classrooms to illustrate conic sections, and drawing ellipses using a string taped to the chalkboard are but a few examples. the use of interactive software on a microcomputer linked to a large monitor or overhead projection palette creates a perfect set-up for class exploration.

3. *Classroom Discussions:*In order to generate a lively discussion, ask probing questions and seek out and test conjectures. Try to create a spirit of group exploration and cooperative learning. After exploring a concept using examples and nonexamples, get the class to generate a definition through a refinement process.

4. *Laboratory activities:*Here again computers play a vital role. students can explore individually with the instructor or lab assistant nearby. Laboratory assignments, whether within scheduled classes time or as outside assignments, can be made progressively more challenging so as to build the students' explanatory power. Assignments can also be individualized to provide challenge or remediation.

5. *Homework Problems:*Assign fewer problems and include some that require sustained effort. For example, you could assign set of problems in series where you ask: Does each series converge or diverge? If it converges, can you prove it? How fast, of slow, does it converge? Explore the speed of convergence using a programmable calculator or a computer. If it diverges, in what sense does it diverge? It is interesting to explore or a computer. If it diverges, in what sense does it diverge? It is interesting to explore on a hand-held or micro computer such as series $10! + 1!2! + 13! + \ldots$ and $1 - 13 + 15 - 17 + \ldots$ to see how dramatically different their rates of convergence are. The first series converges to e very rapidly, and the second converges to $\pi 4$ very slowly. (See Simmons, 1985, pp. 715-716, 720-721, for further exploratory ideas relative to the second series.)

5. *Test Questions:*Again, less can be more. Ask fewer but more probing questions, such as, on a Trigonometry test: Classify each equation as an identity, a conditional equation, or a null equation. If an identity, prove it. If conditional, solve it. If null explain why. Such a group of items will sharpen students' understanding of trigonometric equations and the relationships among the various types.

Conclusion. This article advocates and exploratory approach to teaching and to learning. Too often students view mathematics as a collection of facts and procedures to be memorized, and textbooks exercises and even our own test questions frequently reinforce this shallow view. Overcrowded syllabi and the associated rapid-fire "covering" of the material do not help matter either. An exploratory approach requires extra preparation and effort on the part of the instructor, but pay off in an improved understanding and appreciation of mathematics on the part of the students. So let's encourage our students to explore and teach them how to explore as they do mathematics. Let's get them to realize that exploration is a natural and integral part of mathematics.

The Ohio State University, Columbus OH 43210- 1172

References

Bodner, G.M. (1986) Constructivism: A theory of knowledge. *Journal of Chemical Education*, **63**, 873-878.

Commission on Standards for School Mathematics. (1987) *Curriculum and evaluation standards for school mathematics* (working draft). Reston, VA: National Council of Teachers of Mathematics. (Available from Thomas A. Romberg, Chair, NCTM Commission on Standards for School Mathematics, 1906 Association Drive, Reston, VA 22091.)

Douglas, R.G. (Ed.) (1986) *Toward a lean and lively calculus: Report of the conference workshop to develop curriculum and teaching methods for calculus at the college level*, MAA *Notes*, No. 6, Washington, D.C.: Mathematical Association of America.

Foley, G.D. (1987) Future shock: Hand-held computers. *The AMATYC Review*, 9(1), 53-57.

National Council of Teachers of Mathematics. (1980) *An agenda for action: Recommendations for school mathematics of the 1980s.* Reston, VA: Author.

Patterson, J.H. & Smith, M.S. (1986) The role of computers in higher-order thinking. In J.A. Culbertson & L.L. Cunningham (Eds.) *Microcomputers and education: 85th yearbook of the National Society for the Study of Education – Part I* (pp. 81-108). Chicago: University of Chicago Press.

Pedersen, J. & Polya, G. (1984). On problems with solutions attainable in more than one way. *College Mathematics Journal*, 15, 218-228.

Polya, G. (1950) Let us teach guessing. In *Etudes de philosophie des sciences, en hommage a Ferdinand Gonseth* (pp. 147-154). Neuchatel: Griffon.

Polya, G. (1950) *How to solve it: A new aspect of mathematical method* (2nd ed.). Princeton, NJ: Princeton University Press.

Simmons, G.F. (1985) *Calculus with analytic geometry.* New York: McGraw-Hill.

Steen, L.A. (Ed.). (1988). *Calculus for a new century: A pump, not a filter* [MAA *Notes* No. 8]. Washington, D.C.: Mathematical Association of America.

Suydam, M.N. (1987). Inidcations from research on problem solving. In F.R. Curcio (Ed.). *Teaching and learning: A problem-solving focus* (pp. 99-114). Reston, VA: National Council of Teachers of Mathematics.

Websters ninth new collegiate dictionary. (1986). Springfield, MA: Merriam-Webster.

Constructivism in Mathematics Education: A View of How People Learn

A. Selden and J. Selden, UME Trends 2:1 (March 1990)

There are at least four views on how students learn mathematics: spontaneously, inductively, constructively, and pragmatically. Each has implications for teaching. If students learn mathematics spontaneously, there is little one can do directly to help them, except perhaps provide them a good environment, that is, good explanations and materials. If students learn inductively by working examples and extracting common features and important ideas, then one might present many carefully structured examples to facilitate this. If students learn by making mental constructions to handle mathematical ideas, then one might want to study these mental constructions to discover how to help students make them. If students learn pragmatically as a response to real world problems, one might search for interesting applications. In regard to calculus, for example, most mathematicians *say* they believe students learn inductively through examples, whereas they often *teach* as if calculus were learned spontaneously by listening and watching. (Ed Dubinsky, paper presented at the St. Olaf Conference, October 1989.) Like many researchers in mathematics education today, Dubinsky takes the constructivist view.

The term "constructivist" might be misleading for mathematicians who are likely to associate it with constructive mathematics along the lines of L.E.J. Brouwer, but it means something very different for those in mathematics education research. For Jeremy Kilpatrick, former editor of the *Journal for Research in Mathematics Education*, the constructivist view involves two principles: (1) knowledge is actively constructed, not passively received, and (2) coming to know is an adaptive process of organizing one's experiences and *does not* involve discovering an independent pre-existing world outside the mind of the knower. The first principle, which is generally accepted by those in mathematics education research, is sometimes called *simple* constructivism, whereas acceptance of both principles implies a certain epistemology called *radical* constructivism by some although it is essentially the contribution of J. Piaget. Kilpatrick asserts that certain consequences, claimed to follow from the second view, fit equally well with other philosophical positions. These are: (1) teaching should be sharply distinguished from mere training; (2) students' thought processes are more interesting than their overt behavior; (3) linguistic communication is for guiding students' learning, rather than for transferring knowledge; (4) students' misconceptions provide clues to their attempts at understanding; and (5) in-depth interviews can be used, not only to infer cognitive structures, but also to modify them.

The implication for mathematics teaching is that the emphasis should shift from ensuring that a student can correctly replicate what he or she has been shown to concentrating on helping her or him organize and modify her or his "mental schemas." For mathematics education research it means in-depth studies of students solving problems or coming to terms with concepts-descriptive studies, as well as predictive studies. See, for example, Schoenfeld, et al's 18-month study of seven hours of video tape of a 16-year-old student taking calculus during the summer at Berkeley, as she attempts to come to grips with the concept of slope.

For an easy-to-read, yet extremely informative, introduction to constructivism, see von Glasersfeld's plenary paper, "Learning as a Constructive Activity" (*Proceedings of the Fifth Annual PME-NA*). Von Glasersfeld, currently at the Scientific Reasoning Research Institute, University of Massachusetts, is described by Kilpatrick as the major exponent of radical constructivism. Divided into three parts, the paper first gives an historical overview of mankind's often hotly debated ideas on what constitutes knowledge from the Pre-Socratics to the recent writings of Karl Popper as well as Jean Piaget, whom Schoenfeld calls "the most famous of the constructivists." Next, von Glasersfeld considers what he views as the misguided impression that when a concept is "grasped" its meaning is somehow extracted from the printed words, whereas he contends it is built or constructed by the individual using her or his own experiences and previous knowledge as a guide. Thus, concepts reside within each of us and have subjective representations. We note, however, that this does not necessarily imply that a concept such as "continuity" can mean whatever a student wishes. It must be *consistent* with experience, the meaning of related concepts and the meaning constructed by others. Gerald Goldin of the Center for Mathematics, Science, and Computer Education at Rutgers points out that one need not adhere to a constructivist epistemology to view learning as a constructive activity or advocate guided discovery in the classroom. Finally, von Glasersfeld discusses mathematical knowledge as a product of conscious reflection and he considers the mathematics teacher as a facilitator of student thought processes. The more abstract the concepts the more reflection needed—it is not easy to modify one's mental schemas unless they fail or surprise one. A teacher's role is to *suggest* appropriate critical examples or counterexamples for student reflection; the actual cognitive conflict resolution must be left to the student.

While much mathematics education research along constructivist lines has been with young children, there are notable exceptions, for example, Schoenfeld's extensive studies on problem solving and Dubinsky's work. Dubinsky extends Piaget's ideas to undergraduate students, and his terminology, like that of Piaget, is not easily comprehended by the uninitiated, being replete with terms such as "interiorization" and "encapsulation." This is to be expected, however, as his ultimate goal is a general theory of mathematics learning, which will surely not be simple. He has already investigated how students learn such topics as functions, mathematical induction, compactness, and quantification Using his

general theory as a guide, each concept is analyzed a priori, from his vantage point as a mathematician and teacher. Problem situations are designed to help students construct the concept, and then students are observed at length as they try to make sense out of these situations. Finally, the "instructional treatment" is refined and the experiment is repeated until a satisfactory understanding of the learning process for the concept (a genetic decomposition) has been obtained. ("Constructive aspects of reflective abstraction in advanced mathematical thinking," L. P. Steffe, ed., *Epistemological foundations of mathematical experience.*) For Dubinsky, the primary goal of teaching is to help students construct and reflect on mathematical processes and objects, much like nascent mathematicians. He uses computers and small-group problem solving to implement this approach.

All this suggests structuring courses in which students' activities are more closely related to what mathematicians actually do than is now the case. Practicing mathematicians are generally very creative, constructing or modifying concepts, often after much cognitive conflict. Yet, as Davis and Hersh note, in the classroom professors sometimes view their task as a clear-cut one of "problems to solve, a method of calculation to explain, or a theorem to prove," with an unexpected question from a student seen as causing the "progress of the class to deviate from what the instructor intended" (*The Mathematical Experience.*) The constructivist view, current research in mathematics education, and initiatives like calculus reform may eventually bring the teaching of mathematics closer to its practice.

Constructivism and Mathematics Education

D.A. Crocker
AMATYC Review 13:1 (Fall 1991)

This article[1] is an exploration of the theory of learning that has become known as constructivism. The beginnings of the theory currently labeled constructivism, its current status, and the possible changes in mathematics education that might occur if this perspective is taken are presented.

Beginnings

Tracking the beginnings of constructivist ideas is no easy task. The term has only recently become popular, but the ideas have been around for hundreds of years. Von Glasersfeld (1989b) points out that constructivist ideas started in pre-Socratic times when people questioned whether knowledge of the real world is certain or whether it is derived from experience. The idea of cognitive construction was first discussed in a Latin treatise

---

[1] Parts of this article appear in the author's doctoral dissertation, completed at The Ohio State University in 1991 under the direction of Alan Osborne, major advisor.

on epistemology, by the Italian philosopher Vico in 1710 (Von Glasersfeld, 1989b).

The question of how knowledge is possible, or what is knowledge, moved through several philosophical views during the 17th and 18th centuries (Fabricius, 1983). The three major views were: (a) rationalism, (b) empiricism, and (c) romanticism. Rationalism was based on the idea that the entire world was rational and patterned after mathematical systems. Certain accepted premises enabled people to deduce conclusions that were considered facts about the world. The entire world was considered a mathematical model.

Empiricists did not use deductive reasoning in their view of the world, but rather inductive reasoning (Fabricius, 1983). They tried to generalize and predict based only on what was observed or experienced. Knowledge was based on perceptions of sight, sound, feel, and other senses.

Hume revealed the problems of empiricism and rationalism (Fabricius, 1983) in their inability to explain the structure of knowledge. If knowledge had to be deduced or induced, its structure could not be explained. This paradox is explained by Von Glasersfeld (1989b) by the fact that if true knowledge is representative of the world, then true knowledge cannot be tested. He writes:

If experience is the only contact a knower can have with the world, there is no way of comparing the products of experience with the reality from which whatever messages we received are supposed to emanate... The paradox then is this: to assess the truth of your knowledge you would have to make to know what you come to know before you know it (cited in Narode, 1987, p. 22).

Rationalism and empiricism both place the structure of knowledge in objects. Rousseau (Fabricius, 1983) suggested that the structure of knowledge might not be far from the subject. He felt these structures could be found from within. Rousseau's view represented the opposite extreme. Knowledge was within the subject, not the object. This theory was romanticism.

In the late 18th century, Kant looked at a combination of subject and object (Fabricius, 1983), in particular, how objects appeared to the subject. He argued that the structure of knowledge was in neither subject nor object, but in the experience created by their interaction. His ideas conflicted with rationalism, empiricism, and romanticism. Kant was the first to conceive of this idea. It was a major breakthrough in what has become known as "constructivism."

Current Status

More recently, the work of Piaget has been studied with attention to the constructivist nature of his theories. Piaget studied learning theory from a developmental point of view and stated that experience is assimilated and accommodated in order to be known (Piaget, 1954). The subject, or learner, seeks a state of equilib-

rium and knowledge is created in the process. As Von Glasersfeld (1989a) explains:

Knowledge for Piaget is never (and can never be) a representation of the real world. Instead it is a collection of conceptual structures that turn out to be adapted or viable within the knowing subject's range of experience. (p. 125)

This leads to current attempt to define or explain constructivism. Von Glasersfeld (1989b) distinguishes between radical and trivial constructivism. The radical constructivist, as Von Glasersfeld calls himself, gives up all ideas that knowledge must somehow match the world. The radical constructivist claims: "perception and all forms of seeing... are the result of operations that have to be carried out by an active subject" (p. 22). The trivial constructivist talks about construction of knowledge, but still believes it represents reality. This causes confusion when reading about and discussing constructivism. The reader must be aware that both types of constructivist viewpoints exist and learn to distinguish one from the other.

Narode (1987) says constructivism is the process of individuals defining their own, constructed, worlds. Blais (1988) says "knowledge is something learners must construct for and by themselves... Discovery, reinvention, or active reconstruction is necessary" (p. 627). He claims constructivists distinguish information, that can be told or given to someone, from knowledge, that requires gaining expertise. Piaget (1973) states a similar idea:

To understand is to discover... The goal of intellectual education is not to know how to repeat or retain ready-made truths. It is in learning to master the truth by oneself (p. 106).

Kamii (1985) sees constructivism as the theory that a child builds his own knowledge and it comes from inside. Brooks (1987) thinks constructivism explains how viewers "come to know their own world" (p. 66).

Orton (1988) explains that constructivists consider the processes learners go through to create representations or relationships between ideas and that process involves reflection. Procedural knowledge is active and therefore preferred by constructivists. Von Glasersfeld (1988a) says the test of knowledge "is not whether it matches the world... but whether or not it fits the pursuit of our goals" (p. 2). Von Glasersfeld (1988a) also brings to our attention the following:

Constructivism is a way of thinking which, at this stage in the development of our ideas, seems the most adequate; like all we call knowledge, it is not, and cannot be the description of an ontological reality and therefore it may change as our ways of experiencing and our purposes change (p.5)..

So, constructivism has undergone and will undergo many changes.

Possible Effects on Mathematics Education

The National Research Council (1989) summarizes research on learning as follows:

Research on learning shows that most students cannot learn mathematics effectively by only listening and imitating... that students actually construct their own understanding based on new experiences that enlarge the intellectual framework in which ideas can be created... Much of the failure in school mathematics is due to a tradition of teaching that is inappropriate to the way most students learn. (p. 6)

The Commission on Standards for School Mathematics of the National Council of Teachers of Mathematics (1989) notes that:

...learning does not occur by passive absorption... individuals approach a new task with prior knowledge, assimilate new information, and construct their own meanings. This constructive, active view of the learning process must be reflected in the way much of mathematics is taught. (p. 10)

The Commission on Teaching Standards for School Mathematics (1991) describes the learning of students with the following:

All students engage in a great deal of invention as they learn mathematics; they impose their own interpretation on what is presented to create a theory that makes sense to them. Students do not learn simply a subset of what they have been shown. Instead, they use new information to modify their prior beliefs. As a consequence, each student's knowledge of mathematics is uniquely personal. (p. 2)

All of the above incorporate a constructivist viewpoint on learning and teaching. This viewpoint requires a change in the way mathematics is taught that will, in turn, lead to changes in the curriculum.

The Mathematical Sciences Education Board (1990) states, as one redirection of the mathematics curriculum, the following, "The teaching of mathematics is shifting from an authoritarian model based on transmission of knowledge to a student centered practice featuring stimulation of learning" (p. 5). This redirection will involve drastic changes in methods of teaching and the role for the teacher in the classroom.

Brooks (1987) says the role of the teacher in a constructivist classroom is one of mediator, not implementor. Teachers are developers and deliverers of curriculum. This changing, multi-dimensional role of the teacher is also presented by the Mathematical Sciences Education Board (1990) as follows:

◇ A role model who demonstrates not just the right way, but also the false starts and higher-order thinking skills that lead to the solution of problems;

◇ A consultant who helps individuals, small groups, or the whole class to decide if their work is keeping to the subject and making reasonable progress;

◇ A moderator who poses questions to consider, but leaves much of the decision making to the class;

◇ An interlocutor who supports students during class presentation, encouraging them to reflect on their activities and to explore mathematics on their own;

◇ A questioner who challenges students to make sure that what they are doing is reasonable and purposeful, and ensures that students can defend their conclusions. (p. 40)

◇ Assuming these new roles in the classroom will not be an easy task, changes in methodology and teacher's roles must occur at all levels. The teaching of mathematics in the first two, as well as subsequent, years of college must change as well. Prospective teachers will long remember the ways they were taught rather than the lectures on the ways they should teach. As mathematics educators we must model these changes in methodology and the role of the teacher.

Conclusion

While many mathematics educators may agree with the basic tenets of constructivism, few will take action to change the teaching of mathematics. Agreement that our predominant methods of teaching mathematics are in conflict with the ways students learn is not sufficient. Attention must be directed to the research results on students' learning. Projects to test new methods of teaching and changes in the curriculum must be pursued. Mathematics educators must serve as role models in the classroom. Prospective and in-service teachers must be exposed to research results involving teaching and learning along with ways to incorporate changes into the classroom.

No purpose is served in agreeing that the students construct or invent their own mathematics if we, as mathematics educators, do not capitalize on these theories and change our approaches to teaching. We must encourage our students to appreciate and experiment with mathematics. We must recognize the varied ways they think and use mathematics. The constructivist viewpoint toward mathematics education will mean changes in the role of the teacher, a new respect for the learner, different approaches to mathematics education research, new teaching methods in the classroom, and changes in the curriculum.

Miami University Oxford, OH 45056-1641

References

Blais, D.M. (1988). Constructivism: A theoretical revolution for algebra. *Mathematics Teacher*, 81, 624-631.

Brooks, M. (1987). Curriculum development from a constructivist perspective. *Educational Leadership*, 44(4),63-67.

Commission on Standards for School Mathematics. (1989). *Curriculum and evaluation standards for school mathematics*. Reston, VA: National Council of Teachers of Mathematics.

Commission on Teaching Standards for School Mathematics. (1991). *Professional standards for teaching mathematics.* Reston, VA: National Council of Teachers of Mathematics.

Fabricius, W.V. (1983). Piaget's theory of knowledge: Its philosophical context. *Human Development*, 26, 325-334.

Kamii, C. (1985). *Can there be excellence in education without knowledge of child development?* Paper presented at the Annual Conference of the Chicago Association for the Education of Young Children, Chicago, IL (ERIC Document Reproduction Services No. ED 254 333)

Mathematical Sciences Education Board. (1990). *Reshaping school mathematics: A philosophy and framework for curriculum.* Washington, DC: National Academy Press.

Narode, R.B. (1987). *Constructivism in math and science education.* (ERIC Document Reproduction Services No. ED 290 616)

National Research Council. (1989). *Everybody counts: A report to the nation on the future of mathematics education.* Washington, DC: National Academy Press.

Orton, R.E. (1988, April). *Constructivist and information processing views of representation in mathematics education.* Paper presented at the Annual Meeting of the American Educational Research Association, New Orleans, LA. (ERIC Document Reproduction Services No. ED 297 934)

Piaget, J. (1954). *The construction of reality in the child.* New York: Basic Books.

_____. (1973). *To understand is to invent.* New York: Grossman Publishers.

Von Glasersfeld, E. (1988a). *An outsider's first approach to Bogdanov's theory of knowledge.* Paper presented at the Annual Meeting of the US-USSR Seminar on Systems Theory and Cybernetics, Tallin, Estonia, USSR. (ERIC Document Reproduction Services No. ED 295 814)

_____. (1988b) *Environment and communication.* Paper presented at the ICME-6, Budapest, Hungary. (ERIC Document Reproduction Services No. ED 295 850)

_____. (1989a). Cognition, construction of knowledge, and teaching. *Synthese an International Journal for Epistemology Methodology and Philosophy of Science 80*, 121-140.

_____. (1989b). *Knowing without metaphysics: Aspects of radical constructivist position.* (ERIC Document Reproduction Services No. 304 344)

Linear Algebra Curriculum Reform: A Progress Report

D.H. Carlson, SIAM News 26:3 (May 1993)

Linear algebra is one of the most heavily enrolled lower-division mathematics courses. Because its concepts and techniques are fundamental to analytical and numerical mathematics and have wide applications in computer science, engineering, statistics, and the physical and social sciences, those of us who teach linear algebra have a superb opportunity to help our students learn important and useful material that also conveys the spirit and beauty of mathematics.

How are we doing?

First, we share in the renewed interest in mathematics teaching that has grown out of calculus reform. In the belief that we can effect improvement, we have begun to devote serious attention to the process by which students learn linear algebra, discussing general and specific problems and formulating and testing solutions.

Moreover, because computing appears to have a significant effect on how and what we teach in linear algebra, many of us are not only incorporating computing into our classes, but also participating in workshops on computing in the classroom and writing materials or software. All these efforts at curriculum reform, together with the enthusiasm and excitement they generate, will surely have positive effects.

Second, a number of specific efforts are under way. The Linear Algebra Curriculum Study Group (organized by Charles Johnson of the College of William and Mary, David Lay of the University of Maryland, Duane Porter of the University of Wyoming, and myself) developed a core syllabus for the first semester of linear algebra at a workshop in August 1990 in Williamsburg, Virginia. The syllabus, which emphasizes the concrete, applied, and geometric aspects of the subject, recommends the use of computing and attention to the learning needs of the student. The group presented its curriculum recommendations to large audiences at the San Francisco and Baltimore Joint Mathematics Meetings and at various other regional and special meetings, eliciting significant audience response each time.

With the assistance of the National Science Foundation (which also sponsored the Williamsburg workshop and has consistently supported the reform effort), the study group is compiling and plans to publish a collection of "gems" of linear algebra — expositions of specific topics, short and insightful proofs, and problems and problem sets (including true-false, open-ended, and challenge problems). We welcome contributions of such gems from other teachers and hope that these useful teaching aids will eventually find their way into textbooks and classroom presentations.

Faculty members can learn about aspects of linear algebra or linear algebra teaching at several upcoming workshops and meetings, such as the June 28-July 9 Conference on Matrix Analysis for Applications at the University of Wyoming in Laramie. Charles Johnson will be the principal speaker at this conference, which is being organized by Duane Porter and sponsored by the Rocky Mountain Mathematics Consortium.

Many of these workshops involve the use or creation of courseware for linear algebra. The most ambitious is the NSF-funded ATLAST (Augment the Teaching of Linear Algebra through Software Tools) project of the International Linear Algebra Society. Participants in the summer ATLAST workshops design undergraduate-level computer exercises, then class-test them at their home institutions. The project is directed by Steven Leon with assistance from Richard Faulkenberry (both of the University of Massachusetts, Dartmouth). In 1993, the program's second year, workshops will be held at five U.S. locations during June and July. (See the *SIAM News* calendar for details.) Selected exercises from the 1993 workshops will be published in an ATLAST project book.

Donald LaTorre of Clemson University has been speaking about the use of supercalculators in linear algebra at various mathematics meetings.

The January 1993 issue of *College Mathematics Journal*, which is devoted to the teaching of linear algebra, contains articles on specific topic expositions and applications, the historical role of linear algebra in the curriculum, teaching issues, computing, and a student project. Ann and Bill Watkins of California State University, Northridge, co-editors of *CMJ*, organized this special issue.

Linear algebra curriculum reform has begun to move beyond the undergraduate (and primarily lower-division) level. Through its Education Committee, chaired by Frank Uhlig (Auburn University), ILAS is preparing a report that will contain recommendations for graduate-level linear algebra courses. Reaching out in the other direction, ILAS invited high school teachers to a panel discussion on teaching linear algebra at its annual meeting in March in Pensacola, Florida. The panel featured Paul Halmos (Santa Clara University) and was organized by James Weaver (University of West Florida).

As a result of all these activities, teaching linear algebra is more fruitful and exciting than ever before; we invite everyone who teaches or is interested in the subject to join the reform effort.

David H. Carlson (carlson@math.sdsu.edu) is a member of the Department of Mathematical Sciences at San Diego State University.

Writing about Linear Algebra: Report on an Experiment

G.J. Porter, UME Trends 3:3 (August 1991)

Last semester, I taught a course on linear algebra for nonmath majors. The text I chose was very strong on applications, but less thorough on the theoretical underpinnings. This was appropriate for my class of engineering, finance, and economics majors. To present the material as I wished, I needed to supplement the book with lectures on the theory. Many students have a difficult time understanding theoretical concepts and consequently ignore this material. While teaching the course, I received a copy of *Using Writing to Teach Mathematics* (number 16 in the MAA *Notes* Series) edited by Andy Sterrett. In thirty years of teaching I had never given a writing assignment, but I was so intrigued by the material in this volume that I decided to try this approach to help my students master the basic concepts of linear algebra. The purpose of this article is to share the reactions and results of what to me (and to my students) was a different way of teaching and learning mathematics with the hope that others will consider similar experiments.

The Assignment (abridged). The material on subspaces, spanning sets, basis, dimension, etc., is not covered as thoroughly in the text as in class. This is also true of the related material on lines, planes, and hyperplanes. Your assignment is to write a short chapter to supplement the material in the text. Take as your motivation the need to describe the solution space of a system of linear equations. Use this to motivate the idea of a subspace that in turn will motivate the idea of basis. (How else do you explicitly describe a subspace?) This leads to the notion of dimension and brings us back to lines, planes, and hyperplanes.

Your chapter should be approximately ten pages long. If you can express your ideas more succinctly, that is fine. Remember that it costs money to publish books, so don't be too long winded. You must send your chapter for review to another student in the course. Please feel free to discuss the mathematical material with your classmates and with me, but do not refer to texts on the subject other than the one we are using.

Reactions from the class. After I collected the assignment, I asked my students to write an evaluation describing their reactions to the assignment and detailing its strengths and weaknesses. Many of their responses had to do with the fact that because they were unaware (as was I) that such an assignment was coming, they hadn't allocated time for it or taken as detailed notes in class as they would have (in retrospect) liked. I do agree that it would have been better to have the nature, if not the details, of the assignment announced earlier.

There was concern that access to reference material was strictly limited. The students do need a reference on the material, since they are not being asked to invent the subject. Yet, I fear that unlimited access to other texts would lead to regurgitation of the contents of those texts. I welcome suggestions of other ways to solve this dilemma.

Student reaction to the assignment ranged from shock to disbelief. The following three comments indicate the original reaction and how that evolved.

◇ I have to admit when the writing assignment was announced, I was less than pleased. As I began working on it, by formulating ideas and trying to organize them, I hated the assignment more and more. The reasons why I loathed the assignment are precisely the reasons that such an assignment should be given.

In order to write this chapter, I had to think; something I try to avoid doing at all costs. I was forced not just to know the material but to understand it. I had to ponder each concept and approach it from several directions to be sure I understood enough both to explain it and to relate it to other material. This was hard.

Overall, I think the assignment has good learning aspects, despite the fact that students will inevitably hate it.

◇ Write a chapter to supplement the material in our text book! I was shocked to read the assignment sheet. It was the first time I had a writing assignment in a technical course. The assignment was a good experience. It gave me the opportunity to learn the material by organizing it to present to others.

◇ Having a writing assignment in a math course came as quite a shock. It was something I never expected or even heard of in a math course. However, it proved not to be as difficult as I thought it would be. Although I would not welcome another writing assignment in math, I did learn something and now understand the subject better.

Aspects of the assignment that students found beneficial are reflected in the following comments.

◇ By identifying and arranging the material I learned the subtleties, as well as reinforced the basics which I already knew. The chapter organization forced me to realize relationships and categorize concepts, which in turn made the material more logical and easily understood.

◇ Let me start by saying that I understood this material better than any topic we have had in this class. Even so, this assignment really brought things together for me.

◇ To write the paper I had to understand and digest all the concepts before I began. By doing this I acquired a good understanding of the material presented in class.

◇ It forced students to break away from the "plug and chug" approach and focus on the theory behind linear algebra.

The response was not universally positive, the assignment did take time and made life more difficult for some students. Comments about this included:

◇ Even though I learned the material, I pretty much had to disregard the new material we were supposed to be learning.

◇ I think there must be a better way to learn the material without so much work. There is no real need for us to exert so much energy into the presentation.

◇ I have a stronger knowledge of vector space theory because I had to organize my thoughts for this chapter. I still would like, however, to concentrate on applications where the benefits of the mathematics are readily apparent. I'm not really interested in theory.

The students in my class worked very hard on this assignment and by their statements feel they have learned a great deal. My evaluation is that they made substantial progress toward understanding the material covered; however, there was still a great deal of confusion in their papers. Virtually all the students would have improved their comprehension of the material by rewriting the paper after I commented upon it. One student did rewrite the paper, and his understanding of the concepts was significantly improved.

I graded the papers myself. It took a great deal of time because there were forty students in the course. All of the papers were done on word processors, and the resubmitting and rereading would have been much harder on the reader than on the writers. Most of the papers in *Using Writing to Teach Mathematics* are by faculty in schools smaller than mine and with heavier teaching loads. Surely, this epitomizes their dedication to teaching.

One student wrote, "This was a valuable assignment because it required learning how to do technical writing. Most of us have never written a technical paper. The style must be more concise and comprehensible than the prose that most students are accustomed to. It would have been helpful to have been given guidelines and instruction in scientific writing." I agree and, in the future will ask our writing center to assist me in this task.

Comments about peer review were mixed, and there was no consensus.

◇ The review process stinks. While working in groups and getting others to give advice is very helpful, it is not fair for a person to get a completed paper for review at 2 am the night before it is due.

◇ The review was really a good idea. I wish that I could have started earlier so that I could have had my paper reviewed several times. It was also good to see what approach the person whose paper I reviewed took. Everyone comes up with different ways of looking at things.

◇ Although I got a little out of reading someone else's chapter, I gained little by having another student review my chapter.

It was especially true that in an environment where many papers don't get written until the last minute (and some after that), peer review is difficult. In general, I found the peer reviews were nowhere near as critical as I had hoped they would be. Next time I will encourage, but not require, peer review.

Conclusions.

◊ The assignment succeeded in getting students to take seriously the theory in an applied course. More generally, students tend to ignore material presented in class that is not in the text. Writing about this material certainly focuses student attention on it.

◊ The assignment needs to be as highly structured as possible since the students are unfamiliar with this type assignment. It is my impression that this assignment improved student learning of theoretical aspects of linear algebra more than any other assignment I have ever given in this course. Further improvement would have resulted from having the students rewrite the paper after I commented on it.

◊ Instruction in technical writing should be included or, at a minimum, the students should be asked to read material on scientific writing.

A student sums up the experiment better than I can:

"I believe that the assignment was absolutely invaluable. Not only did it force me to learn the material, it made me learn it to an extent where I could explain it to someone else–which is a true test of knowledge. Oh, what a feeling of satisfaction I got when I finally understood concepts that I couldn't understand last year (and understood them well!) As far as I know, assignments such as this one are not general practice in the math department. However, after completing this one, I feel that they should be made a requirement. Again, it was invaluable–both in writing practice and as a learning exercise."

I want to thank Andy Sterrett and the contributors to *Using Writing to Teach Mathematics.* They have given my students a new way to learn mathematics. It works, and I would not have tried it if I had not read this book.

Jerry Porter is at the University of Pennsylvania.

Structures and Proofs First, Algorithms Later

L. J. Gerstein, UME Trends 2:5 (December 1990)

Pressure for calculus reform comes in part from the desire to make room for discrete math in the first two years of college. The 1986 Douglas calculus report observed that many students "have taken a watered down, cookbook course in which all they learn are recipes, without even being taught what it is they are cooking." If we reform calculus only to introduce cookbook discrete math courses, then much of the point of the reform will have been lost. What should a noncookbook discrete math course be? Here are some observations based on my university's experience.

At UCSB we include a one-quarter discrete mathematics course in our premajor packages in mathematics and computer science; students usually take it after at least two quarters of calculus. There is no agreement as to what a discrete mathematics course should include. Some say, "Algorithms are what the students want and need and are therefore what the students must get." We do *not* stress algorithms, yet our course has been very successful in helping prepare students for upper-division courses. Many seniors have told me that the course clinched their decision to major in mathematics or computer science, and that what they learned in the course was invaluable to them in subsequent courses.

How can this be so, if students want and need algorithms, algorithms, and more algorithms? I wish to argue against this assumption, to say something about students' actual wants and needs, and to indicate some ways in which our course addresses these matters. I do not claim algorithms are unimportant; I just think other things are *more* important at this stage of the educational process.

Algorithms are the core of a typical calculus course: our students use algorithms to compute limits, derivatives, and integrals ad nauseam. Unfortunately, *it is possible for a student to get excellent grades in calculus courses without ever having to give a coherent explanation of anything.* In calculus most definitions are technical and are soft-pedaled. We give a geometric sketch of the basic idea, followed by a relevant test or algorithm. The mechanical part is all students are likely to need for exams. For example, students learn to check injectivity by testing the sign of the derivative, and the definition is essentially forgotten. But in a discrete context, where derivatives are not available, testing for injectivity can usually be done only by applying the definition. Learning to handle functions effectively is a primary goal of the lower-division years. Students planning to major in mathematics or computer science should be able to prove, before they reach upper-division courses, that the composition of two bijections is a bijection. Reasoning must not be sacrificed on the altar of algorithms.

I have the computer science major in mind as much as the mathematics major. Computer science is not just *programming.* Anyone who has seen the definition of "finite state automaton" will immediately agree that computer science majors will not be well served by a diet of algorithms. Likewise, by the junior year a mathematics major should have had nontrivial experience proving theorems. Many mathematically talented students *lack* that level of maturity, and we need to attend to their intellectual needs during their lower-division years. Our discrete mathematics courses for these students need to be as different in spirit from standard calculus courses as they are in content. *Algorithms are not enough.*

One of the dangers of not addressing these needs at the lower-division level is that we run the risk of overwhelming our students with the twin burdens of abstraction and proofs when they start their junior year. If the shock is too great, we will continue to drive many able students away from mathematics.

What do lower-division students want? "Algorithms," many educators say, and again I disagree. This is a matter of some importance, because if we force-feed students with material that makes them uncomfortable or gives them little satisfaction, we will send them scurrying for other majors. Students will tolerate abstraction if they can see a payoff. The payoff does not have to be a numerical answer or a cute algorithm. It may instead be a clarification of a mathematical matter of intrinsic beauty. Questions about cardinality, both finite and infinite, provide excellent raw material. As for finite counting concerns, the so-called sum and product rules are central, but our course doesn't present these as self-evident truths. Indeed, an important moment occurs when students learn that to *count* a nonempty finite set S means to construct a bijection $\{1 \ldots m\}_{\overrightarrow{S}}$ for some positive integer m. [Since childhood, students have believed that counting is a mathematical process; now at last they see where it fits in.] We use this definition and mathematical induction to prove the theorems of set theory underlying the sum and product rules, we point out that those rules are interpretations of the theorems just proved, and *then* we go on to the usual applications, the "How many license plates ...?" sort of thing. When we link the set-theoretic effort to the consequent combinatorial ideas, students experience set theory as a foundation for useful mathematics, and they get a healthy dose of modeling along the way. There is a major difference between what I have described here and a course that labels combinatorial principles "self-evident" and then plunges into applications. Sure students in the latter course get problem-solving experience, as they do in calculus. But are we preparing them for upper-division work?

I have found that students are fascinated with infinite sets. On the first day of class I ask for thoughts on the relative sizes of $\mathbf{N}$, $\mathbf{Q}$, and $\mathbf{R}$; students don't know the answer, and intuitions vary. The curiosity lasts, and when the answers come later in the course, students find them remarkable. Indeed, when they learn that there are levels of infinity, they see that mathematics has a richness they did not anticipate. I point out that the ideas we use to compare infinite sets are a natural extension of our intuitive ideas, but students also learn that careful use of definitions and mathematical constructions allow them to go beyond intuition. This payoff, the clarification of concepts that "common sense" was too weak to handle is at least as rewarding as that from any algorithm, and it is a payoff that will serve them well in more advanced courses.

Our course has an introduction to logic (one week), a substantial (4 weeks) dose of set theory, a week on functions, a week on cardinality (including countability questions), one to two weeks on counting problems and an introduction to permutations, and about two weeks of elementary number theory. No topic is sacred: the

spirit of the course counts more than anything. Indeed the course cannot succeed, however artfully the syllabus may be designed, unless it is taught by enthusiastic faculty who are committed to its fundamental goals: to enhance students' mathematical maturity, to stimulate their imaginations, and to give them a taste of the excitement of mathematics.

Larry Gerstein is at the University of California, Santa Barbara.

Proofs and Algorithms: A Reply to Gerstein

S. B. Maurer, UME Trends 2:6 (January 1991)

In discrete math courses there is an increasing emphasis on algorithmics. In his article "Structures and Proofs First, Algorithms Later" (*UME Trends*, December 1990), Larry Gerstein objects. He argues that students don't need more algorithms – they have enough rote algorithms in calculus – rather they need to learn how to reason and prove. While I have a lot of sympathy with this specific statement, I think it misses the mark as a criticism of current discrete math courses. His criticism is based on two confusions: confusing algorithmics with doing algorithms and confusing discrete math with "transition to higher math."

Consider the first confusion. By algorithmics I mean a point of view in which algorithms play an important role in *thinking* about problems. If the problem, say, is how to determine when a graph has an Euler circuit (can be traversed without covering any edge twice), you don't just try to prove a criterion by existential means, but rather you try to devise an algorithm and a criterion simultaneously, so that a proof of algorithm correctness becomes a constructive proof of the criterion. Thus algorithmics involves strategies for finding and validating algorithms at least as much as it involves learning and using specific algorithms. Algorithmics also involves methods for evaluating algorithm efficiency, helpful when you find more than one correct algorithm for a task. Algorithmics is at least as good for developing mathematical maturity – the ability to discover, reason, and prove – as classical mathematics. In no way is it rote! For instance, to devise a recursive algorithm you need the same agility with basis case and inductive step that you need to do proofs by mathematical induction.

In short, algorithmics is thinking *about* algorithms, not thinking *like* algorithms. For more discussion of this point, see my article "Two Meanings of Algorithmic Mathematics" (*Math. Teacher*, September 1984, pp. 430-435).

Now on to "transition" courses. As Gerstein points out, many students who choose to major in math (or CS) have trouble adjusting to the proof orientation in upper-level courses. This is a serious problem, especially if our profession is to replenish itself. One solution is to offer a course, say to spring-term sophomores, whose purpose is to develop "proving muscle." The course Gerstein describes is of this sort. Another approach (supposedly what happened in the good old days) is to have the whole collection of freshmen-sophomore courses provide a ladder of gradually increasing emphasis on proof. (In this second approach, the nature of algorithmics makes discrete math a very useful rung.)

There are many parts of mathematics one can use for developing proving muscle. Some of the books published for transition courses emphasize the real number system and build the foundations of continuous mathematics. I have taught a transition course using the interplay of convexity with linearity and topological notions. Whatever the topic, so long as you can devise a sequence of exercises whose proofs start short and easy and gradually get longer, you can (if, as Gerstein points out, you are enthusiastic) give a good transition course. It may be somewhat easier to create such a course out of discrete material, but it is hardly necessary.

Now my key claim: Even if one gives a transition course using discrete material, that no more obviates the need for an introductory discrete math course than giving a transition course using analysis obviates the need to teach calculus.

In fact, freshman calculus, junior real analysis, and analysis-based transition courses share the same underlying theory. So why have a calculus course? My answer: Because the calculus course gives a broad audience a good grasp of what the key concepts in continuous mathematics are, why they are very applicable within mathematics and without, how you model problems using these concepts, and (last in importance) how to apply some standard mechanical techniques. In short, a good calculus course is problem oriented, with emphasis on concept development and problem-solving methodology. This orientation does not mean you can't learn any proofs. But formal proofs are not the centerpiece, and you don't try to prove everything from scratch. Instead you prove illustrative results or chains of results. The real analysis course, by contrast, is for math majors (and a few others) and emphasizes the structure of analysis, with practically everything proved from some base point.

In a similar vein, there are two types of discrete math courses, a concept and problem oriented course, and a discrete structures course. (Include a discrete transition course, and you have three types.) I believe that the first type is appropriate, indeed, very important, for students with many interests. Like calculus, it elucidates concepts that any mathematically capable person ought to know. I include concepts of induction and algorithmics (discussed above); I would also include (at the least) concepts about graphs (networks) as models and about recursively defined sequences as a way of treating discrete change. There is almost no overlap between this type of course and Gerstein's, either in content, goals, or

intended audience.

So, should proofs come first, then algorithms? I don't think so. The two need not be mutually exclusive; it's a question of emphasis. If put in the framework of algorithmics, algorithms get my vote for greater emphasis first.

Steve Maurer is at Swarthmore College.

Calculator-Generated Questions: Numerical Analysis

T. Giebutowski, UME Trends 2:1 (March 1990)

Instructors in undergraduate numerical methods courses might make use of the scientific calculator to construct exercises on fixed-point iteration which will give their students nontrivial questions to address, while at the same time leading students to generate their own exercises. Specifically, I generated some interest by asking students to analyze their calculator's behavior when a fixed sequence of function keys is iterated.

As a simple first example, have students punch sin, then cos keys repeatedly (starting from any entry x_0 in radian mode), observe the results (two converging subsequences), and analyze them using whatever theorems in the course apply to the $x_{n+1} = g(x_n)$ algorithm. At this level, just finding Lipschitz constants by max-min methods, graphing the various composite functions which yield the subsequences, finding the number of iterations necessary to insure a certain level of accuracy, and approximating the limit points of the sequence provide meaningful examples of "analysis at work." (After all, any result can be "checked" by punching more buttons.)

The students can generate their own exercises after that (cos, then tan is a more interesting one), and the questions can get more challenging (e.g., generate a sequence with one finite and one infinite limit point). In addition, since in most cases machine error "self-corrects" until the last iteration performed, the answers are usually satisfactorily precise. I've used this technique in our senior level numerical analysis course which usually has an enrollment in the neighborhood of ten, but it might be even more stimulating for students in a lower level methods course. Students like to fiddle with things, and this gets them fiddling in a constructive way. The one bone of contention is with calculator manufacturers who insist on making the degree-radian-grad option a button rather than a switch, but that problem is not limited to this technique.

Ted Giebutowski is at Plymouth State College.

Symbolic Computation: A Revolutionary Force

Z. A. Karian, UME Trends 2:3 (August 1990)

Computer Algebra Systems (CAS) are powerful tools that combine symbolic and numeric computation with graphic capabilities. Although such systems have been commercially available for more than a decade, until recently widespread use was prohibited by high cost. Now, however, as more powerful and less expensive CAS packages become available, they are revolutionizing both mathematics instruction and the practice of mathematics in much the same way as numerical computation impacted numerical analysis.

The general availability of CAS allows mathematics faculty to improve instruction in a variety of ways. At the very least, these systems provide a vehicle through which realistic applications of mathematics can be brought into the curriculum. We no longer need to consider the algebraic complexities of a problem as a limitation for its instructional use. This movement, away from drill on closed-form solutions to problems with real-world consequences, will give students a better sense of the vitality and importance of mathematics. In some courses (for example, calculus, differential equations, numerical analysis, statistics) this approach will lend itself naturally to discussions of the role of approximations and the need to consider error bounds.

There is also a growing conviction among many mathematics educators that computer algebra systems can help expose the underlying principles of mathematics more convincingly than exercises found in current standard texts. Using a CAS to promote discovery of mathematical principles is another scheme through which mathematics instruction can be significantly improved. We can engage students more directly with mathematics by asking them to determine, through experimentation, the form of $D_n[f(x)g(x)]$ or to give the reasons for the presence of vertical asymptotes of rational functions. Learning, as well as attitudes about learning, will certainly be enhanced if students become more active in the exploration of ideas.

As they are currently taught, many elementary mathematics courses are so burdened by symbolic manipulations that instructors find it difficult to spend adequate time on fundamental concepts. In some cases, much of what is done in standard courses can be done by creative use of CAS; such use would free students from unnecessary rote calculations, thus enabling both instructor and student to give greater thought to underlying principles and their applications.

The mathematics community, and to some extent the larger scientific community, finds itself at a crossroads. While many individuals are convinced that CAS should be fully exploited, there continues to be considerable uncertainty and apprehension regarding how symbolic computation systems function; what type of equipment is needed; how the hardware, software, and users inter-act; and most importantly, how such systems can be used effectively. National interest in the impact of symbolic computation on the broader science and engineering curriculum is also growing. For example, the American Society for Engineering Education organized several sessions on the pedagogical uses of CAS in mathematics and engineering courses at its June 1989 conference. Similarly, through various funded projects, physicists at several institutions have begun to investigate the potential impact of CAS on their discipline.

To address these issues, CUPM established the Subcommittee on Symbolic Computation to explore the potential instructional uses of computer algebra systems and to keep the mathematical community informed of emerging developments. With support provided by the Alfred P. Sloan Foundation, the subcommittee has supported a wide range of activities. Anyone interested in contributing to the work of the subcommittee or in receiving its communications should write to its chair, *Zaven Karian, Department of Mathematical Sciences, Denison University, Granville, Ohio 43023.*

For the summer MAA/AMS meeting, the subcommittee is sponsoring a poster session (organized by Joan Hundhausen of the Colorado School of Mines), a panel session (chaired by Robert Lopez of the Rose-Hulman Institute of Technology with panelists William Boyce, John Harvey, Michael Henle, and Jeanette Palmiter), and a special presentation by Paul Zorn of St. Olaf College entitled "Symbolic Computation in Undergraduate Mathematics: Symbols, Pictures, Numbers, and Insights."

Additionally, the Subcommittee on Symbolic Computation has begun work on a set of annotated problems and related materials that take advantage of CAS. (Anyone who has materials that could be considered for this publication should send them to *Arnold Ostebee, Department of Mathematics, St. Olaf College, Northfield, MN 55057.*) The subcommittee also intends to publish an MAA Notes volume on the use of CAS in undergraduate mathematics education. (A solicitation of abstracts for this volume appeared in the May issue of *UME Trends.*) For the January 1991 Annual Meeting in San Francisco, the subcommittee will sponsor a variety of activities, including a minicourse for faculty who are interested in starting to use a CAS in their courses.

Finally, subcommittee member Donald Small from Colby College continues to organize NSF-supported workshops for college faculty; this activity is independent of formal subcommittee sponsorship, but fits in well with our broad agenda.

Zaven Karian of Denison University is chair of the CUPM Subcommittee on Symbolic Computation and a former member of the Board of Governors of the Mathematical Association of America.

Alternatives to the Lecture Method in Teaching Geometry

M.K. Smith, UME Trends 4:4 (October 1992)

One way I can influence the education of future mathematics teachers is by teaching the undergraduate geometry course they are required by state law to take. (Most states have similar requirements.) In Texas, students preparing to teach elementary school with a mathematics concentration are also required to take this course. Since it is not required by mathematics majors, the enrollment is usually about two-thirds future teachers, so I feel justified in tailoring the course to their needs.

I see several such needs which are not met in most of our courses. First, students need to develop spatial skills and intuitive understanding. Second, they need to experience teaching methods other than the usual lecture. Third, they need exposure to a wide variety of topics in geometry that can serve as enrichment material for their students, the kinds of things that turned me on as a high school student. Finally, they need to develop an appreciation for geometry. Indeed, part of my original motivation for teaching the course was that I heard so many future teachers say they hated geometry.

The bulk of the course is on transformational geometry. As time permits, I may do a little on axiom systems or constructions. I also spend about two weeks on an intuitive introduction to non- Euclidean geometry. I carefully resist the temptation to cover a lot, aiming instead for quality and depth in the main part of the course. Three days toward the end of the semester are devoted to student projects, which expand the breadth of the course.

I have not found a good text for the transformational geometry; linear algebra is not a prerequisite for the course, which restricts the possibilities. I have used Smart's *Modern Geometries* twice and Brown's *Tranformational Geometry* twice, but am not satisfied with either, so I use handouts as needed to supplement the text: for example, one on conditional statements, and one with supplementary problems. For the non-Euclidean geometry, I use Petit's *The Adventures of Archibald Higgins: Here's Looking at Euclid*, translated by Ian Stewart, which serves my purposes very well.

Students are assigned readings in the text. I pass out a study guide to help them, but do not lecture on text material. I may ask them questions on their reading, or answer their questions, possibly after having them discuss the reading in small groups.

The first day of class, I hand out the geometry portions of the NCTM *Curriculum and Evaluation Standards*. I mention what the *Standards* are. I have the students write an essay due the next week comparing and contrasting their geometry background with what is recommended in the *Standards*. This serves three purposes. It gets the students to think back about the geometry they learned in high school, it gives me an idea of what their backgrounds are, and it introduces them to the *Standards*.

Also on the first day of class, I have students work on a problem in groups. I choose a problem from *Get It Together* (Equals). These use the gimmick of having clue cards, so each person only has part of the problem; students are thus forced to interact. The gimmick works and gets the class off to a good, participatory start.

I try to have some sort of group work on at least half of the class days. It may be discussing a concept from the reading or working on a problem. When we cover transformations of space and non-Euclidean geometry, group work is hands-on, with student-made models or balls that the students or I bring in. At first I let the students choose their own groups. After a couple of weeks, I try to make up groups that are somewhat but not too mixed in ability level, and which will promote maximum participation of all students. For example, I put aggressive students in groups with others who I think can handle them, and quiet students together so they will be more likely to participate actively.

The bulk of most classes is spent having students present problems at the board. I consider an important part of my job to be to establish and maintain a respectful and courteous atmosphere where students are willing to risk making mistakes. I try to point out what is right about a solution before pointing out what is wrong. No putdowns are allowed, either by me or by students.

Presenting problems is not a cut and dried do-this-one-then-go-on-to-the-next matter. I try to ask, "Can anyone do it another way?" when a student has presented a correct solution. I play it by ear, taking whatever comes from the students as an opportunity to explore or expand. I may compare and contrast different solutions, or ask students to do so, comment on a point of logic or exposition, or lead them to explore a related problem.

Getting maximum student participation takes effort. In addition to "Can anyone do it another way?", I often ask, "Does anyone have anything else to add?" I wait a few seconds for hands to go up before calling on someone, and choose the student who has participated least recently. I interpret "raising a hand" liberally; if someone who hasn't gotten up to the board much looks like they're considering raising their hand, I may call on them over the regulars. Often good students don't volunteer. I put notes on their homework encouraging them to do so. Often it works.

The class usually has slightly more female than male students, but males usually tend to dominate the class discussion at first. I work at getting class participation more representative of the gender distribution of the class. For example, I try to assign the non-participating women to all-female groups until they get accustomed to participating actively, and women get the bulk of the notes on homework.

The projects usually go quite well, thanks to advice from Phyllis Chinn of Humboldt State, whom I met at a MER workshop. They are presented in a science fair display format. This encourages emphasizing the visual aspect of geometry and allows up to thirty projects to be presented in three 75 minute class periods. I give students a lengthy handout the first day of class, explaining the rationale and rules of the game for the projects, giving suggestions for project topics, and how to research them. I have three intermediate due dates. Title and at least two references are due at the end of the third week of class. A one-page description of the project and the broad limits within which the student will work is due the sixth week of class. A one-page outline of what will be included, how it will be presented, what will be left out, and why, is due the ninth week of class.

Students vary in their ability to research their topics. Some have little or no library experience. I have compiled a bibliography for projects, but do not give out items from it until students have found at least two references on their own or unless they have a topic that is particularly difficult to research. Some students come to ask what I want. I tell them that there is no particular format or content I want, but that their job is to decide what fits the topic, audience, and space limitations. I hand out project evaluation criteria to help. Students grade projects on the days they are not presenters. I determine the project grade by averaging my grade and the average of the students' grades. I also count grading projects as a homework grade: uniform "outstanding" ratings indicate a poor job of grading; thoughtful comments backing up the grade assigned indicate a good job of grading.

I cover non-Euclidean geometry just before projects are presented. I do not assign any homework beyond the reading on this but spend a couple of days in class having students investigate spherical geometry in groups. This gives them a chance to spend less preparation time on their projects.

The last day of class I hand out a list of resources for teachers.

Student reactions to the course have generally been favorable. Some students have difficulty with the nontraditional structure, but most appreciate it. My favorite comment is from a student who already had some teaching experience:

"I've learned geometry in this class, certainly, but more than anything I learned about being a good teacher. Seeing these techniques in practice is better than any lecture I ever heard over in [the education building]."

I would be glad to send copies of any of the handouts in the course to people wishing to teach a similar course.

Martha Smith is at the University of Texas at Austin.

Avoiding the Algebra Trap

J.J. Woeppel, UME Trends 5:1 (March 1993)

The majority of entering first year students at many colleges and universities now take some developmental (remedial) courses. At Indiana University Southeast, we have two developmental mathematics courses, Elementary Mathematical Skills and Elementary Algebra. Neither of these courses earns credit toward graduation. There are two additional courses, Intermediate Algebra and College Algebra, whose credit does count for graduation, but they do not satisfy the mathematics distribution requirement for a Bachelor of Arts degree. Thus a student can spend as much as two years preparing to take the mathematics that her or his major requires; e.g., Finite Mathematics and Brief Survey of Calculus 1. There must be a more effective way to do this.

First, what are we trying to do? We are, of course, trying to give students a thorough background in algebra, which they need to be able to take the more advanced mathematics courses required for their major discipline and to be able to take the more analytic courses in their major discipline. This statement is a true statement for any students going into the hard sciences (chemistry, physics, etc.) or mathematics. However, if the student intends to go into other fields such as Biology, Business, Social Sciences, etc., the statement should be examined more closely.

In particular, the word algebra needs to be examined. It has three different meanings for people in different positions. These are:

i) The manipulation of various symbols according to a set of rules.

ii) The application of an algorithm to solve equations or relations made up of well formed expressions involving symbols from a particular set.

iii) The ability to solve a problem given a written description of a situation.

Mathematicians think of algebra in the sense of (i), a generalized arithmetic. Most algebra texts treat it in the sense of (ii), although they profess in the preface to treat it in the sense of (iii). Most individuals in nonscientific application areas expect algebra courses to provide students with the ability to solve problems in that particular area, (iii), which they do not. Most instructors suppress their tendency to teach algebra as a set of rules for manipulating symbols and follow the text's algorithmic approach. Unfortunately, the way most instructors deal with problem solving is to ignore it. The text makes this easy since they isolate "applications" in special sections (the last section of the chapter). Some more daring/naive instructors follow the text and present some template problems.

Using the algorithmic approach (ii), the student is presented with an algorithm to solve a particular type of equation or relation. The algorithm is described as a set of steps displayed in a box in the text. Then several examples are presented to illustrate the algorithm. To be able to solve an equation or relation using this approach the student must recognize the type of equation or relation, recall the algorithm producing the desired result, and properly apply the algorithm to the equation or relation. Often some manipulation of the equation or relation is required to be able to recognize the type of problem, and then various manipulations are usually required in applying the algorithm. Thus the student still must master (i) to some extent. This is a hard way to learn this subject; hundreds of algorithms must be memorized and associated with the proper problem type; the rules of algebra must be deciphered from the examples; and little or no motivation is given for learning all this "nonsense." Even the student who is successful in memorizing all this soon forgets much of what has been learned. Since a symbolic manipulation program can easily do all this, it may be more effective (and possibly more efficient) to have the students buy lap-top computers and teach them to use one of these programs.

There has to be a better way to teach "algebra?" Before we try to answer this question, we must operationally define algebra and more generally determine the purpose of algebra courses in the college curriculum. There are two major functions of algebra as it now exists. One is as a prerequisite for college level mathematics courses and other courses requiring algebra; the second is as a general education requirement. An algebra course is not a good choice for a general education requirement. Primarily, it is not a college level course. Also, as it is customarily taught (see point (ii), above), it is a skills or tools course and does not impart any general knowledge about mathematics. Mathematicians should develop a mathematics course (or courses) which would serve as a good general education requirement, but it is not the purpose of this paper to discuss that issue. Instead we will look at the purpose of algebra in the college curriculum.

As a prerequisite, algebra has a valid purpose. However, this purpose must also be investigated. Providing a filter to screen out students who will have difficulty in analytic courses in their major is is not a valid purpose. The dubious honor of screening out unqualified students often seems to fall on the shoulders of the mathematics courses. This is expected of the mathematics curriculum while the remedial mathematics courses are supposed to "pump" students who are not prepared for college level courses into the college student population to make up for the 18-20 year old students who are not there. I often remind my colleagues that the "real world" example of a pump with a filter is a *meat grinder*! Unfortunately, the mathematics filter is more dependent on background than ability, and thus this process represents de facto discrimination against minority students who are often poorly prepared in the area of mathematics. *We mathematicians should not accept the rap and should stop being the "fall guys" for the other disciplines.*

Mathematics departments have been providing service courses for the hard sciences and engineering for many years. This has not always been an ideal situation but mathematics departments have some understanding of the mathematics background that these departments require for their majors. Similarly, these departments have some understanding of the preparation their majors are receiving. Unfortunately this mutual understanding does not exist to the same extent with Biology, Business, Social Sciences, etc. *Mathematics departments must develop an understanding of the mathematical needs of nonscience majors.* In addition, mathematics departments need to begin a dialogue with the nonscience departments concerning the needs of their majors and what they can reasonably expect mathematics courses to teach their majors. This will be difficult since there is not always a common basis for the discussion. Often these departments will leave it to the mathematics department to decide what is to be taught in the service courses. Then they criticize the preparation their students are getting. It is a better strategy to encourage them to have more say about what is taught. Mathematics departments must be willing to be innovative and develop new courses that do meet the needs of nonscience majors - not merely collect various material that should meet their needs in a course.

The mathematics courses for the nonscience major should not only provide these students with a basis for further study of mathematics (in their major courses or in the mathematics department courses) but also provide tools that are likely to be used throughout the person's professional career. After all, this is what we strive to provide for the science major. However, we seem to fail to provide nonscience majors with either the basis for present studies or the tools for their future endeavors!

The appearance of the programmable graphics calculators should make the task of providing nonscience majors with reasonable and powerful tools much easier. These calculators allow students to graph complicated functions with reasonable ease. The points of intersection that are displayed can be observed. One can "zoom" in on a point to obtain the value to the desired number of decimal places. To be sure there may be problems with the accuracy of this value, but this is not usually a major difficulty for text book problems. Picking the proper domain and range for the function can also be critical to locating the solutions. Calculator skills learned in an introductory course can be expanded in later courses in calculus, linear algebra (matrices), statistics, and other major courses. Also, instilling the concept of working with graphs is not necessarily a bad idea for these students. The applications in their fields often deal with empirical data.

What is being proposed here is:

i) All entering students be given a mathematics placement test. This test should cover arithmetic through pre-calculus (functions).

ii) Students who possess a good knowledge of arithmetic (fractions and decimals) but do not have a sufficient basis in algebra to take college level courses would be required to take a remedial course, *Beginning Algebra*. The Beginning Algebra course would teach the rules of algebra stressing the distributive law and laws of exponents. It would omit or treat lightly rational expressions and radicals. Stress would also be put on graphing linear equations including the traditional analytic geometry for linear equations. The algorithm for solving a linear equation in one variable would be presented for continuity. Graphing and manipulating linear inequalities would also be treated. Absolute value would most likely be postponed to the next course.

iii) The second course would use the graphing calculator extensively. For most programs, there would need to be two courses, College Algebra and Pre-Calculus. Both courses would cover graphing: polynomial functions, rational functions, principal root functions (functions with rational exponents), exponential and logarithmic functions. The Pre-Calculus would also study the graphs of the trigonometric functions. (The standard modern trigonometry course would need to be provided the Pre-Calculus students if they did not previously have trigonometry.) Roots (and other points of intersection) of all these functions would be obtained from their graphs. We would actually tiptoe into the twentieth century mathematics.

While both groups of students, College Algebra (non-science majors) and Pre-Calculus (science majors), would profit from the exposure to the use of a graphing calculator, the Pre-Calculus students need a much higher level of facility with algebraic expressions than the college Algebra students. The College Algebra student would investigate complicated algebraic expressions by graphing them. The charge that a student taking the college algebra course that is proposed above was taught less algebra could be made. However, they have been taught more mathematics, and probably for the "average" student learned more algebra that they will retain for more than a semester.

This approach places students in courses that present material that is new to them. These courses provide students with a basis for further studies in mathematics dealing with models that more closely approximate the reality of their subject area. The ability to use a tool such as a graphing calculator in itself is a skill that students will use throughout their careers. Such a curriculum avoids the "algebra trap" that ensnares many students at the outset of their college program.

James Woeppel is at Indiana University Southeast.

An Alternative to Traditional Tests

J. Eidswick, UME Trends 4:3 (August 1992)

At the Third Annual International Conference on Technology in Collegiate Mathematics, a lively on-stage debate between Ed Dubinsky and John Kenelly reinforced my long held belief that exams stand as a major obstacle to the entire teaching and learning process. The debate, however, did nothing to satisfy my hunger for reasonable alternatives. The best the debaters could offer was to say that resources should be provided to research the subject. After a long inner debate, I finally hit upon an idea that seems to hold some hope.

Let us start by looking at the basics. Why do we give tests? The answer: to assess and to evaluate.

Assessment. Students need to know how they are doing and how they can improve. Presumably this is what happens when we return tests with scores on the top and red marks and comments throughout. Unfortunately, many students never see beyond the score at the top and those who do look at the red marks often see only red. A more meaningful way to assess students is to do it one on one with the student sitting in your office. If you look at their work, tell them what is right, what is wrong, how they can and must improve, give them custom assignments, and have follow-up discussions, you will have a better understanding of what you want them to do.

Evaluation. What do we expect in return for the various letter grades? We may tell our students that it takes 90-100 points to get an "A", 80-89 for a "B", etc., but that is not very informative since we are the ones who write the tests and do the grading. What do we really expect? Let's start with the so called passing grade of "C". For this grade, I suspect that most of us would be satisfied with evidence that the student can do most of the routine stuff without too many blunders. For this grade, we do not expect brilliance, neatness, or even good grammar.

There is a big difference in our expectations between the grade of "C" and the grade of "A". For the grade of "A", we expect performance beyond the call of duty and we expect it regularly. We expect students to perform like we did when we were students. For the grade of "B", we expect something in between.

The two main components of the system to be described are "check-off lists" and "special problems".

Check-off Lists. What is the bare minimum you want all students to learn in a course? What are the things you want students to remember after the course is over and done with? What are the big ten? Make a list, keep it terse, and keep it short. Here is my check-off list for a one-quarter differential calculus course: (1) Functions, (2) Derivatives, (3) Calculation of Derivatives, (4) Zeros, (5) Curve Sketching, (6) Applications, and (7) Terminology.

Special Problems. These are challenging but do-able

problems, given a good understanding of the material and a willingness to expend some effort. Some are abstract, some require the use of a computer, some require cleverness, all require more than five minutes.

The system works like this. To pass the course (= "C"), the student must get through the check-off list. To get a "B", the student must, in addition, work m special problems and to get an "A", n special problems. For check-off, mastery is required, perfection is not. For special problems, nothing short of excellence is acceptable.

That is basically it. The rest depends on how many and what kind of students you have, your style, etc. You set the criteria for check-off, you choose the special problems, pick your own m and n , you be the judge of mastery and excellence. Plan to spend more time in your office with students, less time grading papers. What follows is a summary of my efforts to implement the system.

My efforts have involved four different courses (Calculus I, Linear Algebra, Mathematics for Elementary School Teachers, and Number Theory) over a two-quarter period at two different institutions (California State University at San Bernardino and the University of Montana). Class sizes have ranged from seventeen to twenty-five, students were not preselected, and nobody knew in advance that the system would be used.

My check-off procedure involves a strong cooperative learning component. Groups of three (two if necessary) are randomly selected for each check-off topic and group members work together both before and during check-off quizzes as described below.

Check-off Procedure. For most topics, there are two possible ways to get checked-off: (1) satisfactory completion of a check-off quiz; or (2) convince me.

How do check-off quizzes differ from traditional tests? For one thing, they are taken by groups instead of individuals. Each group turns in only one paper. A group either gets checked-off or it does not. If a person belongs to a group that does not get checked-off, it means that person must spend extra time working on the topic, then convince me that he or she understands it. Another difference is that check-off quizzes are very well focused. There is no attempt to see if the student has mastered all of the material. My quiz on derivatives, for example, focuses on the main ideas of derivatives, nothing more. A student must demonstrate an understanding of the definition of the derivative and its interpretations.

What does "convince me" mean? Convince me possibilities include: (1) make up some relevant problems/questions of your own and then work them correctly; (2) prove that you have correctly worked several relevant problems from the text; or (3) have me make up some problems for you. My experience is that students do not show much imagination. They will take the check-off quiz and hope for a favorable result. If that does not happen, they make excuses, like it was the fault of the others in their group, and ask me to make

up some more problems for them. It is interesting that when (1) is suggested as a possibility, the student typically stops and thinks awhile, then grins and asks for (3). It might not be a bad idea to insist on (1)!

Special Problems Procedure. Students are given the following guidelines for special problem work: (1) all special problem work is optional; (2) all special problem work is to be done individually; (3) the usual time limit for the first draft is one week; (4) after you have solved a problem, write up your solution the best you can, make an appointment to see me, and bring your work with you to the appointment; (5) if you did a great job, fine, if not, you will be sent back to the drawing board with things and/or suggestions on how to improve.

I am a stickler on the requirement of excellence, and students seldom get by without being sent back to the drawing board at least once. It is interesting that most students seem to appreciate this kind of criticism, and if and when they finally get a solution accepted, it is gratifying to see their satisfaction.

Student Reaction. I asked students to share their reactions with me concerning the various aspects of the system and received additional information at the end through questions like "Would you take another course that was based on the same or a similar format?" and "What advice would you give a student taking such a course?"

Most students approved the system and many heaped praise on it. Some asked for other ways to get an "A" or a "B" including a traditional final exam, others said they did not like the idea that they might fail if they could not get checked off on one topic.

The small group work was popular in the elementary classes, less popular in the more advanced classes.

Almost all of the students responded "yes" or "yes, definitely!" to the question about whether they would take another such course. Advice that students would give other students: do the work early, keep up, do not depend on your group to do the work for you.

I anticipate that most reader concerns will regard time, cheating, and free-loading.

Time. I make myself available ten to fifteen hours a week. That's about twice my normal office hours. By the second quarter, I learned to put sign-up sheets outside my office. Would I do it with forty students? Probably. It is a lot of work, but keep in mind that grading time is substantially reduced and so is the time spent determining course grades.

Cheating. Can an unscrupulous student end up with an undeserved "B" or "A"? While this might be possible, it would not be easy. When I read student solutions to special problems, I do it with the student sitting in my office. I usually read slowly and carefully and give the appearance of having trouble understanding what the student has done (often this is no pretense). I think out loud and ask questions. An unethical student who had obtained outside help would have to go to a lot of trouble to fool me, and, even if it did happen, it is unlikely that it could happen often enough to lead to an "A". The point is that you get to know these problem-solvers quite well.

Free-loading. It is easy to be overly concerned about this aspect. While it is true that, through the luck of the draw, a student could "float" through the course and get a "C" on the strengths of others, it is unlikely. For one thing, peer pressure is a powerful force. Most students will want to do their fair share and it may even cause some students to pull themselves up to the passing level. Peer pressure can also shame a person out of a group. Most people do not want something for nothing, especially if others know that they are getting it.

Shortcomings. The system is not perfect. One difficulty is dealing with student frustration over the unfamiliar demands of mastery and excellence. When students work hard and still cannot get checked-off or cannot get special problems accepted, they grow frustrated and those accustomed to getting by on partial credit can cause trouble. Another difficulty is that effective one on one work with students calls for people skills that do not come with a Ph.D. degree in mathematics. Honest, two-way communication can be tricky, especially if mathematics problems are compounded with personality problems. Finally, it would be nice to have a good no-test system for precollege students. The system described here would probably work well enough, but would require more one-on-one meeting time than most precollege teachers could spare.

Jack Eidswick is at the University of Montana.

Retesting with Oral Quizzes

A. Brown, UME Trends 3:1 (March 1991)

Like Joann Bossanbroek (UME TRENDS, August 1990), I believe that asking students to rework the problems they missed on a test can be an excellent exercise. In my upper-division mathematics courses, I use a variation that I find easier to monitor and grade.

First, instead of requiring them to work on their own, I encourage students to ask their peers for help. According to my ground rules, any student is permitted to ask a peer with correct solutions to tutor her on the questions she missed, but she may not copy another's work. When the student presents her corrections to me, she acknowledges by name those who tutored her. This informal but well-defined collaborative learning experience benefits both parties, and is a natural extension of the study partner relationship that we encourage in our department.

Second and more important, rather than simply submitting the written corrections, each student comes to see me during my office hours for a fifteen minute oral quiz. I choose some of the items from the corrected test

and, without using her notes, the student presents the requested definitions, solutions, or proofs on the chalkboard. We engage in a dialogue in which I ask her questions that clarify whether or not she understands the details. My grading scheme allows a student to raise her test grade by at most one letter grade. This gives me the options of leaving the grade unchanged for a poor oral quiz, raising it one third for a minimally passable job, two thirds for a flawed but adequate attempt, and a full grade for a completely satisfactory session.

Besides providing the student with valuable immediate feedback, the oral quiz is also a useful diagnostic tool. When the students talk about their solutions, I often learn more about the source of their confusion than I could by reading their papers. The oral quiz is particularly helpful for weaker students, who otherwise might hesitate to come in and ask questions about their errors on the test. It is an added opportunity to talk with them about their difficulties in the class and to give encouragement. Interestingly, I have seen very few poor presentations from these students. Most show an improved understanding of the material, and a successful oral quiz helps repair some of the damage done to their confidence by the poor test performance.

Anne Brown is at St. Mary's College.

Small Study Groups in Summer Courses

R.J. Maher, UME Trends 3:3 (August 1991)

Over the past five summers I have taught nine sections of Precalculus, Calculus I, Multivariable Calculus, and Calculus- based Statistics by lecturing for less than one-half of each 100 minute period, using the remaining time for problem sessions conducted in small study groups. Such groups have been discussed extensively, particularly in relation to adaptations of the **Professional Development Program** (University of California, Berkeley). What is different here is that the small group work is done within the clock hours assigned to the course; resources like extra hours for study or quiz sessions, tutors, and TA's are not available.

Initially, classes are divided into groups of three-to-five at the first meeting. Each group is assigned part of the classroom for its work; two rooms are used if needed. The groups work on an assigned collection of problems; the teacher circulates among the groups, offering comments and soliciting or asking questions. While some students are naturally suspicious of anything different, feedback in general has been quite favorable and student concerns usually disappear after the first few days.

The structure of the problems used each day is critical; they must be both cyclic and progressive. Each set should begin with one basic problem for each new concept introduced that day. New concepts should then be covered a second time, with these problems intermixed with review problems for ideas introduced during the the previous class. This process should be repeated two more times; the review should cover material from, respectively, two and three classes earlier. Problems from even earlier classes also can be included. The students should be able to complete most of these problems during the 50+ minutes available, so problem sets of a similar nature can be assigned for homework. A number of standard texts for these courses contain the variety of problems needed to construct the problem sets.

The cyclic nature of the problems seems to assist learning and retention. The group activity not only indicates topics that need reinforcement but also identifies students having difficulties. Initially, it also identifies students who are in the wrong class, so that referrals can be made without waiting for the first test.

Summer classes compress one semester's work into a five- to six-week period. My experience indicates that small groups, properly used, minimize the learning and retention problems caused by this compression.

Note: The classes discussed here enrolled from eleven to thirty- one students, contained 0 percent-80 percent non-Loyola students, and met four or five times a week for six weeks. Loyola students in these courses performed at the same level as academic year students in the same courses; non-Loyola students performed at the same level as Loyola students. Performances of department majors in the statistics course were consistent with those of majors taking this course during the academic year; performances in follow-up courses also were similar.

Richard Maher is at Loyola University, Chicago.

Working in Small Groups

F. Gass, UME Trends 1:5 (December 1989)

Last fall was the most satisfying experience I've had teaching calculus in at least a decade, and one significant aspect of the course was the occasional use of students working together in small groups during class. Since most of my students had studied some calculus in high school (but had not received college credit), one purpose in trying this approach was to create a format different from the one they were used to. The course information sheet I had given my two classes contained some description and words of encouragement about group work, and I announced these sessions a day or two in advance but without mentioning any particulars.

Because the class was arranged in groups for this effort, students had immediate sources of help and gained some other benefits as well: they got to know each other, and through working together, identified more closely with the class; they broke the pattern of sitting passively in their accustomed seats by actually moving elsewhere and talking about mathematics. To encourage participation by all group members, my assignments either required

full-scale cooperation to achieve results (see example below) or were linked with a subsequent quiz in which individual scores were somehow tied to group success (e.g. everyone gets more credit if each group member does sufficiently well on the quiz). Besides trying to insure that each group had a mix of high-, medium- and low-achievers (relative to homework) and at least one "idea" person, I did some mixing and matching of personalities as I came to know the students better.

As an example, before integrals and Riemann sums came up, I asked my classes to find the approximate length of the graph of $y = x^3$ for $0 \leq x \leq 2$, using their calculators as necessary and submitting one answer from each group of 5 or 6. (I allowed use of the textbook but will probably disallow it next time, to compare results.) At stake was a fairly miniscule reward — a few "bonus" homework points that depended on the accuracy of their answer, with extra points for the group having the most accurate result. I expected that the idea of group competition would provide an additional drive to succeed, but now I would say that the satisfaction and novelty of working together were more significant factors.

To my surprise I found that during most of these sessions, students were so absorbed in discussion as to seem oblivious of me and the other groups. During the length problem, a few people tried to show off advanced knowledge by talking about integrals but were unsuccessful for various reasons. (In the first place, this particular problem is not amenable to the Fundamental Theorem of Calculus.) One of them gave up when a fellow prodigy recalled that integrals give only area, not length. That comment about area illustrates well the value of eavesdropping on groups, especially as it reveals fundamental misunderstandings.

In the end, 11 of the 12 groups in my two classes handed in solutions that were approximating sums based upon partitions of $[0,2]$, although in some cases the summands were incorrect (e.g. $\sum \Delta \, y_i$ rather than $\sum (\Delta \, x_i^2 + \Delta \, y_i^2)^{1/2}$). The 12th group declined to hand in its calculated result because it was less than the straight-line distance they had calculated from $(0,0)$ to $(2,8)$. Another group had used a graph and a piece of string to check its work for gross errors.

In addition to strictly mathematical ideas, the lessons of that day included: the wisdom of being sure to understand what other group members propose to do, the value of concerted brain-storming as opposed to idle page-flipping in search of formulas, the need to cooperate in dividing up work, the need to check your work and the reasonableness of answers, and the need to plan ahead so as to finish on time. Of course, these are lessons that bear repeating.

Fred Gass is at Miami University of Ohio.

Grading the Work of Peers

V. P. Schielack, UME Trends 1:1 (March 1989)

One of the more frustrating problems encountered by the mathematics teacher at the upper undergraduate level is the plight of students making the transition from computation-based introductory courses to theory-oriented courses. Most students simply are not prepared by prior courses for the level of abstraction required in a rigorous course, and a substantial amount of the instructor's efforts must be spent imparting an appreciation for precision in mathematical language. This difficulty is particularly acute in courses for nonmathematics majors such as those desiring teacher certification.

The method presented here is an attempt to facilitate the acquisition of critical-thinking skills necessary in upper-level undergraduate mathematics by having students grade the mathematical work of their peers. The fundamental premise of this technique is that analysis of another's work will make the student more critical of her or his own. When I first taught a theorem-proof geometry course for elementary and secondary mathematics education majors, the first examination results were very discouraging. Many students had extreme difficulty writing logically correct proofs and using quantifiers properly, although considerable class time was spent on these topics; especially disheartening was their inability to write grammatically correct mathematical sentences. I was compelled to make students realize that mathematics is a language that must be used properly, and felt the most expedient means of achieving this goal was having them confront improper usage.

After the first examination was returned and discussed, the class was given a pop quiz; students were told they could earn a bonus of 5 points toward the examination grade. (I wanted an honest attempt at answering the questions.) The quiz consisted of stating definitions and theorems, negating quantified statements, and writing a short proof; all questions were very similar to questions from the examination. Quizzes were then collected and redistributed; no one received her or his own paper. Students were then told that quiz performance was immaterial; the bonus would be achieved by accurately grading the papers of their fellow students. They were instructed to grade the papers on a scale of 20 points (according to a predetermined point distribution per problem) and return them to me the following class period. I evaluated this work by grading the quiz myself, computing the difference of the grader's point assignments and my own on each problem, and awarding the grader the point values of the problems minus the differences. There were thus 20 possible points, and the total was quartered for a maximum bonus of the promised 5 points.

Without any empirical measure of the effects of this procedure, I feel that subsequent course tests showed a sizable improvement in student awareness of mathematical language. I have used this technique several times in the aforementioned geometry course and a real analysis course for secondary mathematics education majors; enrollments are usually 35–45 students. I feel the technique should be beneficial in any introductory theory-oriented course, but is particularly useful in courses for education majors who perceive grading as a skill they must eventually master.

Vincent Schielack is in the Mathematics Department at Texas A&M University.

Peer Grading

B. J. Winkel, UME Trends 1:5 (December 1989)

I let *all* the students in my calculus class grade homeworks from their calculus class! I grade the first week's homeworks to determine starter graders from among the good students and to give the class a model on how one could grade. I show up in class with a key to the homework and a roster. I give the roster and key to a student and say, "You're on tonight." When I return the graded papers the next day, I give each student a complete key. I continue this process *of choosing a new student each day and working my way through the entire class roster* throughout the course. *If needed I go through the top of the grader roster again. All students get a chance to grade homeworks.*

Of the 60 students in two sections this winter, not one thought it a bad idea! At the end of the quarter I asked, "What did you think of the student grading of homework?"

Some immediate responses were: "We got homeworks back faster." "I liked to see how other people worked the problems." "It helped the person who graded to understand the material better." "It gave me the feel of a problem as I worked through different approaches." "Since the same grader did not grade every night, an 'assembly line grading' didn't occur." "It showed me how sloppy and unorganized some people could be and how structured others could be."

The grading usually takes no more than one hour. I simply say, "10 points per problem and be consistent." We discuss the difference between a significant calculus error and a lesser arithmetic or copying error and they each set their own standard. Some are harsh and others are more forgiving. It is obvious that some student graders take it more seriously than others. There is never any sign of vindictive behavior. There is a sense of involvement in the process. Occasionally (but rarely) there are inconsistencies in grading. I address those on an individual basis. The problems seem very small and the benefits are great from the perspective of the teacher *and* the student.

This winter, when I had finished the first round of grading, (i.e., *when all students in the class had a chance to grade*) I began to go through the list of graders again.

In one section I had not finished the roster and at the start of class the next morning the one student who had not been asked to grade came up and said, "How come I didn't get a chance to grade homeworks?" When I handed him the roster sheet with the homework key for that night, his face lit up and he said, "Thank you, Dr. Winkel." *He was thanking me for letting him grade.* Shades of Huck Finn and fence whitewashing! The students are proud to be asked to grade homeworks. And they take pride in the work. While I am not sure how peer grading could be used in your class, I do know that the students who grade like it and they believe it helps them learn better. Thus I commend the method to you.

Brian Winkel is at Rose-Hulman Institute of Technology.

Challenging the Better Students I

R. Czerwinski, UME Trends 3:3 (August 1991)

Instructors in undergraduate mathematics courses are almost always faced with the dilemma of teaching students who are at different levels of preparedness. This is especially true in freshman-sophomore courses. We must challenge the stronger, better-prepared students. We must encourage and motivate the students in the middle level of preparedness, who are willing to work hard in order to learn. At the same time, we must meticulously and patiently work with the weaker students who require a different approach. How does the instructor accomplish these diverse tasks in the same course? There is, of course, no single answer; however, I use a grading device that I feel helps in this situation.

I indicate in my syllabus that there is a starred problem requirement for a good grade in the course. The student must solve a fixed number of problems, picked from among problems (both my own and the textbook's) that require a clearer, more in-depth understanding of the concepts covered in the course. I usually designate from thirty to forty starred problems and require that an A student solve any ten of these problems along with earning 90-100 percent of the available points on tests and homework in the course. For a B, the student may have to solve four of these starred problems, along with earning 80-89 percent of the points. for a C grade, I might require only 1 starred problem. I don't require any for a D. I try to provide enough starred problems that a C student can solve to encourage everyone to at least try one.

During the course of the semester, the student may turn in more than one attempt at a solution for the same problem. I check the attempts and keep all correct solutions until the end of the course. If the solution is incorrect or incomplete, I return the attempt, along with some indication of what is wrong. The students are encouraged to interact with me individually outside of class when they are hot on the trail of a starred problem solution.

This technique does introduce extra overhead in the grading process, and some students object to any type of nontraditional grading in their course. Although the process is susceptible to cheating, I find that student pride in accomplishing something extra and the ease of spotting copied solutions for this type of problem minimize this disadvantage.

I find that the grading technique keeps the stronger, more serious students challenged and makes them even stronger and more serious! It encourages perserverance and exposes the better-prepared students to more in-depth material than can be covered with the class as a whole. I also like how it adds elements of independent, nonprogrammed, and open-minded pursuit of mathematical knowledge. It also adds prestige and pride to good grade accomplishment. Finally, the technique allows me to feel less anxious about keeping some students challenged while I go slower and exercise more patience during regular class meetings.

Ralph Czerwinski is at Millikin University.

Challenging the Better Students II

B. Gold, UME Trends 3:3 (August 1991)

Czerwinski's article ("Challenging the Better Students I," above) prompts me to mention a somewhat similar activity I have directed here at Wabash for the last six or so years. A handout of supplementary problems, keyed to the syllabus, some taken from the starred problems in the textbook, some taken from old actuarial exams, some from assorted other sources (Apostol's calculus book, colleagues, etc.) is given to every calculus student at the beginning of the semester. Students are invited to come Tuesdays at Chapel period (no classes are held then, a residue of the time when there was compulsory chapel), to discuss these supplementary problems. This allows students from *all* the sections of calculus to participate, so that good mathematics students from across the campus will get to know each other early in their career here, and also, to get students used to discussing problems with each other. Students are encouraged to come whether or not they have managed to work any of the problems. They are then set to working on the problems on the blackboard in groups of two or three. The faculty present move from group to group, observing the progress, suggesting that the group look at something more carefully when an error has been made, giving a hint when the group is completely stumped, encouraging students who are following a reasonable path (even if it will not lead to the correct answer eventually). Students who are busy with other activities during Chapel period are invited to submit solutions in writing.

In terms of how it counts towards a grade, the borderline between A and A- is significantly lower for students who participate regularly in the supplementary problem session–students who do not participate can still get an A, but it's much harder. I simply require participation for it to count for the better grade, not the quality of the participation, for I feel that it is beneficial for everyone to try to work on hard problems, even if they don't always succeed. I count it in a less well-defined fashion for students with lower grades. I feel this differential for the top grade is not unfair because my tests are fairly routine–I dislike asking students to think creatively in the limited time (fifty minutes for six to ten problems) there is for a test, but feel that an A should reflect more than routine work.

Students who participate fairly regularly increase in mathematical sophistication significantly more than their peers do over the course of the year. They get excited about the problems and frequently leave the session still discussing them. However, because my colleagues have not chosen to reward participation in any significant way (although most have come to help run the session), students from other sections only rarely participate, and so these sessions have not served their secondary purpose in forming an active core of students interested in mathematics. Still, it is a way of getting students to work on hard problems with relatively little effort on the part of the faculty.

Special Assignments

A. Riskin, UME Trends 1:5 (December 1989)

The biggest difficulty I had in teaching lower division students how to write proofs and to solve relatively difficult conceptual problems was finding an appropriate medium in which to assign these problems. Assigning them on the nightly homework is not effective because the student only tries the problem once, and if he or she can't do it, the answer is heard the next day, and so the student never learns how to do it her or himself. The type of assignment I designed to counter this difficulty seems to be effective, although it is admittedly not perfect.

I assign between three and five proofs in a quarter, which I grade on clarity of expression as well as mathematical content. The students can revise these and turn them in as many times as they want throughout the quarter to get new comments from me and/or to improve their scores. I make these problems, which I call "special assignments", collectively worth 15% of the class grade in order to give the students incentive to work hard on them. Even though there technically is no due date, I make the students turn them in for the first time about a week after I assign them. When I didn't do this, many students would put off doing them until the end of the quarter, which defeats the purpose of the assignments. In the past I have made each assignment worth a

straight ten points. However, for the first time this quarter, I am giving each assignment two ten points scores; one for math and one for writing, in order to reflect more accurately the dual criteria on which I grade.

Special assignments have two main advantages. First, they allow a student to work and re-work relatively difficult problems until he or she solves them. No one learns by seeing a problem solved after having tried it only once. Second, the special assignment lets the student practice mathematical writing skills without being penalized for not catching on right away. Very few people, including professional mathematicians, can get the notation and language of their proofs in an acceptable form without help and comments from other people, and the special assignment takes this fact into account. This type of assignment also gives the student an idea of what mathematical research is like, because no one ever tells the answer; it has to be worked out by the student, her or himself.

Because the special assignments are worth a high percentage of the class grade, I have had some trouble with cheating. At first I passed the assignments back with the homework, but people would steal the ones with high scores and copy them. Now I pass them back individually. Insisting on solutions written in prose makes cheating harder because if a student doesn't understand the solution being copied, he or she will tend to copy it word for word, which is pretty easy to spot. I don't know how to completely eliminate cheating on these assignments, but I think it is still worthwhile to assign them, because the serious students benefit from them. The other drawback is that it is a lot of work to grade the papers over and over with extensive comments, but that seems unavoidable.

I have used this type of assignment in all three quarters of freshman calculus, as well as in sophomore linear algebra, always in classes of 40–60 students. If you would like copies of my special assignments, or if you have any questions or comments, please feel free to write me.

Adrian Riskin is at the University of California, Davis.

Pedagogical Post-its^{TM}

J. Pedersen, UME Trends 2:5 (December 1990)

Have you ever put a problem on the board, laboriously and carefully worked out the solution, and *then* had a students ask "Didn't you copy the problem wrong?" Or, having developed a mathematical result–for example, that it is possible to factor $x^n - y^n$ into at least two factors, namely, $(x-y)$ and $(x^{n-1} + x^{n-2}y + x^{n-3}y^2 + \ldots + y^{n-1})$ –you ask if there are any questions and the only question you get is, "Is this going to be on the next test?"

Although the lack of response (in the first case, above) and the response (in the second case) annoy me, what disturbs me more is that most students simply don't participate in large classes. I have for some time been puzzled as to how I could get students (a) to tell me about my errors before they eat up a lot of class time and (b) to begin thinking about the *mathematical* ideas that are involved in the material I am presenting.

What I really wanted was a system that would reward the student for good classroom behavior, but if I were to take the roll to every class, ask the student who participated her or his name, and record bonus points it would take up valuable class time and – more important – it would break the natural flow of the classroom discussion.

Last quarter I decided to try something that I hoped would be an effective way to deal with this problem. I stamped each page of a Post-it^{TM} pad (those wonderful little blocks of paper with sticky stuff on one edge) with the rubber stamp that has my name and address on it (any mark that can't be easily forged would have done) and the next day when a student timidly corrected an error I had made at the board (*before* it had serious consequences) I gave her a Post-it^{TM} and told her to stick it on her next test and she would be awarded bonus points for it. This didn't disturb the natural flow of the discussion. (When the tests were turned in I simply added the bonus points to their test score, crossing out my name on the Post-it^{TM} so that it couldn't be used again.) On the contrary, the response was almost electric. I could hear the students sitting up in their chairs and, within a few minutes, they had asked more questions than I had heard in all of the previous class periods. The students even began to ask questions about steps they had trouble understanding (and, of course, they were rewarded for those questions, too).

Then someone said, "During the last class period you said that we can *always* factor x^n-y^n and today you have shown that we can factor x^n+y^n when $n=3$, but not when $n=2$. Is there a way to tell when you can factor $x^n + y^n$?" I was so delighted that I gave this student two Post-its^{TM} (and, of course, the class then realized that I placed a higher value on *mathematical* questions).

I have since refined the system so that yellow Post-its^{TM} (for catching mistakes or asking clarifying questions) are worth one point, and blue Post-its^{TM} (for asking mathematical questions) are worth two points – and, of course, there is *no reward* for asking "Will this be on the next test?" (In fact, I jokingly told a student who asked that question that his question would *cost him* a Post-it^{TM}!)

Perhaps my students are only behaving the way they do because they want the extra credit points, but now they keep me from making costly mistakes, they ask many more questions about what they don't understand, and they are learning the habit of asking some genuinely mathematical questions – and, not least of all, they have stopped asking questions about what will be on the next test. Since I started using the Post-its^{TM} I feel better after class.

Jean Pedersen is at Santa Clara University, Santa Clara, California.

Seminar-style Learning

L.M. Berard, UME Trends 3:1 (March 1991)

In some of our upper-level mathematics courses (Topology, Functions of Several Variables, Abstract Algebra II), I have employed a combined lecture/seminar format. That is, I lecture on new material once per week, and during the remaining weekly meeting each student presents solutions to problems assigned specifically to her or him. Because enrollments in these classes are small (usually 3 to 5 students), and because the material is sophisticated, this approach translates into more than occasionally doing homework problems at the board. For one thing, problems are sometimes assigned in advance of pertinent lecture material, so that students must spend a week or two "researching" the material on their own time in order to be prepared in time for their presentation. Furthermore, students have available the extended class time needed to prove complicated results, and they have many opportunities to make presentations (generally, each student makes at least one presentation per week). Although students are shy and nervous at first, a seminar atmosphere soon develops: each student knows in advance the proofs for which he or she is responsible and is motivated to work toward a polished presentation; students question each other about the mathematical correctness and clarity of presented solutions; and the instructor provides helpful comments to improve a student's style of writing and explaining mathematics. Furthermore, much collaboration occurs among students outside class; they begin to "talk mathematics" frequently among themselves.

Certainly this approach differs from a typical graduate-level seminar in that the core material is presented by the instructor rather than by the students. For undergraduates, with more limited mathematical backgrounds, this approach is "safer": the core material is presented in a consistent manner by someone experienced in doing so. Yet students have the opportunity to get much more involved in mathematics than if they were solely listening to lectures and submitting written work; they learn to feel comfortable making formal presentations to their peers, and they become adept at writing clear and concise mathematical arguments.

If you are interested in this method, you will probably find it worthwhile to read an article by D. Cohen, "Modified Moore Method for Teaching Undergraduates" in the 1982 MonthlyBG.

Louise Berard is at Wilkes University.

Lesson Guides —
An Aid and Model for Learning

D. Small, UME Trends 4:5 (December 1992)

Helping students learn how to learn mathematics is an instructor's most important objective, i.e., responsibility. If we expect a student's active mathematical life to extend beyond the short time span of our courses, then we must provide the student with tools for doing so. In particular, if we expect our students to prepare the day's lesson (containing new material) *before* coming to class, then we must help them learn how to do so. Lesson guides provide students with both an effective study aid and also a model for how to approach new material. The following is an example showing the format that I use.

Lesson #12: Limit Concept

Primary Reading:

Small & Hosack: *Calculus, An Integrated Approach*

Sections 2.3 & 2.4 (pp. 107-112)

Homework:

p. 106/ 4, 5, 6, 8, 12; p. 120/1(d), 2(a,h,l,p)

Essential Terminology:

Limit of a Function

Objectives:

a. Understand the *Algebra of Limits*.

b. Understand the sequence approach to limits. (Same approach as in Discrete Dynamical Systems.)

c. Understand how composing a function with a sequence yields a sequence.

d. Understand the process of evaluating the limit of a function by composing the function with an appropriate sequence and then evaluating the limit of the resulting sequence.

e. Understand the six observations concerning Definition 2.4.6, Limit of a Function. (In particular, understand observation 6 concerning the *epsilon-delta* definition of limit.)

Study Questions:

a. Does the phrase "limit of a sequence" always refer to the limit as n approaches infinity? Does it make any sense to speak of the limit of x_n as $n \rightarrow 10$? Why?

b. In evaluating a limit, why must we consider all possible approaches to the point in question?

c. When is it permissible to find the limit of a function by substitution?

Comments:

The limit concept is the basic concept on which all of the calculus rests. This concept is probably the most difficult to understand of all the concepts in calculus. Although the cofounders of calculus, Isaac Newton and Gottfried Leibniz, had strong intuitive (geometric) understanding of the limit concept, an analytical definition was not rigorously formulated until over a hundred years later. Augustin-Louis Cauchy in his 1829 text *Lecons sur le calcul differentiel* gave the essential statement of our present definition.

In Section 2.2, we encountered the second example of an *Algebra* theorem (the first was in Section 1.6). We will encounter several more *Algebra* theorems in the future. A purpose of these theorems is to allow us to simplify computing by not having to explicitly apply the definition each time. Generalizing a concept would be very, very difficult if there were no *Algebra* theorem for the concept.

In the first part of the course, each of the categories in the lesson guides are filled out in detail. As the course progresses, less and less detail is provided until only the outline headings are presented at the end of the course. Part of the students' homework responsibility is to fill in the lesson guide. Thus by the end of the course, students are organizing a full guide on their own for each lesson.

Class discussions on the Objectives that students have entered into their guides provide excellent opportunities to address the question of determining relative importance among theorems, definitions, and examples. This is particularly effective when students are given the chance to "defend" their choices. The Study Questions that students list in their guides offers a similar opportunity for lively class discussions. Learning how to question is a difficult but essential object of becoming an independent learner. Lesson guides are a helpful tool in learning how to question as well as providing a model for addressing new material.

Don Small is at the U. S. Military Academy, West Point.

Writing Proofs

R.J. Hendel, UME Trends 4:1 (March 1992)

Introduction: This article studies the challenging task of teaching a diverse student population to write proofs in a real analysis course preceded by three semesters of calculus. Typical examples are (dis)proofs of (uniform) continuity of rational functions, defined on (closed)(open) subsets of **R**, whose numerators and denominators have small degrees. Transference of the methods presented to other undergraduate areas of proofs seems desirable.

The principle innovation is the request for three *versions* of every proof: the (i) *detailed*, (ii) *motivational*, and (iii) *polished* versions. This *three-version* approach is fruitful because an instructionally specific, seven-step procedure produces the *detailed* version, from which the *polished* version can be produced by a simple transformation rule.

The Seven Steps: To counteract the student complaint of "... but where do I begin," **Step 1**, requests citation of a *textbook definition* of continuity, while **Step 2**, *functional substitution*, requests rewriting this definition with each occurrence of "$f(x)$" replaced by the specific function to be dealt with.

In **Step 3**, students list the *variables* in the definition and decide on appropriate quantifiers, *all or some*. **Step 4** uses the *mechanical* rules of simplification. The following "dependence" rule, useful in continuity proofs, does not explicitly appear in textbooks.

A variable in the prefix of a formula is functionally dependent only on the variables preceding it in the prefix, that are bound by universal quantifiers.

Steps 5 and 6 use an "*assume-exists-and-solve*" strategy to complete the proof.

Example: The following are **Steps 3–6** in the disproof on the uniform continuity of $1/x$ on $(0,1)$.

Step 3: ε—*for all*; δ—*for some*; a—*for all*; x—*for all*.

Step 4: Not (For all ε)(For some δ)(For all x, a) $[|x - a| < \delta \text{ implies } |(1/x) - (1/a)| < \varepsilon] \leftrightarrow$ (For some ε)(For all δ)(For some x, a) $[|x - a| < \delta \text{ and } |(1/x) - (1/a)| \geq \varepsilon]$.

Dependencies: ε is constant. $x = x(\delta)$ and $a = a(\delta)$ depend only on δ.

Step 5: Students now attempt to "find values" (numbers of functions) for the (three) *essentially quantified* variables by *assuming* that a real ε *exists*, that a real δ *(universally quantified)* is *given*, and that functions $x = x(\delta)$ and $a = a(\delta)$ *exist*. An algebra "review," uniform for most proofs, shows $x = a(\delta/2)$ suffices for $|x - a| = \delta/2$, which implies $|x - a| < \delta$ (the first matrix conjunct). Consequently (second matrix conjunct), $(1/x) - (1/a) = ((\delta/2)/a(a + (\delta/2)))$.

Step 6: The attempted "finding" of e and a is completed by seeking both a function, $a = a(\delta)$, and a corresponding numerical lower bound, ε, of the right hand side of this last expression. Students, using a graphing calculator, experimenting with a "standard" increasing sequence of functions, $a = 1, \ln \delta, \delta, \delta^n, ...$, easily discover that the function $a(\delta) = \delta$, with corresponding numerical lower bound, $\varepsilon = 1/3$, suffices as long as $\delta < 1$. (The assumption $\delta < 1$ appears reasonable to students since the problem statement assumes the domain of $1/x$ restricted to $(0,1)$.)

Step 7, which by itself constitutes the *polished version*, *summarizes*, from the previous steps, the variable assignments, (let $\varepsilon = 1/3$. Given δ, let $a = \delta$, ...), and consequent matrix (in)equalities. Students are surprised to learn that good mathematicians prepare *motivational versions*, consisting only of Steps 5–6, prior to writing the *polished* versions found in texts.

Reactions: The *detailed version's* quasi-mechanical, slow-paced nature (i) allows diagnostic specificity, thereby encouraging step-by-step improvement of poorer students' proofs, (ii) facilitates instructor allocation of partial credit, and (iii) sheds perspective on which components of a proof are essential, thereby aiding good students to produce better *polished versions*.

Russell Hendel is at Dowling College.

A Number Theory Course for Liberal Arts Students

R.L. Hatcher, UME Trends 3:5 (December 1991)

Four years ago, when I was on the faculty of St. Olaf College, I was asked to propose a liberal arts mathematics course. The course would be limited to fifteen students, predominantly non-science majors with an interest in satisfying their core mathematics requirement. I wanted to propose a course in which the students would develop the confidence and ability to solve problems which require some creative thought.

With some apprehension, I decided to propose a course in elementary number theory. Although number theory is normally taught at the sophomore mathematics major level or above, I felt that it could be done at a lower level. The topics we covered in the course included: divisibility, prime numbers, modular arithmetic, applications of modular arithmetic to tournament scheduling, calendar days and cryptography, the Chinese Remainder Theorem, Pythagorean triples, rational and irrational numbers, and magic squares. Since I could not find a text at the appropriate level, the students depended solely on lectures for the material. The students were assigned homework almost every class period. Although some of the problems were routine, I always tried to include at least a few problems which required original thinking and which they would not necessarily solve completely. The following are examples of the more difficult problems: (1) Find one thousand consecutive composite integers. (2) Using the fact that $r(6)$ is irrational, prove that $(2) + r(3)$ is irrational. Rather than writing up the homework problems and turning them in to me, the students discussed the problems at the beginning of the next class. For any problems which were not well understood, a student who had solved or made significant progress on the problem would present it on the board for the entire class. To encourage participation, I allowed the weaker students to present the easier problems and I focused attention on what student did right rather than on their mistakes. If the student at the board had not completely and correctly solved the problem, the entire class would discuss it and work together to solve the problem in class with limited help from me. Even reluctant students participated since twenty-five percent of the course grade was based on class participation, with the remainder based on three examinations and a final problem set which consisted of answering one hard problem from a choice of several different problems.

I taught the number theory course described above for three years, and it was quite successful. The course seemed to work well for two main reasons. First of all, number theory has a wealth of interesting problems which students can approach in an exploratory manner. However, there are other subjects which would work well. More important to the success of the course was its structure. By providing a noncompetitive, interactive class atmosphere, the students saw mathematics as a lively and surprisingly social subject. As I had hoped, most of the students left the course with an improved ability and willingness to solve difficult mathematical problems and with a more positive attitude toward mathematics.

Rhonda Hatcher is at Texas Christian University.

Statement of Purpose of UME *Trends*. ed. Research, Teaching, Industry

E. Dubinsky, UME Trends 1:1 (March 1989)

Mathematics is all three of these, and each is essential to our discipline. Research constantly expands, revises, and renews the content of mathematics. Teaching inspires young people to join our ranks, refreshing our efforts with new ideas and new energy. And our relationship with industry establishes us as part of a modern technological society which stimulates us with practical problems whose solutions we seek.

Today, there is a crisis in mathematics education, from Kindergarten through Graduate School. In particular, with respect to undergraduate mathematics, industry is questioning the mathematical knowledge of our graduates, the mathematics major at some universities is fading away, engineers and scientists are expressing dissatisfaction with our service courses, and many of our best graduate students will not do mathematics research in this country.

In short, both the infrastructure and cutting edge of American mathematics is in an early state of decay.

Fortunately, there is widespread recognition of the problem which, though critical, is not unsolvable. As we see our community generate and implement a multitude of ideas and projects, it is particularly encouraging to observe that our response has been creative and energetic with a potentially unifying effect. The three professional societies — AMS, MAA, and SIAM — are demonstrating their understanding that the fundamental problems of the profession cut across research, teaching and industry; and that the appropriate response to a widespread danger is a united effort to move in ways that will set things right.

It will be the role of *UME Trends* to report on, stimulate, and contribute to that movement. Created through the efforts of all three professional societies, partially funded by the NSF, and established as a program of the Joint Policy Board for Mathematics. *UME Trends* is already an example of that unity and we hope it will make a contribution towards alleviating the crisis by fostering improvement in undergraduate mathematics education.

The content of *UME Trends* will include feature articles, regular departments, and news items. Opinions expressed in signed articles will be those of the author. Opinions in unsigned articles will be avoided. Editorial policy, established by the editor with the advice of the Editorial Committee, will occasionally have the effect of making a statement. For example, this first issue contains several articles about minorities and women in mathematics. We hope that this expresses the importance of paying serious attention to these groups both because they form an essentially untapped resource for the renewal of mathematics and because it is right to do so.

We have great hopes that *UME Trends* will perform a service to the mathematics community. We are firmly convinced that these hopes can only be realized through your participation. Please send me your comments, criticisms, praise, suggestions for improvement, and offers to help.

UME Trends will be distributed free during the first year, on request during the second, and at a subsidized subscription rate for the third.

E̲ditor's N̲ote: *One of the regular sections of UME Trends is called* **Research Sampler.** *These are very short descriptions of research in mathematics education and a selection is reprinted below. The editors of the Research Sampler section state the following about their intent:*

This department will sample current research in tertiary mathematics education. The aim is to interest and enlighten by providing capsule commentaries on noteworthy articles. Our remarks will be neither complete nor objective hopefully, they will be provocative. References will be included.

Research Sampler
Edited by A. Selden & J. Selden

The Diversity of Knowing
UME Trends 4:4 (October 1992)

Using their research on programming styles, Sherry Turkle and Seymour Papert of MIT make a case for epistemological pluralism, and in particular, for acceptance and re-evaluation of concrete approaches to knowledge, which they term *bricolage*. Bricoleurs prefer rearrangement and negotiation of their materials, whether musical notes, words in a sentence, or elements in a program, think of mistakes as mid-course corrections, and get close to and sometimes even anthropomorphize the objects they work with. This way of working contrasts with abstract, analytical approaches to knowing, which have become the canonical scientific style, causing bricoleurs, including many women, to feel left out. [*Journal of Mathematical Behavior* 11 (1), 3-33; *Humanistic Mathematics Network Journal* 7, 48-68.]

Support for the superiority of abstract approaches may derive, in part, from Piaget's developmental theory of intelligence, in which formal reasoning is seen as the final, most mature stage. Turkle and Papert reject rigid stage theory, noting that most people adopt aspects of both styles, while feeling more comfortable with one. They see the computer with its graphics, sound, and animation as a facilitator for those who relate to the world through intuition, visual impression, and movement. Computational objects, of which the icon is the simplest example, can be manipulated directly, thereby rendering the abstract concrete. Yet the computer culture values the formal approach almost exclusively and insists on structured, top-down programming. Lorraine, a successful computer science graduate student, thinks about what a program "feels like from inside," but does not want others to know her way of working. Lisa and Robin, two Harvard classmates in an introductory programming course, are instructed to program "the right way," dutifully "fake it," but feel alienated, preferring to work as they do when writing a poem or composing music.

As further evidence for the validity of concrete approaches, Turkel and Papert point to studies like Jean Lave's *Everyday Cognition*, in which adults were found to use effective mathematical strategies very different from those taught in schools, to ethnographies showing ad hoc scientific discoveries which were only later formalized, and to feminist scholarship, which has documented the power of concrete contextual reasoning in a wide range of domains. [See, for example, S. Traweek, *Beamtimes and Lifetimes: The World of High Energy Physicists* and M. K. Belenky, *et.al, Women's Ways of Knowing*.]

Turkle and Papert think icons, object-oriented programming, and biologically based "emergent AI" all challenge the idea that formal, abstract methods are the only legitimate ones for mathematics, science, and engineering. They suggest that, while discrimination blocks access to such fields, the dominant ways of thinking may also make bricoleurs, including many women, reluctant to join in.

Personal Teaching Styles Resist Change

UME Trends 4:4 (October 1992)

In a recent survey of research on K-12 teachers' beliefs, Dona Kagan of the University of Alabama suggests that strongly held, highly personal pedagogies develop because there are many alternative explanations of the nature of learning and cognitive growth, of why students behave the way they do, and of the best way to teach a topic. Teachers work in fluid and uncertain settings with immediate practical classroom problems to solve, and their belief systems provide them with a sense of control and a way to make decisions. [*Educational Psychologist* 27(1), 65-90.] Teachers' beliefs may be largely unconscious or unexamined, yet can affect both classroom behavior and the approach to specific academic content. In elementary science, teachers' beliefs determine whether they present facts, laws, and formulae to be memorized, demonstrate the scientific method, provide hands on activities for discovery, or challenge students to predict and explain. [D. Smith and D.C. Neale, *Teaching and Teacher Education* 5 (1), 1-20.]

Beliefs are highly resistant to change and preservice teachers tend to leave college with their beliefs about teaching unaffected by the research they read, their education courses, and their student teaching. Since teachers are not immune from seeing what they expect to see and since their work is often done largely in isolation, their beliefs act as the dominant filter unless challenged. Over time, however, personal pedagogies evolve; longitudinal research is needed on how.

In an article on the difficulties of getting didactic engineering products, from decimals and ratios to calculus, used in mathematics classrooms, M. Artigue and M. J. Perrin Glorian of the University of Paris note that teachers tend to reduce the distance between their customary teaching practice and the teaching sequences proposed by researcher/reformers. They found an incompatibility between teachers' beliefs and their own use of Brousseau's theory of didactic situations with its constructivist base. Such beliefs include: errors are always dangerous, need to be prevented by warning signals, and, failing this, should be systematically corrected; learning is a continuous progression best handled by the teacher via small increments in difficulty; pupils should be actively involved, but conceptual tasks are too difficult, so active participation is limited to exercises. Practicing basic skills for external assessments takes priority over all else, especially researchers' teaching sequences, which may be useful for introducing a topic, but are of secondary importance. [*For the Learning of Mathematics* 11 (1), 13-18.]

Do college teachers develop similar idiosyncratic, robust, personal pedagogies? Perhaps one should study how curriculum reform projects affect college teachers' beliefs.

Now It's Reading

UME Trends 4:2 (May 1992)

Writing to learn mathematics is now a familiar idea whose raison d'être is the stimulation of learning through writing. The emphasis is on process, rather than product. For example, students keep journals in which affective, as well as cognitive, observations are made. This reconceptualization of writing as process entailed a break with the past: a paradigm shift. Reading comprehension is currently being rethought in a similar constructivist way.

Traditionally, reading has been interpreted as the art of extracting information from written text. However, recent transactional models portray readers as actively constructing new meanings while transforming text based on their experiences, knowledge, and beliefs. Instead of emphasizing vocabulary and correct inferences, readers are asked to engage in *transactional activities* such as transforming the text, building on analogies, and raising questions.

Rafaella Borasi, a mathematics education researcher, together with Marjorie Siegel, a reading researcher, implemented such ideas in a semester-long graduate course on non-Euclidean geometries for secondary mathematics teachers at the University of Rochester. The goal of their NSF-funded study was to have students experience reading to learn, while building up a collaborative team for the preparation of secondary school materials. (*For the Learning of Mathematics*, November 1990.)

They began with the exploration of a simple non-Euclidean geometry, taxi geometry, embedded in a participant's story about the search for an ideal apartment in Manhattan. (For the story and additional observations, see Borasi, Sheedy, Siegel, *Language Arts*, February 1990.) Students engaged in generative reading of the story in pairs, stopping to say something at self-regulated points of interest or difficulty. Then, for homework, the students were assigned Krause's text, *Taxicab Geometry* (Dover, 1986). As a transmediation activity between the two, the students were asked, as a group, to extend the story using results from that text. Subsequently, they read an historical account of the development of non-Euclidean geometry (Kline, 1980) and *Flatland*. As a follow-up, each student used the cards to create a map of conceptual relationships for presentation to the class. Finally, they came up with metaphors for alternative geometries, drawing a parallel between their initial resistance to studying this material and resistance to employing alternative teaching methods.

The authors are positive about this experience. The students found and resolved anomalies (often internal), reflected on content, thought about representations, and engaged in problem solving. They feel that reading to learn can contribute to students' redefinition of what it means to learn and understand mathematics.

Although this study was done with secondary students

as the ultimate target, such a course might be suitable for liberal arts students in colleges.

Whats the Effect of Taking High School Calculus?

UME Trends 4:2 (May 1992)

A full year of high-school calculus, whether AP or not, improved traditionally-taught first semester college calculus students grades one notch, from C- to B-, when compared with students who had taken only a brief introduction to calculus or none at all. Students with less than a semester of high school calculus performed only slightly better than those with no calculus background on the procedural questions appearing on the first test and the final exam. However, there was no noticeable difference in grades for any of those continuing on to second-semester college calculus. Conducted with 751 students at a mid-sized state university in 1987-1988, these results were statistically significant even after adjusting for the effects of mathematics SAT scores. (Ferrini-Mundy and Gaudard, *JRME*, January 1992, 56-721.) Currently, considerable time is spent on calculus topics in 12th-grade precalculus courses. Might not students be better off studying functions or other topics? The authors caution that their results may not generalize to calculus reform projects, where the emphasis is often more conceptual, applications-oriented, and technology intensive.

Are We in the Business of Changing Students' Beliefs About Mathematics?

UME Trends 3:6 (January 1992)

If we aren't, we ought to be. Beliefs influence mathematical behavior and lie somewhere between primarily cognitive factors (one's knowledge of mathematical facts and procedures and the effective use of strategies and techniques) and primarily affective factors (anxiety, motivation, etc.) [Alan Schoenfeld, *Mathematical Problem Solving Academic Press*, 1985, Chapters 5 & 10]. Clearly, what students know about mathematics determines what they can do with it and their attitudes can affect their performance. In addition, whether overtly expressed or not, their views about the nature of mathematics can influence how they tackle mathematical problems. The Institute on Problem Solving and Thinking, an NSF-funded project at Georgia State, changed inservice teachers' beliefs from nonproductive ("I'm good in math if I can do problems fast.") to productive ("It's okay if math takes time.") [Karen Schultz, *Proceedings PME*-15, Vol. III, 1991]. While studying both high school geometry and college cognitive science students,

Schoenfeld observed beliefs such as that formal mathematics has little or nothing to do with thinking, all mathematics problems can be solved in at most ten minutes, only geniuses can create mathematics, and mathematical procedures are passed down "from above." Behavioral consequences were that the students did not invoke formal mathematics when constructions or discoveries were called for, using instead a *naive empiricism*, with students giving up quickly when they forgot a formula ("either you know it or you don't").

In a study of elementary education majors in mathematics and mathematics methods courses, Dina Tirosh and Anna Graeber found that although only ten percent explicitly believed "multiplication always makes bigger," more than fifty percent explicitly believed "the quotient must always be less than the dividend" [*Educational Studies in Mathematics* 20, 1989]. Many of these students still thought of division in terms of the partitive model in which an object (like a pizza) is divided into equal parts. This means the divisor *must* be a whole number and the quotient *must* be less than the dividend. It may have been the source of their belief that "division always makes smaller." Despite these students' mechanical proficiency with the division algorithm (eighty-seven percent correctly divided 5 by .75), this model dominated their thinking when solving word problems. Given the problem, "The price of one bolt of silk fabric is $12,000. What is the cost of .55 of the bolt?", one student argued, " ... you want to know the price of this part, a part, of the bolt. So you are going to divide .55 into 12,000 to find out what that part is." After executing the division algorithm and obtaining 21,818, the student continued, "I moved it [the decimal point] over here [in dividend], so I move it back two places. About 218." According to Fischbein, et al [*Journal for Research in Mathematical Education* 16 (1985)], "models become so deeply rooted in the learner's mind that they continue to exert an unconscious control over mental behavior even after the learner has acquired formal mathematical notions that are solid and correct." Such beliefs (misconceptions) are often robust, but can be addressed by inducing cognitive conflict for class discussion and self resolution and by suggesting the consideration of other models. For division, this might be the measurement model, in which one determines how many times a given quantity is contained in a larger quantity.

Metamathematical and mathematical beliefs affect our students' problem-solving behavior. They are often abstracted from past and present classroom experiences. Is it possible we are "accidentally teaching" things we do not believe?

What Kind of Talk is Effective for Learning Mathematics in Small Groups?

UME Trends 3:6 (January 1992)

In a meta-analysis of eighteen prior studies linking task-related verbal interaction in small groups with effective mathematics learning, Noreen Webb concludes that giving elaborate content-related explanations is positively related to one's own achievement [JRME, 22(5), 1991]. Mixed results were obtained regarding the receiving of responsive feedback, such as having an error corrected along with an explanation. However, nonresponsive feedback, such as receiving only the correct answer, was definitely negatively related to achievement. It was postulated that in order for an explanation to be effective, the receiver must understand it, internalize it, and use it to correct misunderstandings. Typically, achievement was measured by tests taken individually at the end of instructional units. Although these studies dealt with early elementary through late secondary mathematics students, perhaps some implications, and certainly some questions, can be drawn for the use of small groups in undergraduate mathematics instruction.

Other findings include that explicit requests for help were more likely to elicit explanations than general requests, and boys and extroverted students were more likely to ask direct questions and receive help. The most effective compositions for engendering active participation were homogeneous medium-ability and moderate-ability range (highs and mediums or lows and mediums) groups. Homogeneous high-ability or low-ability and heterogeneous groups of highs, mediums, and lows were detrimental to the performance of at least some students. Training students in the art of giving explanations was effective.

Although the precise nature of effective explanations was not addressed and needs study, Webb suggests ten strategies from observations of expert adult teachers and student tutors: (1) activate students' prior knowledge when explaining, (2) use multiple representations (symbols, numbers, graphs), (3) show how to translate between representations, (4) provide detailed justifications of the reasoning used to solve a problem, (5) use specific examples to illustrate a general concept, (6) become "tuned into" target students' problem-solving processes so as to correct errors immediately, (7) translate unfamiliar terms into familiar language, (8) assess students' understanding by giving them opportunities to solve problems on their own. (9) respond to errors and confusion by providing detailed explanations, instead of brief responses or only correct answers, and (10) encourage students to disagree with others, to say they don't understand, and to control the pace and content of discussion. Good suggestions for any teacher, not just peer tutors. But how would one implement them?

Do Students Have a Sense of Function?

UME Trends 2:6 (January 1990)

If one means anything more than manipulative ability with functions seen as formulas, the answer is, regrettably, almost always *no*. To go beyond this and discuss recent work on students' conceptions of function, about twenty mathematics education researchers from Poland, Israel, France, Canada, and the United States met on the Purdue campus in October. The intimacy and intensity of the discussions and the perfect fall weather were reminiscent of conferences at Oberwolfach, although perhaps the setting was not so bucolic. Many participants were mathematicians, intrigued by the complexities of the teaching/learning process. They ranged from experimentalists who gather data on students' understanding of function to designers of educational software to builders of learning theories.

A sampling from papers and discussions may help convey the flavor and variety of the interchanges. Steve Monk of the University of Washington related recent research revealing undergraduate students' inability to see and express functional relationships in a physical model of the classic rate of change problem involving a ladder sliding down a vertical wall. Michèle Artigue of the University of Paris reported on a three-year study of some 300 first-year French university students' difficulties learning differential equations, especially in the graphical setting.

Judah Schwartz of Harvard and Education Development Center outlined an ambitious software project to avoid the asymmetry of current graphing packages he calls "half-tools." Such packages only give the graphical results of manipulating symbolic representations of functions. Schwartz is adding the ability to manipulate graphs directly while producing the corresponding symbolic representations. Joel Hillel of Concordia University considered strengths and weakness of CAS vis-a-vis functions, arguing for their thoughtful use. However, "retrofitting" commercial software to learning situations may not be ideal pedagogically.

Anna Sierpinska of Warsaw, who has done extensive studies of students' understandings of function, described numerous *epistemological obstacles*. An example is the historical, and also currently widespread, notion that functional relationships are *causal* – a change in x causes a change in y. Perhaps the difference between a physical law and a function should be stressed in high school and beginning college.

The Purdue contingent, including Guershon Harel, Keith Schwingendorf, Ed Dubinsky, Daniel Breidenbach, and Devi Nichols, described positive results of instructional treatments using ISETL to elevate college students from an action to a process conception of function. An *action* conception includes the ability to plug numbers into algebraic expressions and calculate, whereas a *pro-cess* conception includes thinking of functions as transformations and being able to compose them. In many U.S. high schools, students merely engage in activities for manipulating formulas–this leaves them with a poorly formed function concept. A student has reached an *object* conception of function when he or she can act on functions using operators and consider them as legitimate elements of vector spaces.

In schools and colleges, one linearly organizes complex mathematical concepts in order to teach them – a difficult task that Brouseau has termed the "didactic transposition" of knowledge. This is especially difficult for concepts like function, which can be understood in several different ways, none of which is adequate by itself. Furthermore, teaching is often based on the fiction that a good teacher can "smooth out" the learning process so that it is a gradual accumulation of knowledge without difficulties or errors. This view fails to account for resistant, and seemingly unavoidable, learning difficulties such as epistemological obstacles.

A recurring theme of the conference was students' difficulties with visualization. Students need explicit teaching as graphs contain concise information that needs "unpacking." Like Dienes blocks for representing base 10 to young children, graphs are only useful *after* one understands how to interpret them. Naive students expect a close resemblance between the shape of a graph and the "real world" situation the graph refers to. For example, when given information about an individual cycling up and over a hill (including a diagram), adolescent students incorporated contours of the hill into their speed versus position graphs. [Paper presented at the 1986 AERA Annual Meeting by Schultz, K, Clement, J., and Mokros, J.] Michèle Artigue's differential equations students had great difficulty using graphs to demonstrate whether a solution intersects a given curve or has an infinite branch. She conjectured the students had not yet encapsulated graphs as *objects* for reasoning. How can knowledge of graphs be linearly organized for teaching?

Many current calculus reform projects use the "rule of three," that is, they rightly consider it important for students to work with functions in their numerical, graphical, and symbolic representations. Perhaps such projects should also pay attention to the difficulties of understanding these diverse representations and building bridges between them.

How Should Proofs Be Taught?

UME Trends 2:4 (October 1990)

One prerequisite for becoming a mathematician is the ability to recognize a proof when you see one, and presenting proofs clearly is how you teach them. Or is it? For some in mathematics education and the philosophy of science, the acceptance of new mathematics, including the proof of a new theorem, is a process of social negotiation. Pedagogical consequences are drawn from this view, and almost everyone writing on the subject references, quotes, interprets, or misinterprets Imre Lakatos' seminal work, *Proofs and Refutations: The Logic of Mathematical Discovery*. Using a closely reasoned Socratic dialogue based on the historical evolution of the proof of the Descartes-Euler conjecture that for polyhedra, $V - E + F = 2$, Lakatos develops the method of "proofs and refutations." It has four basic stages: (1) a primitive conjecture, (2) a rough proof, (3) emergence of counterexamples, and (4) re-examination of the proof and improvement of the conjecture by redefining the concepts involved. He indicates that the teaching of mathematics should better reflect its practice. There seems to be more to say about creating mathematics, and we wonder how Lakatos's work might have evolved had it not been truncated by his untimely death.

Gila Hanna of the Ontario Institute for Studies in Education also uses an argument based on the examination of the role of proof in mathematics. She draws the pedagogical conclusion that rigorous proof should be deemphasized in the secondary school curriculum, preferring *proofs that explain* to *proofs that only prove*. She uses the word *explain* for a proof that makes use of the motivating mathematical ideas. An example of such a proof is Gauss's demonstration that the sum of the first n positive integers is $n(n+1)/2$ by listing the integers in descending as well as ascending order, thereby displaying the n identical summands $n+1$. A proof of the same result by mathematical induction is seen as a proof that merely proves. [*Rigorous Proof in Mathematics Education*, OISE Press, Toronto, 1983; "Some Pedagogical Aspects of Proof," *Interchange* 21 (1), 6-13.]

Hanna considers rigorous proof as only one element in the acceptance of a theorem. More important in her view are: (1) understanding the theorem and its implications, (2) significance of the theorem in its relation to various branches of mathematics, (3) compatibility of the theorem with other accepted mathematical results, (4) the reputation of the author, and (5) a convincing argument of a type encountered before, whether rigorous or not. In advancing her argument, she cites the views of P. Kitcher that mathematical knowledge is always sensitive to peer challenges and is sustained by community approval, of T. Tymoczko that proving theorems is a public activity, and of P. J. Davis that a proof is never complete as evidenced by the inclusion of phrases like "it is easy to show that." How many mathematicians would agree?

Hanna goes on to draw implications for mathematics education: (1) Competence in mathematics is not "synonymous with the ability to create the form, a rigorous proof," although careful reasoning and the building of arguments needs to be conveyed to students. (2) Teaching beginning students the formalities of proof omits the "crucial element," namely, the significance of the idea and its implications. (3) Students need help in maintaining the level of concentration required to negotiate

a line of reasoning. (4) Sometimes pictorial explanations or analogies are pedagogically better than precision. Proofs that explain are to be preferred to proofs that only prove. (5) A balance between rigor and intuition needs to be conveyed to students. These views could, it seems, be held without reference to the preceding philosophical position.

It seems reasonable that the nature of mathematics, as well as the nature of learning, should influence how one teaches proof, and professional mathematicians might make some useful contributions.

Incorporating Learning Research into the Classroom
UME Trends 2:4 (October 1990)

Bernard Cornu of the University of Grenoble surveys this topic in "Didactical Aspects of the Use of Computers for Teaching and Learning Mathematics," *Educational Computing in Mathematics*, North-Holland, 1988. A computer can be used in many ways from classroom demonstrations where it is essentially an "improved blackboard" to labs where it is a resource for homework or projects. However, hardware and software are not enough; there needs to be a pedagogical strategy or a "didactical approach" for their use. Cornu mentions several possible approaches: (1) Tall and Vinner have put forth the idea that each student has *concept images* consisting of collections of examples, nonexamples, intuitions, etc., which determine how he or she reacts, and it is the teacher's job to act on these images by completing them, enriching them, and destroying contradictions. With the computer one can provide "microworlds" or environments in which students can manipulate different aspects of concept images. ["Concept Image and Concept Definition in Mathematics with Particular Reference to Limits and Continuity," *Educational Studies in Mathematics* 12, 151-169.] (2) Dubinsky analyzes the different steps in the acquisition of a concept and describes what kind of abstraction is necessary at each step, obtaining a concept's "genetic decomposition." He uses the programming language ISETL to help students through the necessary steps. [*Learning Discrete Mathematics with ISETL*, Springer, 1989 and "ISETL– Interactive Set Language," AMS Notices 37(3), 276-279.] (3) G. Bachelard and A. Sierpinska study epistemological obstacles, which must be encountered to acquire a whole concept. Many of these have been encountered historically by mathematicians in the development of the concept. Perhaps appropriate computer experiences would help here. ["Humanities students and epistemological obstacles related to limits," *Educational Studies in Mathematics* 18, 371-397.] (4) Errors are symptoms of a student's cognitive structure–they are logical consequences of the student's present state of knowledge. Analysis of errors is difficult and requires time, but there are experiments where errors are analyzed, and appropriate computer exercises are given to allow a student to readjust her or his knowledge. (5) The computer can be a partner in problem solving.

Cornu warns that the computer is only a tool; it has no automatic effect on knowledge and must be used with a definite pedagogical strategy in mind. This is the hard job of what the French call "didactical engineering," which consists of research, followed by designing "didactical situations" and appropriate software. This is so time-consuming as to be impractical for ordinary classroom teachers. However, as such products and approaches become available, we should consider taking them up.

The Derivative as the Slope of the Graph Itself

UME Trends 2:2 (May 1990)

David Tall of Mathematics Education Research Centre, University of Warwick, a mathematics education researcher as well as a mathematician found that calculus students in Britain have great difficulty interpreting phrases like "tends to a limit" and "as close as we please." ("Understanding the Calculus," *Mathematics Teaching* 110 1985, 49–53.) In a similar vein, Bernard Cornu reported that French students have the idea that the limit is a value that cannot be surpassed, whereas the limiting process is seen as a dynamic one of going on forever. "Apprentissage de la notion de limite: conceptions et obstacles," doctoral thesis, Grenoble, 1983.) For example, they said that the limit of the sequence $0.9, 0.99, 0.999, \ldots$ was 1, but the sequence tended to $0.a(,9)$, which they considered to be just less than 1. Students told Tall that secant lines approach the tangent line, but never really get there.

Like most mathematics education researchers today, Tall takes the constructivist view. "Constructivism in Mathematics Education — An Underlying View of How People Learn," (*UME Trends*, March 1990). That is, he considers calculus from the students' viewpoint and seeks to help them build up an understanding of the subject from their current state of knowledge. Using data from several hundred questionnaires in which students with no knowledge of calculus were asked about secant lines and tangent lines, Tall concluded that, although the limiting process may be intuitive in the mathematical sense, it appears *not* to be so in the psychological sense.

Tall sees great use for microcomputer experiences here, but *not* by having secant lines approach tangent lines. Instead, he proposes that the derivative be introduced as the *slope of the graph itself*. When students zoom in on a portion of the graph of a differentiable function, they easily see that the graph tends to a straight line, making the notion of the slope of the graph a reasonable one

for them. One merely magnifies the graph sufficiently and observes the slope of the resulting (approximately) straight line. By looking at points where functions are not differentiable, such as $- x^2 - x -$ at $x=1$, one clearly sees upon zooming in, a different slope to the right from that to the left. In their software *A Graphic Approach to Calculus* (available from Sunburst), Tall and coauthors Piet Blokland and Douwe Kok include the graph of a nowhere differentiable function. When one tries to make this function straight, one cannot. Tall contends such computer experiences of a geometric sort, before studying calculus more formally, help students in their cognitive development.

Gem on Teaching Probability Goes Unnoticed

UME Trends 2:2 (May 1990)

After reading our January column, "Can We Improve People's Judgments Under Uncertainty?", David Carlson of San Diego State pointed out J. Michael Shaughnessy's fine article, "Misconceptions of Probability: An Experiment with a Small-Group Activity-Based, Model Building Approach to Introductory Probability at the College Level." (*Educational Studies in Mathematics* 8, 1977, 295–316) Shaughnessy describes nine experimental group activities for the probability course usually taken by business majors. These range from asking students to discover how many distinct words one can write using the letters L A Z K, if every arrangement is considered to be a word, to designing an experiment to test the truth of the statement, "Pulse rates go up when taken by a member of the opposite sex." At the end of the course, he compared his two experimental sections with two traditional lecture sections and found that his students not only enjoyed the course more, but also were more successful in overcoming beliefs such as the idea that outcomes which are more familiar are more likely to occur. Did this gem go unnoticed because it was published in a journal so few of us regularly see? Perhaps there should be a journal on collegiate mathematics education research.

Undergraduates Cannot Solve Algebra Word Problems

UME Trends 2:2 (May 1990)

Eighty-five junior and senior computer science undergraduates enrolled in an artificial intelligence course were asked to solve problems such as, "Mary can do a job in 5 hours and Jane can do a job in 4 hours. If they work together, how long will it take to do the job?" Many found these problems nonroutine, failing to solve them or solving them with difficulty. (Hall, Kibler, Wenger, and Truxaw, "Exploring the episodic structure of algebra story problem solving," *Cognition and Instruction*, 6(3), 1989, 223–83.) The students were definitely not expert problem solvers, as they did not exhibit the use of schema or rule automation. ("When experts solve problems, what do they do that novices don't?" *UME Trends*, January 1990.) However, since a prerequisite was three calculus courses and a corequisite was discrete mathematics, they were also not naive problems solvers, and the authors coined the term *competent* problem solvers to describe them. In this thorough study supported by the Office of Naval Research, students' written protocols were divided into coherent problem-solving episodes, each of which was examined for strategic purpose, tactical content, and conceptual content. Most of the students used nonalgebraic reasoning within the situational context to comprehend a problem, to construct a candidate solution, to gather evidence for checking its plausibility, and to recover from false starts. Algebraic formalism made up only a small portion of their problem-solving behavior, yet current high school and college algebra curricula emphasize formal manipulation. The authors, all of the Department of Information and Computer Science, University of California at Irvine, want to exploit this natural tendency to reason within a problem's immediate situational context. They suggest algebra students might be taught to make pictorial representations of linear relationships, similar to static force diagrams in physics, to help them analyze such problems before going on to formal manipulation. Surely something needs to be done to improve students' problem-solving skills, but we suspect such specialized methods, which are applicable to only very restricted domains, do little to improve general mathematical problem-solving abilities. Should we introduce what appears to be yet another aspect of formal manipulation into the algebra curriculum?

Can We Improve People's Judgements Under Uncertainty?

UME Trends 1:6 (January 1990)

In his popular book Innumeracy: Mathematical Illiteracy and its Consequences, John Allen Paulos amusingly and disturbingly describes many everyday results of people's ignorance of elementary probability and statistics. Researchers Tversky and Kahneman, whom Paulos at times quotes, found that when the role of chance is salient and the sample space is small, most adults reason probabilistically. However, when a situation is complicated, even people with some training in probability, tend to use a limited number of heuristic principles which can lead to severe and systematic errors. The majority of those answering a questionnaire at a meeting of the Mathematical Psychology Group responded as if there were a "law of small numbers," that is, as if a randomly drawn sample, no matter what its size, would necessar-

ily be highly representative of the parent population. (Judgment under Uncertainty: Heuristics and Biases, Cambridge University Press, 1982.) A "representativeness heuristic," in which the probability of observing a particular sample is estimated by noting the degree of similarity with the parent population, is used by many who indicate that the sequence MMMMMM of male and female births in a family is less likely than the sequence MFFMMF. When asked to estimate the probability of the most likely outcome of rolling a loaded die, a significant number of undergraduate subjects at the University of Massachusetts preferred inspecting the die to using information on 1,000 trials. (Konold, "Informal Conceptions of Probability, Cognition and Instruction 6(1), 59–98.) In his closing section Paulos acknowledges that "an appreciation for probability takes a long time." Others have noted that although traditional probability instruction is effective in producing correct answers to formal test questions, it is less effective in changing actual behavior. We know of one attempt to design a curriculum aimed at removing probabilistic misconceptions in junior high school students' judgments. (Beyth-Marom and Dekel, "A Curriculum to Improve Thinking under Uncertainty," Instructional Science 12, 67–82.) Perhaps there should be more.

When Experts Solve Problems, What Do They Do That Novices Don't?

UME Trends 1:6 (January 1990)

There has been a good deal of research on this question in the past few years. A survey of the literature suggests that experts: (1) have a better memory for relevant problem details; (2) classify problem types according to underlying principles, rather than surface structure; (3) work forwards towards a goal, rather than backwards from it; and (4) use well-established procedures or rule automation. The first three of these can be viewed in terms of *schemas*, cognitive structures that suggest the category to which a problem might belong, as well as appropriate solution strategies. Just how mathematics students, or anyone else, might be helped to acquire schemas is of considerable interest, yet little research on this seems to have been done. Owen and Sweller have observed that solving a large number of conventional exercises is largely ineffectual (JRME 20, 322–328). Both schemas and rule automation reduce memory load, allowing an expert to handle familiar aspects of a problem routinely, while freeing cognitive capacity for investigating the novel aspects of a problem. We note that even if students understand relevant procedures, they may not have fully automated them, and even if they have automated them correctly, they may not know how or when to access them. Is that for lack of appropriate schemas?

Teaching Away Misconceptions Is Hard, But Not Impossible

UME Trends 1:6 (January 1990)

Current research points out that traditional instruction often has little influence on students' prior conceptions when these are at odds with concepts central to a subject, as for example in photosynthesis, Newtonian mechanics, and probability (see above review). Often students do not realize that a straightforward text is *supposed* to change their conceptions of a topic. Roth obtained positive results with a text on photosynthesis that presented and refuted common misconceptions, along with the scientifically accepted view. ("Conceptual Change Learning and Student Processing of Science Texts," paper, Annual Meeting of the American Educational Research Association, Chicago, March 1985.) In physics, where students often come into the course with an Aristotelian point of view and go out with it virtually unchanged, labs have been designed to demonstrate counterintuitive results and promote discussions. Problems requiring nonquantitative answers have been assigned with good effect. (Clement, J., *Proceedings of the Second International Seminar, Misconceptions and Educational Strategies in Science and Mathematics*, Cornell University, 84–97; Hake, R.R., *American Journal of Physics* 55, 878–884; Minstrell, J., *Observing Science Classrooms: Perspectives from Research and Practice*, A.E.T.S. Yearbook XI, ERIC, 1984.) Marked changes in student perceptions of mechanics occurred after students carried out tasks using a computer simulation which incorporated Newton's laws without friction or gravity. (White, B.Y., *Cognition and Instruction* 1, 69–108.) How could we go about confronting misconceptions in mathematics?

The Pedagogy of Educational Software: Good News, Bad News

UME Trends 1:4 (October 1989)

Both research and informed opinion support the promising view that the focus in calculus can be shifted away from mechanics to central ideas using technology (Richard Shumway in "Symbolic Computer Systems and the Calculus," *AMATYC Review*, September 1989). This article provides a useful guide to the current literature on symbolic computer systems, concept learning, and the research linking the two. It is about equally divided in length between a concise survey and an extensive bibliography. For Shumway, computers and supercalculators are the wave of the future, freeing students and teachers from drudgery for an exhilarating exploration of concepts, examples, counterexamples, and proofs.

Euphoria needs caution, however. Conventionally de-

veloped graphing software, together with its uninformed use, can blur, or even obscure, concepts according to E. Paul Goldenberg of Education Development Center, Inc. ("Mathematics, Metaphors, Human Factors: Mathematical, Technical and Pedagogical Challenges in the Educational Use of Graphical Representations of Functions," The *Journal of Mathematical Behavior* 7(2), 1988, 135–173). Using an early prototype of the Function Analyzer component of EDC's Algebra Visualizing, Goldenberg observed algebra students' numerous misinterpretations of graphs. His findings suggest a veritable Murphy's Law of student-software interaction.

The upper of two congruent parabolas, one merely a vertical translate of the other, was seen as more obtuse—a phenomenon related to optical illusion. A parabola can resemble a vertical line when the x-axis is sufficiently compressed. The scale and position of a graph, as well as the shape of the viewing window, affect how it is perceived; yet there is a near-total lack of research on these issues. Linear zooming may be suitable for lines, but a "variable zoom" might be better for polynomials—a cubic might be changed by a factor of m along the x-axis and a factor of m^3 along the y-axis, thereby preserving its usual appearance. The concept of variable, always difficult for students to grasp, can be further obscured by conventional software which encourages students to vary parameters, as for example, a, b and c in ax^2+bx+c. Goldenberg sees promise in the thoughtful design and use of software and speculates on implications for the curriculum. He notes that continuity, usually considered a more advanced topic, may eventually arise quite naturally in algebra, as students easily zoom and follow graphs.

Although a bright future may lie ahead, current software is no panacea. More research is needed on how software effects students' conceptual development.

Calculus Reform at a French University

UME Trends 1:4 (October 1989)

Many students only mimic what the teacher writes, use algorithms mindlessly, and regard proofs as meaningless formal exercises. Sound familiar? This is how Daniel Alibert and his colleagues at the University of Grenoble see their first-year students and why they developed the method of "scientific debate" in which students construct their own concepts, understand the need for proof, and sense the need for mathematics. Certain "customs" for the classroom are laid down: (1) Uncertainty (conjecture) is part of the learning process. (2) Proofs are for convincing others of the truth of a conjecture. (3) Mathematical tools are for solving complex problems, often from the physical sciences. The complexity justifies the effort needed to develop the tool. (4) Students need to reflect on their own thought processes. A sample debate on the Riemann integral is included. The method is being refined so as to be replicable ("Towards New Customs in the Classroom," *For the Learning of Mathematics* 8(2), June 1988, 31–35).

Compare and Contrast
Euclid, Bourbaki, Lakatos, Et.Al

UME Trends 1:4 (October 1989)

Axiomatic and discovery approaches to mathematics and mathematics education seem to swing back and forth pendulum-like in a rather difficult-to-read philosophical-cum-historical essay by Hans-Georg Steiner ("Two Kinds of 'Elements' and the Dialectic between Synthetic-deductive and Analytic-genetic Approaches in Mathematics," *For the Learning of Mathematics* 8(3), November 1988, 7–15).Characterized by van der Waerden as the greatest schoolmaster in the history of mathematics, Euclid dealt deductively with eternal truths in the *Elements*. Criticisms of his approach go back to the 16th century when Ramus advocated a more discovery oriented approach, influencing French mathematicians through his disciples for some 50 years. Clairaut, two centuries later, criticized texts written for beginners for starting from definitions and postulates, promising the reader nothing but dry learning. In 1934 the Bourbaki group purposely chose the axiomatic method with structures (group, topological space, etc.) as basic, thus profoundly influencing present day university level mathematics education. More recently, Lakatos criticized this authoritarian type of presentation for hiding the struggle and adventure of discovery, proposing instead a hybrid heuristic approach (*Proofs and Refutations*, Cambridge, 1976).

What are Prospective
Math Teachers Learning from Us?

UME Trends 1:4 (October 1989)

Mathematics education research indicates that children construct their own mathematical knowledge and that elementary school teachers should organize instruction to facilitate this. The old idea of learning by being told is seen as an inaccurate model for arithmetic learning. Unfortunately, from their own experiences in college, future teachers usually come to believe their knowledge of mathematics devolves from the combined authority of teacher and text, rather than evolving from their own reasoned thought and effort. Thus, challenging preconceptions of both teaching and learning is the objective when Deborah Ball explores permutations with prospective elementary teachers ("Unlearning to Teach Mathematics," *For the Learning of Mathematics* 8 (1), February 1988, 40–48). Furthermore, when first-grade

teachers think pedagogically about mathematics — how best to present topics, etc. — they tend to make extensive use of word problems, are aware of their pupils' problem-solving strategies and get better results (Peterson, Fennema, Carpenter, and Loef, "Teachers' Pedagogical Content Beliefs in Mathematics," *Cognition and Instruction* 6(1),1989, 1–40). Not enough is known about the effects of the teacher education curriculum on those who experience it. The National Center for Research on Teacher Education at Michigan State is asking questions like: What goes on in college mathematics courses? How do they influence attitudes towards the teaching and learning of mathematics? Hmmm? What sort of role models are we?

Do You Know the
Students-and-Professors Problem?

UME Trends 1:2 (May 1989)

In the mathematics education literature this problem is so well discussed that it might be considered an "old saw". The original research was done by Peter Rosnick and John Clement at the University of Massachusets, Amherst. They took 150 first-year engineering majors and asked them to translate English sentences into algebraic equations. The prototype problem was:

Write an equation using the variables S and P to represent the following statement: "There are six times as many students as professors at this university." Use S for the number of students and P for the number of professors.

Easy, eh?

Remarkably, only 63% of these students and 43% of social science students got it right. The most common error was the reversal error, i.e., writing 6S=P. Furthermore, based on tutoring interviews they found this misconception very resilient and not easily taught away. This and similar results have appeared in the *Journal of Mathematical Behavior* (3,p. 3-28), the *Monthly* (88,p.286-290), and the *Mathematics Teacher* (74, p.418-420). Curricular roots of the problem were explored by Sims-Knight and Kaput in *Focus on Learning Problems in Mathematics* (5,p.63-79). Soloway, Lochhead, and Clement have reported some success in getting freshmen and sophomores to write correct computer programs in BASIC for variants of this problem ("Does Computer Programming Enhance Problem Solving Ability? some Positive Evidence on Algebra Word Problems," *Computer Literacy,* R. Seidel, Ed., Academic Press (1982), p. 171-185).

Most recently one of us saw Kathleen M. Fisher's "The Students-and-Professors Problem Revisited" in the *Journal for Research in Mathematics Education* 19 (3) May 1988, p.260-262. Using an introductory physics course at the University of California, Davis, she gave half the students the original problem and the rest a variant, using the variable names Ns and Np for number of students and number of professors. She conjectured this more literal notation might improve performance. It didn't; only 43% answered this version correctly, as opposed to 60% for the original problem. The reversal error accounted for three-fourths of the wrong answers in both groups.

Not wanting to believe so many of our best and brightest freshman would make this error, one of us gave the problem to our own better than average engineering calculus class and obtained similar results. Disturbing, isn't it? And we haven't heard the last of this problem – additional research is in the pipeline.

Metamorphosis: From Mathematician to Mathematics Education Researcher

UME Trends 1:2 (May 1989)

Alan Schoenfeld has written a short intellectual autobiography ("Confessions of an Accidental Theorist," For *The Learning of Mathematics* 8,1). He relates his initial excitement as a young mathematician in 1974 upon discovering Polya's *How to Solve It,* followed by disillusionment upon being repeatedly informed by mathematics faculty who coach students for the Putman Exam that it is of no use whatever. Schoenfeld set out to discover why Polya's widely admired strategies didn't work for students. He began by looking at cognitive science and artificial intelligence, e.g., General Problem Solver, a program to solve problems in symbolic logic and chess, described in Newell and Simon's *Human Problem Solving* (1972). He relates his progression through three phases as he made ever deeper observations of actual student problem solving using video-tapes and interviews.

This entertainingly written article provides a concise introduction to Schoenfeld's *Mathematical Problem Solving* (Academic Press 1985), in which he discusses mathematical problem-solving performance in terms of four categories: cognitive resources (one's basic knowledge of mathematical facts and procedures, heuristics (strategies and techniques one has), control (or *metacognition,* i.e., how one uses what one knows), and belief systems (or *weltanschauung,* e.g., all math problems can be solved in ten minutes or less if one understands the material).

Do End-of-Section Exercises Prevent Learning?

UME Trends 1:2 (May 1989)

Quite possibly, argues John Sweller of the University of New South Wales in "Cognitive Load During Prob-

lem Solving: Effects on Learning" (*Cognitive Science,* 12 (1988), 257-285). Sweller's research suggests that, since human short-term memory may be severely limited, conventional problem (textbook exercise) solving may use a large amount of cognitive processing capacity which is thus unavailable for the acquisition of schemas.

Experts seem to group problems based on their solution mode whereas novices use surface structures. In physics, an expert might say, "This is a conservation of energy problem," whereas a novice might say, "This is a lever problem" (Chi, Glasser, and Rees, "Expertise in problem solving," *Advances in the psychology of human intelligence,* R. Sternberg, ed. , Erlbaum, 1982).

In one of Sweller's experiments a group of trigonometry students was asked to find as many lengths of sides as possible (nonspecific goal) while another group was asked to find a specific side (conventional problem solving); the former performed better. Sweller argues that the latter suffered from the heavy cognitive load of means-ends strategy.

Critical Thinking About Modeling

UME Trends 1:2 (May 1989)

In addition to giving students the mathematical tools and the technical ability to develop models, Ole Skovsmose advocates developing a critical pedagogy in which students are encouraged to reflect on the uses, misuses, and societal consequences of particular models [Mathematics as a Part of Technology, *Educational Studies in Mathematics* 19(1988) 23-41]. He has in mind models like SMEC (Simulation Model of the Economic Council) or the North Sea Model and argues against the pragmatic trend in mathematics education, according to which the only important thing is to apply and learn how to apply mathematics. This article is basically philosophical in nature, contains an certain amount of jargon, but definitely provides food for thought., Skovsmose contends that "Technology not only provides solutions, to social problems, it also creates new social problems. As supervisor of students at Aalborg University Centre he has developed projects that promote critical student reflection, however, these are not provided. Perhaps those interested in details could write him at the Department of Mathematics and Computer Science, Aalborg University Centre, Strandvejen 19, DK 9000 Aalborg, Denmark.

You're the Professor, What Next?

Author Index

Anderson, B.J., B-II-19

Askey, R., B-III-30

Berard, L.M., B-III-84

Berriozábal, Manuel P., 106

Blackwelder, M. Annette, 15, 25

Blum, D.E., B-III-19

Bogart, Kenneth P., 60

Bradley, C., B-II-23

Brawley, Joel V., 52

Brown, A., B-III-79

Bushaw, Donald W., 108, commentary

Carlson, D.H., B-III-67

Case, B.A., 9, 15, 25, 32, B-III-36, commentary

Clapp, M.H., B-III-17

Cole, Donald R., 112

Corzatt, Clifton, 90

Crannell, A., B-I-2

Crocker, D.A., B-III-64

Cuoco, A.A., B-III-14

Curjel, C.R., B-III-42

Czerwinski, R., B-III-82

Davis, Paul, B-I-15

Davis, Philip J., B-I-24, B-III-53

Doblin, Stephen A., 113

Douglas, R.G., B-I-37

Downs, F., B-III-32

Dubinsky, E., 114, B-III-23, B-III-87

Earles, G.A., B-III-41

Effros, E.G., B-III-12

Eidswick, J., B-III-77

(Escalante, Jaime, see Gillman p. 121)

Evangelista, F., B-I-14

Ewing, J., B-I-14, B-III-12

Fisher, N., B-III-22, B-III-25

Foley, G.D., A-1, B-III-61

Frandsen, Henry, 82; illustrations: see iv

Freeman, Michael, 119

Fusaro, B.A., B-I-40

Garfunkel, S.A., B-I-33, B-III-51, B-III-56

Garity, Dennis, 77

Gass, F., B-III-80

Gerstein, L.J., B-III-71

Giebutowski, T., B-III-72

Gillman, Leonard, 121

Gold, B., B-III-82

Goroff, Daniel L., 71

Gray, Mary W., 92

Groszek, Marcia J., 60

Guerrieri, Bruno, 127

Halmos, P.R., B-III-44

Harrison, Jenny, B-II-3, B-II-8

Hassett, M., B-III-32

Hatcher, R.L., B-III-86

Hendel, R.J., B-III-85

Henrion, C., B-II-9

Herman, R.H., B-I-6

Herrmann, Diane, 93

Hilton, P., B-III-46

Humke, Paul D., 90

Hunt, F.Y., B-I-33

Jackson, A., B-I-9, B-I-11

Jaco, W., B-III-21

Jenkins, J., B-III-32

Jones, J.A., B-II-26

Jordan, Sam, 82

Karian, Z.A., B-III-73

Kary, M., B-II-8

Kasube, Herbert E., 95

Keith, S.Z., B-III-41

Kennedy, Judy, 66

Keny, S.V., B-III-44

Keynes, H.B., B-II-25, B-III-8

Krantz, Steven G., 86, 128

Lasser, S.J.S., 136

Lazarus, A.J., B-I-34

Lekander, Brian, 130

Levermore, C.D., B-I-17

Lewis, J., B-I-35

Lucas, W.F., B-III-59

Luchins, Edith H., 96

Lutzer, D., B-III-40

Madison, B.L., B-III-43

Maher, R.J., B-III-79

Maurer, S.B., B-III-70

McCaffrey, J., B-II-29

McDowell, Robert, 86

Merkes, Edward P., 46

Millman, Richard S., 11

Morrell, J.H., B-III-50

Nashed, M. Zuhair, 66

Newman, R.J., B-II-22, B-II-28

O'Leary, M.E., 136

Palais, Eleanor G., 133

Patterson, Wanda M., 135

Pearson, R.W., B-III-58

Pedersen, J., B-III-46, B-III-83

Pengelley, D., B-III-55 (see also Gillman p. 124)

Phelps, Robert R., commentary

Pincus, S, B-I-25

Poiani, E.L., B-III-29

Porter, G.J., B-III-68

Proctor, T.G., 136

Ringeisen, Richard D., 100

Riskin, A., B-III-83

Rodi, Stephen B., 10, 34, 43, 157, commentary

Roitman, J., B-II-14 (see also Harrison p. B-II-6)

Ross, A.E., B-III-27

Ross, S.C., B-I-27

Rossi, H., B-III-10

Rota, G.-C., B-I-19

Schafer, Alice T., B-II-16

Schielack, V.P., B-III-81

Selden, A., 138, B-III-39, B-III-63, B-III-87

Selden, J. , 138, B-III-39, B-III-63, B-III-87

Shell, T., B-I-28

Shure, Pat, 102

Small, D., B-III-85

Smart, J., B-I-38

Smith, D.A., B-III-48

Smith, M.K., B-I-22, B-III-74

Snelshire, R.W., 136

Stakgold, Ivar, 12, 66

(Stephens, Clarence F., see Gillman p. 124)

Sward, M.P., B-I-40

Tapia, Richard, 102

Thurston, W.P., B-III-2

(Treisman, Uri, see Gillman p. 123 and McCaffrey p. B-II-29)

(Uhlenbeck, Karen, see Harrison p. B-II-6)

Walker, Janice B., 103

Walvoord, Barbara E., 141

Watkins, Ann E., 142

Watkins, B.T., B-III-26

Weiss, Guido L., 43, B-III-20

Williams, Roselyn E., 145

Willoughby, S., B-I-38

Winkel, B.J., B-III-81

Woeppel, J.J., B-III-75

Wood, C.S., B-I-36

Young, G.S., B-III-56